THIRD EDITION

REMOTE SENSING FOR GEOSCIENTISTS

IMAGE ANALYSIS AND INTEGRATION

THIRD EDITION

REMOTE SENSING FOR GEOSCIENTISTS

IMAGE ANALYSIS AND INTEGRATION

GARY L. PROST

CRC Press
Taylor & Francis Group
Boca Raton London New York

CRC Press is an imprint of the
Taylor & Francis Group, an **informa** business

CRC Press
Taylor & Francis Group
6000 Broken Sound Parkway NW, Suite 300
Boca Raton, FL 33487-2742

First issued in paperback 2019

© 2014 by Taylor & Francis Group, LLC
CRC Press is an imprint of Taylor & Francis Group, an Informa business

No claim to original U.S. Government works

ISBN-13: 978-1-4665-6174-8 (hbk)
ISBN-13: 978-0-367-86757-7 (pbk)

Library of Congress Cataloging-in-Publication Data

Prost, G. L.
 [Remote sensing for geologists.]
 Remote sensing for geoscientists : image analysis and integration / Gary L. Prost. -- Third edition.
     pages cm
 Includes bibliographical references and index.
 ISBN 978-1-4665-6174-8 (hardback)
 1. Geology--Remote sensing. 2. Prospecting--Remote sensing. 3. Remote sensing. 4. Geology--Geographic information systems. 5. Image analysis. I. Title.

 QE33.2.R4P76 2013
 681'.755--dc23   2013016498

Visit the Taylor & Francis Web site at
http://www.taylorandfrancis.com

and the CRC Press Web site at
http://www.crcpress.com

To Nancy, with love

Contents

Section I Initiating Projects

Section II Remote Sensing in Geologic Mapping and Resource Exploration

Section IV Environmental Remote Sensing

Section V   Astrogeology

17. Mapping Planetary Structure

18. Mapping Planetary Stratigraphy

19. Planetary Resources

Section VI   Remote Sensing, Geoscience, and the Public

20. Public Relations, the Media, and the Law

Preface

The third edition of this text has a new title. The previous *Remote Sensing for Geologists: A Guide to Image Interpretation* is now *Remote Sensing for Geoscientists: Image Analysis and Integration*. The title change reflects (1) that this edition applies to a broad spectrum of geosciences, not just geology; (2) remote sensing has become more than photointerpretation; and (3) an emphasis on integration of multiple remote sensing technologies to solve Earth science problems. Since publication of the second edition in 2001, new remote sensing systems have been acquiring not only visible, infrared, and microwave images, but also have been detecting atmospheric gases, ocean temperatures, wind speeds, and mapping minerals on the moon and planets, as well as recording potential fields and digital elevation data. The purpose of this text is to review systems and applications, explain what to look for when analyzing imagery, and stress the importance of integration of multiple remote sensing and field-based tools. Case histories are included to show how these systems are being used to best effect in exploration, engineering design, and environmental monitoring.

Remote sensing technology can be traced at least to the thirteenth century invention of eyeglasses by Roger Bacon. This technology literally got off the ground in 1858 when Gaspard Tournachon took photographs of Paris from a balloon to produce topographic maps (Reeves et al., 1975). During World War I, aircraft-based photography was used extensively for the first time as a means of gathering intelligence, particularly the disposition of enemy troops. Modern remote sensing can be traced to airphoto surveys begun in the 1930s and 1940s to map topography, geologic features associated with petroleum accumulations, and as a basis for construction projects. During World War II, color infrared film was used to detect and circumvent camouflage, and radar was invented to detect enemy aircraft. In fact, the technology has often been advanced by the demands of the military and intelligence communities. With the advent of satellites and multispectral scanners, the science of remote sensing became increasingly useful in geologic exploration, engineering and logistical planning, and environmental monitoring. An objective of this book is to show the interested reader how to analyze, interpret, integrate, and extract information from remote sensing imagery.

In general terms, "remote sensing" is defined as technologies and techniques used to obtain information about distant objects using reflected or emitted electromagnetic radiation (Figure 0.1), acoustic energy, potential fields (gravity, magnetics), or geochemical measurements. This book concerns itself primarily with analysis of electromagnetic images, supplemented by acoustic (sonar), potential field, and geochemical images and maps. Data that are measured at points and then contoured (geochemistry, gravity) or along profiles (Lidar, aeromagnetics, seismic) will be touched on to a lesser extent. "Image analysis" (also known as "image interpretation") can be defined as the process of extracting useful information from remote sensing images, whether they are digital or analog, hardcopy (paper, film) or on a computer screen. This imagery is usually integrated, either digitally or by comparison with other data types during the analysis, and may be displayed or compiled as an image, map, report, or some combination thereof.

The process of interpretation draws heavily on field experience; that is, it is important for the interpreter to have observed features on the ground, preferably in three dimensions, to understand what is being observed on images. In geologic remote sensing, this draws heavily on structural geology and geomorphology; in environmental remote sensing, it is important to understand plant communities (composition, density, distribution) and to

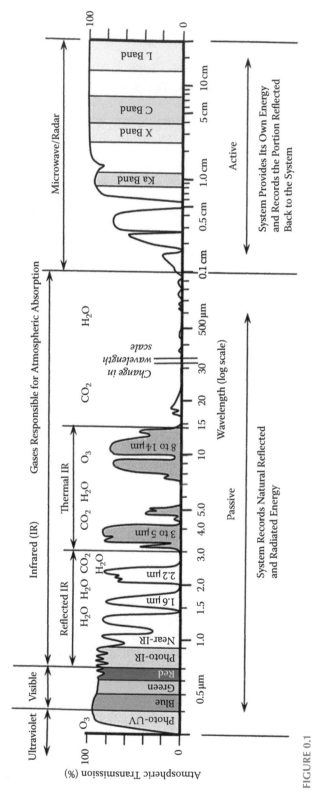

FIGURE 0.1

The electromagnetic spectrum as referred to in this text. (Data from Sabins, F.F. *Remote Sensing Principles and Interpretation*, 2nd ed. W.H. Freeman & Company, New York, 1987.)

know the distribution and effect of soil type, surface water and groundwater, contaminants, infestation, wildlife behavior, and climate; and in engineering and logistics planning, it is useful to know about slopes and slope stability, soil type, bedrock (composition, bedding dips), and groundwater conditions. Whereas "photogrammetry" is the precise measurement or surveying of ground features from photos to make, for example, topographic maps, image analysis deals with the discrimination and identification of stratigraphy, structure, landforms, soil cover, plant cover, surface composition, and cultural features and determining their significance. Analysis and integration provide information about surface and subsurface geology, environmental sensitivity, or the suitability of an area for building structures.

Image analysis also requires knowledge of the instruments used to acquire the image and the image processing techniques that went into generating the picture. You do not really know what you are seeing unless you know how the image was made. This book deals with instrumentation and processing as they apply to interpreting imagery: for a more complete understanding of the physics of remote sensing, remote sensing instrumentation, and digital image processing, the reader is referred to the *Manual of Remote Sensing* (Rencz and Ryerson, 1999) as well as books by Campbell and Wynne (2011), Petrou and Petrou (2010), Varshney and Arora (2010), Clark and Rilee (2010), Canty (2009), Lillesand (2007), Woods et al. (2007), Sabins (2007), Schowengerdt (2006), Jenson (2004, 2006), Richards (2005), Drury (2004), and others (Gupta, 2003; Condit and Chavez, 1979).

Remote Sensing for Geoscientists is organized into six sections: Section I (Chapters 1 through 3) deals with initiating a project, that is, determining the objective, choosing the right tools, and selecting imagery. Section II (Chapters 4 through 10) describes techniques used in geologic mapping and mineral and hydrocarbon exploration. Section III (Chapters 11 through 13) describes image analysis as used during mine development and petroleum exploitation, for structural or civil engineering, site evaluation, groundwater development, surface water monitoring, geothermal resource exploitation, and logistics. Section IV (Chapters 14 through 16), on environmental concerns, demonstrates how imagery is used to establish environmental baselines; monitor land, air, and water quality; map hazards; and determine the effects of global warming. Section V (Chapters 17 through 19) provides examples of geologic mapping on other planets and moons. In these chapters, we show how to analyze planetary surface processes, map stratigraphy, and locate resources. Section VI (Chapters 20 and 21) deals with remote sensing and the public: geographic information systems and Google Earth, how imagery is used by the media, in the legal system, in public relations, and by individuals. The reader should come away with a good understanding of what is involved in image analysis and interpretation and should be able to recognize and identify geologic features of interest. Having read this book, a student or project geologist should be able to effectively use imagery in petroleum, mining, groundwater, surface water, engineering, and environmental projects.

References

Campbell, J.B., R.H. Wynne. 2011. *Introduction to Remote Sensing*, 5th ed. New York: Guilford Press: 667 p.

Canty, M.J. 2009. *Image Analysis, Classification, and Change Detection in Remote Sensing: With Algorithms for ENVI/IDL*, 2nd ed. Boca Raton, FL: CRC Press: 471 p.

Clark, P.E., M.L. Rilee. 2010. *Remote Sensing Tools for Exploration: Observing and Interpreting the Electromagnetic Spectrum*. New York: Springer: 360 p.

Condit, C.D., P.S. Chavez, Jr. 1979. *Basic Concepts of Computerized Digital Image Processing for Geologists; U.S. Geological Survey Bulletin 1462*. Reston, VA: U.S. Geological Survey: 16 p.

Drury, S.A. 2004. *Image Interpretation in Geology*, 3rd ed. Malden, MA: Blackwell Science: 304 p.

Gupta, R.P. 2003. *Remote Sensing Geology*, 2nd ed. Berlin: Springer-Verlag: 656 p.

Jenson, J.R. 2004. *Introductory Digital Image Processing*, 3rd ed. Upper Saddle River, NJ: Prentice Hall: 544 p.

Jenson, J.R. 2006. *Remote Sensing of the Environment: An Earth Resource Perspective*, 2nd ed. Upper Saddle River, NJ: Prentice Hall: 608 p.

Lillesand, T. 2007. *Remote Sensing and Image Interpretation*. New York: Wiley: 804 p.

Petrou, M., C. Petrou. 2010. *Image Processing: The Fundamentals*, 2nd ed. New York: Wiley: 818 p.

Reeves, R.G., A. Anson, D. Landen. 1975. *Manual of Remote Sensing*, 1st ed., Chap. 2. Falls Church, VA: American Society of Photogrammetry: 27 p.

Rencz, A.N., R.A. Ryerson. 1999. *Manual of Remote Sensing: Remote Sensing for the Earth Sciences*, Vol. 3, 3rd ed. New York: Wiley: 728 p.

Richards, J.A. 2005. *Remote Sensing Digital Image Analysis: An Introduction*, 2nd ed. Berlin: Springer-Verlag: 464 p.

Sabins, F.F. 2007. *Remote Sensing: Principles and Interpretation*, 3rd ed. Long Grove: Waveland Press: 512 p.

Schowengerdt, R.A. 2006. *Remote Sensing: Models and Methods for Image Processing*, 3rd ed. Burlington: Academic Press: 560 p.

Varshney, P.K., M.K. Arora. 2010. *Advanced Image Processing Techniques for Remotely Sensed Hyperspectral Data*. Berlin: Springer-Verlag: 338 p.

Woods, R., R.C. Gonzales, P.A. Wintz. 2007. *Digital Image Processing*, 3rd ed. Upper Saddle River, NJ: Prentice Hall: 976 p.

Acknowledgments

Thanks are directed to Amoco Production Company (now BP) and ConocoPhillips for allowing publication of this manuscript, and to Jean Munshi, Morton Lovestad, Stephen Hansen, and Tiffany Cortez for drafting many of the diagrams in this work. I am grateful to the many individuals and companies that did the image processing and allowed me to use their work to illustrate the concepts presented here. I am indebted to Larry Lattman, Keenan Lee, and Eric Nelson for my background in geomorphology, remote sensing, and structural geology, respectively. Finally, many of the ideas put forward here are the result of discussions and many hours of head-scratching with coworkers over the years, including Steve Nicolais, Bill DiPaolo, Dave Cole, Glen Steen, Don Erickson, Dave Koger, and John Berry. Thanks to Fred Kruse, Joseph Boardman, Bob Agar, Sandra Feldman, Jim Ellis, Ronald Marrs, and Rebecca Dodge for contributing case histories to and/or reviewing the second edition. The assistance of Ralph Baker (independent consultant), Matt Hall (Agile Geoscience), Cara Hollis (NEOS), Dave Coulter (Overhill Imaging), Phoebe Hauff (Spectral International Inc.), Sandra Perry (Perry Remote Sensing), Doug Peters (Peters Geosciences), Deet Schumacher (independent consultant), Bill Hirst (Shell Global Solutions International), David Haddad (Arizona State University), Elspeth Robertson and Juliet Biggs (University of Bristol), Michael Henschel (MacDonald Dettwiler & Associates, or MDA), David Ferrill (Southwest Research Institute), Richard Eyers (Spatial Energy; The Geologic Remote Sensing Group), Rebecca Dodge (Midwestern State University), and Lito Cillo (Cenovus Energy) in providing examples, case histories, and helpful discussions for the third edition is gratefully acknowledged.

Illustrations have been credited to the company or agency that originally acquired and/or processed them. There are many service companies that provide remote sensing surveys and build and operate remote sensing instruments. It is not the intention of this book to promote any particular company or vendor, nor would it be possible to list them all due to the large number and because the list is constantly changing. Several but by no means all government agencies and principal distributors of data are listed in this text. For information regarding companies that can provide services, the reader should consult local professional photogrammetric societies and journals or the internet under "Remote Sensing Service Providers."

Author

Gary L. Prost earned a BSc in Geology from Northern Arizona University and MSc and PhD in Geology at Colorado School of Mines. Over the past 35 years, he has worked for the U.S. Geological Survey (mapping coal), the Superior Oil Company (mineral and oil exploration), Amoco Production Company (worldwide oil exploration), Gulf Canada (international new ventures), and ConocoPhillips Canada (frontier exploration and development). While at Superior Oil and Amoco, he spent more than 20 years working as an image interpreter in the search for oil and minerals in more than 30 countries. He is presently working for ConocoPhillips Canada on field development in the Canadian oil sands. Dr. Prost is a Registered Professional Geologist in Wyoming, and Alberta and the Northwest Territories, Canada. He is also author of the *English–Spanish and Spanish–English Glossary of Geoscience Terms*.

Section I

Initiating Projects

All remote sensing projects begin with a problem we are trying to solve: we may want to find out where the oil is, where to drill, the best way to get to a drill site, and where to build a pipeline or a gas plant for the least cost and with minimal environmental disruption. We may wish to lay out a seismic program most efficiently or find out where a competitor has shot a seismic program. We require a source of water for drilling, for a coal slurry pipeline, or for keeping dust down in a mining operation. If we are involved with mineral exploration, we will be looking for any evidence of mineralization in a new mineral province, or which direction to extend a known deposit. We need to know the state of the terrain before mining so that we know how to restore it to its premining condition. Was there natural acid drainage before mining or is it coming from the tailings ponds?

In order to determine the best imagery to evaluate, we must know what we are looking for. Is the area large or small? Does our problem require us to see fine details (centimeters up to 10 m resolution), moderate detail (20–100 m resolution), or regional features (100 m to 1 km resolution or more)? What scale do we wish to work with? Do we need to detect color changes (e.g., lithology, alteration) or vegetation stress? Is the area always under clouds? Is the area in a polar region that has an extended dark season? Are we looking for changes in moisture conditions? Is the area under water? Do we require or want a certain date or specific time of year? Do we need multitemporal (repetitive) coverage or historical coverage? Finally, how much time do we have and what kind of budget do we have to work with? Should we go to a vendor, the government, or process the data ourselves? The answers to these questions will determine the products that are acquired and the types of analyses that are possible.

1

Project Flow and Obtaining Data

Chapter Overview

There are many factors to consider when starting a new remote sensing project. These include, among others, clearly defining the problem to be solved, the type of imagery needed, the size of the area requiring analysis, the amount of detail that is required, the time frame for resolving the problem, and the budget available. The answers to these questions will determine, for example, whether the imagery is color or black and white (B/W), on film or digital, the required scale, whether it is from an aircraft or satellite, historical imagery or real time, and whether it is in the visible, thermal, or microwave (radar) portion of the spectrum.

There are several providers of imagery and image databases that are available once the type of imagery required is known. These can be quickly found online by going to a search engine and typing in the type of imagery or data required.

Choosing the Right Imagery for the Job

The first step in any project is to clearly understand the problem that needs to be solved and to determine the best approach to solving it (Figure 1.1). Although it sounds obvious, too often not enough thought goes into this step and imagery is obtained that does not answer the question we have. A large number of data types are available to suit various needs. If we are studying thrust belts, we may wish to order 1:1,000,000 B/W single-band Landsat multispectral scanner (MSS) images with 80 m resolution and large area coverage (185 × 185 km) to make a mosaic of part of a continent. If we are doing a basin analysis, then a color MSS image at 1:250,000 or 1:500,000 should be acceptable. If we are mapping details of lithologic and facies changes, or vegetation patterns and wildlife habitats, we can use color SPOT multispectral (XS) imagery with 20 m resolution and 60 × 60 km (vertical view) coverage at a scale of 1:100,000 or 1:50,000. Color Landsat Thematic Mapper (TM) imagery, with 30 m resolution, and covering 185 × 185 km at a scale of 1:100,000 would also work. For mapping alteration associated with mineral deposits, the ideal choice is high resolution (1–10 m) airborne hyperspectral imagery. If that is not available or is too expensive and the area is large and remote, we may wish to use Landsat TM for a reconnaissance look. If we want to know where there are trails that can be used to access remote areas, or where a well was drilled several years ago in a poorly mapped part of the world, we chose a SPOT P high resolution (10 m) panchromatic image or a Soyuzkarta KFA-1000 photograph with

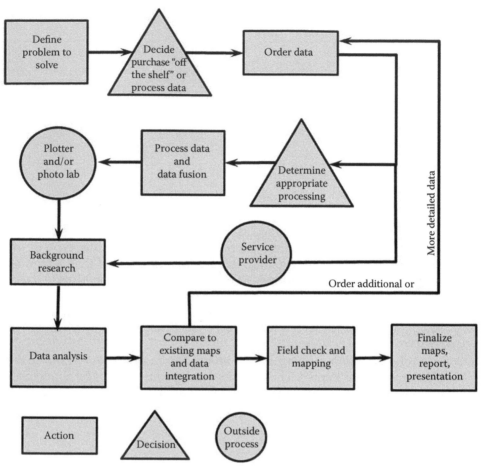

FIGURE 1.1
Project flowchart for a typical remote sensing study.

5 m resolution. For very fine detail, GeoEye or WorldView images have approximately 1 m resolution. These can be enlarged to 1:25,000 and still appear clear and sharp.

In order to determine the type of imagery that best suits the needs of a project, the following factors should be taken into account when ordering data (Dekker, 1993).

1. Cost

   If cost is no object, the best procedure, which provides the most flexibility in end products, is to purchase a digital image and process it. However, this requires one to have an image processing system, the knowledge to operate it and keep it updated, and the time to do the processing. For those with some latitude in the amount they can spend, there are a variety of high-quality custom products available from vendors with a wide range of costs. If the budget is limited, the least expensive option is to purchase imagery off the shelf from a government agency or primary distributor.

   One factor to keep in mind is that the smaller the area covered, the higher the cost per unit area. Airphotos or airborne imagery will almost always cost more per unit area than satellite images.

2. Timing

If the imagery is needed immediately, one can screen grab free but rather low-quality imagery from internet sources such as Google Earth (subject to their License Agreement) or purchase off-the-shelf data from a government agency or vendor. Large image archives exist, and data can often be obtained quickly. The cost will increase if a "rush" job is requested. Purchasing custom images from vendors and consultants, or processing digital data in your own shop can take up to several weeks.

3. Coverage

Large area coverage can be obtained using weather satellites such as GOES (covers a full hemisphere), the Advanced Very High Resolution Radiometer, which covers a 2700 km swath on the Earth's surface, or the SeaWIFS instrument, with a 1502–2801 km swath width. Moderate size areas can be covered using some handheld Shuttle (and other mission) photos (variable area coverage), as well as Landsat images (MSS and TM), which cover 185 × 185 km. Systems that cover 50 × 50 to 500 × 500 km include MK-4 photos (120–270 km), KATE-200 photos (180 × 180 km), KFA-1000 photos (68–85 km), and the SPOT systems (XS and P) that cover 60 × 60 km. Satellite images generally cover larger areas than airborne photos or images, and the synoptic view is one of their greatest advantages. For regional studies, one usually requires large area coverage; for field or local studies, airborne surveys or small area satellite images will save cost and/or provide more detail. Recent satellites, such as Worldview 2 and GeoEye-1, have very high resolution (up to 41 cm for the panchromatic sensor and 1.65 m for the multispectral instrument) and cover correspondingly smaller areas (16.4 and 15.2 km swaths, respectively).

4. Resolution

Because fine resolution sensors create very large digital files, larger areas tend to be imaged by coarser resolution systems, whereas smaller areas are imaged in finer detail. Historically, satellites available to civilian users always have less resolution than airborne photos and images, which are flown closer to the ground. This has changed in the past decade. Photographs almost always had better resolution than digital images, since the grain of the film is finer than the array of picture cells or "pixels" produced by the spacing of detectors. Digital sensors now compete successfully with photography with respect to resolution, both on satellite and airborne platforms.

The scale of the final image will to some extent be a function of the sensor system resolution, in that one cannot enlarge, say, an image with 80 m resolution to a scale of 1:100,000 without the image becoming "pixelated," that is, breaking up into the individual resolution elements that appear as an array of colored squares.

5. Stereoscopic Coverage

Stereoscopic viewing is a technique that creates the illusion of depth. Two overlapping images are viewed through a stereoscope and combined in the analyst's brain to give the perception of a three-dimensional representation of the Earth.

If stereoscopic coverage is needed for mapping the dip of geologic structures, for example, one could use airphoto or satellite images with overlapping ground swaths. Airborne imaging scanners can provide stereo using "frame-grabber" technology. This technology utilizes video cameras to acquire images at intervals along the flight path such that one obtains the overlap required for parallax. Satellites such as Landsat provide stereo only where adjacent orbits overlap, known as sidelap. Thus, there may be only 10% overlap near the equator, but up to 60% or more overlap in polar regions where orbits converge. SPOT and some other satellites offer stereo imagery by using one forward-looking sensor and another near-vertical one.

6. Color versus Black/White

All multispectral data and most films can be processed to provide color images. B/W airphotos, SPOT P, radar, and thermal data consist of a single band or channel generally displayed as B/W images. Color often reveals subtle details regarding lithologies or vegetation, mineral alteration, or water depth that are not readily available from panchromatic images. However, any B/W data can be artificially colored by digitally assigning colors to various densities (gray levels) or by merging it with other types of imagery during image processing. This will often enhance features not otherwise obvious on the original B/W image.

7. Haze Suppression

As humidity increases, moisture condenses on atmospheric particulates and increases the amount of light that is scattered. Junge (1963) showed that as humidity increased from 70% to 95%, horizontal visibility decreased by a factor of six. Longer wavelengths penetrate haze more than shorter wavelengths (which is why we enjoy red sunsets). Thus, long wavelength microwave (radar) images penetrate cloud over, while color infrared or B/W infrared images, created using wavelengths longer than visible light, show less dust or humidity-related haze than true color or visible panchromatic images. Haze can also be diminished by flying close to the ground and by using haze-reducing (generally "minus blue") filters (Reeves et al., 1975).

8. Cloud Cover

Only radar imagery can penetrate clouds as a result of the long wavelength of microwaves with respect to water vapor particle size. This is especially useful in tropical climates or areas known to have cloudy periods. Other ways to minimize clouds include timing overflights to occur early to mid-morning, before clouds begin building, or after passage of a cold front, when skies tend to be clear.

9. Nighttime Surveys

Thermal and radar surveys can be flown effectively at night because neither system relies on reflected sunlight: the radar instrument illuminates the surface by providing its own energy source, and thermal energy is radiated from the surface. Predawn thermal imagery reveals, among other things, lithologic contrasts related to differing rock and soil densities or color tones (light versus dark). Nighttime thermal imagery can reveal shallow groundwater and moist soil (generally warmer than background), can detect oil spills on water, and can help map underground coal mine fires. Because radar illuminates the ground with microwaves, it can be flown at night to map oil spills, for example, or during polar night to map the movement of ice floes that could threaten an offshore oil rig or platform.

10. Seasonal/Repetitive Coverage

Certain seasons are better for specific surveys. For example, a geologic mapping project in an area covered by temperate forest would see more of the ground in spring before deciduous plant leaf-out or in the fall, after leaves have dropped. High sun elevation angle (summer) provides images with the best color saturation, which can be useful when mapping lithologies in low contrast areas. On the other hand, low sun angle images (flown during the morning or in winter), especially with a light snow cover, enhance geologic features in low-relief terrain.

If repetitive coverage is needed to monitor natural (e.g., flooding, ice floes) or man-made (e.g., drilling, roads) changes, it is often most cost effective to use satellites because of their regular repeat cycles. Repeated aircraft surveys provide more detail but are much more costly.

11. Relief

Low-relief terrain may require low sun angle or grazing radar imagery to enhance subtle topographic and structural features. Although most airphoto surveys are flown with a 15.25 cm (6 inch) lens, airphotos obtained with a 7.6 cm (3 inch) focal length lens have increased vertical exaggeration and thus amplify subtle features. On the other hand, high-relief terrain poses the potential problem of large shadowed areas that can obscure important areas or details. These areas should be flown during mid-day or using radar with a steep depression angle to minimize shadows. Visible imagery will always have some information in shadowed areas due to scattered light; radar shadows have no data as there is no information reflected from the shadowed area.

12. Vegetation Cover

Color infrared images are very sensitive to changes in vegetation type or vigor, since the peak reflection for vegetation is in the near infrared region. Combinations of infrared and visible wavelengths have been used to map changes in vegetation related to underlying rock types and even hydrocarbon seepage (Abrams et al., 1984). Longer radar bands will penetrate a few layers of leaves (long microwaves penetrate farther than short microwaves), but radar is not particularly well suited for mapping vegetation. However, it will faithfully map topography in vegetation-covered areas where the canopy follows topographic contours. In this respect, it is much like Lidar, which will also provide an image of the top of the vegetation canopy that looks like topography. There are some image processing methods, called "vegetation suppression" techniques, which appear to remove vegetation and reveal subtle changes in the underlying soil or bedrock. These algorithms tend to remove the reflectance attributed to vegetation and enhance the remaining wavelengths.

13. Water-Covered Areas

In areas with clear water, uniform and light-colored bottom material, little or no suspended sediment, and overhead illumination, one may be able to map submerged features to 20 or 30 m by measuring the amount of light reflected off the bottom. Dark bottom material, suspended sediment or algae, and low sun angle would limit water penetration by light. Shorter wavelengths (blue and green light) penetrate water farther than longer wavelengths. The euphotic or light-penetrating zone is known to extend to 30 m in clear water (Purser, 1973).

Infrared light, which has longer wavelengths than visible light, is absorbed by water and does not provide information on bottom features. Landsat TM, with its blue band, is excellent for mapping shallow water features such as shoals, reefs, or geologic structures. Likewise, true color and special water penetration films such as Kodak Aerocolor SO-224 have excellent water penetration capabilities (Reeves et al., 1975). Side-scan sonar is available for shallow and deep water mapping, and produces images of the sea bottom reflectance using acoustic energy, much like radar uses microwave energy to produce an image.

Acquiring Imagery

After the appropriate type of imagery has been determined, the user must obtain it. As mentioned earlier, low-quality images can often be obtained on the internet. There are three sources of high-quality imagery: government agencies, service companies, and universities. A list of service companies can be obtained through technical literature, the phone book, or internet under the heading Aerial Surveys, Photogrammetry, Radar, Thermal, or other survey type (see, e.g., http://virtual.vtt.fi/virtual/space/rsvlib/ or http://virtual.vtt.fi/virtual/space/rsvlib/other.html#satdat), or from various photogrammetric organizations. A list of government agencies can be obtained from the same sources. Agencies generally have archives that can be searched for images that meet the user's requirements. University libraries, such as the Berkeley Earth Science and Map Library, can be very useful (http://cluster3.lib.berkeley.edu/EART/aerial.html). Service companies will fly a survey to the user's specifications. This usually requires a greater investment of both time and money than ordering imagery off the shelf. If a survey is being planned, consult your local photogrammetric society for standard survey parameters or refer to the U.S. Bureau of Land Management Aerial Photography Specifications (U.S. Bureau of Land Management, 1983).

On receiving the imagery, it is necessary to check the quality and confirm that the area you expected to be covered is correct. One is then ready to begin image analysis.

References

Abrams, M.J., J.E. Conel, H.R. Lang. 1984. *The Joint NASA/Geosat Test Case Project Sections 11 and 12.* Tulsa: AAPG Bookstore.

Dekker, F. 1993. What is the right remote sensing tool for oil exploration? *Earth Obs. Mag.* 2: 28–35.

Junge, C.E. 1963. *Air Chemistry and Radioactivity.* New York: Academic Press: 382 p.

Purser, B.H. 1973. *The Persian Gulf.* New York: Springer-Verlag: 1–9.

Reeves, R.G., A. Anson, D. Landen. 1975. *Manual of Remote Sensing,* 1st ed, Chap. 6. Falls Church: American Society of Photogrammetry.

U.S. Bureau of Land Management. 1983. *Aerial Photography Specifications.* Denver: U.S. Government Printing Office: 15 p.

Additional Reading

Abrams, M.J., G. Asrar, R. Balstad, P. Minnett, M.T. Chahine, V. Salomonson, V. Singhroy et al., eds. 2011. *Encyclopedia of Remote Sensing, Encyclopedia of Earth Science Series*. Berlin: Springer-Verlag: 1000 p.

Berger, Z. 1994. *Satellite Hydrocarbon Exploration*, Chap. 1. Berlin: Springer-Verlag: 3–34.

Lawrance, C., R. Byard, P. Beaven. 1993. *Terrain Evaluation Manual*, Chaps. 1–2. London: Transport Research Lab, Department of Transport State of the Art Review 7, HMSO Publications Centre: 1–12.

Morain, S.A., A.M. Budge, eds. 1999. *Earth Observing Platforms and Sensors*, **1**, *Manual of Remote Sensing*, 3rd ed. New York: John Wiley & Sons: CD-ROM.

Qu, J.J., W. Gao, M. Kafatos, R.E. Murphy, V.V. Salomonson, eds. 2007. *Earth Science Satellite Remote Sensing*. Berlin: Springer-Verlag: 1066 p.

Vincent, R.K. 1997. *Fundamentals of Geological and Environmental Remote Sensing*. New York: Prentice Hall: 480 p.

Webber, V.L., ed. 2009. *Environmental Satellites: Weather and Environmental Information Systems*. Hauppauge: Nova Science Pub. Inc.: 112 p.

Wigbels, L., R.G. Faith, V. Sabathier. 2008. *Earth Observations and Global Change: Why? Where Are We? What Next?* Washington, DC: Center for Strategic and International Studies: 48 p.

2

Photointerpretation Tools and Techniques

Chapter Overview

In this chapter, we discuss equipment needed to analyze imagery and the different approaches and techniques commonly practiced. Historically, image analysis has been done on film products. Today, the extraction of data from imagery is done primarily on a computer screen. Both of these techniques will be addressed here.

There are several tools of the trade that make photo analysis possible and others that make it easier (Table 2.1). Among those items generally considered indispensable are stereoscopes, magnifying glasses, and light tables. Other items include various types of marking pens, clear mylar overlays, filters, rulers at various scales, and machines for changing scales from photos to maps (commonly called "transferscopes"). For a more complete discussion of devices for measuring vertical and horizontal distances, refer to Ray (1960) and Miller and Miller (1961).

Image interpretation is, by definition, a subjective process based largely on an individual's experience and training. There are, however, image interpretation and image processing techniques such as image enlargement, filtering, and foreshortening that allow an interpreter to see and extract more information from photos and images.

Tools

Light Tables

Light tables are used to backlight positive transparencies, a common format of airphotos or space imagery. When used with paper or Duratrans (a color transparency film) prints, backlighting makes it easier to see detail otherwise hidden in dark areas of the image, and colors also appear more vibrant. Backlighting also helps prevent eye fatigue.

Light tables are available in many sizes and with varying types of illumination. It is desirable to get the illumination as even as possible to avoid distracting light bands, which can be caused by fluorescent tubes, for example. This is best achieved by having an "opal" glass or "frosted" glass cover on the light box to diffuse the light. If the table is used for cutting and splicing, it is wise to have a sheet of common glass over the frosted glass to avoid scratching the diffuser, which can be expensive to replace.

Magnifying Glasses

Magnifying glasses come in an assortment of sizes and powers, some illuminated. Illuminated magnifiers are particularly convenient when a light table is not available to

TABLE 2.1

List of Some Basic Equipment Needed to Begin Photointerpretation

Large work surface with good lighting or a light table
Colored pens that write on film
Colored pencils that write on photos
Clear film for overlays
Polymer erasers (soft enough that they do not scratch photographic emulsion)
Magnifying glasses (usually between 3× and 10× magnification)
Stereoscopes (pocket or mirror)
Rulers (metric or English)
Scale rulers (with distances given for various scales)
Planimeter (for measuring areas inside a closed polygon)
Topographic or planimetric maps as paper or film (to use as a base for the interpretation)
Weights and tape (to secure images and maps)

backlight prints. The most common types magnify images 3–10 times. Users should be aware that prolonged use of magnifying lenses can produce eye strain.

Stereoscopes

Stereoscopes allow three-dimensional viewing of the earth's surface and provide the image analyst with the best remote way to estimate the dip of units or the relief in an area (particularly when topographic maps are not available). Using vertical exaggeration, one can recognize dips as low as 1°–2° that otherwise would not be evident. Without stereo viewing, for example, it could be difficult to determine in some cases whether an area is a ridge or a valley, or whether a dark area consists of dark surface material or is just an area in shadow. Although some workers are capable of seeing three-dimensional images without using the stereoscope, for most analysts the stereoscope makes it a lot easier.

Stereoscopes come in pocket or table models, have various magnifications available, and make the interpretation of stereo images easier and more accurate. Stereoscopic viewing requires parallax, the apparent displacement of an object as a result of viewing from two different locations. This is achieved in aerial photography by overlapping successive photos acquired as the airplane moves along its flight path. Satellite parallax can be acquired using one of three techniques: (1) by imaging an object from adjacent orbital paths using a vertical scanner, (2) by imaging an area from different viewing angles using a nonvertical scanner, or (3) by acquiring overlapping photos in the case of orbital photography. Stereo viewing generally relies on 60% endlap of photos along a flight line and on 10%–20% sidelap of photos from adjacent flight lines (Figure 2.1).

Pocket stereoscopes are convenient for field work, but may require bending of one photo to see the entire area of overlap. Mirror stereoscopes are larger and more bulky, but allow the viewer to see the entire overlap area without moving the photos (Figure 2.2). Combined with magnifying lenses, these tabletop scopes can provide a maximum of detail with a minimum of eye strain.

Transferscopes

The purpose of these scale-changing devices is to transfer an interpretation from the image to a planimetric or topographic map base. This is achieved in several ways, but commonly this is by projection or split optics.

Projectors illuminate the image and the image is then projected through a lens onto a glass surface (screen). The base map is on clear or frosted film and is positioned on the

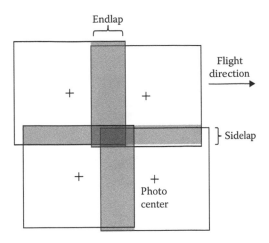

FIGURE 2.1
Diagram illustrating the difference between endlap and sidelap.

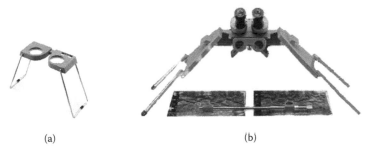

(a)   (b)

FIGURE 2.2
Stereoscopes. (a) Pocket stereoscope. (b) Mirror stereoscope.

glass screen. The scale is matched by varying the size of the projection by moving the original photo or an intermediate mirror closer or farther from the lens. One must have at least two landmarks or "ground control points" on both the image and base map to match scales. Distortion on these instruments is radial, increasing outward from the center of the projected image, so that only the central portion can be used with any degree of accuracy. Examples of such instruments include the double-reflecting projector (P.B. Kail Associates, Denver, CO) and reflecting projector (Keuffel and Esser "Kargl," NY).

Scale-changing instruments that use split optics are built by Bausch and Lomb (Rochester, NY), among others. These zoom transferscopes, both monoscopic and stereoscopic, use varying illumination intensity on the photographs and base map to allow the viewer to see either the photo, base map, or both simultaneously (Figure 2.3). The images are viewed through binocular lenses that can enlarge the photos, and other lenses can be used to enlarge the base map. These lenses impose a radial distortion on the transferred image. Flat or gently rolling topography is less affected by the distortion of lenses than high-relief areas.

All scale-changing equipment is more accurate than "eyeballing" an interpretation from an image to map base. Problems with data transfer from image to map arise out of the radial distortion inherent in airphotos, differences in cartographic projection between satellite images and base maps, and the amount of relief. If a transferscope is not available, the best technique to minimize distortion is to superimpose a grid on the image and match scales and landmarks within each grid cell.

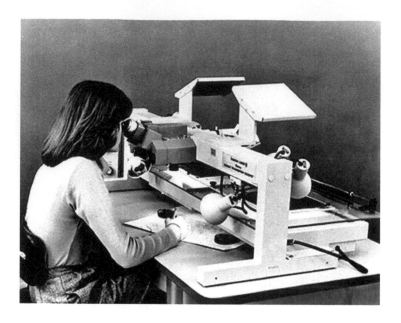

FIGURE 2.3
The zoom transferscope has split optics that allows the user to view both the photos and a base map simultaneously.

Scales and Planimeters

Engineering supply stores carry not only metric/English rulers but also often carry "scale" rules that give measurements in meters/kilometers, feet/miles, chains, rods, and so on for a given map or photo scale. These can be convenient for reading distances directly from the images when the scale is known. Most drawing, drafting, and mapping programs have distance tools that provide lengths and distances in any unit system.

Planimeters are devices, either mechanical or electric, that measure the circumference of an object and provide the area of that object. This is often faster and more accurate than overlaying a grid of known size and counting squares.

Parallax Bars

Parallax bars, or stereometers, are used to determine the height of features on photographic stereo pairs, generally for 23 × 23 cm (9 × 9 in.) frames. Consisting of a bar with two small glass plates, the instrument is positioned such that dots inscribed on the plates fall over identical ground points on the two overlapping photos. These dots are fused stereoscopically into a single dot that appears to float in the three-dimensional stereo image. By placing the dot on the top of a feature (e.g., a hill or strike ridge), then at the base of the feature, one can determine its height or differential parallax. This technique is described by Ray (1960) and Miller and Miller (1961). Of particular interest, this instrument can be used to help calculate stratigraphic thickness, determine dip magnitude, or generate topographic profiles.

Overlays

Photointerpreters have historically used soft lead or grease pencils to mark directly onto photographs. This has the disadvantage of covering the feature being mapped and making corrections difficult (erasing can remove the emulsion). These drawbacks can be overcome by using clear film overlays. Overlays can be cut to fit the photo or image, taped along one edge, and then annotated with special pens that write on film. The marks can be erased without damaging the photo, and the film can be raised to see the original image beneath. Overlays can be removed and the photo reinterpreted if a second opinion is required.

Base Maps

Interpretations are easiest to plot and compile on topographic maps because of the large number of corresponding features on the image and map. Planimetric maps generally have only roads and waterways to use as landmarks. Regardless of which type map is available, they should be in the form of a frosted film, or mylar, so that images can be projected on them, or so they can be directly placed over an image and the annotation traced or transferred to the map. Frosted film takes drafting ink better than clear film and maintains true scale better than paper. In areas with great relief or on cluttered maps, it is often helpful to have the map reproduced as a halftone or as a screened print. This makes the base map lighter than the inked lines of the interpretation and makes annotation easier to see and read.

Filters

Filters are used to help view images and map lineaments. Filters can be physical or digital. Digital filters are discussed in the section "Image Processing." Among physical filters are films that allow high-frequency information to be preserved at the expense of low-frequency data (high-pass filter) and enhance linear features. A coarse (79 line pairs per cm) diffraction grating on film, known as a Ronchi filter, can be used to view an image. Slowly rotate the filter and lineaments perpendicular to the grating become sharp, whereas lineaments at angles to the ruling appear diffuse (Pohn, 1971). This filter is an analog edge-enhancement technique.

A low-tech low-pass filter that allows the interpreter to concentrate on gradual tonal or color changes and exclude sharp or abrupt changes consists of simply removing one's glasses (or contact lenses). This is only effective when the interpreter requires corrective lenses and works because the high-frequency information in an image becomes unfocused. The use of filters is discussed further in the section "Mechanics of Image Analysis."

Image Characteristics

Photo Finish

Most photos have a semigloss or matte finish to reduce glare and allow one to annotate on the photo using color pencils. Some workers, however, prefer a glossy finish because details appear sharper. Photos carried into the field are often laminated to protect the emulsion from scratches and moisture and still allow annotation.

Paper versus Film

Paper prints are generally more convenient to carry into the field and annotate but must be protected from moisture and rough handling. Positive transparencies are often easier on the eyes when used with clear film overlays on a light table. They are, however, difficult to use and easily damaged in the field.

Hard Copy versus Digital

Traditionally, information extraction from imagery was performed on photographs, either monoscopic or stereoscopic, using a work table. It is probably safe to say that today much if not most image analysis is done digitally on a computer screen. Whether analysis is done on hard copy or digitally may ultimately come down to the analyst's preference, much as reading a book in digital or hard-copy form is determined by user preference.

The advantage of a paper print is that it can be easily carried to the field and it forms a more-or-less permanent record of the earth's surface at a point in time. It may be easier to hold a hard-copy image proprietary or confidential. The advantages of digital data analysis include the ability to duplicate an original quickly and inexpensively, send it anywhere quickly via the Internet, and allowing the analysis to be done by any number of interpreters or by the same analyst multiple times. Digital images can be digitally enhanced or filtered and are easily integrated with other types of data such as topography, gravity, or cultural data. Increasingly data are being carried to the field in digital form on pad-type devices.

Mechanics of Image Analysis

Image Interpretation

Image interpretation, or image analysis, is the process of viewing imagery and attributing meaning to what is seen. The analysis involves pattern recognition, recognizing targets based on experience in the field and prior training, what has been termed "the calibrated eyeball." It is essential to know not only what may be on the ground but also how the imagery was generated to properly understand what one is seeing.

Different people have different spatial cognitive skills (Kastens and Ishikawa, 2006; Kastens et al., 2009; Titus and Horsman, 2009; Kastens and Manduca, 2012). In the geosciences, it is helpful to think and perceive objects in three dimensions. This is the ability to recognize that a chair is the same object whether it is facing toward you or away from you or is hanging upside down from the ceiling. In this way one can recognize what is physically plausible, and what is not (Figure 2.4). Some people see only what they expect to see, and some people just see things differently than the rest of us. Having said that, we are all naturally programmed to focus on certain shapes such as the human face (e.g., the "face" reported in the Cydonia region of Mars seen on a 1976 NASA Viking image; Figure 2.5) and to tease order out of randomness (e.g., the Martian canals of Percival Lowell; Figure 2.6).

Our brains work using feedback loops to impress objects in our memory.

> The initial input – a sight or sound – becomes altered by the downstream conception of what that sight or sound may be. Our higher level [experiential] processing actually modifies our initial processing. Perception is not pure and direct; it is affected by our learned expectations. So our prior experience with [objects] – what they look … like, where they occur, when we've seen them before – all can quite literally affect the way we perceive [them] today.
>
> **Lynch and Granger, 2008**

For this reason those who have seen bedding features, folding, or mineral alteration in the field will be better at recognizing these things on imagery. As a former teacher put it, "reviewing by viewing promotes learning while relearning" (Dinkel, 1959). In other words, cognition (learning) is a function of recognition (experience).

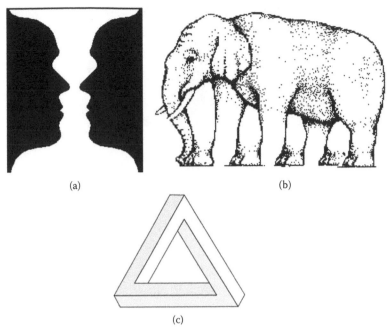

(a)   (b)

(c)

FIGURE 2.4
Several illusions. (a) Faces or vase? Two analysts looking at the same image may see different objects. (b) Elephant looks feasible at first glance, but on closer inspection is physically impossible. (c) Penrose triangle. Any given vertex is feasible, but taken all together it is another physical impossibility.

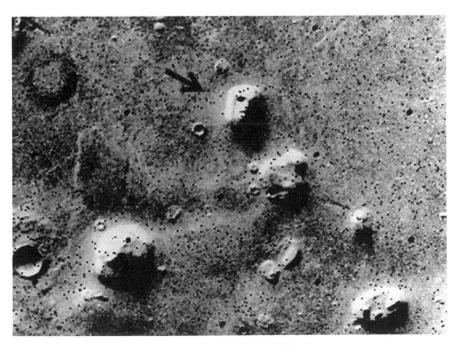

FIGURE 2.5
NASA Viking image of Cydonia region, Mars. This shows the infamous "Martian face" that some people are convinced is a cultural artifact. (Image courtesy of NASA.)

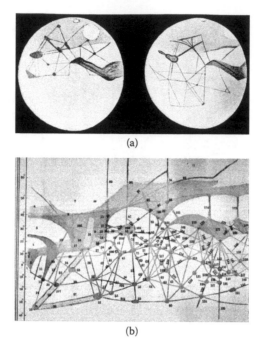

FIGURE 2.6
Martian canals as depicted by the renowned astronomer Percival Lowell. (a) Hemispheres. (b) Detail of Lowell's canals. These drawings are the product of a respected astronomer seeing what he expected to see rather than what is there.

As an aside, it is not necessary to see in stereo to analyze imagery, but it probably helps in the case of structural interpretation because recognizing dip is often the key to a correct interpretation.

Interpretations are prone to several kinds of error. They can miss a target entirely; just not see something that is actually there. They can see something on imagery that is not there. They can see something that is there, but identify it as the wrong thing. Or they can recognize and identify an object correctly and attribute the wrong significance to that object, such as identifying a landslide scarp, but feeling it is not important enough to warn the homeowner living downslope of potential danger.

Ideally the image analyst will either correctly identify the object of the exercise or provide a short list of possible alternative interpretations for what is seen. One can provide a scale of confidence in the interpretation of an object, much as a Geological Survey will mark heavy or solid lines on a map where a fault or contact is clear, and thin or dashed or dotted lines where an interpretation is less clear or speculative.

A proper image analysis is a result of multiple interrelated factors, including training and experience, the appropriate scale for a particular interpretation, knowing the context, recognizing and filtering out noise (the "signal-to-noise ratio"), and understanding the unique perspective, color, contrast, and texture of the imagery.

Influence of Scale

The scale of the image has a great influence on the amount of detail observed and mapped. Landsat Multispectral Scanner (MSS) images, for example, in the early 1970s were often interpreted at scales of 1:1,000,000. Considering the amount of detail visible at 1:250,000, let alone the ease of mapping, it is not surprising that new interpretations of the same images at an enlarged scale can provide a wealth of new information. In the author's experience, larger scales yield better (more detailed) maps. If only regional features are required, it is always possible to back off and view the image from a distance.

Often one has a choice whether to use satellite images or more detailed airphotos. Not only do most airphotos offer stereo viewing, they also provide a wealth of detail generally not available from orbital altitudes. The main drawback when dealing with airphotos is that large numbers of photos may be required to cover an area that only one or a few small-scale satellite images would cover. For example, it takes four photos at 1:25,000 to cover the same area covered by one photo at 1:50,000. The interpreter must then consider the trade-off between detail versus time to interpret and the added cost for more photos.

Influence of Color

A similar argument can be made when considering the use of color versus black-and-white imagery. Some have claimed that an experienced interpreter can glean as much information from black-and-white photos as from color. Most geological interpreters disagree. Color photography not only makes interpretation quicker and easier but also makes it possible to recognize lithologic units of interest, trace them laterally with greater confidence, and recognize repeated section or missing section in faulted and folded terrains. Estimates of additional information over black-and-white images range from 10% to 25%, but that information can be critical in the final analysis. Again, the main trade-off is one of cost. Color photography can cost two to three times as much as black-and-white photography. Digital imagery is almost entirely in color or multispectral.

Black-and-white satellite imagery has some advantages over color. It is often quicker and less expensive to generate one band of data than the three required for a color composite. Black and white of the near-infrared or red spectrum (e.g., Landsat MSS band 7, Thematic Mapper [TM] band 4, or SPOT P) can be an excellent choice for a "quick look." Interpreters generally see different features on different bands (wavelengths). In some cases, a black-and-white image can highlight certain features lost in the background of a color scene. As each project has unique objectives, each case must be considered on its merits.

False color images, particularly color infrared, have specific advantages over true color. Although it is easy to recognize units using true color photos, color infrared images will present geology quite well, eliminate most atmospheric scattering (haze) by eliminating the blue band, and emphasize variations in vegetation (species, growth stage, seasonal variations, stress). False color combinations can emphasize other features, from mineralogical changes to moisture changes. The primary disadvantage is that the interpreter must become familiar with each new color combination and what the colors represent under the unique conditions of image acquisition and processing. It may be necessary to obtain control points in the field to improve the accuracy of the interpretation.

Technique

Every interpreter has a preferred technique for working through a photo or image. Generally, one begins in a known area, an area with field mapping or control points, or where a complete section is exposed and there are good outcrops. One then works outward into lesser known or poorly exposed areas. This is the process of extrapolating the known into the unknown.

Often there are large areas of poor or monotonous exposure, or none of the area is well known. In such cases, one can impose a grid over the image and work in one grid cell at a time. This has the effect of concentrating one's attention on small areas rather than the whole scene. The entire image can have more information than is readily absorbed at a glance.

Effect of Viewing Direction or Perspective

An image often yields different information depending on the direction it is viewed. For this reason, many workers will rotate an image and observe it from multiple angles. The direction that shadows fall often has an effect on the perception of topography. A stream can appear either low (as a valley) or high (as a ridge) depending on how the shadows fall. Generally, if an image is oriented such that shadows fall toward the viewer, the relief will appear correctly (Drury, 1987).

Foreshortening

A technique that is often used by seismic interpreters is viewing the image at a shallow or glancing angle. The effect, known as foreshortening, enhances slight changes in shape and detection of continuity. It is especially effective for joining discrete segments of a fault zone into a regional lineament, for example.

Filtering

Filtering is done to improve the "signal-to-noise ratio," that is, the information content of the imagery. Physical filters such as the Ronchi grating can enhance lineaments without image processing. Viewing black-and-white images through red film provides the illusion of depth.

When viewing imagery on-screen, there are a multitude of digital filters that can be applied, including those that sharpen edges or remove random noise (despeckling). Some filters smooth out high or low spatial frequencies or filter out specific wavelengths. For a discussion of filter types, see the "Filtering" section under "Image Processing."

Viewing with "Fresh Eyes"

Several passes at an image may be required before all the pertinent information is extracted. In the first stage, for example, one might trace all bedding and look for repeated section, missing section, and other irregularities. The second stage might involve mapping all traces of fracturing. A third phase might include looking for tonal or vegetation anomalies, and a final pass would be mapping and interpreting drainage patterns.

A variation on this theme is having another person take a look at your image (a second opinion). The process of image analysis is extremely individualized, as each person has a different set of experiences and training and natural ways of looking at things. The old adage of "two heads are better than one" more often than not proves to be the case.

Multiple Images of the Same Area

When several images overlap in an area, a comparison of the area of interest on each of the images often reveals new information due to changes in illumination, plant cover, or other feature (snow cover, moisture change) visible on imagery acquired at different times. In addition, more details of the surface may be obvious due to the stereo capability that derives from the overlap.

Image Processing

Image processing takes a digital file such as a photograph or satellite image and creates an image or manipulates an image to make it easier to analyze. Digital images consist of "picture elements," or "pixels," equivalent to the grain in film, that have a brightness value in each of the colors (also known as channels, bands, or wavelength ranges) that make up the pixel. It is not the intent of this text to go into detail on image processing techniques. Excellent texts describe image processing: some of the more recent include Petrou and Petrou (2010), Parker (2010), Canty (2009), Croitoru and Agouris (2009), Gao (2008), Nixon and Aguado (2008), O'Gorman et al. (2008), Chen (2007), Gonzalez and Woods (2007), Burger and Burge (2007), Richards and Jia (2006), Jenson (2004), and Varshney (2004). Many software packages exist, including ENVI, ArcView, SPRING, MultiSpec, GRASS, PCI Image Works, TNT Lite, ER Viewer, DLGView, DEM3d, MrSID GeoExpress View, OpenEV, and Photoshop (see http://www.cof.orst.edu/cof/teach/for421/Software.html for free software).

Image processing can now be done almost instantaneously on commonly available programs such as Photoshop and on standard personal computers with average capabilities. A word of caution is in order, however. Despite what you see in the movies, an image cannot be "improved" if there is no information content in the data. No amount of processing or filtering can create data: the information has to be there for it to be enhanced. Likewise, digital enlargement just makes more or larger pixels with the same information content. It does not add new data.

The following is a brief rundown of some of the more common image processing techniques and what they are used for.

Contrast Enhancement (and Suppression)

Contrast enhancement allows the user to maximize the contrast between brightness values in an image. For eight-bit data (still the most common format), this means redistributing the 256 brightness values, usually in a near Gaussian histogram that uses only part of the available range of "digital numbers" or brightness values. Contrast enhancement, or stretching, focuses on getting the most information from where most of the brightness values are. This has the effect of altering the histogram of brightness values such that instead of a normal (Gaussian) distribution they are now "stretched" to fill the entire available dynamic range (0–255). An unfortunate side effect is that areas which appear light on the original image may get washed out and original areas which are dark may go black. Alternatively, the contrast enhancement may focus on only the light areas to the detriment of the middle and dark tones or only on the dark areas and lose middle and light area contrast (Figure 2.7).

Contrast suppression, the inverse of contrast enhancement, can be useful in areas where there is too much contrast, say in areas with white sand over basalt. In this case, the contrast could be altered to enhance information in the dark and light areas of the image (e.g., histogram equalization filtering).

Filtering

As mentioned in "Filtering" under the section "Technique," digital images can be filtered to achieve various effects (e.g., Photoshop Elements online help). For example, the "equalize" filter redistributes the brightness values of the pixels in an image so that they more evenly represent the entire range of available brightness levels (Figure 2.8).

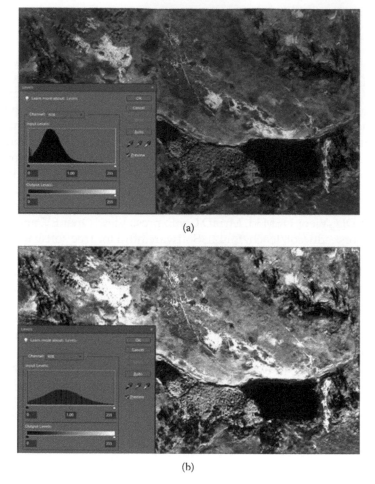

(a)

(b)

FIGURE 2.7
Airphoto and corresponding histograms showing (a) no contrast enhancement and (b) contrast stretch. Note how some light areas appear "washed out" on the contrast-enhanced image. (Figures enhanced using Adobe Photoshop.)

The "threshold filter" allows you to specify a certain level as a threshold: all pixels lighter than the threshold are converted to white and all pixels darker than a threshold are converted to black.

"Blur filters" soften all or part of an image and are useful for retouching. They smooth transitions by averaging the pixels next to the hard edges of lines and shaded areas in an image. "Gaussian blur" blurs a selection by an adjustable amount. Photoshop applies a weighted average to the pixels that results in a Gaussian histogram. This filter adds low-frequency detail and produces a fuzzy effect. "Box blur" blurs an image based on the average value of neighboring pixels. By adjusting the radius of affected pixels, you can increase the amount of blurring.

"Noise filters" generally remove noise or randomly distributed color or tone levels. They blend a sharp contrast pixel into the surrounding pixels. The "despeckle filter" detects the edges in an image where significant color or tone changes occur and blurs all the image except those edges. Noise filters remove noise while preserving detail. "Median filters" reduce noise in an image by blending the brightness of pixels within a selected area.

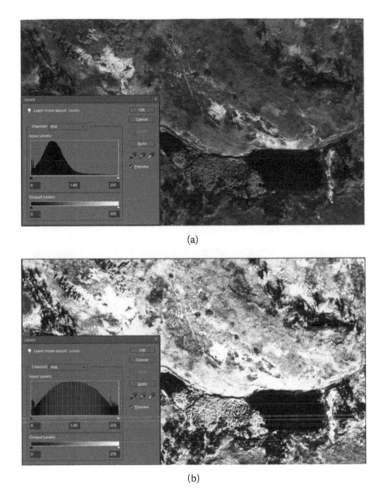

(a)

(b)

FIGURE 2.8

Airphoto and corresponding histograms showing (a) no contrast enhancement and (b) histogram equalization stretch. (Figures enhanced using Adobe Photoshop.)

The filter searches the area for pixels of similar brightness, discarding those that differ too much from adjacent pixels and replacing the center pixel with the median brightness value of the searched pixels.

"High-pass filters" retain edge details in a specified radius around each pixel where sharp color or tone transitions occur and suppress the rest of the image (Figure 2.9). This filter retains high-frequency information and removes low-frequency detail. It has the opposite effect of a blur filter and can cause an image to appear speckled. "Edge enhancement" or "sharpening filters" identify areas of the image with significant transitions and emphasize the edges. They highlight edges or abrupt color or tonal transitions with a line of darkened pixels against a lighter background. These filters can be modified to sharpen only edges while preserving the overall smoothness of the image. They are useful for finding structural alignments such as faults or intersections of bedding with the surface. A drawback to this type of filter is that it will enhance any linear noise in the data. For this reason, one should remove noise before applying a sharpening filter.

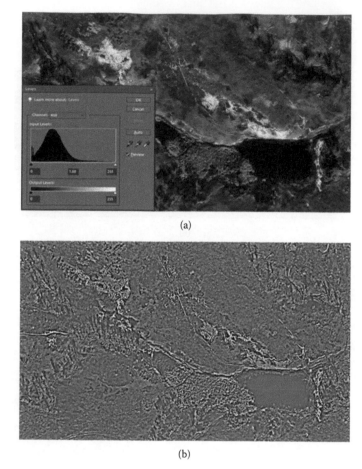

(a)

(b)

FIGURE 2.9
Airphoto showing (a) no contrast enhancement and (b) high-pass filter. (Figures enhanced using Adobe Photoshop.)

A "low-pass filter" will remove high-frequency data while preserving the low frequency information. It is the inverse of the high-pass filter.

A "band-pass filter" allows only a specified continuous frequency range to remain and suppresses or deletes the remaining brightness values.

Ratioing

Ratioing is a processing technique where brightness (reflectance) values in one channel are divided by values for the same pixel in another channel. Ratios enhance the contrast between materials with different reflectance at specific wavelengths and suppress the effects of shadows (topography), since low reflectance values are divided by low values in shadowed areas (Figure 2.10). Although the range of possible ratio values goes from zero to infinity, most ratio values fall between 0.5 and 2.0 and are then stretched, or scaled, back to the normal range of 0–255 (for 8-bit data such as Landsat and SPOT). If three sets of ratios are combined and each is assigned a different color, a color composite ratio image is generated. The ratio technique is one of many image processing methods that use discrete, multiband multi-spectral data to generate images to show the distribution of specific mineral assemblages.

FIGURE 2.10

Clementine ratio image of the Moon. Bands used are 750/415 nm displayed as red; 750/950 nm green; 414/750 nm blue. Blue indicates iron-rich high titanium areas; yellow-orange indicates iron-rich low titanium. Note how crater topography has been suppressed. (Image courtesy of NASA. http://www.mapaplanet.org/explorer/ help/data_set.html#moon_clementine_ratio.)

Transforms and Spectral Band Merging

Some techniques ought to be mentioned briefly when discussing methods for enhancing the spectral information content of images. "Principal components (PCs)," "intensity–hue–saturation (IHS)," and "decorrelation stretch transformations" make it easier for the interpreter to distinguish between materials with low spectral contrast.

The PC transformation is used to decrease the amount of correlation between bands of data and maximize the differences between bands (Loe've, 1955). In most images, the information content of adjacent wavelength bands is nearly identical (it is highly correlated). An object that is highly reflective in the red region of visible light is probably also highly reflective in yellow or even green light. A three-dimensional plot of reflectance values in three bands (displayed as red, green, and blue and known as "RGB space") would generate an ellipsoidal cluster of points. This ellipsoid is usually *not* parallel to one of the reflectance axes. The tightness of this cluster is an indication of the redundancy of the data sets: the tighter the cluster (the more it approximates a straight line), the more the correlation between bands. The PC transformation generates a new set of orthogonal (noncorrelated) axes in space such that the principal axis is oriented through the long dimension of the ellipsoid, and the secondary and tertiary axes define correspondingly smaller distributions or densities of data points. This transformation can be performed on any number of bands and will generate axes in *n*-space where the *n* axes correspond to the number of input bands. Generally speaking, the first PC contains the brightness information of the image; the second, third, and so on components contain progressively less information; and the last components tend to contain first the systematic and then the random noise in the image (Figure 2.11). One can create a color image using any three components. Usually the first three components are used. The second and third components tend to reveal subtle differences in surface materials. One drawback of this technique is

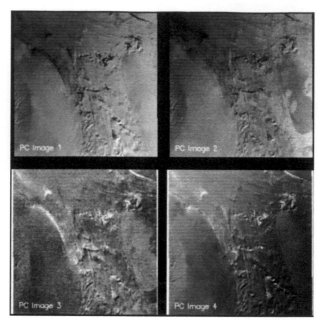

FIGURE 2.11
First four principle components derived from Landsat bands 2, 3, 4, and 5. (Image of Antarctic ice courtesy of Coyote's Guide to IDL Programming, Copyright© 2007–2008 David W. Fanning and the National Snow and Ice Data Center, University of Colorado, Boulder, CO.)

that, unlike standard color images or ratios, it is not possible to predict what colors will correspond to specific surface materials, since each PC image is a function of the unique distribution of brightness values within the original image. This also means that the color combinations are not consistent from one image to the next, that is, the colors generated for a limestone in one image are not necessarily the same colors as those in the same limestone in a later image, or for the same limestone in an adjacent image with a different distribution of brightness values.

A variation of the PC image is the decorrelation stretch (Kahle et al., 1980). In the decorrelation stretch, a PC image is generated, then is contrast enhanced using, for example, a histogram equalization stretch. This new data are then subjected to an inverse PC transformation that returns it to the original RGB color space. The resulting data has the same color combinations as the original color image, but the colors tend to be highly saturated and spectral differences are intensified (Figure 2.12).

The IHS transform takes an image from the familiar red–green–blue color space to a new system of IHS (Short and Stuart, 1982). Intensity is a function of brightness; hue represents the color of the object; and saturation is the purity of the color. An image is first transformed into IHS space, where the IHS components are individually contrast enhanced. They are then transformed back to the original RGB color space and result in an image with greater dynamic range in colors and tones. There are a number of variations of this transformation. In one example, known as "IHS with band substitution," the original Landsat TM image has bands 7-3-1 as red–green–blue. These are transformed to IHS space and stretched. Then the intensity band is removed and replaced by the TM band 4 and the transformation is reversed to return to RGB space (Figure 2.13). This technique has been shown to amplify subtle spectral differences in the original image.

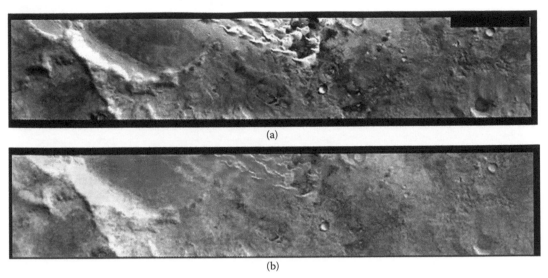

(a)

(b)

FIGURE 2.12

Decorrelation stretch, Kaiser Crater, Mars. (a) Gray scale image showing surface temperature. (b) False color composite made from three THEMIS bands using a decorrelation stretch to emphasize composition. Basaltic sand dunes are pink/magenta. 100 m pixels. Mars Odyssey mission. (Courtesy of NASA/JPL.)

FIGURE 2.13

Landsat Thematic Mapper (TM) image over the Hodna field area, Algeria. The image was generated using an intensity–hue–saturation transform on bands 7-3-1, and then replacing intensity with TM band 4 and transforming back to red–green–blue space. (Processed by Amoco Production Company, Houston, TX.)

A transform that has been used to minimize noise, especially noise introduced by atmospheric gases, is called "minimum noise fraction analysis" (Green et al., 1988). A correlation matrix is created to estimate the noise in the imagery. The transformation decorrelates and scales the noise fraction, creates components with variance information, and then transforms these components back to the original space and number of channels using only the nonnoise components.

Color Manipulation

Color manipulation can be as simple as arbitrary assignment of red, green, or blue to various wavelength bands. It can involve manipulating the intensity, hue, and saturation of individual and collective bands. It may be advantageous to change the contrast and brightness of one specific color band. One can remove a specific color or replace it with another. One can adjust the color balance, making, for example, an image with a greenish cast (shows vegetation better in true color) into one with a reddish cast (shows sulfide alteration better on a true color image).

Density Slicing

Density slicing, or gradient mapping, maps pixel brightness to colors or gray shades in a preselected gradient (Figure 2.14). The gradient can be continuous or incremental (e.g., a continuous gray scale can be broken into 8 or 10 discontinuous gray steps). A color gradient can be useful, for example, in visualizing warm to cool areas on an originally black-and-white gray scale thermal image.

Vegetation Indices

Vegetation indices have been generated to enhance specific features of vegetation. A "vegetation index" is "the reduction of multispectral scanning measurements to a single value for predicting and assessing vegetation characteristics" (NRCan, 2008). This could include values for healthy versus stressed vegetation, leaf area, biomass, chlorophyll content, plant height, or percent ground cover. For healthy vegetation, this could be as simple as dividing the near-infrared band by the red band, since healthy plants are highly reflective in the near-infrared and much less so in the red. The equation for this would then be

$$VI_{healthy} = \frac{NIR}{R}$$

where NIR is reflectance in the near-infrared and R is reflectance in red wavelengths.

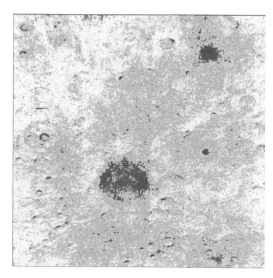

FIGURE 2.14
Density slice image, Kaiser Crater. (Courtesy of NASA.)

The Normalized Difference Vegetation Index (NDVI) reveals the extent to which reflection is influenced by photosynthesis, that is, parameters such as vegetation quantity, biomass, and productivity. It is calculated:

$$NDVI = \frac{(NIR - R)}{(NIR + R)}$$

There are some limitations to the use of vegetation indices. The NDVI should not be used to investigate short-term events like fires; they are more appropriate for long-term processes such as vegetation growth throughout a season. Atmospheric scattering increases the amount of red light measured by a sensor, which tends to reduce the NDVI values. Scattering increases when the sensor is pointed off-nadir. NDVI pixels rarely cover homogeneous vegetation, so the response is an average of all natural and agricultural vegetation in the area covered by a pixel. Likewise, soil and rock reflectance effects NDVI values: greater soil or rock reflectance gives lower NDVI values, particularly when there is between 40% and 75% plant cover. Finally, sensor calibration can have a significant impact on NDVI values. Since radiometers drift and degrade over time, the calibration must be adjusted as well (University of Reading, 2002).

Other techniques, including spectral unmixing algorithms, have been developed as "vegetation suppression methods" to minimize the effects of vegetation cover and enhance the remaining information content of the image. These can be extremely useful for mapping soil or bedrock in heavily forested areas (see the "Geobotany and Geological Mapping" section in Chapter 15).

Classification

The objective of digital image classification is to classify objects in the image. We wish to subdivide a landscape into categories to help understand the distribution of surface features. The surface features can be rocks, mineral alteration, soils, vegetation, water bodies, or man-made features (mines, tailings, clear cuts, etc). The classification is based on the knowledge that different materials have different spectral characteristics. If we have a sensor with n spectral bands, we should be able to get the spectral signature (brightness range) for the surface materials in each of those bands or in n spectral dimensions. If we can determine the spectral signature for each surface material, we can then classify all the materials in an image.

Classification can be either unsupervised or supervised. "Unsupervised classification" is based entirely on an analysis of the image statistics with no prior knowledge of what exists on the ground. In an unsupervised classification, a program looks at each pixel in multiple channels or wavelengths (n-dimensional space), establishes a region in n-dimensional space characterized by a cluster of brightness values, and assigns that cluster of values to one of a number of "classes." The program determines statistically which clusters have the best separation in n-dimensional space. It is then up to the interpreter/analyst to determine exactly what each class represents.

A "supervised classification," on the other hand, is "a procedure for identifying spectrally similar areas on an image by identifying "training" sites of known targets and then extrapolating those spectral signatures to other areas of unknown targets" (NRCan, 2008). Supervised classification requires a training set, or areas on the ground where the surface materials are known. The range of brightness values for a surface material is quantified in n-dimensional space, where n is the number of wavelengths or bands available. The classification program scans the remainder of the image looking for surface materials with brightness characteristics similar to those of the training areas and assigns them to known classes of materials.

Once a supervised classification has been performed, it is necessary to determine the accuracy of the classification. This is usually done by verifying the classes during a field program.

Geographic Information Systems

A "geographic information system" (GIS) is "a computer-based system designed to input, store, manipulate, and output geographically referenced data" (NRCan, 2008). Any data with a location (*x*–*y* coordinates) is geographic information. All imagery is geographic information, as is cultural data (roads and other infrastructure, land use information, and land boundaries). Geologic maps, hazard maps, vegetation cover, soil maps, topography, bathymetry, and geochemical sample locations all qualify as geographic information. These different types of information must be in digital format and rectified to a given coordinate system before they can be digitally overlain, merged as different colors or densities, or mathematically fused into new kinds of maps.

A GIS can be descriptive, predictive, or prescriptive (NRCan, 2008). It can describe the surface at a point in time under a specific set of conditions. Conclusions about the surface can be compiled, as in making a geologic or soil moisture map. A GIS can help the analyst predict future surface conditions based on knowledge of the present surface cover and active processes. This could include hazard maps showing where landslides or floods are likely to occur. Finally, a GIS can incorporate present surface information and processes into multiple scenarios, such as suggesting the most economic versus safest versus least environmental impact maps for mine and tailings development plans. Examples of data integration using GIS are provided throughout the text.

TABLE 2.2

Commonly Used Photogeologic Map Symbols

Symbol	Description	Symbol	Description
⊤	Strike and dip of bedding; dips 1–3°	⇌	Strike-slip fault, displacement shown
⊤⊤	Strike and dip of bedding; dips 4–10°	▼▼▼	Thrust fault, teeth on upper plate
⊤⊤⊤	Strike and dip of bedding; dips 11–25°	——	Fracture or lineament: fault, joint, or shear zone, displacement undetermined.
⊤⊤⊤⊤	Strike and dip of bedding; dips 26–45°	⊣⊢	Anticline
∇	Strike and dip of bedding; dips greater than 45°	⊣⊢	Syncline
⊕	Strike and dip of horizontal beds	⇕	Geomorphic high or arch
⊣—	Strike and dip of near-vertical beds	⇕	Geomorphic low or trough
⤢	Strike and dip of overturned beds	- - - - -	Contacts
⊤	Geomorphic strike and dip (no outcrops)	··········	Structure form lines
⊤	Normal fault, ball on downthrown side; dashed where approximately located; dotted where concealed or extrapolated, querried where questionable.		

Each Earth science problem has a unique set of objectives, and it is worthwhile pausing during the analysis to review these objectives and ensure that the work stays on track. It never hurts to get a coworker to provide a second opinion: more often than not they will notice something important that was missed.

Some of the more common photointerpretation symbols are shown in Table 2.2. These should be used in conjunction with the standard mapping symbols given in field mapping texts such as Compton (1962) or Lahee (1952).

References

Burger, W., M.J. Burge. 2007. *Digital Image Processing: An Algorithmic Introduction Using Java*. New York: Springer, 586 p.

Canty, M.J. 2009. *Image Analysis, Classification, and Change Detection in Remote Sensing: With Algorithms for ENVI/IDL*, 2nd ed. Boca Raton, FL: CRC Press, 471 p.

Chen, C.H. 2007. *Image Processing for Remote Sensing*. Boca Raton, FL: CRC Press, 400 p.

Compton, R.R. 1962. *Manual of Field Geology, Appendix 4*. New York: John Wiley & Sons, 334–337.

Croitoru, A., P. Agouris. 2009. *Next Generation Geospatial Information: From Digital Image Analysis to SpatioTemporal Databases*. Leiden, the Netherlands: A.A. Balkema Publishers, 198 p.

Dinkel, R.E. 1959. Reviewing by viewing promotes learning while relearning. *The Math. Teach.* 52: 459.

Drury, S.A. 1987. *Image Interpretation in Geology*, Chap. 2. London: Allen & Unwin, 26–27.

Gao, J. 2008. *Digital Analysis of Remotely Sensed Imagery*. New York: McGraw-Hill Professional, 674 p.

Gonzalez, R.C., R.E. Woods. 2007. *Digital Image Processing*, 3rd ed. Upper Saddle River, NJ: Prentice Hall, 976 p.

Green, A.A., M. Berman, P. Switzer, M.D. Craig. 1988. A transformation for ordering multispectral data in terms of image quality with implications for noise removal. *IEEE Trans. Geosci. Rem. Sens.* 26: 65–74.

Jenson, J.R. 2004. *Introductory Digital Image Processing*, 3rd ed. Upper Saddle River, NJ: Prentice Hall: 544 p.

Kahle, A.B., D.P. Madura, J.M. Soha. 1980. Middle infrared multispectral aircraft scanner data analysis for geological applications. *Appl. Opt.* 19: 2279–2290.

Kastens, K.A., T. Ishikawa. 2006. Spatial thinking in the geosciences and cognitive sciences: A cross-disciplinary look at the intersection of the two fields. *GSA Spec. Pap.* 413, 53–76.

Kastens, K.A., C.A. Manduca. 2012. Earth and Mind II: A synthesis of research on thinking and learning in the geosciences. *GSA Spec. Pap.* 486, 210 p.

Kastens, K.A., C.A. Manduca, C. Cervato, R. Frodeman, C. Goodwin, L.S. Liben, D.W. Mogk et al. 2009. How geoscientists think and learn. *EOS Trans. AGU.* 90: 265–272.

Lahee, F.H., 1952. *Field Geology*. New York: McGraw-Hill Book, 615–620.

Loe've, M. 1955. *Probability Theory*. Princeton, NJ: Van Nostrand, 515 p.

Lynch, G., R. Granger. 2008. *Big Brain: The Origins and Future of Human Intelligence*. New York: Palgrave MacMillan, 259 p.

Miller, V.C., C.F. Miller. 1961. *Photogeology*, Chap. 3. New York: McGraw-Hill Book, 51–58.

Nixon, M., A.S. Aguado. 2008. *Feature Extraction & Image Processing for Computer Vision*, 2nd ed. Amsterdam, the Netherlands: Academic Press, 424 p.

NRCan. 2008. Glossary of Remote Sensing Terms. Available at http://www.nrcan.gc.ca/earth-sciences/geography-boundary/remote-sensing/kids/1776?destination=node%2F1776 (accessed September 9, 2012).

O'Gorman, L., M.J. Sammon, M. Seul. 2008. *Practical Algorithms for Image Analysis with CD-ROM*. Cambridge: Cambridge University Press, 360 p.

Parker, J.R. 2010. *Algorithms for Image Processing and Computer Vision*, 2nd ed. Indianapolis, IN: Wiley Publishing, 504 p.

Petrou, M., C. Petrou. 2010. *Image Processing: The Fundamentals*, 2nd ed. Chichester, UK: John Wiley & Sons, 818 p.

Photoshop Elements online help. Available at http://helpx.adobe.com/photoshop.html (accessed September 10, 2012).

Pohn, H.A. 1971. Analysis of images and photographs by a Ronchi grating, Part 1, Remote sensor application studies progress report, 1968–69. National Technical Information Service, PB 197 101: 9 p.

Ray, R.G. 1960. *Aerial Photographs in Geologic Interpretation and Mapp, U.S. Geol. Survey Prof. Pap. 373*. Reston, VA: U.S. Geological Survey, 41–49.

Richards, J.A., X. Jia, 2006, *Remote Sensing Digital Image Analysis: An Introduction*, 4th ed. Berlin: Springer-Verlag, 468 p.

Short, N.M., L.M. Stuart. 1982. *The Heat Capacity Mapping Mission, NASA SP 465*. Washington, DC: U.S. Govt. Printing Office.

Titus, S., E. Horsman. 2009. Characterizing and improving spatial visualization skills. *J. Geosci. Educ.* 57: 242–254.

University of Reading, Department of Meteorology. 2002. Normalised Difference Vegetation Index. Available at http://www.pvts.net/pdfs/ndvi/3_3_ndvi.PDF (accessed September 9, 2012).

Varshney, P.K. 2004. *Advanced Image Processing Techniques for Remotely Sensed Hyperspectral Data*. Berlin: Springer-Verlag, 338 p.

Additional Reading

Desjardins, L. 1950. Techniques in photogeology. *Bull. Am. Assn. Pet. Geol.* 34: 2284–2317.

Lawrance, C., R. Byard, P. Beaven. 1993. *Terrain Evaluation Manual*, Chap. 3. Transport Research Laboratory, Department of Transport State of the Art Review 7. London: HMSO Publications Centre, 1–12.

Ray, R.G. 1956. Photogeologic procedures in geologic interpretation and mapping. *U.S. Geol. Survey Bull.* 1043-A: 1–21.

3

Remote Sensing Systems

Chapter Overview

Most of us have used cameras to take family or vacation photos. In the not too distant past we used cameras and film. Today, the majority of us have become adept at using digital cameras. Viewers of the evening news have seen, perhaps without knowing it, weather maps produced from thermal infrared (TIR) imagery acquired by dedicated weather satellites. The maps show the distribution of clouds over the Earth's surface. The satellites detect clouds because they are cooler than the background. Weather radar images indicate areas of light to heavy rainfall, and Doppler radar can warn the public of the location and movement direction of tornados. Medical instruments use thermal scanners (e.g., mammography), and energy efficiency consultants use thermal imaging to show where buildings require insulation. There is a bewildering array of instruments, from cameras to scanners and antennas, both active and passive, on satellites and aircraft, which are capable of recording a wide range of the electromagnetic (EM) spectrum. These sensors are available for exploration, environmental and engineering work, research, and public information purposes.

This chapter is divided into two parts. In Instrument Systems, the physics behind sensors, and their useful characteristics and applications are described. This is not meant to be an exhaustive discussion; references are provided so the interested reader can learn more about the instruments. The second section, Instrument Platforms, enumerates and describes the airborne and satellite platforms themselves. Although some airborne surveys are flown by governments, many are flown by commercial service providers. Some providers are mentioned. Although satellites were initially the exclusive domain of governments, there are now many satellites that are purely commercial or developed as government–private partnerships. Earth observation satellites (including those that observe the oceans and atmosphere) are listed chronologically by the primary sponsoring country or agency. The purpose of the satellite and the instruments on board are described.

Instrument Systems

Cameras and Photography

"Airphotos" are photographs acquired by aircraft using cameras specially adapted so that the film advance and shutter exposure is synchronized to flight speed. This is to ensure that each frame has at least the 60% overlap (endlap) required for stereoscopic viewing.

Several terms should be familiar to the user working with airphotos:

"Scale" is the ratio of image distance to ground distance and is a function of altitude and camera focal length:

$$\text{Scale} = \frac{\text{Focal length}}{\text{Camera height}}$$

For example: 6 in. (0.5 ft.) focal length lens/24,000 ft. altitude = 1:48,000 scale.

A 15.25 cm (6 in.) focal length is used on most airborne cameras. Larger scales mean more detail is available on the photo, but less area is included in each frame. For example, standard 22.9 × 22.9 cm (9 × 9 in.) frames at a scale of 1:63,360 (flown at 9,748 m, or 31,680 ft.) each cover an area ~210 km² (81 mi.²), whereas the same size frame at a scale of 1:24,000 (flown at 3,692 m, or 12,000 ft.) covers only 30 km² (11.6 mi.²). Halving the scale from 1:24,000 to 1:48,000 quadruples the area covered, and uses a quarter of the number of frames. This should be considered from a cost standpoint when planning surveys.

"Vertical exaggeration" occurs when the vertical scale of a stereo image is not the same as the horizontal scale. This can help with the interpretation of slopes, dips, fault displacement, and unit thickness. Exaggeration is influenced by changes in focal length, camera height, separation of the photo pair, distance between adjacent photo centers, viewing distance, and eye separation. A 7.6 cm (3 in.) focal length lens is commonly used to increase exaggeration and distinguish dips more clearly in low-relief terrain. Exaggeration is increased when the photos are acquired farther apart, or at lower altitudes (lower camera height), with increased separation of the photos on the light table, or with increased viewing distance.

Airphotos have "radial distortion," meaning that objects appear to lean outward from the center point of a frame (the "photo center") due to the bending of light as it passes through the changing thickness of a camera's lens. In high-relief terrain the distortions can become severe. The only area essentially free of distortion on a stereo pair is the point halfway between the two photo centers, called the "perspective center."

Distortion can also be introduced by gradual or abrupt changes in altitude (scale distortion), by tilting of the aircraft (roll and pitch changes that cause oblique distortion), and by crabbing (frames at an angle to flight direction, usually due to flying in a crosswind).

"Spectral distortion" is caused by radial falloff of light intensity outward from the center of the photograph and is known as "vignetting" (pronounced "vin-yetting"). This is again due to the curvature of the lens, and can be compensated for by the use of special filters. More information on airphotos and distortion can be found in Miller and Miller (1961) and Colwell (1983).

Airphotos are acquired along flight lines, which generally have a minimum of 10% side overlap (sidelap) between adjacent lines. Flight lines are usually laid out north–south or east–west for easy navigation, but can be oriented along bedding strike or structural grain to take advantage of geology, or along pipeline corridors, and so on, as needed.

Most airphotos are vertical to minimize distortion. "Vertical airphotos" are nadir-viewing, that is, they are aimed directly at the surface below the aircraft. Under some circumstances, however, it is useful to work with "oblique photos," those that view the surface at some angle other than vertical. These can be "low oblique" (near vertical) or "high oblique" (looking out toward the horizon). Oblique airphotos can be particularly useful when mapping exposures along cliffs or for orientation purposes.

Low Sun-Angle Photos (LSAPs) are acquired early or late in the day, or during the winter when the sun is low in the sky. LSAPs have the advantage of exaggerating slight topographic irregularities in seemingly flat landscapes by lengthening shadows and increasing the contrast between shadowed and illuminated slopes. This type of photography is useful for enhancing structures in interior basins, coastal plains, and other low-relief areas. They are also useful to enhance subtle topographic changes in areas with monotonous cover such as uniform soil, vegetation, or snow.

"Airphoto mosaics" are often useful when large area satellite images are not available. Joining photos into flight lines and merging lines into mosaics often provides the big picture that cannot be seen or deduced from individual frames. The drawback is that this removes photos from stereo viewing, and the resulting mosaic often displays a distracting, quilt-like patchiness due to variations in print exposure and vignetting.

Mosaics can be either controlled or uncontrolled. In "controlled mosaics" an attempt is made to fit the photos together so as to minimize distortion. Generally, only the center part of each frame is used, and all the frames fit together without obvious offsets along the edges. "Uncontrolled mosaics" make no attempt at geometric fidelity. Frames are simply overlapped and pasted together. Typically, uncontrolled mosaics are used only as index maps, to quickly determine which frames cover a specific area. Controlled mosaics are used to view and evaluate a large area. In many cases, this role has now been fulfilled by large area satellite and high-altitude aircraft images.

Black-and-White Photography

The principal advantage of black-and-white (B/W) or panchromatic photography is its low cost, often a fraction of the cost of color photos (Table 3.1). Most older (analog film) surveys were routinely flown using B/W photographs; these have been and still are used in exploration programs to great effect (Figure 3.1). The ability to map structure is usually not compromised. Units are distinguished on the basis of tone (light or dark) and texture (smooth, mottled, etc.). The inability to identify soils, lithologic units, or mineral alteration on the basis of color is the main disadvantage. To the extent that one may not recognize missing or repeated section due to a lack of color, mapping fault type would be more difficult without color. Differences in vegetation species and vigor are difficult to distinguish.

Panchromatic B/W film, such as Kodak 3414 High Definition Aerial film, covers the spectral region from 400 to 700 nm, or the visible range. Atmospheric scattering is generally removed using a minus blue filter (e.g., Wratten 12). In contrast, infrared black and white (IR B/W) films such as SO-289 are available that record reflected light in the range from 400 to 900 nm. Used

TABLE 3.1

Film Characteristics

Film Type	Spectral Range	Common Filters	Relative Cost	Ability to Map			Haze Penetration	Water Penetration
				Lithology	Structure	Vegetation		
Panchromatic (B/W)	0.4–0.7 μm	Wratten 12	Low	Fair	Good	Fair	Moderate	Moderate
B/W Infrared	0.4–0.9 μm	Wratten 89B	Moderate	Fair	Good	Good	Good	Poor
Color	0.4–0.7 μm	Wratten 12	Moderate	Good	Good	Good	Moderate to good	Moderate
Color infrared	0.4–0.9 μm	Wratten 12	High	Good	Good	Excellent	Good	Poor

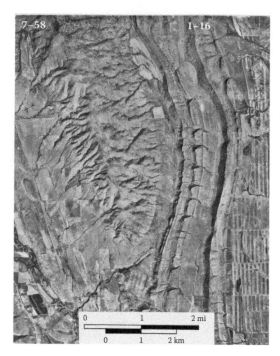

FIGURE 3.1
Black/white U.S. government airphoto of Milner Mountain anticline, near Fort Collins, Colorado. (Image courtesy of USGS.)

with a Wratten 89B filter, only wavelengths from 700 to 900 nm are recorded (Colwell, 1983). This film allows the detection of subtle vegetation changes since the peak reflectance of vegetation is near 800 nm, just beyond the visible range. This film does not record atmospheric haze (mostly in the blue part of the spectrum), and is sensitive to bodies of standing water (black) that absorb IR light. The cost is less than that of color-infrared (CIR) films.

The advent of low-cost, high-resolution satellite and airborne scanners has rendered B/W airphotos to a large degree obsolete.

Color Photography

The chief advantage of true color imagery (analog or digital) is the ability to recognize geologic units in areas where the stratigraphy is known, and the increased ability to interpret rock units where the geologic section is unknown (Figure 3.2). Correlation from area to area is improved, particularly if a unique sequence of units is recognized. Alteration minerals, often associated with ocher to red gossans or bleaching, may not be recognizable without color. Color images also have an aesthetic quality that makes them particularly appealing for public relations work. A disadvantage of color photography is its cost relative to B/W film. The user must weigh the cost versus benefits to determine the need for color. A disadvantage of film-based photography is the inability to manipulate it digitally and view it on a computer. Consequently most color imagery is now digital.

Color films such as Ektachrome Aerographic 2448 or SO-242 cover the spectral region from 400–700 nm. CIR film ("false color" or "camouflage detection film") such as Aerochrome Infrared 2443 covers the range from 400 to 900 nm and is used with a Wratten 12 filter to remove blue wavelengths (Colwell, 1983). It is particularly sensitive to subtle changes in

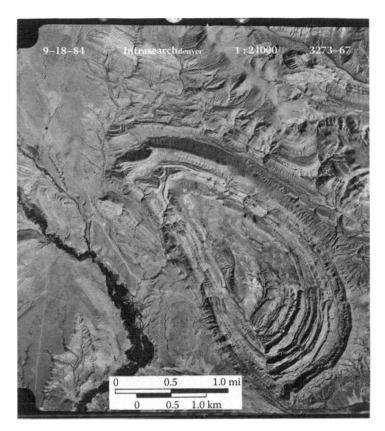

FIGURE 3.2
Color airphoto of Circle Ridge anticline, Wind River basin, Wyoming. (Photo by IntraSearch, Littleton, Colorado.)

vegetation and soil moisture and diminishes the effects of atmospheric haze. The cost is generally higher than that of true color film, and there are problems associated with handling the film: it must be stored in a cool or refrigerated area, and is sensitive to exposure setting and slight changes in photo lab processing chemistry.

The interpretation of CIR photographs requires an adjustment in the way the analyst looks at color on a photo. The sensitivity of the film's pigments has been shifted so that visible blue light is not recorded, and green light is assigned the film pigment blue. Likewise, red light is recorded on film as green, and IR light is recorded as red. The effect of this shift is to eliminate haze (most scattered light in the atmosphere is blue), and cause deep water to appear black. Green rock and shallow or turbid water appears blue, red rocks and gossans appear in shades of yellow to green, and vegetation appears as shades of red (Figure 3.3). Most analysts quickly get accustomed to this color scheme.

There is a widespread misconception that CIR photography shows the distribution of heat, and that red areas are warm. In fact, this film records only reflected light in and slightly beyond the visible range. Thermal energy, or emitted radiation, has wavelengths longer than 3.0 μm and requires special crystal detectors discussed in the section "Thermal Scanners and Imagery." The reason that vegetation appears red on CIR photography is that plants are many times brighter in the near-infrared (NIR, 700–900 nm) than in the green region (500–600 nm). Since human eyes are not sensitive to IR light, we see plants as green, which is where they are brightest in the visible range (Figure 3.4).

FIGURE 3.3
Color-infrared NASA airphoto of Milner Mountain, Colorado.

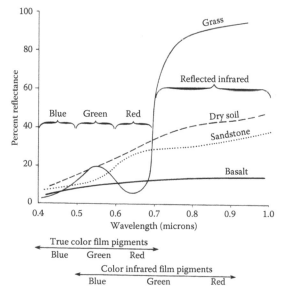

FIGURE 3.4
Comparison of color and color-infrared film pigments and reflectance curves of some typical surface materials. (From Condit, H.R., *Photogram. Engr. and Rem. Sens.*, 36, 955, 1970; Moran, D.E., *Geology, Seismicity, and Environmental Impact*. University Publishers, Los Angeles, CA, 1973; and Goetz, A.F.H., *Remote Sensing Geology, Landsat and Beyond*. SP43-30, Jet Propulsion Lab, Pasadena, CA, 1976.)

With a little practice, an analyst will adjust to the differences between CIR and true color film. Photointerpretation will be improved by the sharpness, clarity, and vegetation information that is available with CIR imagery.

Satellite Photography

Satellite photography began with military reconnaissance missions in the late 1950s and early 60s. The first publicly available satellite photography comprised hand-held, generally oblique true color photos from the U.S. Mercury, Gemini, Apollo, and then Shuttle programs (catalogs may be obtained from the National Aeronautics and Space Administration [NASA] Johnson Space Center, Houston, or from the Earth Data Analysis Center [EDAC], University of New Mexico, Albuquerque [http://edac.unm.edu/]). Satellite photography, while still available, has been largely superseded by government and commercial high-resolution panchromatic and color (multispectral and hyperspectral) digital imagery.

The main advantage of satellite photography over most digital images is stereo capability. In the past, the resolution was generally better for photographs than digital images: that advantage has virtually disappeared, and some satellites acquire digital stereo images using two cameras. Disadvantages of photography include the difficulty involved in making geometric and spectral corrections, the difficulty in filtering high- or low-frequency data and performing contrast enhancements, and the inability to apply mathematical transformations that enhance lithology, structure, or other features. Provision must be made to replenish and retrieve film lifted into orbit. Much of this older photography, particularly Russian data, is now available in digital form.

Thermal Scanners and Imagery

Radiometers indirectly measure the temperature of the surface by recording thermal energy emitted from the surface. These passive systems measure thermal emissions from the surface of the Earth (both land and water).

Thermal scanners operate in two spectral regions: from 3.0 to 5.0 µm and from 8 to 14 µm. This is a function of the composition of their detectors and the fact that there is a strong atmospheric absorption of TIR radiation, primarily due to water molecules, in wavelengths between 5 and 8 µm. These instruments do not measure the temperature of objects, but rather the heat energy radiated from the object. This radiant flux (F) is related to the temperature (T) by the equation:

$$F = \varepsilon S T^4$$

where ε is the emissivity of the material and S is the Stefan–Boltzman constant, 5.67×10^{-12} W/cm^2 $^\circ$k^4.

"Emissivity" is a function of a material's ability to absorb and reradiate energy. It is defined as the ratio of radiant flux from a body to that from a blackbody at the same kinetic temperature (a "blackbody" is a perfect absorber and a perfect emitter of radiant energy). Objects that absorb and reradiate large amounts of energy have an emissivity near but always less than 1.0. Objects with low emissivity do not easily absorb and reradiate energy. Pure water has the highest natural emissivity (0.993); an offshore petroleum slick has an emissivity of 0.972, asphalt has an emissivity of 0.959, granite has an emissivity of 0.815, and a polished aluminum surface has an emissivity of 0.06 (Sabins, 1997).

Objects facing the sun warm faster than objects in shadows. Therefore, daytime thermal imagery tends to emphasize topography by showing warm and cool slopes. This can be useful for mapping structures, but other types of images (such as B/W airphotos) are better for mapping structure and are less expensive to acquire. On flat, homogenous surfaces, a thermal image will be most sensitive to changes in moisture, since water has a high emissivity. This can help locate sources of water needed in drilling programs, sources of leaks in buried pipelines, or faults that impound groundwater in alluvial fans.

The most useful thermal images for geologic mapping are flown just before sunrise, since the lack of illumination at night gives all slopes and surface materials a chance to cool to equilibrium. This is when thermal images most accurately show temperature variations due to different surface materials or moisture content. Temperatures in predawn images are chiefly a result of rock or soil color and albedo. "Albedo" is a material's reflection coefficient, that is, the ratio of reflected radiation to incident radiation. The brighter (lighter) a material is, the greater its albedo. With respect to thermal radiation, darker rocks get warmer and stay warmer than lighter rocks. Rock density is also a factor in that dense rocks cool more slowly (have a higher thermal inertia) than less dense rocks. "Thermal inertia" is the rate at which the temperature of a body approaches that of its surroundings and is a function of absorptivity, the material's specific heat, its thermal conductivity, its dimensions, color, density, and other factors. A basalt, for instance, should appear warmer in predawn thermal imagery than a granite because of its dark color, and a dense gray sandstone would be warmer than a gray shale in predawn imagery because of the density contrast (Figure 3.5). In this way, thermal imagery can discriminate lithologic units and help map structure (Figure 3.6).

Many materials have unique spectral curves in the thermal part of the spectrum. When several narrow bands of thermal data are acquired, one can match the measured

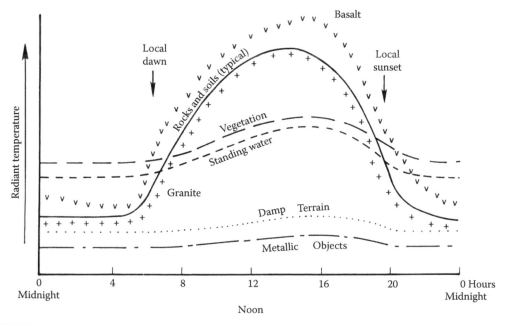

FIGURE 3.5

Schematic diagram of daily radiant temperature curves for some common materials. (From Sabins, F.F., *Remote Sensing—Principles and Interpretation*, Long Grove, IL, Waveland Press, 1997.)

FIGURE 3.6
Predawn thermal image of the Wolverine Creek syncline (arrow) developed in Cretaceous Aspen Formation shales, Caribou Range, eastern Idaho thrust belt. (Flown by Mars, Inc., Phoenix, AZ.)

spectral curve to a reference curve. In the thermal region, one can begin to discriminate between silicate minerals on the basis of these spectral curves (Khale and Goetz, 1983; Hook et al., 2005).

Thermal detectors (such as InSb [Indium Antimonide] or HgCdTe [Mercury-Cadmium-Tellurium]) usually have sensitivities from 0.1°C to 1.0°C. There generally are three gratings in the instrument that record energy in the visible and near-infrared (VNIR) (from 500 to 900 nm), short-wave thermal infrared (SWIR) (from 2.0 to 2.3 µm), mid-wave thermal infrared (MWIR) (3.0 to 5.0 µm), and long-wave thermal range (LWIR) (from 8.6 to 11.3 µm). Airborne thermal scanners are generally flown at altitudes that provide 15–20 m resolution or better. The scanners are gyro-stabilized for roll, pitch, and yaw, so the only distortion is the result of crabbing. Satellite radiometers generally have a ground resolution on the order of tens to hundreds of meters.

Radar

Radio detection and ranging (Radar) is an active system in that it illuminates the surface with a beam of microwave radiation. Radar is most sensitive to surface roughness; shorter radar wavelengths are most sensitive to microtopography, whereas long wavelengths are more sensitive to macrotopography (Sabins, 1997; Drury, 1987; Gupta, 1991). Radar wavelengths are designated as shown in Table 3.2.

A useful equation for converting radar frequency to wavelength is

$$\text{Wavelength (cm)} = \frac{30}{\text{Frequency (GHz)}}$$

TABLE 3.2

Radar Band Designations

Band	Frequency (GHz)	Wavelength (cm)
L band	1–2	15–30
S band	2–4	7.5–15
C band	4–8	3.75–7.5
X band	8–12	2.5–3.75
Ku band	12–18	1.7–2.5
K band	18–26.5	1.13–1.67
Ka band	26.5–40	0.75–1.13
Q band	30–50	0.60–1.0
U band	40–60	0.50–0.75
V band	50–75	0.4–0.60
E band	60–90	0.33–0.50

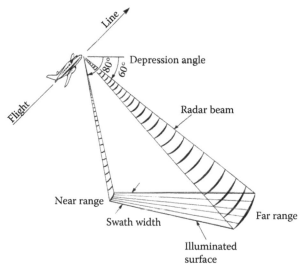

FIGURE 3.7

The side-looking airborne radar system.

Radar is extremely sensitive to surface moisture differences, which affect the complex dielectric constant, a measure of the electrical properties of surface materials.

Early imaging radars were called side-looking airborne radar (SLAR), and resolution was a function of antenna length. Radar energy is beamed outward from an aircraft perpendicular to the flight direction, this being the "look" direction (Figure 3.7). "Polarimetric radar" has the ability to transmit and receive polarized microwaves. Radar beams can be transmitted and received either "like polarized" or "cross polarized." "Like-polarized waves" are transmitted horizontal and received horizontal (HH) or transmitted vertical and received vertical (VV). "Cross-polarized waves" are either transmitted horizontal and received vertical (HV) or transmitted vertical and received horizontal (VH).

With the development of onboard computers came synthetic aperture radar (SAR), which synthetically creates a larger antenna, thus achieving greater resolution. Imagery is available from a number of sources including the U.S. Geological Survey (USGS) EROS Data Center, the Canada Centre for Remote Sensing, the Earth Remote Sensing Data Analysis Center (ERSDAC) of Japan, and the European Space Agency (ESA), among others. Radar images have been generated by the U.S., Canadian, German, French, Japanese, and Russian satellites, among others.

The radar image consists of three types of radar return signal: from corner reflectors (bright), diffuse reflectors (shades of gray), and from specular reflectors and shadows (black, no return).

Advantages of radar imagery over visible and thermal satellite images or airphotos include the ability to penetrate clouds (important in the Tropics) and darkness (important in the Arctic and at night). Contrary to common belief, most radar systems cannot penetrate a vegetation canopy. However, where the canopy follows topography it generally provides the same topographic information as if there were no vegetation cover (Figure 3.8). Radar only penetrates the surface microlayer in leaves or soils, with the exception of sand in hyperarid environments. Under unusually arid conditions, microwaves have been shown to penetrate from one to two meters below the surface. Ground-penetrating radar (GPR), a ground-based instrument, can penetrate to between 20 and 50 m depending on soil type, moisture content, and wavelength (GPR systems are described in the section "Ground-Penetrating Radar").

Radar enhances topography and therefore structurally controlled landforms. In rugged terrain, one can plan a survey with a steep "depression angle" (the angle between the horizontal and the radar beam) to minimize shadows; in flat terrain, a shallow or small depression angle highlights subtle topographic changes such as gentle folds or small faults.

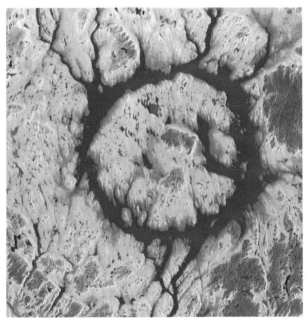

FIGURE 3.8

Colorized RadarSat synthetic aperture radar image of the Manicougan impact structure, Quebec. This area is covered by dense black spruce boreal forest, yet the image looks like scoured bedrock. (Image courtesy of UCSB and NASA. http://www.geog.ucsb.edu/~jeff/wallpaper2/canada_manicouagan_reservoir_radarsat_dec142003_landsatwall.jpg.)

Airborne radar flight lines can be laid out so that structural trends are accentuated. The best results are obtained when flying at about 30° to the strike of structures, so that the beam hits the trend at 60° to strike.

Some companies have experimented with "multispectral color radar," assigning colors to different radar wavelengths, but it is not common. Experiments by NASA and Jet Propulsion Lab (JPL) have demonstrated that multipolarization radar images are sensitive to differences in vegetation size, density, and distribution as well as differences in soil moisture and, in some cases, composition (Evans et al., 1986). These tests used L band radar over several sites and combined like- and cross-polarized images in color to generate surface feature maps. They found that like-polarized waves are most sensitive to topographic effects and surface water, whereas cross-polarized waves respond to changes in vegetation cover and soil moisture.

Disadvantages of radar include the high cost relative to airphotos and VNIR satellite imagery and the lack of color to assist in lithologic mapping. "Layover" and "foreshortening" are unique geometric distortions that make it difficult to estimate dip. These distortions are caused by radar beams returning from elevated objects before beams reflected off low objects (layover) or radar beams returning from the base of a hill before returns from the top (foreshortening). Layover will cause hills, for example, to appear displaced toward the sensor, whereas foreshortening will cause the slope to appear compressed. This distortion is minimized in flat terrain. Most vendors attempt to rectify the imagery to minimize geometric distortions.

In most cases, stereo overlap is not available with radar imagery. A kind of pseudo-stereo can be obtained by keeping all parameters the same but flying again at another attitude, or by flying with two opposing look directions. The first of these options provides low relief because of the low base-to-height ratio that controls vertical exaggeration. The second technique generates shadows on both sides of features, which makes viewing difficult. Unlike photography, in which shadows contain scattered light, radar has no useful information in shadowed areas, since these are true "no data" zones. In moderate to high-relief areas, full coverage would necessitate flying a survey with opposing look directions to fill in the shadows.

Radar Profiling/Microwave Sounders

Radar profiling is used to map clouds, vegetation canopies, and even wind. The instrument can be located on the ground, in an aircraft, or on a satellite. The first Millimeter-wave Cloud Radar (MMCR) began operating at the U.S. Department of Energy's Atmospheric Radiation Measurement Cloud and Radiation site in northern Oklahoma in November 1996. This unit uses a 3 m diameter antenna in a protective radome; the beam width is 0.2°. The radar's Doppler processor provides estimates of reflectivity, vertical velocity, and spectral width simultaneously at multiple elevations, typically from 0.1 to 15.1 km above the ground. In addition to revealing the layer heights and thicknesses of clouds, the MMCR, when combined with data from radiometers or LIght Detection And Ranging (Lidars), provides information for estimating particulate sizes and concentrations (National Security Space Road Map, online).

A "wind profiler" uses radar or sound waves (SOund Detection And Ranging [SODAR]) to detect wind speed and direction at various elevations above the ground. The profiler records samples along each of five beams: one aimed vertically to measure vertical velocity, and four off vertical and orthogonal to one another to measure the horizontal components of the wind. The energy scattered by wind eddies and received by the profiler is orders of magnitude smaller than the energy transmitted. However, if the scattered energy can be clearly identified above background noise, then the mean wind speed and direction can be determined. Pulse-Doppler radar wind profilers operate using EM signals to remotely sense winds aloft. The 4D radar transmits short radio frequency pulses along each of the five antennas to detect a target's location and velocity. The length of the pulse emitted by the antenna corresponds to the volume of air encountered by the radar beam. Delays are built into the data processing so that the radar receives scattered energy from discrete altitudes (range gates). The Doppler frequency shift of the backscattered energy is determined, and is used to calculate the location and velocity of the wind moving toward or away from the radar along each beam as a function of altitude. The source of the backscattered energy is small-scale wind turbulence. The radar is most sensitive to scattering by turbulent eddies whose spatial scale is half the wavelength of the radar, or ~16 cm for an ultrahigh frequency (UHF) profiler (U.S. EPA, 2000).

"Microwave sounders" are radiometers on aircraft or satellites that passively record surface or atmospheric microwave radiation in multiple wavelengths. The Advanced Microwave Sounding Unit (AMSU), for example, is a multichannel microwave radiometer (MWR) installed on National Oceanographic and Atmospheric Administration (NOAA) and ESA weather satellites. The instrument performs atmospheric sounding of temperature and moisture levels by recording several bands of microwave radiation generated by the atmosphere.

Radar Altimeters

A radar altimeter (RA) measures the distance between an aircraft or satellite and the terrain below. A radar transmits radio waves toward the ground and records the time to receive the reflection. Since the velocity of the radio waves is known, the two-way travel time is used to determine the distance from the reflecting surface. The altimeter on aircraft or satellites is nadir-pointing radar (vertical, as opposed to side-scanning). It is used to map the land and ocean surface, map and monitor sea ice distribution, and measure wave and land heights, among other things. RAs normally work in the E band, Ka band, or S band. RAs flying over land measure "absolute altitude," the height above ground level, as opposed to "true altitude," the height above mean sea level (MSL).

A second radar altimetric technique is "frequency modulated continuous-wave radar." The frequency shift of the radio waves is measured and the greater the shift, the greater the distance traveled. This method, which is the current industry standard, achieves better accuracy than timing radar reflections. Most RAs are used by civilian aircraft for approach and landing and for proximity warning systems (Radar Altimeter Tutorial, online).

The decision whether to use radar or Lidar altimetry depends on a number of factors, including the size of the area of interest and required resolution. Satellite radar will always be cheaper, but has a large resolution cell. The Terra SAR X is currently making a 30 m worldwide digital elevation model (DEM). The SPOT HRS is generating DEMS with 30 m resolution over limited areas. Airborne SAR provides two to ten meter DEMs but is project driven and is more costly due to mob/demob and remote or foreign access issues. Lidar has the higher resolution (mm to cm vertical and one to five m horizontal, depending on flight altitude), but is currently the more expensive option, so is usually used for limited areas.

Radar Interferometry

As just stated, knowing the velocity of the radar signal and the time from sending to receiving makes it possible to calculate the distance between the radar and the surface. This allows us to generate DEMs from radar imagery and create topographic maps (Figure 3.9).

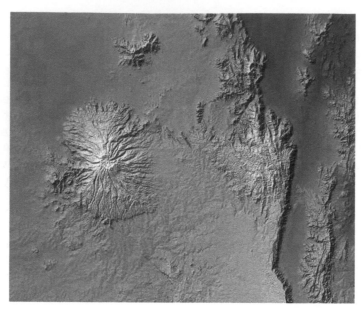

FIGURE 3.9
Color digital elevation model of Mt. Elgon and central African rift, Uganda–Kenya derived from the Shuttle Radar Topography Mission. (Courtesy of NASA/JPL.)

"Radar interferometry" uses SAR images recorded at different times to measure small variations in surface elevation (on the order of 1 mm to 1 cm) over large ground swaths. The two images are registered to an accuracy of about 1/8th pixel, the phases of the corresponding pixels are subtracted, and the result is an interferogram. Phase differences in an interferogram have cycles of 360° and must be unwrapped to derive absolute phase. After unwrapping, the absolute phase is used to derive altitude information. The closer the images are acquired in time and space, the more accurate the interferogram will be. Surface variations from one image acquisition to the next result in a phase change known as "temporal decorrelation." There should be no phase changes unless there has been a change in surface elevation that changed the distance from the surface to the antenna (Gabriel et al., 1989; Ehrismann et al., 1996). Using a C band radar interferometer formed by a pair of antennas displaced across track, Zebker (1991) reports that he could measure the topographic height of each point on the ground with accuracies between 2 and 12 m. This error could be improved by removing systematic noise due to the aircraft motion and increasing the signal-to-noise ratio. This technique has been applied to both satellite and airborne radar.

An example of a remote sensing system that uses interferometric radar is the instrument aboard ESA's CryoSat-2 satellite. SIRAL-2 is the SAR/Interferometric Radar Altimeter. This program uses an interferometric radar-range finder to (1) determine the spacecraft's altitude and (2) to measure the elevation of ice and open water in and near the polar ice caps (Encyclopedia Astronautica, online).

Passive Microwave Radiometry

Microwave radiometers (MWRs) measure spectral radiance emitted from the Earth's surface in the microwave region, that is, wavelengths from 0.15 to 30 cm (or frequencies from 1 to 200 GHz). These passive systems record multiple frequencies and polarizations and require a large instantaneous field of view (IFOV) and wide bands because the energy source is weak. The emitted spectral radiance is called the "brightness temperature." "Spectral radiance" is related to the kinetic temperature by the Rayleigh–Jeans Approximation:

$$L_\lambda = \varepsilon\ 2kcT_{kin}/\lambda^4$$

where L_λ is spectral radiance, k is Planck's constant, c is the speed of light, ε is emissivity, T_{kin} is kinetic temperature, and λ is wavelength.

Kinetic temperature is related to brightness temperature by the relation:

$$T_b = \varepsilon T_{kin}$$

Emissivity is a function of a material's dielectric constant. The dielectric constant of water is 80, whereas that of ice is 3.2, vegetation is 3.0, and air is 1.0. Thus, soil moisture has a large effect on brightness temperature and spectral radiance.

Passive microwave radiometry is used to map atmospheric water vapor at frequencies where there are atmospheric absorption bands, such as at 85 GHz. Passive MWRs are also used to monitor surface temperatures (land and ocean) and soil moisture (Plevin, 1974).

Passive MWRs have been available on NASA aircraft since 1972. The Passive Microwave Imaging System was an experimental instrument flown on a NASA NP3A aircraft that recorded at 10.69 GHz. The AMSU has been flown by NOAA on meteorological satellites from 1978 to present. The Scanning Multichannel Microwave Radiometer (SMRR) has been in operation since 1987; the Tropical Rainfall Measuring Mission (TRMM) has been operational since 1997; the Advanced Microwave Scanning Radiometer (AMSR-E) has flown since 2002.

Spectrometers, Spectrophotometers, and Interferometers

"Spectrometry" and "spectrography" measure radiation intensity as a function of wavelength. Spectral measurement devices are referred to as spectrometers or spectrophotometers. Types of spectroscopy are distinguished by the interaction between EM energy and a material. These interactions include

- Absorption: when energy from a source is absorbed by a material.
- Emission: radiated energy is emitted by a material. Emission can be induced by other sources of energy or by EM radiation (in the case of fluorescence).
- Reflection spectroscopy: examines how incident radiation is reflected or scattered by a material.
- Impedance spectroscopy: the ability of a medium to impede or slow transmitted energy. For optical applications, this is characterized by the index of refraction.
- Inelastic scattering phenomena: involve an exchange of energy between the radiation and matter that shifts the wavelength of the scattered radiation.

The unique energy state of molecules provides unique spectra or "absorption," the lack of spectra.

Multispectral and Hyperspectral Scanners

Spectrometers measure the reflectance of the Earth (surface and atmosphere) in the solar spectral range (390–1040 nm) and include multispectral scanners (MSS) and hyperspectral scanners flown on aircraft and satellites. These are the most common imaging instruments on aircraft and satellites. The imagery produced by these instruments has been given the name "multispectral" if there are a relatively few rather broadbands, or "hyperspectral" if there are many narrow bandwidth channels. These systems are also available for video (videography, video grammetry, and video remote sensing). Video multispectral surveys are generally integrated with GIS (Thomas, 1997).

Commercial companies involved in remote sensing data acquisition are available from several sources, including the American Society of Photogrammetry (ASPRS) and online.

Imaging scanners provide digital information recorded on a hard drive, although one can request that the vendor provide hardcopy in the form of transparencies or paper prints. The spectral range may be adjustable or preset, with the number of "channels" (discrete wavelength intervals) and "bandwidth" (wavelength range) varying considerably.

Some scanners have the ability to configure the sensor to record spectral features of particular interest, such as clays, carbonates, or vegetation stress (Figure 3.10). Airborne systems can adjust mission timing to avoid adverse weather or atmospheric conditions, or to respond to a crisis, whereas satellite systems provide repeat coverage at specific

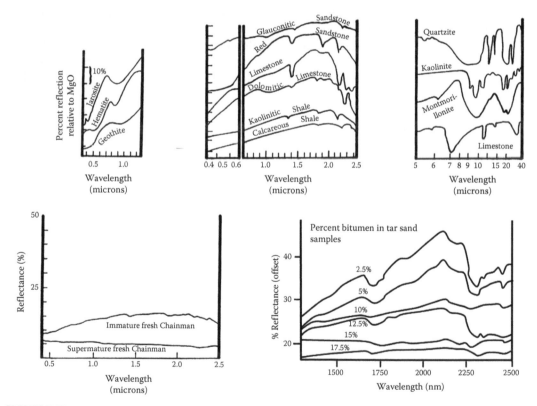

FIGURE 3.10

Typical spectral reflectance curves of some common geologic materials (From Hunt, G.R., *Mod. Geol.*, 5, 211, 1976; Hunt, G.R., *Econ. Geol.*, 74, 1613, 1979; Khale, A.B., *Geology*, 8, 234, 1980; Andreoli et al., *Institute for the Protection and Security of the Citizen*. European Commission Joint Research Centre. Luxembourg: European Communities: 36 p., 2007; Rowan, L.C., *Bull. Am. Assn. Pet. Geol.*, 76, 1008, 1992.)

time intervals. One can adjust airborne mission parameters to vary the resolution, scale, and so on. Airborne scanners must be synchronized with flight speed, or a skewing of the image occurs. This distortion can be corrected by means of image processing.

Flying an MSS survey is now cost competitive with flying airphotos. The output will require postflight image processing rather than photo processing. Stereo is not available without further processing and a DEM. On the other hand, working with multiple narrow bands of data and custom processing allows the user to identify, classify, or otherwise map features not readily detected on standard airphotos.

Imaging scanners are available with spectral bands throughout the ultraviolet (UV), VIR, SWIR, and TIR range. NASA has made available ER-2 and C-130B aircraft operating from NASA Ames Research Center in Moffet Field, California, and a Learjet operating from Stennis Space Center in Bay St. Louis, Mississippi. Instruments carried include the NS-001 Thematic Mapper Simulator (TMS), Zeiss mapping cameras, a Thermal Infrared Multispectral Scanner (TIMS), and the Advanced Solid-State Array Spectroradiometer (ASAS), among others. These are used for research and commercial ventures. For more information in North America, contact JPL, NASA Ames Research Center, NASA Goddard Space Flight Center (SFC), NASA John C. Stennis Space Center, or the Canada Centre for Remote Sensing. In Europe, contact ESA. Besides those instruments developed for research by NASA and JPL, several vendors build scanners for airborne and satellite remote sensing (contact ASPRS or consult the Internet).

The "spatial resolution" (smallest feature that can be resolved, detected, or discriminated) of imaging scanners is a function of the instrument's IFOV and the platform (aircraft or satellite) altitude. Generally it is in the range of 1.0–20 m. The spectral resolution (the spectral width of a band or channel) is also an instrument parameter, and can be designed narrow (e.g., 10 nm or less) to wide (4 μm or more). Most airborne scanner surveys have a "swath width" (the width of the ground imaged by the instrument) in the range 5–10 km. Satellite-based scanners generally have swath width in the tens to hundreds of kilometers.

The simultaneous acquisition of many narrow bands allows the analyst to construct spectral reflectance curves (or emittance curves in the thermal range) for the ground area covered by any picture element. The spectral curve can then be matched against a set of reference curves for various minerals, plants, or soils in an attempt to identify the surface materials (Figure 3.11; see Chapter 6). The greater the spectral resolution (narrower the bandwidth), the more accurate the spectral curve, and the better the chance that the material in question can be identified.

The advantage of multispectral and hyperspectral scanners over photography is their ability to generate spectral curves and discriminate or identify certain minerals, soils, vegetation, and other surface materials. Image processing allows many more options than photo processing. The resulting images are used to help identify hydrothermal alteration, oil seeps, near-surface water, source rock, or soil and rock alteration associated with seeping hydrocarbon gases. Other advantages of digital image processing include

1. Ability to filter, enhance contrast, and apply transformations to maximize geologic information content
2. Ability to correct geometric and spectral distortions

The advantages of satellite imagery include

1. Repetitive coverage can provide cloud-free images of multiple seasons (high and low sun elevation, varying plant cover, snow cover, moisture conditions, etc.)
2. Synoptic view (large area coverage) at relatively low cost per unit area

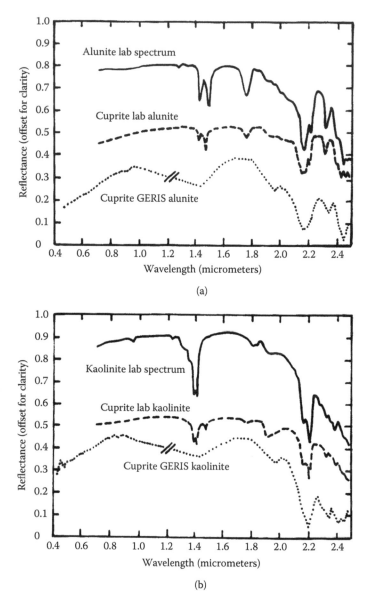

FIGURE 3.11

Laboratory reference curves and field spectra should be similar, but differences can be caused by variations in grain size, weathering, moisture, soil and plant cover, and atmospheric and illumination conditions. These curves compare pure lab sample reflectance against field measurements of the same mineral and airborne GERIS (a 63-channel imaging spectrometer) reflectance of the mineral. (a) Alunite spectra and (b) kaolinite spectra. (From Kruse, F.A., *Photogramm. Eng. and Rem. Sens.*, 56, 83, 1990.)

3. Access to remote areas and, in the case of satellites, restricted areas without regard to terrain, climate, or politics

4. Confidential exploration and scouting activity

On the other hand, multispectral and hyperspectral scanners generate a large amount of data that require processing to obtain a usable image. Identification of surface materials is complicated by mineral mixing, alteration, weathering, and by vegetation cover.

The imagery tends to be most useful in arid or boreal climates where weathering and vegetation are minimal.

Imaging Absorption Spectrometers

Imaging absorption spectrometers compare light coming from the sun to light reflected by the earth, thus providing information on the atmosphere. The objective is to map the concentration of gases and aerosols. An absorption spectrometer measures light that is transmitted, backscattered, and reflected by the atmosphere. The SCanning Imaging Absorption spectroMeter for Atmospheric CartograpHY (SCIAMACHY) on ESA's Environmental Satellite (ENVISAT) uses this type of spectrometer to measure atmospheric trace gases including CO and O_3.

Occultation Spectrometers

Occultation spectrometers detect light from distant stars and measure the depletion of that light in specific spectral regions by gases in the earth's atmosphere. The most common gases and materials identified using this technique include nitrogen dioxide, nitrogen trioxide, ozone, and aerosols (Giroux et al., 2010). The Atmospheric Chemistry Experiment-Fourier Transform Infrared Spectrometer (ACE-FTS) solar occultation spectrometer acquires data in the IR spectral range. It is flown on SCISAT-1 and is used to monitor ozone and trace atmospheric gases.

Michelson Interferometers

The Michelson interferometer is a spectrometer that produces an interference pattern by splitting a light beam into two paths, bouncing the beams back and recombining them. The different paths may be of different lengths or transmitted through different materials to create alternating interference fringes on a back detector. Interferometers passively sound the atmosphere for IR radiation and can provide atmospheric pressure and temperature profiles. They also profile stratospheric trace gases including nitrogen dioxide, nitrous oxide, methane, nitric acid, ozone, and water (HyperPhysics, online).

Lasers, Lidar, and Lidar Altimeters

Light Detection and Ranging (Lidar) uses UV, visible, or NIR lasers to measure the distance to, or properties of a target by illuminating the target and measuring the reflected or backscattered light. "Lidar altimeters," much like RAs, are downward-looking lasers flown on aircraft or satellites (e.g., the Geoscience Laser Altimeter System [GLAS] flown on the Ice, Cloud and Land Elevation Satellite [ICESat]). They can be used to measure surface topography (roughness, slope aspect, and inclination), measure surface elevation, or the distribution and volume of sea ice and MSL. They can also be used to scan outcrops, measure bathymetry, cloud heights, aerosols in the atmosphere, snow cover, reflectivity, and other surface characteristics.

Haddad et al. (2012) describe airborne and terrestrial Lidar platforms that are used in studies of active tectonics. Two types of platform are typically used in research that uses Lidar data. "Airborne Lidar" systems use an aircraft-mounted laser scanner that scans topography in side-to-side swaths at rates that range between tens and several hundreds of kilohertz (Figure 3.12). The orientation (yaw, pitch, and roll) of the aircraft is monitored by an onboard inertial navigation unit, and its location is determined by a high-precision kinematic global positioning system (GPS) (El-Sheimy et al., 2005; Carter et al., 2007; Shan et al., 2007). Postprocessing places the Lidar data in a global reference frame as a point

(a)

(b)

FIGURE 3.12

Airborne Lidar instrumentation and setup. (From Haddad, D.E., *High-Resolution Lidar Digital Topography in Active Tectonics Research*, Arizona State University, Tempe, AZ, 2012.)

cloud of laser returns with typical shot densities >1 m^{-2}. "Terrestrial Lidar" platforms use a tripod-mounted laser scanner that can be operated from various near-field positions to ensure complete scan coverage of the feature of interest (Figure 3.13). Reflective targets with known geographic coordinates are needed to align point clouds from the different scan positions into a final point cloud within a global reference frame. Point cloud densities for terrestrial Lidar can reach up to >10^4 m^{-2} depending on the scanning distance. Furthermore, acquisition geometry of terrestrial Lidar systems provides a true three-dimensional (3D) representation of the scanned feature or outcrop. As a result, complete 3D representations of features scanned by terrestrial Lidar can be accomplished, as opposed to airborne Lidar platforms that scan topographic features in "2.5D."

The utility of airborne and terrestrial Lidar datasets for the visualization and analysis of topographic data is showed by gridded DEMs and how they reveal tectonic information (Figure 3.14). These DEMs are generated from the spatially heterogeneous point clouds

FIGURE 3.13
Terrestrial Lidar system. (From Haddad, D.E., *High-Resolution Lidar Digital Topography in Active Tectonics Research*, Arizona State University, Tempe, AZ, 2012.)

that were detected by the Lidar scanner. Where the point spacing is less than the desired resolution of the DEM, a local binning algorithm is applied to compute values within a specified search radius at each node using a predefined mathematical function (e.g., mean, minimum, maximum; El-Sheimy et al., 2005).

The following is a list of suggested Web resources where publicly available Lidar data and processing capabilities are available for airborne and terrestrial Lidar datasets:

- http://Lidar.asu.edu
- http://www.opentopography.org
- http://www.ncalm.org/
- http://facility.unavco.org/project_support/tls/tls.html#interface/
- http://Lidar.cr.usgs.gov/
- http://lvis.gsfc.nasa.gov/

Lidar altimeters are used by commercial firms to generate DEMs for precise engineering work (powerlines, pipelines, dams, mine surveys, excavations) and for environmental and hazard monitoring (landslides, slope creep, magma dome growth, flood plain surveys, oil field subsidence). Lidar technology offers one of the most accurate and cost-effective ways to derive elevation information. Lidar data are commonly processed to produce DEMs in LAS, ASCII, ESRI and CAD formats. Horizontal accuracy for airborne Lidar surveys is typically in the range between 20.0 and 100.0 cm. Vertical accuracies generally range between 9.0 and 18.5 cm for fixed-wing aircraft and 1.0 and 3.0 cm using helicopter Lidar. Products in addition to DEMs include digital terrain models, slope maps, topographic maps, planimetric maps, tree height analyses, image rectification, shallow-water bathymetry, and excavation and landfill modeling. Lidar also provides an opportunity to detect

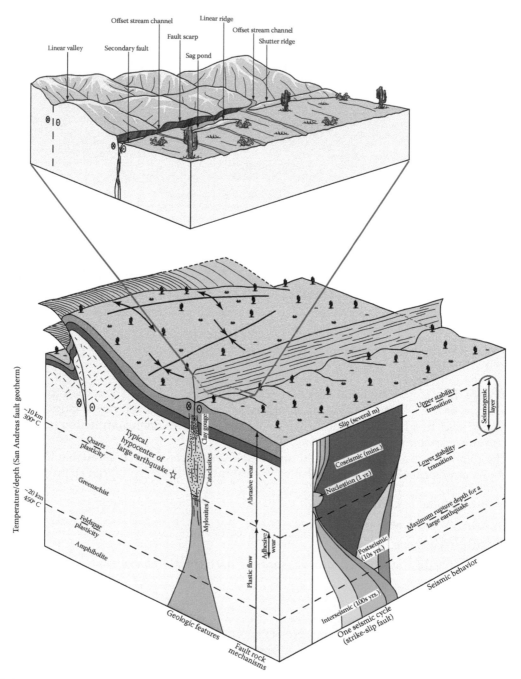

FIGURE 3.14

Synoptic overview of fault zone processes as manifested at the earth's surface and interior. In Earth's interior, geologic and seismic properties of fault zones transmit strain through the seismogenic layer (outlined in red) to Earth's surface. On Earth's surface, the geomorphology of fault zones provides clues to seismicity at depth by displaying faulted geomorphic elements. Derived from concepts developed by Vedder and Wallace (1970), Sylvester (1999), and Scholz (2002). (From Haddad, D.E., *High-Resolution Lidar Digital Topography in Active Tectonics Research*, Arizona State University, Tempe, AZ, 2012.)

and measure atmospheric gases, most typically CO_2, by measuring the spectral absorption of the reflected or backscattered light.

Commercially flown Lidar surveys are provided by a large number of survey companies. A quick check of the Internet under "Lidar surveys" and "airborne Lidar mapping" brought up the following North American and European service providers, including Helica srl, Fugro EarthData, LSI (Lidar Services International), Merrett Survey Partnership, Scandinavian Laser Surveying, Optech, SGL, Leica Aerometric, LaserMap Image Plus, and Credent Technology (Lidar News, online; Lidar data web directory, online).

Backscatter Lidar

"Backscatter Lidar" (also Atmospheric Backscatter Lidar) involves transmitting a high-power pulsed laser beam into the atmosphere or into an aerosol plume and detecting the backscattered signal. The system transmits at one wavelength and detects changes in the backscattered wavelength because of aerosols and dust in the atmosphere. Backscatter Lidar has been used primarily for atmospheric research and meteorology. Wavelengths in a range from thermal (about 10 µm) to the UV (250 nm) are used. Transmitted light is reflected by "backscattering," the diffuse reflection of the signal due to scattering by molecules or aerosols. Combinations of wavelengths allow remote mapping of atmospheric constituents by measuring wavelength-dependent changes in the intensity of the returned signal.

Differential Absorption Lidar

Differential absorption Lidar (DIAL) is used to detect specific gases in the atmosphere, particularly ozone, CO_2, and water vapor. The Lidar transmits two wavelengths, one that is absorbed by the gas of interest, and another that is not absorbed. The differential absorption between the two wavelengths is a measure of the concentration of the gas as a function of the range (distance to the target). Molecules studied using DIAL include SO_2, NH_3, O_3, CO, CO_2, HCl, NO, N_2H_4, NO_2, and SF_6 (Figure 3.15) (Considine, 1988; Spectrasyne, online).

Fluorescence Lidar

Fluorescence Lidar uses two wavelengths and a spectrometer to separate the wavelength-shifted fluorescence from the strong atmospheric (Rayleigh) backscatter. The laser is tuned to the absorption band of the molecule of interest, and reradiated fluorescence is detected by spectral filtering of the returned radiation. Fluorescence is greater in the UV than in the IR, but for some applications this limits the effectiveness of the system because the detector is overwhelmed by normal solar background radiation. Thus, the system works best at night and when tuned to wavelengths less than 1 µm. Fluorescence Lidar has been used to detect atmospheric trace metals including Na, K, Li, Ca, and the hydroxyl ion (Considine, 1988).

Doppler Lidar

Doppler Lidar is used to measure temperature and wind speed along the beam by measuring the frequency of backscattered light relative to transmitted light. The Doppler shift provides information about air speed, turbulence, and wind shear. ESA's Atmospheric Dynamics Mission (ADM-Aeolus), to be launched in 2013 or 2014, will have a Doppler LIDAR to provide global measurements of vertical wind profiles (NOAA a, online).

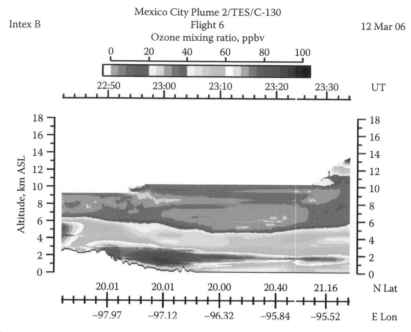

FIGURE 3.15

NASA DIAL-derived atmospheric ozone over Mexico City. This cross section was acquired by Intex B Flight 06/ Mexico City Plume 2–TES–C130/March 12–13, 2006. High ozone values are shown as red to black. (Courtesy of NASA. http://asd-www.larc.nasa.gov/Lidar/intex-b/dial_06.html.)

Broadband Lidar

NASA's Goddard SFC, Greenbelt, MD, has been flying a "broadband Lidar" instrument on their DC-8 before installing it on the Active Sensing of Carbon dioxide Emissions over Nights, Days, and Seasons (ASCENDS) satellite mission. "ASCENDS" is a NASA satellite to be launched in the 2018–2020 time frame. The goal of the ASCENDS mission is to measure the sources, distribution, and variations in CO_2 with high precision worldwide. The majority of CO_2 variability occurs in the first 100 ft. above the surface of the earth. To measure abundance from a satellite, an instrument must look through Earth's entire atmosphere. NASA's broadband Lidar uses an IR laser aimed at the earth's surface. CO_2 molecules in the atmosphere absorb some of the laser beam as the light passes through the atmosphere before and after reflecting off the ground. Standard Lidar systems emit light at specific wavelengths. The CO_2 molecule, however, absorbs light at several IR wavelengths. The broadband laser emits light with a range of wavelengths that are susceptible to CO_2 absorption. Since the amount of absorption is a function of CO_2 concentration, this technique can be used to measure and map global CO_2 (NASA a, online).

Potential Fields Systems (Gravity and Magnetics Instruments)

Airborne magnetometers and gravimeters are used primarily by the mining industry, but have been flown as well to search for hydrocarbons, for monitoring surface water and groundwater, and for environmental surveys. Many commercial systems are available, and many companies exist to acquire these data, including IIT, GEM Systems, Airborne Petroleum Geophysics, Fugro, ARKeX, Sander Geophysics, Gravitec, SGL, and ARETECH

Group, to name a few. For more information, consult online directories that provide lists of companies and describe their products to varying degrees (GeoExplo, online; Open Directory Project, online).

Gravimeters

Variations in terrain elevation (such as the presence of mountains) and geology (such as the density of rocks) cause variations in the earth's gravitational field. For example, a mid-ocean ridge causes a measurable bulge in MSL over the ridge. Gravity maps are usually adjusted for the effects of topography and rock density, and the resulting maps show the corrected gravity. Dense material (e.g., mafic igneous rock) causes a high local gravity field, as do mountains, whereas oceanic trenches and less dense (e.g., sedimentary) rocks cause gravity lows.

Gravity variations are measured by gravimeters. A gravimeter is a specialized accelerometer for measuring the downward acceleration of gravity. Though the design is the same as in other accelerometers, gravimeters are designed to measure minute changes in the earth's gravity caused by changes in rock density, by the shape of the earth, and by tidal variations. Gravimeter measurements are in units of "gals," where the gal is defined as 1 centimeter per second squared ($1 \ cm/s^2$).

Airborne gravimeters are a subset of terrestrial gravimeters: modifications are required to isolate the instrument from aircraft accelerations, and postprocessing is required to remove high-frequency noise. Ground-based gravimeters have been adapted for use in airborne systems. For example, the German Research Centre for Geosciences (Potsdam) flies the modified LaCoste-Romberg air/sea gravity meter Model S124b, a highly damped, spring-type gravity sensor mounted on a gyro-stabilized platform (LaCoste & Romberg, 2004). This is a spring-based gravimeter, basically a weight on a spring, and local gravity is measured by observing the amount the weight stretches the spring.

Another type of airborne scalar gravimeter is the Airborne Inertially Referenced Gravimeter system (AIRGrav) flown by Sander Geophysics Ltd. (Ferguson and Hammada, 2001). This platform-type inertial system has three navigation-grade accelerometers and two gyroscopes with two degrees of freedom mounted on a block in a temperature-controlled environment. The accelerometers are fully isolated from aircraft angular motions by three gimbals controlled by servo motors that use the gyroscopes to adjust for aircraft motion. The system is designed to allow the accelerometers to be aligned with the gravity vector and 180° from it. Gyroscope drift is monitored and corrected. The system was tested over flat ground and compared to ground gravity measurements. Corrections are made for upward continuation of the gravity field to flight altitude (averaging 575 m, but as high as 1150 m), and gravity readings are subjected to low-pass filters to remove high-frequency noise. GPS data are used to remove kinematic accelerations and the Coriolis effect. Results show agreement with ground measurements to within 1.0 mGal for a 2 km half-wavelength spatial resolution. Flight line spacing for these tests was between 1 and 10 km.

Satellites use gravimeters to measure the pull of Earth's gravity as it affects a satellite's orbit to determine gravitational potential along an orbit. The strength of this pull is then plotted as a gravity map (Figure 3.16). A second way satellites can measure gravity is to record small changes in the distance between two identical satellites in the same orbit. When the first satellite passes over a region of slightly stronger gravity, it is pulled slightly ahead of the trailing satellite. This causes the distance between the satellites to increase. The first spacecraft then passes the anomaly and slows down again; meanwhile the following spacecraft accelerates then decelerates over the same area. By carefully measuring the

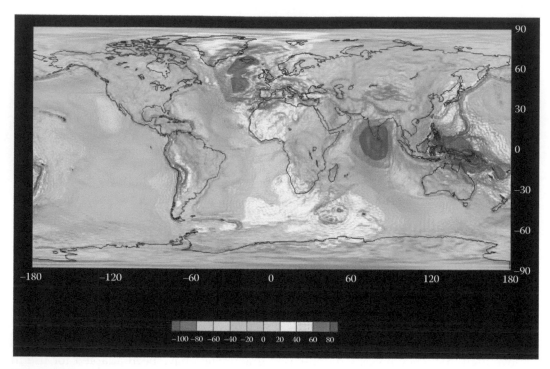

FIGURE 3.16
Global gravity map acquired by ESA's GOCE satellite. Colors show variations in the geoid in cm. (Courtesy of ESA. http://www.esa.int/esaLP/ESAYEK1VMOC_LPgoce_0.html.)

changing distance between the two satellites (using GPS) and combining that data with precise positioning measurements, one can construct a detailed map of earth's gravity. The NASA Gravity Recovery and Climate Experiment (GRACE) satellites derive gravity from measurements of the change in intersatellite distance between two satellites in identical orbits at an altitude of 200 km. The GRACE satellite mission measures the long wavelength structure of gravity and the geoid with high precision, and can therefore detect changes in gravity over time due to mass redistribution (ice movement, sea level changes).

ESA's Gravity field and Ocean Circulation Explorer (GOCE) mission, on the other hand, has higher spatial resolution, which can be used to model the geoid as a surface for studies of ocean circulation (e.g., bulges associated with currents) or for geodynamics (uplift, subsidence, fault movements). The GOCE satellite uses "gravity gradiometry," which measures variations in the acceleration due to gravity. The gravity gradient is the spatial rate of change of gravitational acceleration. Gravity gradiometry is used to measure the rate of change of rock types or densities. From this, one can build a map of gravity variations that can then be used to target hydrocarbon and mineral deposits, to determine water column density, to locate submerged objects, or to map bathymetry.

An example of an airborne gravity gradiometer is described by Murphy (2004). A 3D-Full Tensor Gravity gradiometer, designed and built for the U.S. Navy by Bell Geospace (now Lockheed Martin), uses a set of three rotating disks, each containing two opposing pairs of orthogonally mounted accelerometers. The gradient is measured as the difference in readings between the opposing pairs of accelerometers on each disk. Taking the difference of the gravity field sensed by each pair of accelerometers allows the Air-FTG™ to compensate for most of the aircraft turbulence. The gravity field is composed of three orthogonal vectors, and each vector contains three gradients (the rate of change of the vector). Thus, there are nine gradients to fully describe the accelerations. The entire assembly is rotated at constant speed about a vertical axis to minimize bias related to orientation or movement direction of the instrument. Rotational accelerations are minimized by positioning the gradiometer near the center of roll, pitch, and yaw of the aircraft. Accelerations that remain are measured and noise is removed by applying low-pass filters during postprocessing. Since the signal strength decreases with the cube of the distance to the target, the surveys are flown close to terrain, typically 80 m above the ground surface at line spacing from 50 to 2000 m. When compared to ground surveys, the difference has a standard deviation of about 5.5 Eo (Eötvös, or 0.1 mGal/km). The system is considered to have resolution of about 5 Eo when filtered to remove wavelengths less than 400–600 m.

A third way that satellites measure gravity is using laser altimeters. The satellite laser reflector (SLR) is a system that collects data on the long wavelength gravity field and its temporal changes. It is possible to determine the tidal responses of the solid Earth and to improve knowledge of the long wavelengths of the ocean tides by accurately measuring the distance between a satellite and the surface using satellite-borne lasers and surface reflectors. In this case, the lasers are distributed as part of an international network on the ground while the satellite passively carries the reflector. The precise measurement of the two-way time of an ultrashort (150 picoseconds) laser pulse between an SLR ground station and a retroreflector-equipped satellite is then corrected for atmospheric refraction using ground-based meteorological sensors. Satellite altitudes vary from 400 to 20,000 km (GPS, GLONASS) and measurement accuracy is within 1–2 mm (Degnan, online; Pearlman et al., 2010).

Gravimeters provide one way to determine the Geoid. The "Geoid" is a theoretical equipotential surface that would coincide with the mean ocean surface and land surface in equilibrium, a smooth but irregular surface that is not the actual Earth's surface but one reflected by the earth's gravity. It may be considered a surface of equal or constant gravitational potential at zero elevation. One can conceptualize this by thinking of a hypothetical Earth that is perfectly spherical. Gravity would still change from location to location due to variations of rock and water density. If that perfect sphere were covered in water, the water depth would not be the same everywhere because areas with greater density would cause a thicker water column than areas with low-density bedrock. That water surface would be a good approximation of the geoid.

Research institutes such as the German Research Center for Geosciences (GFZ, Potsdam) fly gravity meters to measure the geoid, map fault zones, salt domes, volcanic structures, ore deposits, or map tectonic shifts. In the case of GFZ's GEOHALO mission, they fly both LaCoste & Romberg and Chekan-AM gravity meters (GFZ a, online).

Airborne gravimeters are available from a number of geophysical service providers. A quick review of the Internet provides a far-from-complete list that includes Aeroquest Ltd., Bell Geospace, Edcon PRJ, GeoData Solutions, GPX Surveys, Fugro Airborne Surveys and Fugro LCT, Helica srl, Sander Geophysics Ltd., and Universal Tracking Systems Pty. Ltd. (GeoExplo, online).

Magnetometers

Spacecraft magnetometers are used aboard satellites to measure variations in the strength and direction of the earth's magnetic field. Magnetometers are used, for example, to locate ferrous mineral deposits, to identify geologic structures such as faults, to outline geologic basins, and to help map igneous intrusions.

Magnetometers can be divided into two basic types: "Scalar magnetometers" that measure the total strength of the magnetic field, and "Vector magnetometers," that measure one or more components of the magnetic field in a specific direction. Vector magnetometers measure both the magnitude and direction of the total magnetic field. Three orthogonal sensors are used to measure the components of the magnetic field in three dimensions. Both azimuth and inclination of the magnetic field can be measured using three orthogonal magnetometers. Total magnetic intensity can be calculated by taking the square root of the sum of the squares of the components' total magnetic field strength.

Examples of vector magnetometers are fluxgates, superconducting quantum interference devices, and the atomic Spin Exchange Relaxation-Free (SERF) magnetometer. Vector magnetometers are subject to temperature drift and the dimensional instability of their ferrite cores. They also require leveling to obtain component information, unlike total field (scalar) instruments. Scalar magnetometers measure total magnetic field strength only, and are often used to provide calibration for vector magnetometers. Types of scalar magnetometer include Proton Precession, Overhauser, and alkali vapor (Cs, He, or K) instruments (NASA b, online; Silliman et al., 2000).

A "fluxgate magnetometer" consists of a small, magnetically susceptible core wrapped by two coils of wire. An alternating electrical current is passed through one coil, driving the core through an alternating cycle of magnetic saturation. This constantly changing field induces an electrical current in the second coil, and this induced current is measured by a detector. In a magnetically neutral background, the input and induced currents will match. However, when the core is exposed to a background field it will be more easily saturated in alignment with that field and less easily saturated in opposition to it. Hence, the alternating magnetic field, and the induced output current, will be out of step with the input current. The extent to which this is the case will depend on the strength of the background magnetic field. The current in the output coil yields an output voltage proportional to the background magnetic field.

The "Overhauser magnetometer" measures the resonance frequency of protons in the magnetic field being measured. The precession frequency depends only on atomic constants and the strength of the ambient magnetic field. A low-power radio-frequency field is used to align (polarize) the electron spin of the free radicals in a hydrogen-rich fluid. Electrons couple to protons by means of the Overhauser effect. The current is then interrupted, and as protons realign themselves with the ambient magnetic field they precess at a frequency that is directly proportional to the background magnetic field. This produces a weak rotating magnetic field that is amplified and fed to a frequency counter whose output is scaled and displayed as field strength.

The "cesium vapor magnetometer" consists of a cesium lamp, an absorption chamber containing cesium vapor, a "buffer gas" through which the emitted photons pass, and a photon detector. The technique is based on the fact that alkali atoms (including He and K) can exist at multiple energy levels. When a cesium atom within the chamber encounters a photon from the lamp, it is excited to a higher energy state, emits a photon, and falls to a lower energy state. The cesium atom is sensitive to the photons from the lamp in three of its nine energy states. Assuming a closed system, all the atoms will eventually fall into a state in which the photons from the lamp pass unhindered through the buffer gas and can be measured by the detector. In the most common cesium magnetometer, an alternating magnetic field is applied to the chamber. Since the difference in energy levels of the electrons is determined by the external magnetic field, there is a frequency at which this field will cause the electrons to change states. In its new state, the electron will once again be able to absorb a photon of light, causing a signal on a photo detector that measures the light passing through the cell. The instrument's electronics generate a signal exactly at the frequency corresponding to the external field.

A number of companies fly airborne magnetometers, including Aeroquest Ltd., Airmag Surveys Inc., Edcon, GeoData Solutions Inc., Geophysics GPR International Inc., GPX Airborne Pty. Ltd., Flux Geophysics Ltd., Fugro Airborne Services, Goldak Exploration Technology, New-Sense Geophysics Ltd., Oracle Geoscience International, Pearson deRidder and Johnson, Inc., Precision Geosurveys Inc., Sander Geophysics Ltd., SkyTEM ApS, Terraquest, Tundra Airborne Surveys Ltd., and UTS Geophysics.

A special category of aeromagnetic program, the high-resolution aeromagnetic (HRAM) survey, is usually flown using airplanes (loosely draping over the terrain) or helicopters flying at a fixed elevation above the surface. Fixed-wing surveys are usually flown at 125–150 m above ground, whereas helicopters, used to collect data over rugged terrain, generally fly less than 50 m above the ground. Flight line spacing ranges from 200 to 800 m, with tie line spacing from 600 to 2400 m (Stone, 2008). These surveys are flown by companies such as IIT of Calgary.

The first spacecraft-borne magnetometer was placed on Sputnik 3 in 1958, and the most detailed magnetic observations of the earth to date (2012) have been gathered by the Magnetic Field Satellite (Magsat) (fluxgate and cesium vapor magnetometers, 1979–1980) and Ørsted (fluxgate and Overhauser magnetometers, 1999 to present) satellites (Figure 3.17). Magnetometers were taken to the moon during the Apollo missions. Many instruments have been used to measure the strength and direction of magnetic field lines around Earth. Spacecraft generally carry two magnetometers: a helium ionized gas magnetometer that is used to calibrate the fluxgate magnetometer for more accurate readings. Satellites carrying fluxgate vector magnetometers include Swarm, Ørsted, Challenging Minisatellite Payload (CHAMP), and SAC-C. SAC-C also carries a scalar helium magnetometer (SHM).

Gamma Ray and X-Ray Spectrometry

Airborne gamma-ray spectrometers are passive detectors that measure gamma rays emitted from the surface of the earth, with little or no contribution from the subsurface. "Gamma rays" are high-frequency, high-energy EM radiation: wavelengths are on the

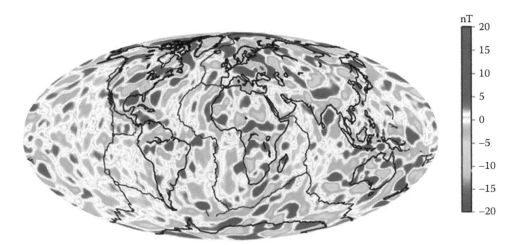

FIGURE 3.17
Earth's magnetic field from Magsat measurements. High values displayed as red and low values as blue. (Image courtesy of NASA. http://www.nasa.gov/centers/goddard/news/topstory/2004/0517magnet.html.)

order of 10^{-11} m or less. All rocks and soils are to some extent radioactive and contain detectable amounts of radioactive elements. The gamma-ray spectrometer is designed to detect gamma rays associated with these radioactive elements and to identify minerals by the energies of the associated gamma rays.

"Radiometric surveys" detect and map natural gamma rays derived from rocks and soils. Gamma radiation from Earth materials come from the decay of three elements: urnaium, thorium, and potassium. Radiometric surveys determine either the absolute or relative amounts of U, Th, and K in surface rocks and soils. These measurements are affected by and must be corrected for meteorological conditions, the topography of the survey area, the influence of cosmic rays, the height of the sensor above ground, and the speed of the aircraft.

Gamma rays are emitted spontaneously by three isotopes. These are Bi^{214}, Tl^{208}, and K^{40}. Bi^{214} comes from the decay of U^{238} and is a uranium indicator. Tl^{208} comes from the decay of Th^{232} and is an indicator of Th content. K^{40} is a minor isotope of K but is the only one that is radioactive.

Gamma rays are classified by their energies, measured in electron volts (eV). One eV is the amount of energy acquired by an electron in moving through an electrical potential difference of 1 V. The gamma rays from Tl^{208} have an energy of 2.62 million eV (MeV). The gamma rays from Bi^{214} are characterized by 1.76 MeV; those from K^{40} have an energy of 1.46 MeV. All three of these energies are constant and form well-defined peaks in the EM spectrum of rocks.

Not all of the gamma rays associated with these energy peaks come from these three minerals. The most important cause of extraneous energy is Compton scattering. "Compton scattering" is a result of gamma rays giving up some of their energy when they collide with electrons. The gamma rays are deflected and continue on with lower energy, that is, lower frequency. The most important non-geologic source of gamma radiation is cosmic rays. Charged particles from space collide with nuclei in the earth's atmosphere and produce 3–6 MeV gamma rays. Radon gas, naturally emitted from the earth, in conjunction with K^{40} dust occurs in layers or clouds, particularly when there is little or no wind, at heights of up to 300 m above the ground. In addition, as the gamma-ray detector moves further from the source fewer gamma rays are detected. Thus, it is necessary to correct for the altitude of the sensor above the ground. After calibration of the system, the corrected count rate in each channel is converted to abundance of the radioactive isotopes at the ground surface. The abundance ratios, U/Th, U/K, and Th/K, are often more diagnostic of changes in rock types, alteration, or depositional environment than the abundance values of the radioisotopes themselves. The U/Th ratio, for example, is often used in exploration for U because it increases near uranium ore. Changes in the concentration of the three radioelements U, Th, and K are often associated with major changes in lithology, and as such this method can be used to help make geologic maps (Figure 3.18). Variations in isotope ratios also characterize primary mineralizing fluids and the resulting veins, gossans, and skarns, as well as supergene alteration and leaching (Moon et al., 2006).

Radiometric abundance ratios are often displayed as ternary images by assigning a primary color to each of the three elements. Commonly Th is assigned red, U green, and K blue. Using this color scheme, green areas would indicate U and blue would show areas of high K (i.e., granites). Surveys are generally flown at elevations of 100 m or less. A number of companies fly radiometric surveys, including Aeroquest Limited, Airborne Petroleum Geophysics, GeoData Solutions, Fugro Airborne Surveys, Goldak Exploration Technology, Novatem Airborne Geophysics, Precision GeoSurveys, Sander Geophysics

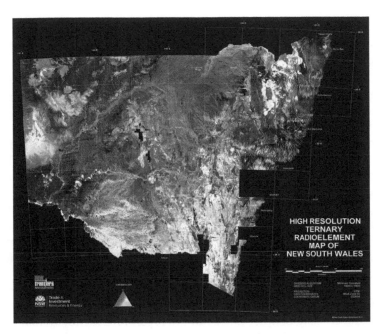

FIGURE 3.18

Radiometric image of New South Wales, Australia. Red–green–blue image of airborne gamma-ray scintillom-eter data shows proportions of the isotopes K40 (red), Th232 (green), and U238 (blue) in rocks or soil at or near the surface. Mixtures of the three radioelements are indicated by the proportion of the three corresponding primary colors, while white equals high amounts of all three radioelements; May 13, 2009. (Courtesy of the government of New South Wales. http://www.flickr.com/photos/landlearnnsw/3526316101/in/photostream/.)

Ltd., Terraquest Ltd. Airborne Geophysics, and Tundra Airborne Surveys Ltd. (Wilford, 2002; GeoExplo, online).

Another type of radiometric remote sensing is used primarily on planetary missions. The alpha particle x-ray spectrometer (APXS) has flown on the lunar Pathfinder mission and on the Martian rovers. The objective of the APXS on the Martian rover Curiosity is to characterize geology and investigate the processes that formed the rocks and soils. The low detection limits, especially for elements like S, Cl, and Br, allow identification of local anomalies. APXS helps characterize samples after they are collected. The elemental data can be used to extract normative mineralogy. A new technique uses backscattered peaks of the primary x-ray radiation to detect x-ray invisible compounds, such as bound water or carbonates, if present in amounts greater than ~5% by weight. The instrument consists of a sensor mounted on the rover's robotic arm. Measurements are taken by positioning the sensor head near a sample, placing the sensor head in contact with or within 2 cm of the sample, and measuring the emitted x-ray spectrum for 15 minutes to 3 hours. The Curiosity APXS has ~3 times the sensitivity for low-atomic-number elements and ~6 times for higher atomic number elements of earlier Martian APXSs. A full analysis with detection limits of 100 ppm for Ni and ~20 ppm for Br now requires 3 hours, while quick look analysis for major and minor elements at ~0.5% abundance, such as Na, Mg, Al, Si, Ca, Fe, or S, can be done in 10 minutes or less. Low-atomic number element x-rays come from the upper 5 μm of the sample, whereas higher atomic number elements such as Fe are detected from the upper ~50 μm. The APXS results average the composition over the sampled area (NASA c, online).

Airborne Electromagnetic Surveys

Airborne electromagnetic (AEM) surveys are flown as a rapid, relatively low-cost exploration tool to locate metallic conductors. The strongest EM responses come from massive sulfides, followed in intensity of the signal by graphite, unconsolidated sediments, and igneous and metamorphic rocks. Graphite, pyrite, and pyrrhotite are responsible for most of the observed AEM responses.

AEM actively creates an alternating magnetic field by passing a current through a coil or along a long wire. The field is measured with a receiver consisting of an electronic amplifier and meter. If the source and receiver are near a conductive anomaly, strong eddy currents are caused within the conductive material and a measurable secondary magnetic field is created. In airborne systems, the receiver coils are usually towed as a "bird" and the transmitter is a large coil encircling a fixed wing aircraft. In helicopter systems there are one or more small coils in the same bird that houses the transmitting coils.

Two systems are commonly used to generate and receive the EM field: transient or "time domain" systems and a/c "frequency domain" systems. Time domain systems contain a transmitting coil, usually around a fixed wing aircraft, energized by a step current. When a conductor is present a sudden change in magnetic field intensity induces current flow in the subsurface conductor, which slows decay of the induced field. Frequency domain AEM systems tend to be helicopter borne. In the typical frequency domain helicopter EM system, both the transmitting coil and the receiver coil are carried in a rigid boom or bird towed beneath the helicopter. Typically, this boom is 3–5 m long and contains two to six coil pairs. The receiver measures the in phase and out of phase, or quadrature, of the secondary field. The different coil orientations provide information about the geometry and depth of the target body.

A quick survey of the Internet indicates a number of companies provide AEM and very low frequency electromagnetic (VLF-EM) surveys, including Aeroquest Ltd., Airborne Petroleum Geophysics, Fugro Airborne Surveys, GeoData Solutions Inc., Geotech Ltd., Goldak Exploration Technology, SkyTEM ApS, and Tundra Airborne Surveys, among others (GeoExplo, online).

Airborne Very Low Frequency Electromagnetic Survey

The "VLF-EM" technique is an active remote sensing system that takes advantage of high-power, extremely low-frequency (3–30 KHz) EM signals generated by military transmitters for the purpose of communicating with submerged submarines. Sensed at a distance of more than a few tens of kilometers from a transmitter, VLF-EM energy acts as plane waves propagating outward horizontally. When these waves intersect a buried conductor, they induce eddy currents that generate a secondary magnetic field concentric around the source of the currents. The strength of the eddy currents is greatest when the long axis of the conductor is oriented parallel to the direction of wave propagation.

Modern VLF-EM instruments conduct measurements at a number of frequencies in sequence to ensure optimum measurement of secondary field signal strength. However, survey lines should still be oriented perpendicular to the expected target trend. The VLF-EM receiver, consisting of coils and supporting electronics towed in a bird behind a fixed wing aircraft or helicopter, measures changes in the propagating waves as they move through anomalous conductive geologic materials and assist in locating and estimating the depth of these bodies. The instrument generally contains three orthogonal coils that measure total field strength and the vertical component of the VLF field. Resolution, the ability of an EM system to recognize and separate nearby conductors, increases with decreasing flight elevation and coil separation. The effective depth of exploration of an EM system, as generally described, includes the elevation of the system above ground. Systems with large transmitter–receiver coil separation, usually referred to as Tx–Rx, have greater penetration. Surveys are generally flown at very low altitudes, for example, at or below 100 m and with 150–400 m line spacing.

VLF-EM surveys involve measuring the orientation of the vector sum of the primary (horizontal) and secondary magnetic field vectors. As the instrument passes over a vertical target, the vector orientation changes from a maximum on one side to a minimum on the other side. The point at which the reading changes from positive to negative lies directly above the conductor. If the conductor dips, then the anomaly shape will be distorted in either the positive or the negative sense. The VLF method is primarily used in mineral exploration work but has also been successfully applied in engineering and groundwater surveys to detect conductive fault zones and other subvertical conductors. Airborne VLF-EM data are used mainly for geological mapping, interpretation of large geological features such as faults and shear zones, and locating conductive graphite, base metal deposits (sulfides, copper, zinc), or kimberlites. VLF-EM surveys tend to not work well in mafic volcanic terrain (conductive background), coastal areas with abundant saline or brackish groundwater, or highly resistive carbonate landscapes (GeoExplo, online).

Airborne Resistivity

Airborne resistivity can be used to explore for sand and gravel deposits, groundwater, oil sands, ice, and permafrost, among others. The equipment is mounted in small planes or helicopters and is used to collect data on the conductivity of rock over large areas. The system consists of a large antenna suspended below a helicopter or around an aircraft. The antenna produces electrical currents focused on the ground whose induced magnetic field is measured to determine resistivity of the subsurface material. Researchers use these data to map the character of the subsurface to depths of ~300 m.

Depending on the type of survey, a significant amount of processing may be required to correct for various factors. The final product is usually a map showing the conductivity (or resistivity) of the area surveyed. The data is interpreted by observing the changes in resistivity and searching for more resistive areas such as carbonate buildup in soil (Walker and Rudd, 2008; U.S. DOT, online; NSF, 2012).

Airborne Induced Polarization

Induced polarization (IP) is a geophysical imaging method that measures electrochemical responses of subsurface material (sulfide ore, clays) to an injected current. A current is introduced through electrodes, and voltage is monitored through other electrodes.

Time domain IP surveys measure the rate of voltage decay once injection of the current is complete. The integrated voltage is used. Frequency domain IP surveys use alternating current to induce an electrical charge in the subsurface. The apparent resistivity is measured at different alternating current frequencies.

Frequency domain measurements are more precise when IP effects increase with depth; time domain surveys are better when IP effects decrease with depth (U.S. EPA, 1993).

A special case is airborne IP. Airborne induced-polarization measurements can be obtained with standard time-domain AEM equipment. The measurements are only valid under limited circumstances, that is, when the ground has sufficient resistivity that the EM response is small and when the ground's ability to be polarized is large enough that the IP response can override the EM response. In order for the response to be recognized the dispersion in conductivity must be within the bandwidth of the EM system (Smith and Klein, 1996).

Ground-Penetrating Radar

GPR is an active terrestrial remote sensing method that uses a towed antenna to pulse microwave EM energy (typically 50–1000 MHz) into the subsurface. As the polarized pulse travels downward, it interacts with materials in the ground and part of the energy is reflected back to the surface antenna. The GPR unit measures the amplitude of the reflected signal and the time delay between the transmitted and received pulses to map the location and depth of subsurface features. The frequency of the antenna can be changed depending on the required depth of investigation and the nature of the expected target. The signal may interact with subsurface materials in a variety of ways (including attenuation, reflection, refraction, diffraction, and scattering). The two most important physical conditions that affect the radar waves are a materials' dielectric properties and its conductivity. The dielectric constant determines the velocity of the EM wave; the lower the dielectric constant, the faster the wave propagation. A sudden change in dielectric constant, such as at a geological boundary, results a velocity change and a consequent reflection of some of the energy. The conductivity of the material is the most important factor determining the rate of signal attenuation. Materials with high conductivities cause rapid dissemination of the transmitted pulse. Signal loss is consequently greatest in clay soils. Signal loss also results from scatter of the transmitted pulse during interaction with different materials in the subsurface.

GPR provides output that appears much like a seismic line (Figure 3.19). The latest GPR can be processed to show either vertical or horizontal slices of the near surface. GPR uses a single, low-power broadband impulse of 1–6 nanoseconds to transmit, followed by a 20,000

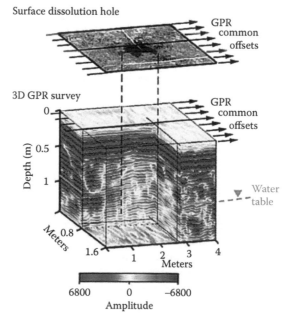

FIGURE 3.19

Ground-penetrating radar image of a small sinkhole in the Late Pleistocene Miami Limestone, Everglades National Park, Florida. (Courtesy of USGS. http://sofia.usgs.gov/cacl/projects/char_mlimestone/.)

nanosecond recording interval. A common frequency is centered at 120 MHz (250 cm). Higher frequencies provide better resolution, but do not penetrate as deep. Penetration depths depend on soil type and moisture content. Maximum depths for low frequencies are on the order of 50 m in dry, sandy soil (Davis and Annan, 1989). Most systems with bands centered at 100 MHz penetrate to about 20 m. Effective penetration depth is usually determined by soil moisture: penetration depth decreases with increasing soil moisture.

GPR can detect a variety of materials, including the water table, bedrock–alluvium interface, clay or sand lenses, pipeline leaks (gas or liquid), and wood or metallic objects, depending on the contrast with background rocks or soils (Battelle Labs, 1983; Inkster et al., 1989; Graf, 1990). The primary factor controlling the amplitude of the reflector is the contrast between electrical properties of the target and surrounding materials. Metallic targets such as oil drums are strong reflectors under most conditions; nonmetallic targets are generally weak reflectors. Filled-in trenches have been detected in the near surface, and law enforcement agencies have used GPR to search for bodies (McCormick, 1999). No target is detectable if covered by a thin layer of moist clay. Saline groundwater also limits the depth of penetration. Clay is both a good reflector and strong absorber of radar waves, and is the least favorable medium for detecting buried objects. Dry sand and gravel, on the other hand, are excellent radar signal transmitters.

GPR data are generally presented as grey scale images showing the amplitude of the reflected radar energy. Two-way time is on the vertical axis and distance along the survey line is displayed on the horizontal axis. Strong reflections appear bright on the image. Typically these are buried metallic objects. Where control is available, the travel time can be converted to approximate depth.

An advantage of GPR over geophysical methods such as EM terrain conductivity measurements is that GPR is not significantly influenced by surface cultural features. One limitation of conventional GPR usage is that the antenna must be towed along the ground, making progress slow. Rough or uneven ground can make surveys difficult. Newer techniques have begun to overcome this problem, but GPR is only effective under a narrow range of conditions. GPR is particularly limited with regard to sensitivity, resolution, and penetration. The data generated can be difficult to interpret where soils have a wide variety of electrical properties as a result of composition and moisture.

For a list of GPR service providers, see Internet directories (MALÅ, online; GSSI, online).

Other EM Systems

A number of products are similar to GPR in that they are ground-based instruments that provide information on the subsurface. These are EM systems, both active and passive. Although it would be difficult to mention every such system, a couple may serve as examples.

"Atomic Dielectric Resonance" (provided by Adrok, Edinburgh, U.K.) is an active EM system that transmits a low-power (less than 1 W), ultrawide band of collimated, pulsed directional beam of radio wave, microwave, and thermal EM energy into the ground. Rather than being absorbed, the energy causes electrons in its path to resonate in a characteristic manner. The resonance releases energy, part of which is transmitted back to the surface detector. Harmonic analysis of the resonant energy compares the signal to a library of known responses and a classification is performed to determine what material is being sensed. Depth is based on the length of the signal return time. To date (2012) this technique reportedly can penetrate up to 4000 m of rock and ~75 m of water. The technique cannot determine porosity or permeability. Adrok is working on making the technology airborne as well as effective in deep water. It is claimed that oil and gas have been detected to within 7 m depth in the range 300 m to 4 km (Stove, 2012).

"Power Imaging" (Wave Technology Group, Houston, TX, and D. Schumacher, Mora, NM) is a passive EM method used for onshore detection of hydrocarbons. This geophysical prospecting method uses the public electric power grid as a continuous source of energy for investigating the earth's subsurface structure, stratigraphy, and hydrocarbon potential. The electric power grid induces EM waves in the earth. These EM waves are at specific frequencies, which are harmonics and subharmonics of 60 Hz (50 Hz in many areas of the world). The waves propagate into the subsurface as plane waves and encounter various geologic boundaries. Those boundaries having dielectric and/or conductivity contrast reflect a portion of the waves back to the earth's surface. With continuous sourcing from the electric power grid, the waves resonate between subsurface boundaries and the earth's surface. The distance and direction to any one power line is not important because the power grid as a whole creates the resonance. The power grid-induced waves are organized such that there is a direct relationship between the resonating frequencies and the depths to various geologic boundaries. An EM signature can be detected by measuring the resonant frequencies at the surface of the earth that result from the electrical contrast between hydrocarbon-bearing rocks and their surrounding formations. Interpretation of this signature yields an EM hydrocarbon indicator that allows the direct detection of hydrocarbons, along with depth and approximate thickness of the hydrocarbon-bearing interval (Schumacher, 2012).

Side-Scan Sonar

Side-scan sonar is an active, imaging marine remote sensing instrument. Imaging sonar presents a display similar to side-looking radar imagery, that is, an oblique view perpendicular to the ship's track. Unlike radar, this system uses acoustic waves instead of microwaves. The sonar instrument is carried in a "fish" that is towed at varying depths below and behind the survey vessel, and can have adjustable depression angles and swath widths. Depths as shallow as 15 m (50 ft.) or as deep as 4,000 m (13,000 ft.) have been surveyed. Generally speaking, resolution becomes coarser as depths increase. The sound frequencies used in side-scan sonar usually range from 100 to 500 kHz; higher frequencies yield better resolution but have a shorter effective range. Sonar is frequently acquired with opposite look directions simultaneously, with a no data zone directly below the towed fish. Overlapping swaths are required to properly mosaic images, prevent shadows from falling in opposite directions, and provide stereo viewing.

Germeshausen & Grier (later E.G. & G., Inc.) developed the first towed, dual-channel commercial side-scan sonar system between 1963 and 1966. Martin Klein helped find Henry VIII's flagship Mary Rose using sonar in 1967. In 1968, Klein founded Klein Associates, Inc. (now L-3/Klein) and developed the first commercial high-frequency (500 kHz) systems, the first dual-frequency side-scan sonar, and the first combined side-scan and subbottom profiling sonar.

The Geological Long Range Inclined Asdic (GLORIA) side-scan sonar was developed by Marconi Underwater Systems for surveying large areas at relatively low frequencies to obtain long range coverage. The National Institute of Oceanography of Great Britain and Institute of Oceanographic Sciences, Wormley, U.K., have been involved in the GLORIA program since the early 1970s and have mapped large areas of the seafloor around the world (Laughton, 1981). Their Mark II system uses a 100 Hz linear FM pulse, usually 2 seconds long and repeated every 30 seconds. A 5° wide transmit beam provides ranges of 7–30 km (900 m swath at maximum range) and can be acquired in two opposite look directions simultaneously at speeds up to 18 km/h (10 knots). Resolution is 45 × 120 m (146 × 390 ft.) at these parameters (Searle et al., 1990).

The USGS has been using the GLORIA Mark III system to map the Exclusive Economic Zone of the United States and its territories since 1984. This system uses either a 20-, 30-, or 40-second pulse repetition rate and images a swath on both sides of the fish 15, 22, or 30 km (9, 13.2, and 18 mi.) wide, respectively. The beam is 2.7° wide and 10° vertical. GLORIA uses a frequency of 6.2 and 6.8 KHz with a 100 Hz bandwidth to image in depths up to 4,000 m (13,000 ft.) (Chavez, 1986).

Sonar imagery can be used at any depth, and has been used not only to locate shipwrecks but also to map slope stability for oil platforms and pipeline and cable routes, pinpoint pipeline damage, and map shallow geologic structures such as anticlines and faults (Jenkinson, 1977; Prior et al., 1981; Clausner and Pope, 1988). In depths up to 100 m, where surfaces were exposed to erosion during lowered Pleistocene sea levels, sonar is particularly good at mapping geomorphic features such as breached anticlines and fault scarps (Figure 3.20). Mudlumps, salt domes, and tar mounds have been identified, and seeps are evident because gas bubbles cause the acoustic signal to break up in a vertical, upward-widening pattern. Gas leaking from the seabed sometimes leaves craters that can be identified. Chemosynthetic communities established around seafloor seeps can be identified as large mounds. Deep sea manganese nodule fields and near-shore placer deposits have also been mapped.

Many companies provide commercial sonar surveys. They are generally listed under Engineering Surveys or Sonar in phone or Web directories. For a list of side-scan sonar service providers, check online directories (SeaDiscovery, online).

FIGURE 3.20
Side-scan sonar image of a submerged, eroded anticline in the Santa Maria basin, offshore California. (Courtesy of GeoCubic Inc., Ventura, CA.)

Side-scan sonar data are frequently acquired along with bathymetric soundings and subbottom profiler data, thus providing a glimpse of the shallow structure of the seabed. The sonar is used to detect debris and other hazards on the seafloor that may be harmful to shipping or to drilling platforms. In addition, the status of pipelines and cables on the seafloor can be monitored, particularly in the arctic where they are subject to sea ice scour. Finally, side-scan sonar is used for fisheries research, dredging operations, and environmental studies. Through the mid-1980s, commercial side-scan sonar images were produced on paper records. Now data are all digital.

"Interferometric sonar" is used to measure water depth, depth changes, and for classification of sea-bottom types (Hadden and Green, 1999). It is possible to determine the acoustic depth and amplitude and analyze these attributes for sediment properties, then assign them to broad classes such as bedrock, gravel, sand, and silt. Applications exist in slope stability studies, locating hazardous wreckage or ordnance when surveying for offshore platforms or wells, for mining marine minerals, cable or pipeline surveys, and shallow marine seismic surveys (bottom hazard mapping).

Instrument Platforms

NASA/JPL in the United States and space agencies in other countries have flown imaging spectrometers on aircraft and satellites. Both multispectral and hyperspectral spectrometers are used. These scanners are also commercially available on aircraft and satellites, as hand-held devices, or as table-top systems to evaluate core from wells. Some websites that list current remote sensing resources include the following:

- NASA Missions: http://www.nasa.gov/missions/current/index.html
- NASA Goddard SFC list of Satellites: http://ilrs.gsfc.nasa.gov/satellite_missions/list_of_satellites/
- Wikipedia list of Earth Observation Satellites: http://en.wikipedia.org/wiki/List_of_Earth_observation_satellites
- Berkeley Earth Sciences and Map Library: http://cluster3.lib.berkeley.edu/EART/aerial.html
- USGS EROS Data Center: http://eros.usgs.gov/error.html
- USGS National Aerial Photography Program: http://eros.usgs.gov/#/Find_Data/Products_and_Data_Available/NAPP
- USGS Landsat Project: http://landsat.usgs.gov/
- Belgian Earth Observation Platform: http://eo.belspo.be/directory/Sensors.aspx
- German DLR Earth Observation Portal: http://www.dlr.de/dlr/en/desktopdefault.aspx/tabid-10376
- Russian Federal Space Agency: http://eng.ntsomz.ru/
- ESA Living Planet Program: http://www.esa.int/esaLP/ESA3QZJE43D_LPswarm_0.html and http://www.esa.int/Our_Activities/Observing_the_Earth

- Indian Space Research Organization Missions: http://www.isro.org/satellites/ allsatellites.aspx
- EO Portal Directory: https://directory.eoportal.org/web/eoportal/home

Airborne Scanner Systems

A handful of historical and recent airborne systems operated by government agencies are listed chronologically and described. For more information on airborne and space-based imaging spectrometers, including commercially available systems, two of the many available websites are provided as follows:

- Michael Schaepman's comprehensive list of Imaging Spectrometers (accessed September 15, 2012) http://www.geo.unizh.ch/~schaep/research/apex/is_list.html
- Hyperspectral Imaging Systems.doc (accessed September 15, 2012) https://docs .google.com/viewer?a=v&q=cache:fU8z6BPZXhIJ:www.scs.gmu.edu/~rgomez/ EOS%2520Lectures/8Lecture%252020%2520Oct%252003/HSI%2520Systems/ Hyperspectral%2520Imaging%2520Systems.doc+AMSS)+Airborne+Multispectra l+ Scanner+-+Geoscan+Pty+Ltd.,&hl=en&gl=ca&pid=bl&srcid=ADGEESh8yDe dQdfQKKn4od1ftZCwVZ-EKCYfyez7OtQW9O1ROR2dRppu1yZx3Y4K1FLJb_O- c-8ipAO7_gliTXh54SQJtJvZNFXPfJBLvTpn9GtzGTp2y-qkMAkk1w0i_SHa_Gw0U wE7&sig=AHIEtbQULrDoMijCTcoNICDhKfSY-SGX2A

Thermal Infrared Multispectral Scanner

Starting in 1982, NASA began flying the six-channel TIMS in the range 8.2–12.2 μm. Bandwidth is 0.4–1.0 μm. Developed by Daedalus Corporation with JPL and NASA, this radiometer has a sensitivity of 0.1°C–0.3°C. The 2.5 mrad IFOV provides a resolution of 50 m and a 31.3 km swath from an altitude of 19,800 m (65,000 ft). Because of the silica absorption features located at the instrument wavelengths, this instrument is particularly well suited to mapping silica-rich minerals and silicification (Khale and Goetz, 1983; Watson et al., 1990). The TIMS multispectral radiometer flies on NASA's C-130, ER-2, and the Stennis Learjet.

Airborne Imagining Spectrometer

The Airborne Imagining Spectrometer (AIS) was developed by JPL and flown by NASA starting in 1983. AIS-1 was the first hyperspectral scanner, with 128 bands in the range 900 nm to 2.4 μm, each band averaging 9.6 nm wide. AIS-2 operates from 800 nm to 2.4 μm. The AIS instrument is generally flown at altitudes from 3 to 18 km (9,000–60,000 ft.): this provides a ground resolution on the order of 2–20 m.

Airborne Visible/Infrared Imaging Spectrometer

The Airborne Visible/Infrared Imaging Spectrometer (AVIRIS), developed by JPL and operated by NASA, was first flown in 1986 and continues to operate in 2012 (Figure 3.21). The first truly operational hyperspectral scanner, AVIRIS has flown on four aircraft platforms: NASA's ER-2 jet, Twin Otter International's turboprop, Scaled Composites' Proteus, and NASA's WB-57. The hyperspectral instrument has 224 bands in the range 400 nm to 2.50 μm. Each band is 9.6 nm wide. With an IFOV of 1 mrad, at an altitude of 20 km it

FIGURE 3.21
AVIRIS airborne image cube of the Louisiana coast. Images like this allow agencies to monitor fragile environments for changes due to the effects of the 2010 Mocambo oil spill. (Image courtesy of NASA/JPL/Caltech/Dryden/USGS/UCSB.)

has a 10 km swath and 20 m pixel (Jansen, 1992; Vane et al., 1993). The main objective of the AVIRIS project is to identify, measure, and monitor constituents of the earth's surface and atmosphere based on molecular absorption and particle scattering signatures. The instrument has been used to test mineral exploration strategies by identifying alteration; to map tailings and acid mine drainage; and to map vegetation and soils, among others (JPL, AVIRIS, online).

Compact Airborne Spectrographic Imager

The Compact Airborne Spectrographic Imager (CASI) is a commercial imaging system built by ITRES Research (Calgary, Alberta). CASI has flown in both aircraft and helicopters since 1988 and can operate in two modes: spatial mode has 19 bands, whereas spectrometer mode has 288 bands, each 1.8 nm wide. It records spectra in the range from 403 nm to 1.00 μm. Resolution can range from 0.25 to 10 m, depending on altitude (Borstad ASL, CASI, online). A typical swath has 1500 pixels cross-track.

Thematic Mapper Simulator

Developed in the mid-1970s, the TMS was built by Daedalus Corporation and flown by NASA on their ER-2 aircraft. The sensors simulate the Landsat Thematic Mapper instruments first launched on Landsat 4 in 1982. The instrument acquires data in 12 spectral bands: Band 1 from 420 to 450 nm, Band 2 from 450 to 520 nm, Band 3 from 520 to 600 nm, Band 4 from 600 to 620 nm, Band 5 from 630 to 690 nm, Band 6 from 690 to 750 nm, Band 7 from 760 to 900 nm, Band 8 from 910 nm to 1.05 μm, Band 9 from 1.55 to 1.75 μm, Band 10 from 2.08 to 2.35 μm, Band 11 from 8.5 to 14.0 μm (high gain), and Band 12 from 8.5 to 14.0 μm (low gain). Spatial resolution is 25 m and swath width is 15.6 km from an altitude of 20 km (65,000 ft.). The instrument was used primarily for geologic mapping, mineral exploration, and vegetation and land cover mapping (NASA d, online).

NS-001

The NS-001 MSS was flown in the late 1980s and 1990s by NASA Ames Research Center. This eight-channel visible to TIR imaging scanner was flown on a NASA C-130B aircraft. Band 1 records visible light from 458 to 519 nm, Band 2 from 529 to 603 nm, Band 3 from 633 to 697 nm, Band 4 collects NIR light from 767 to 910 nm, Band 5 from 1.13 to 1.35 μm,

Band 6 from 1.57 to 1.71 µm, Band 7 from 2.10 to 2.38 µm, and Band 8 records thermal energy between 10.9 and 12.3 µm. Ground resolution is 7.6 m and swath width is 7.26 km from an altitude of 3,000 m (10,0000 ft.).

AIRSAR

Developed by JPL for NASA, AIRSAR has been flying radar missions since 1988. Mounted on a DC-8 aircraft, AIRSAR is a side-looking radar instrument that can collect fully polarimetric data (POLSAR) at three radar wavelengths: C band (0.057 m), L band (0.23 m), and P band (0.68 m). AIRSAR can also collect two types of interferometric data: cross-track interferometric data (TOPSAR), which are sensitive to topography, and along-track interferometric data (ATI), which can be used to measure ocean surface currents (NASA e, AIRSAR, online).

Multispectral Infrared and Visible Imaging Spectrometer

The Consiglio Nazionale delle Ricerche (CNR) of Italy has flown the Multispectral Infrared and Visible Imaging Spectrometer (MIVIS) since 1993. This instrument, built by Daedalus Enterprises, has a 2.0 mrad IFOV and 102 bands in the range 430–830 nm, 1.15–1.55 µm, 1.985–2.479 µm, and 8.21–12.70 µm. Ground resolution varies with altitude (Kruse, 1999).

Hyperspectral Digital Imagery Collection Experiment

The Hyperspectral Digital Imagery Collection Experiment (HYDICE), flown since 1994, was developed by the U.S. Naval Research Laboratory (NRL) and operated by the Environmental Research Institute of Michigan (ERIM, now Michigan Technical University). HYDICE has 210 channels covering 410 nm to 2.504 µm. The swath is generally 1 km at an altitude of 6 km, with better than 4 m resolution (Resmini et al., 1996).

Short Wavelength Infrared Full Spectrum Imager

The Short Wavelength Infrared Full Spectrum Imager (SFSI) has been flown by the Canada Centre for Remote Sensing (CCRS) since 1994. The push broom sensor can simultaneously acquire high spatial (to 20 cm) and spectral (10.4 nm) resolution in 22–115 bands (the number of bands is adjustable) with 0.01 µm (10 nm) bandwidths in the range 1.22–2.42 µm. IFOV is 0.33 mrad, with ground resolution generally less than 10 m (Rowlands and Neville, 1994; Hauff et al., 1996). During initial flights of the instrument hyperspectral imagery was acquired over a calcite quarry and a dolomite quarry. Both of these minerals have a distinctive carbonate absorption feature near 2.3 µm, which this instrument has resolved into two closely spaced features. The absorption feature is centered ~0.1 µm shorter in dolomite than in calcite.

Digital Airborne Imaging Spectrometer 7915

The European Union and DLR (German Aerospace Center) commissioned the 79-channel Digital Airborne Imaging Spectrometer (DAIS 7915), which was built by Geophysical Environmental Research Corporation. This sensor covers the spectral range from the visible to TIR wavelengths at variable spatial resolution from 3 to 20 m depending on the aircraft flight altitude. DAIS 7915 has been in use since the spring of 1995 for environmental monitoring of land and marine ecosystems, vegetation distribution and stress investigations, agriculture and forestry resource mapping, geological mapping, and mineral exploration as well as to provide input data for geographic information systems (Figure 3.22). Six spectral

FIGURE 3.22

DAIS image IL4_1.B, Mt. Sedom area near the Dead Sea, Israel. Channels 76–50–25 (10.5–2.15–0.912 μm) as red–green–blue reveal details of folding related to an active salt diapir in the area. (Courtesy of DLR. http://www.op.dlr.de/dais/israel97/israel97 .htm.)

TABLE 3.3

Spectral Characteristics of the Visible and Infrared DIAS Instrument

Wavelength Range	Number of Channels	Bandwidth	Detector
400–1000 nm	32 Bands	15–30 nm	Si
1.500–1.800 μm	8 Bands	45 nm	InSb
2.000–2.500 μm	32 Bands	20 nm	InSb
3.000–5.000 μm	1 Band	2.0 μm	InSb
8.000–12.600 μm	6 Bands	900 nm	MCT

channels in the 8–12 μm (thermal) region are used to obtain the temperature and emissivity of surface objects. These and 72 narrow band channels between 450 and 2450 nm help investigators examine land surface processes, with a special emphasis on vegetation–soil interactions. Bands, bandwidths, and detectors are provided in Table 3.3 (DLR, DIAS, online).

Portable Hyperspectral Imager for Low-Light Spectroscopy

The U.S. NRL's Portable Hyperspectral Imager for Low-Light Spectroscopy (PHILLS) is a multisensor system designed for ultrabroadband, high-resolution spectroscopy. PHILLS consists of four modules operating in the UV/VIS/NIR range. The PHILLS system includes the imaging sensors and the Optical Real-time Adaptive Signature Identification System (ORASIS) consisting of algorithms that find compositional end members using convex mixing techniques. Several PHILLS missions have been flown, starting in 2000, on the NRL's P-3 aircraft over areas including the polar ice cap and the Gulf of Mexico. PHILLS has been deployed onshore for terrain characterization, target detection, and plume tracking missions. The adjustable PHILLS instrument records with a 16-bit digital camera that obtains up to 1024 wavelengths in the range 200–1100 nm. Each band has 0.5–3 nm resolution and a 5–50 mrad IFOV (Davis, online).

MODIS/ASTER Airborne Simulator

The MODIS/ASTER (MASTER) Airborne Simulator was developed to test sensors to be used in the Advanced Spaceborne Thermal Emission and Reflection Radiometer (ASTER) and Moderate Resolution Imaging Spectroradiometer (MODIS) projects (Figure 3.23). ASTER and MODIS are both instruments on the Terra platform launched in the fall of 1999. JPL and the University of Arizona conducted a joint experiment in December 1998 to test the instrument. Since that time MASTER has been flown by the Department of Energy (DOE). MASTER is an imaging scanner with 50 channels in the visible and IR between 400 nm and 12 μm (Hook et al., 2001; Tables 3.4 and 3.5).

FIGURE 3.23

MASTER (MODIS/ASTER) simulator image of Mt. St. Helen, Oregon. One day before Mt. St. Helens erupted on October 1, 2004, NASA acquired infrared images that indicate heat at and near the surface. "Based on the IR signal, the team predicted an imminent eruption," according to Steve Hipskind, acting chief of the Earth Science Division at NASA Ames Research Center. In this color composite, blue indicates the snow cover, orange and yellow colors characterize the rock type and lava age, and red indicates heat. (Image courtesy of NASA. http://www.nasa.gov/vision/earth/lookingatearth/mt_st_helen_jb.html.)

TABLE 3.4

Summary Characteristics of the MASTER Instrument

Wavelength range	$0.4 \pm 13 \ \mu m$
Number of channels	50
Number of pixels	716
Instantaneous FOV	2.5 mrad
Total FOV	85.92°
Platforms	DOE King Air Beachcraft B200, NASA ER-2, and NASA DC-8
Pixel size	DC-8 10 ± 30 m
Pixel size	ER-2 50 m
Pixel size	B200 5 ± 25 m

Source: Hook, S.J., *Rem. Sens. Environ.* 76, 2001.

TABLE 3.5

Spectral Characteristics of the Visible and Infrared MASTER Channels

Channel	Full Width Half Maximum	Channel Center (μm)	Channel Peak (μm)
1	0.0433	0.4574	0.458
2	0.0426	0.4981	0.496
3	0.0427	0.54	0.538
4	0.0407	0.5807	0.58
5	0.0585	0.6599	0.652
6	0.042	0.711	0.71
7	0.0418	0.7499	0.75
8	0.042	0.8	0.8
9	0.0417	0.8658	0.866
10	0.0407	0.9057	0.906
11	0.0403	0.9452	0.946
12	0.0542	1.6092	1.608
13	0.0526	1.6645	1.666
14	0.0514	1.7196	1.718
15	0.0521	1.7748	1.774
16	0.0506	1.8281	1.826
17	0.0457	1.8751	1.874
18	0.0575	1.9244	1.924
19	0.0504	1.9807	1.98
20	0.0481	2.0806	2.08
21	0.0511	2.1599	2.16
22	0.0508	2.2106	2.212
23	0.0513	2.2581	2.258
24	0.0683	2.3284	2.32
25	0.0641	2.3939	2.388
26	0.1559	3.1477	3.142
27	0.1459	3.2992	3.292
28	0.1478	3.4538	3.452
29	0.1544	3.6088	3.607
30	0.1345	3.7507	3.757
31	0.1524	3.9134	3.912

TABLE 3.5 (*Continued*)

Spectral Characteristics of the Visible and Infrared MASTER Channels

Channel	Full Width Half Maximum	Channel Center (μm)	Channel Peak (μm)
32	0.1548	4.0677	4.067
33	0.153	4.2286	4.224
34	0.153	4.3786	4.374
35	0.1446	4.5202	4.522
36	0.1608	4.6684	4.667
37	0.1521	4.8233	4.822
38	0.1487	4.9672	4.962
39	0.1495	5.116	5.117
40	0.1578	5.2629	5.272
41	0.3645	7.7599	7.815
42	0.4333	8.1677	8.185
43	0.3543	8.6324	8.665
44	0.4253	9.0944	9.104
45	0.4083	9.7004	9.706
46	0.3963	10.116	10.115
47	0.5903	10.631	10.554
48	0.6518	11.3293	11.365
49	0.4929	12.117	12.097
50	0.4618	12.8779	12.876

Source: NASA, http://master.jpl.nasa.gov/sensor/characteristics.htm.

IceBridge

Operation IceBridge is a NASA airborne mission initiated in 2009 that acquires altimetry, radar, and other geophysical measurements to monitor and characterize the earth's cryosphere (the portion of the earth's surface where water is frozen, including sea ice, snow cover, glaciers, ice caps, and permafrost).

Its primary goal is to extend the measurement of ice altimetry initiated by NASA's ICESat. The IceBridge mission will continue until the launch of ICESat-2, estimated for early 2016. The primary instruments are Lidars—including a photon counting Lidar—used to survey the major ice sheets in Greenland and Antarctica, sea ice in the Arctic and Southern Ocean, and glaciers and ice caps in Alaska and Canada that may be significant contributors to sea level rise (Figure 3.24).

IceBridge also acquires data to help understand ice dynamics. It carries snow mapping radars, gravimeters, and cameras. IceBridge currently conducts one Arctic and one Antarctic campaign each year. IceBridge relies on piloted aircraft from NASA and academic sources, and will use unpiloted aerial systems as soon as they are available, estimated to be during 2013. All data collected by Operation IceBridge are made public as quickly as possible through NASA's Distributed Active Archive Center at the National Snow and Ice Data Center (NSIDC), University of Colorado, Boulder (NASA f, g, Operation IceBridge, online).

DISCOVER-AQ

The objective of the Deriving Information on Surface Conditions from Column and Vertically Resolved Observations Relevant to Air Quality (DISCOVER-AQ) investigation,

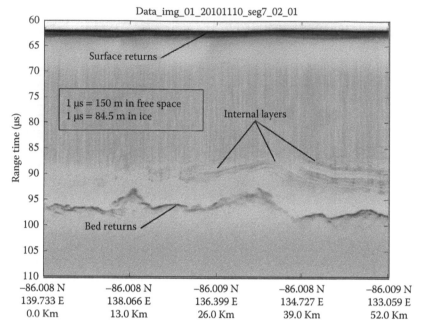

FIGURE 3.24

ICEBRIDGE airborne radar. Echogram from the depth sounder radar on NASA's DC-8 aircraft. Data were collected from an altitude of ~9000 m over the Antarctic ice sheet to a maximum depth of 3 km. The base of ice and internal layers are clearly indicated. (Courtesy of CReSIS/NASA. http://www.earthzine.org/2010/12/10/radar-instrumentation-for-operation-ice-bridge/.)

operated out of NASA Langley Research Center and beginning in July 2011, is to improve satellite observations of near-surface conditions related to air quality. To determine air quality, reliable surface, airborne, and satellite data on aerosols and ozone precursors are needed for use in air quality models. DISCOVER-AQ provides input to an integrated dataset of airborne and surface observations relevant to air quality (NASA h, i, DISCOVER-AQ, online).

DISCOVER-AQ observes and measures column-integrated, surface, and vertical distributions of aerosols and trace gases as they change throughout the day. This is accomplished using two NASA airborne platforms (King Air and P-3B aircraft) that repeatedly sample in coordination with a surface sampling network. One aircraft is used for extensive *in-situ* profiling of the atmosphere while the other conducts passive and active sensing of the atmospheric column extending from the aircraft to the surface.

Key Earth observing instruments and platforms involved in this mission and flown on the P-3B include Langley Aerosol Research Group E (LARGE) for aerosols; Thermal Dissociation and Laser Induced Fluorescence (TD LIF) to detect NO_2, HNO_3, peroxynitrates, and alkylnitrates; Diode Laser Hygrometer (DLH) for H_2O; Differential Absorption CO Measurement (DACOM) to measure CO, CH_4; Difference Frequency Generation Absorption Spectrometer (DFGAS) to observe CH_2O; Chemiluminescence (NCAR 4-channel Chemiluminescence Instrument) to detect O_3, NO_2, NO, and NOy; Proton-Transfer-Reaction Mass Spectrometer (PTR-MS) for non-methane hydrocarbons; and Atmospheric Vertical Observation of CO_2 in the Earth's Troposphere (AVOCET) to measure CO_2. King Air instruments include High Spectral Resolution Lidar (HSRL) to measure aerosol profiles and Airborne Compact Atmospheric Mapper (ACAM) to determine column O_3, NO_2, and CH_2O.

Airborne Microwave Observatory of Subcanopy and Subsurface

The NASA Airborne Microwave Observatory of Subcanopy and Subsurface (AirMOSS) project, initiated in early 2012, uses an airborne polarimetric L band SAR with the ability to penetrate substantial vegetation canopies and soil to depths around 1.2 m. NASA's Uninhabited Aerial Vehicle Synthetic Aperture Radar (UAVSAR) will be tested on a Gulfstream-III aircraft before becoming operational on an Uninhabited Aerial Vehicle (UAV). The instrument will normally operate at 12,497 m (41,000 ft.) with a bandwidth of 80 MHz (2 m range resolution), and a swath on the order of 16 km. Extensive ground and aircraft measurements will confirm root-zone soil measurements and carbon flux model estimates. Surveys are planned at subweekly, seasonal, and annual time scales. AirMOSS data help validate root-zone soil measurement algorithms from the Soil Moisture Active Passive (SMAP) mission, and help assess the impact of fine-scale heterogeneities in its coarse-resolution products. The project is being handled by the University of Southern California, JPL, and Langley Research Center (NASA j, AirMOSS; NASA k, UAVSAR, online).

CARVE

The NASA Carbon in Arctic Reservoirs Vulnerability Experiment (CARVE) mission began in early 2012, and uses a C-23 Sherpa aircraft to fly a payload that includes an L band radiometer/radar and a nadir-viewing spectrometer. The mission delivers simultaneous measurements of surface parameters that influence gas emissions (i.e., soil moisture, freeze/thaw state, and surface temperature) and will detect total atmospheric columns of CO_2, methane, and carbon monoxide. Deployments occur during the spring, summer and early fall when Arctic carbon fluxes are large and rapidly changing and ecosystem sensitivities are maximized. Continuous ground-based measurements provide calibration for the CARVE airborne measurements. The program is managed by JPL and NASA Langley Research Center (NASA l, CARVE, online).

Hurricane and Severe Storm Sentinel

The Hurricane and Severe Storm Sentinel (HS3) is a 5-year NASA mission, beginning in 2012, designed to improve our understanding of hurricane intensity changes in the Atlantic basin. HS3 will determine the extent to which the environment or processes internal to the storm are key to intensity changes.

The project uses two Global Hawk UAVs. The high Global Hawk flight altitude allows overflights of most storm convection as well as sampling of upper-tropospheric winds. The instrument payload includes the scanning High-resolution Interferometer Sounder (HIS), dropsondes, the TWiLiTE Doppler wind Lidar, and the Cloud Physics Lidar. The over-storm payload includes the HIWRAP conically scanning Doppler radar, the HIRAD multi-frequency interferometric radiometer, and the HAMSR microwave sounder. Deployments from NASA's Wallops Flight Facility and 30-hour flight capability allow access to the entire Atlantic Ocean, with on-station times of 10–24 hours depending on storm location. Deployments are planned from mid-August to mid-September 2012–2014, with ten 30-hour flights per deployment. The program is managed by NASA Goddard SFC, NASA Ames Research Center, and the Langley Research Center (NASA m, HS3, online).

Satellites and Their Primary Earth Observation Systems

Some of the more common non-military Earth observing satellites are listed. Satellites often carry more than one instrument system, and these are described and their purpose is given. They can be a single spacecraft or part of a long-running series. Joint missions are projects shared among several countries and agencies, usually for a common purpose such as sea level, pollution, or weather monitoring. Some, such as the European Space Agency (ESA), are multinational organizations with wide-ranging research and practical applications. The following descriptions are for current and planned Earth observing missions, as well as some that are no longer operational but that provided useful Earth science data. These spacecraft are described in chronological order by their primary sponsoring organization.

United States of America (NASA, NOAA)

The National Aeronautics and Space Administration (NASA), founded in 1958, is the agency of the United States that is responsible for the nation's civilian space program and aerospace research. NASA facilities include the John F. Kennedy Space Center, which operates launch facilities at Cape Canaveral. Other major facilities include the Marshall SFC in Huntsville, Alabama; Goddard SFC in Greenbelt, Maryland; Johnson SFC in Houston, Texas; JPL in Pasadena, California; Stennis Space Center in Mississippi; Ames Research Center in California; the Glenn Research Center in Ohio; Langley Research Center in Virginia; and the Dryden Flight Research Center (DFRC) in California.

The NOAA is part of the U.S. Department of Commerce. It is responsible for daily weather forecasts, severe storm warnings, climate monitoring, fisheries management, coastal restoration, and supporting marine commerce.

Orbiting Geophysical Observatory

The Orbiting Geophysical Observatory (OGO) program consisted of six satellites launched by NASA between 1964 and 1969 to study the earth's magnetosphere. Each satellite had between 20 and 25 instruments, but the two primary Earth observation instruments were magnetometers. OGO 1, 3, and 5 (also known as EOGO 1, 3, and 5) were in equatorial orbits: OGO 2, 4, and 6 (also known as POGO 1, 2, and 3) were in polar orbits. The last of these pioneering satellites was turned off in 1972 (NASA n, OGO series, online).

Skylab

The first serious effort at civilian Earth resources photography from space was during the Skylab program from May 1973 to February 1974 (NASA o, Skylab, online). Vertical true color and multispectral B/W CIR photos were acquired from an altitude of 435 km (261 mi.) using two film-based camera systems, the S190A and S190B, both of which provided stereo coverage (Figure 3.25). With a 152 mm (6 in.) focal length, the S190A covered 163 × 163 km (98 × 98 mi.) and had an effective resolution between 60 and 150 m (195–488 ft.). The S190B had a 45.7 cm (18 in.) focal length, covered 109 × 109 km (65 × 65 mi.), and had a resolution between 15 and 30 m (50–100 ft.). Repeat coverage was every 5 days between 50° north and 50° south latitude. Coverage is incomplete because of scheduling and cloud problems. These data were indexed by EDAC, University of New Mexico, and are available from there and the EROS Data Center in Sioux Falls, South Dakota. Much of the photography has changed color or faded with age, but the vertical and oblique images are still useful for a synoptic view of terrain.

Landsat

Civilian satellite imagery in digital form was first made available by the EROS Data Center's Landsat (originally "ERTS") program beginning in July 1972. Landsat satellites

FIGURE 3.25
Skylab Earth Terrain Camera (S190B) true color photograph of the Dolores River area, Uncompahgre Uplift, southwestern Colorado, June 1973. Note the blue cast due to atmospheric scattering. (Courtesy of NASA.)

orbit with either a 16-day or 18-day repeat cycle at an altitude of 918 km (Landsats 1, 2, and 3) and 705 km (Landsats 4, 5, 7). The sun-synchronous polar orbit allows the satellite to pass over mid-latitudes at the same mid-morning time every day. Areas north and south of 81° latitude are not covered. Sidelap between orbital paths varies from 7.6% at the equator to about 70% near the poles. Landsats 1, 2, and 3 contained a Return Beam Vidicon (RBV) and MSS; Landsats 4 and 5 contained the MSS and Thematic Mapper (TM) imaging systems. The Landsat 5 TM operations were suspended in November 2011 due to an electronics problem. Landsat 6 failed during launch. Landsat 7, launched in 1999 and still operational as of 2012, contains the Enhanced Thematic Mapper (ETM) with eight channels (NASA p, Landsat, online).

The RBV on Landsats 1 and 2 consisted of three television cameras that recorded green, red, and NIR bands with 40 m (130 ft.) ground resolution. Landsat 3 contained two RBV cameras, each with a single panchromatic band in an effort to acquire stereo imagery. Images covered 99 × 99 km (60 × 60 mi.). Problems with extreme contrast caused poor-quality images, and the system was not orbited on later Landsats.

The MSS recorded reflected light in four wavelengths: 500–600 nm (Band 4, green light), 600–700 nm (Band 5, red light), 700–800 nm (Band 6, NIR), and 800 nm to 1.1 μm (Band 7, NIR). The MSS image covers an area 176 × 176 km (110 × 110 mi.) with a resolution of 79 m (257 ft.).

The TM system has seven channels: Band 1, 450–520 nm, or blue light; Band 2, 510–600 nm, or green light; Band 3, 630–690 nm, red light; Band 4, 760–900 nm, short-wave infrared; Band 5, 1.55–1.75 μm, SWIR; Band 7, 2.08–2.35 μm, SWIR; and Band 6, 10.4–12.5 μm, LWIR. Images cover 176 × 170 km (110 × 105 mi.), have resolution of 28.5 m (93 ft.) in the reflected channels, and 120 m (390 ft.) in the thermal band (Figures 3.26 and 3.27). Thermal imagery can be acquired both during daytime (mid-morning) and nighttime (late evening).

Landsat 7 contains the ETM. The first seven bands are the same as those in the TM, whereas an eighth, panchromatic channel, covers 520–900 nm with 15 m resolution. The satellite has 16-day repeat coverage and a 185 km swath from a 705 km sun-synchronous orbit.

FIGURE 3.26
Landsat 7 Enhanced Thematic Mapper true color image of Detroit, Michigan, acquired December 11, 2001. (Courtesy of NASA/USGS.)

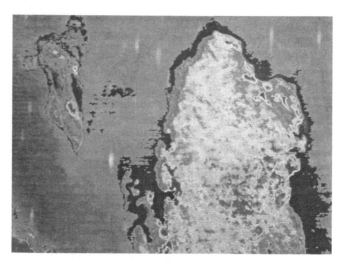

FIGURE 3.27
Color density-sliced Landsat Thematic Mapper thermal image of Qatar and Bahrain. Warm areas are red and yellow; cool areas are light and dark blue. (Processed by Amoco Production Company.)

TM imagery has a combination of moderately high resolution and large area coverage and seven spectral channels that allow good discrimination of surface materials. The blue light band has good water penetration capability and allows one to generate bathymetric maps and true color images. The thermal band is useful in mapping shallow groundwater, faults that channel groundwater, springs, offshore currents, and color/density contrasts between geologic units.

Geostationary Operational Environmental Satellite

The NASA/NOAA Geostationary Operational Environmental Satellites (GOES) are stationary with respect to their position over the earth, orbiting at an altitude of 35,000 km (22,000 mi.) and recording weather patterns over most of North and South America. The primary payload instruments include the Imager and the Sounder. The Imager is a multichannel instrument that senses IR radiant energy and visible reflected solar energy from the Earth's surface and atmosphere. The GOES Imager has a visible band (550–700 nm) with 1 km (0.6 mi.) resolution, and a thermal band (10.5–12.6 μm) with 8 km (5 mi.) resolution. The Sounder provides data for vertical atmospheric temperature and moisture profiles, surface and cloud top temperature, and ozone distribution. Originally launched in 1974, there have been 15 satellites in this series to date (NASA q, GOES-P, online). Data are acquired every 30 minutes and are distributed by the National Weather Service, NOAA, and other federal agencies. These images can be used to monitor not only weather but also winds, snow cover, and ocean currents, and help with the operation of shipping, offshore platforms, and flood prediction. GOES 12 (launched 2001) through 15 (launched 2010) are operational as of 2012. The fourth generation of GOES (designated GOES R, S, T, and U) is slated to be launched starting in 2015.

Laser Geodynamics Satellite 1 and 2

Laser Geodynamics Satellite-1 (LAGEOS 1) was designed by NASA and launched in 1976 into a 5860 km (3600 mi.) circular orbit. It was the first spacecraft dedicated exclusively to precision laser ranging. LAGEOS-2, identical to LAGEOS-1, was built by the Italian Space Agency (ASI) and was launched in 1992 into a 5620 km (3370 mi.) circular orbit. The LAGEOS satellites are completely passive and covered with 426 corner reflectors. Measurements are made by transmitting pulsed laser beams from Earth to the satellites. The laser beams then return to Earth after hitting the reflecting surfaces; the two-way travel times are precisely measured, permitting ground stations to determine their separations to better than 2 cm in thousands of kilometers. The LAGEOS mission goals include: (1) to provide an accurate measurement of the satellite's position with respect to Earth, (2) to determine the planet's shape (geoid), and (3) to determine tectonic plate movements associated with continental drift (NASA r, LAGEOS, online).

The National Oceanographic and Atmospheric Administration and MetOp-A

The Advanced Very High Resolution Radiometer (AVHRR) system is mounted on NOAA weather satellites in a polar orbit at an altitude of 850 km (510 mi.). Fifteen NOAA and European Organization for the Exploitation of Meteorological Satellites (EUMETSAT) MetOp satellites carrying this instrument have been launched since 1978. Each satellite images the entire Earth each day (Figure 3.28). Because they are weather satellites they tend to cover large areas and have coarse imaging resolution. The ground swath is 2700 km wide (1620 mi.), with 1.1 km (0.7 mi.) resolution in five spectral bands: 0.55–0.68 μm (visible), 0.73–1.10 μm (NIR), 3.55–3.93 μm (thermal), 10.5–11.5 μm (thermal), and 11.5–12.5 μm (thermal). The NOAA-15 platform, launched in 1998, carries an additional radiometer centered at 1.6 μm. These data are available from NOAA. NOAA-15 through NOAA-19 and MetOp-A continue to be operational as of 2012 (NASA s, NOAA-N; EUMETSAT a, MetOp, online).

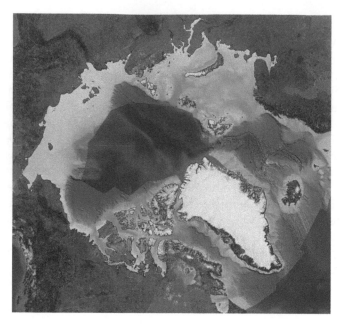

FIGURE 3.28

NOAA AVHRR satellite image (onshore) merged with 90 m bathymetry offshore for a portion of the Arctic Ocean around Greenland and Iceland. (Courtesy of TCarta Marine © 2012 and NASA.)

Heat Capacity Mapping Mission

The Heat Capacity Mapping Mission (HCMM) was launched in 1978 and gathered data until 1980. The objective was to map circulation patterns of marine currents, but large parts of the continents were also imaged to map surface moisture by measuring the soil's thermal inertia. Orbiting at an altitude of 620 km (372 mi.), the Heat Capacity Mapping Radiometer had a ground resolution of 500 m (1625 ft.) in the reflected light range (0.50–1.10 µm) and 600 m (1950 ft.) in the thermal region (10.5–12.5 µm). Each image covered a 700 × 700 km area (420 × 420 mi.), and both day and night images were acquired (Figure 3.29). Coverage is limited mainly to North America, western Europe, north Africa, and east Australia. Index catalogs and computer tapes are available from the National Space Science Data Center, NASA Goddard, Greenbelt, Maryland (Harris, 1987).

HCMM imagery is a good option for regional coverage. For example, it takes 16 Landsat images to cover the same area. Disadvantages include limited coverage and low resolution. It is more useful for plate tectonic reconstructions than for detailed exploration.

SeaSat

SeaSat, designed to monitor the oceans, was launched in June 1978 and failed prematurely in October 1978 (NASA t, SeaSat, online). This satellite carried the first civilian Earth observing radar, an L band (23.5 cm) Imaging Radar. SeaSat orbited at an altitude of 790 km (474 mi.), and had a swath width of 100 km (60 mi.), with depression angles from 67° to 73°. Ground resolution was 25 m (81 ft.). Coverage was limited mainly to North and Central America, Europe, and part of North Africa. Digital tapes can be obtained from the Environmental Data and Information Service of the NOAA, Ashville, North Carolina. Catalogs are available from JPL.

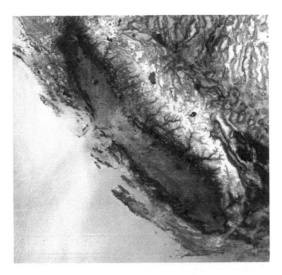

FIGURE 3.29
Heat Capacity Mapping Mission thermal image of central California including the Great Valley and the Sierra Nevada. Warm areas are dark; cool areas are bright. (Courtesy of NASA/USGS.)

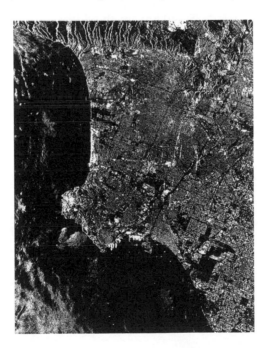

FIGURE 3.30
East-looking SeaSat radar image of the greater Los Angeles area, California. Note the geometric distortion due to "layover" in the Santa Monica Mountains (upper left of the image). (Image courtesy of NASA/JPL.)

SeaSat radar was able to penetrate clouds, especially in tropical areas. It recorded surface roughness changes indicative of changes in soil or lithology, and had the ability to map structure by accentuating topographic relief. Unfortunately, the imagery had limited coverage and extreme layover caused by the steep depression angle (Figure 3.30).

Magnetic Field Satellite

The Magnetic Field Satellite (Magsat) was launched by NASA in 1979 and failed during 1980. The mission was to map the earth's magnetic field. To that end, the satellite was equipped with two magnetometers. The Cesium vapor/scalar magnetometer determined the total magnetic field and was used to calibrate the vector magnetometer. The three axis vector flux-gate magnetometer determined the strength and direction of Earth's magnetic field. The satellite flew in an elliptical 300 × 550 km sun-synchronous near polar orbit (Langel et al., 1982).

Space Shuttle

The Space Shuttle program acquired vertical and oblique black/white, color, and color-infrared photographs using hand-held cameras and the Large Format Camera beginning in 1982 and ending in 2011. This instrument has a 30.5 cm (13 in.) focal length and was used at altitudes from 239 to 370 km (143–222 mi.). Some stereo coverage is available. Contact the EROS Data Center (Sioux Falls, South Dakota) or Johnson Space Center (Houston, Texas).

The Shuttle Imaging Radar-A (SIR-A) was flown on the Space Shuttle in November 1981, and SIR-B in October 1984. Both used an L band radar, but SIR-A orbited at 250 km (150 mi.) and had 38 m (123 ft.) resolution, whereas SIR-B orbited at 225 km (135 mi.) and had 25 m (81 ft.) resolution. SIR-A had a depression angle of 37°–43° (50 km swath), and SIR-B had a variable depression angle of 30°–75° (40 km swath width).

SIR first observed the phenomena, reported in Science in 1982 (McCauley et al., 1982), that radar signals could penetrate through as much as 2 m of sand in the hyperarid deserts of Egypt and the Sudan (Chapter 12). These workers reported mapping paleodrainage channels not evident at the surface. In other areas, such as the Oman–Abu Dhabi border, anticlines have been mapped on SIR-A imagery through a veneer of windblown sand.

The Shuttle Imaging Radar-C (SIR-C), first launched in 1994, was a joint effort between NASA, DLR, and ASI (Figure 3.31). The system comprised a dual-frequency radar with

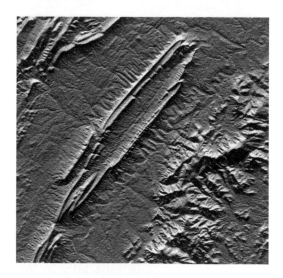

FIGURE 3.31
Shuttle Radar Topography Mission C band radar image of Massanutten Mountain in the Shenandoah Valley of northern Virginia. Rock layers form a synclinal fold. The ridges are capped by a resistant sandstone whereas limestones and shales are less resistant and form the valleys. (Courtesy of NASA/JPL. http://www.jpl.nasa .gov/spaceimages/details.php?id=PIA03382.)

L band (23 cm) and C band (6 cm) wavelengths, and an X-SAR with an X band (3 cm) wavelength. The ground swath is variable, from 15–90 km (9–56 mi.), and resolution varied, depending on the instrument and configuration, from 10–200 m (33–656 ft.). Incidence angles ranged between 15° and 55° off nadir. Flight altitude was 225 km (135 mi.). SIR coverage is extremely limited due to the short duration of the missions and to the mechanical problems in deploying the SIR-B antenna.

SIR-A, B, and C films are available from the National Space Science Data Center, NASA Goddard. SIR-B data, which also are available in digital form, can be obtained from JPL.

The Modular Optico-electronic Multispectral Scanner (MOMS-1) was built by the German AeroSpace Research Establishment (DFVLR) and flown on Shuttle Missions STS-7 (1983) and STS-11 (also known as STS 41-B, 1984). It had two bands in the range 570–920 nm. Swath width is 140 km (84 mi.) with a ground resolution of 20 × 20 m (65 × 65 ft.). This instrument acquired a limited number of images; they are available from the DFVLR (DFVLR a, online).

Geodetic Satellite

The GEOdetic SATellite (GEOSAT) was a U.S. Navy satellite launched in March, 1985 into an 800 km (480 mi.) orbit, with a 23-day repeat cycle. The satellite carried a radar altimeter to measure the distance from the satellite to sea surface within 5 cm. After a classified Geodetic Mission ended, GEOSAT was maneuvered into a 17-day repeat orbit to begin the public Exact Repeat Mission (ERM) in November, 1986. The purpose of GEOSAT was to provide information on the marine gravity field. The ERM ended in January 1990, due to failure of the two onboard tape recorders. More than 3 years of ERM data have been made available to the scientific community (NOAA b, GEOSAT, online). NOAA/National Ocean Service produced the ERM Geophysical Data Records that are distributed by the NODC (National Oceanographic Data Center), Silver Spring, Maryland.

TOPEX/Poseidon

Launched in 1992, TOPEX/Poseidon was a joint satellite mission between NASA and the French Space Agency (*Centre National d'Etudes Spatiales*, CNES) to map ocean surface topography. A malfunction ended normal satellite operations during January 2006. TOPEX/Poseidon carried two onboard altimeters, ALT and Poseidon, as well as the TOPEX instrument (NASA u, TOPEX/Poseidon, online).

- ALT (radar altimeter): a NASA-built nadir-pointing RA with both C band (5.3 GHz) and Ku band (13.6 GHz) for measuring height above sea surface to within 2.4 cm (1 in.).

- Poseidon (also referred to as Single-Frequency Solid-State Altimeter [SSALT]): a CNES-built nadir-pointing Ku-band (13.65 GHz) RA with 2.5 cm accuracy.

- TOPEX (also known as TMR, TOPEX Microwave Radiometer) built by JPL: TMR operates at 18, 21, and 37 GHz to measure the total water vapor content along the altimeter pulse path to correct for water-vapor-induced range delay.

Data produced by the Topex/Poseidon mission are available to the international scientific community through the U.S. data center at JPL called the Physical Oceanographic Distributed Active Archive Center (PODAAC), and through the French data center, Archiving, Validation and Interpretation of Satellite Oceanographic data (AVISO).

Tropical Rainfall Measuring Mission

The TRMM is a NASA satellite launched in 1997 (and still operational as of 2012) that provides information to test and improve climate models (NASA v, w, TRMM, online). TRMM determines rainfall amounts in the tropics and subtropics. These regions make up about two-thirds of the total rainfall on Earth and are responsible for driving Earth's climate. TRMM contributes to a better understanding of where winds blow and how much, where clouds form and rain occurs, where floods and droughts will occur, and the relationship between wind and ocean currents. TRMM does this by providing rainfall data and information on heat released into the atmosphere as a result of precipitation (latent heat of evaporation).

There are five instruments on TRMM. The Precipitation Radar, built by the National Space Development Agency (NASDA) of Japan, provides 3D maps of storm structure. Measurements provide information on the intensity and distribution of rain, on rain type, and on the height where raindrops form. The TRMM Microwave Imager (TMI) is a passive sensor that supplies quantitative rainfall information over a wide swath. TMI quantifies water vapor and rainfall intensity by measuring the microwave energy emitted by the earth and its atmosphere. The Visible and Infrared Scanner (VIRS) is an indirect indicator of rainfall, and ties TRMM measurements to other satellite measurements. In addition, TRMM carries the Clouds and the Earth's Radiant Energy System (CERES) instrument and the Lightning Imaging Sensor (LIS). Data from CERES is used to study the energy exchanged between the Sun and Earth's atmosphere and surface. The LIS detects and locates lightning over the tropics. LIS surveys lightning and thunderstorm on a global scale from a 350 km (218 mi.) geostationary orbit.

Geosat Follow-On

The successor to GEOSAT is the Geosat Follow-On (GFO) mission, launched in 1998. GFO carried a water vapor radiometer as well as an RA, and operated in the same orbit as GEOSAT's ERM. The purpose of the satellite was to measure Earth's marine gravity field using an RA. In late September 2008, the deteriorating state of the spacecraft resulted in its shutdown (NASA x, GEOSAT Follow-On, online).

TERRA (EOS AM)

The Earth Observing System (EOS AM) satellite carries several instruments including MODIS, ASTER, Multi-angle Imaging Spectroradiometer (MISR), and Measurements of Pollution in the Troposphere (MOPITT). The purpose of NASA's MODIS instrument is to measure a variety of biological and physical processes, over both land and oceans and in the atmosphere (Figure 3.32). The satellite occupies a 705 km (425 mi.) sun-synchronous orbit with a 2330 km (1400 mi.) wide image swath. MODIS was launched into Earth orbit by NASA in 1999 onboard the TERRA satellite. The instruments capture data in 36 spectral bands ranging in wavelength from 415 nm to 14.4 μm and at varying spatial resolutions (two bands in the region 620–876 nm with 250 m [750 ft.] ground resolution; five bands from 459 nm to 2.155 μm with 500 m [1500 ft.] resolution; and 29 bands from 405 nm to 14.385 μm with 1000 m [3300 ft.] ground resolution; Table 3.6). Thermal resolution is 0.3°C–0.5°C over water and 1°C over land. Together, the instruments image the entire Earth every 1–2 days (ESA a, MODIS, online). MODIS products are available from several sources. Land Products are available through the Land Processes DAAC at the USGS EROS Data

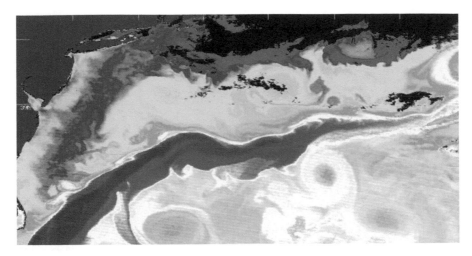

FIGURE 3.32

MODIS thermal image of the Gulf Stream off the eastern seaboard of the United States. This image is from NASA's AQUA satellite, but the instrument is the same as on the TERRA satellite. (Courtesy of NASA and UCSB. http://www.geog.ucsb.edu/~jeff/wallpaper2/gulf_stream_modis.jpg.)

TABLE 3.6

MODIS Instrument Parameters

Primary Use	Band	Bandwidth (nm)
Land/cloud/aerosols	1	620–670
	2	841–876
	3	459–479
	4	545–565
	5	1,230–1,250
	6	1,628–1,652
	7	2,105–2,155
Ocean color/phytoplankton/biogeochemistry	8	405–520
	9	438–448
	10	483–493
	11	526–536
	12	546–556
	13	662–672
	14	673–683
	15	743–753
	16	862–877
Atmospheric water vapor	17	890–920
	18	931–941
	19	915–965
Surface/cloud temperature/atmospheric/temperature	20	3,660–3,840
	21	3,929–3,989
	22	3,930–3,989
	23	4,020–4,080
	24	4,433–4,498
	25	4,482–4,549

(Continued)

TABLE 3.6 (*Continued*)

MODIS Instrument Parameters

Primary Use	Band	Bandwidth (nm)
Cirrus clouds/water vapor	26	1,360–1,390
	27	6,535–6,895
	28	7,175–7,475
Cloud properties/ozone	29	8,400–8,700
	30	9,580–9,880
Surface/cloud temperature	31	10,780–11,280
	32	11,770–12,270
Cloud top altitude	33	13,185–13,485
	34	13,485–13,785
	35	13,785–14,085
	36	14,085–14,385

Source:　USGS and NASA, https://lpdaac.usgs.gov/products/modis_overview.

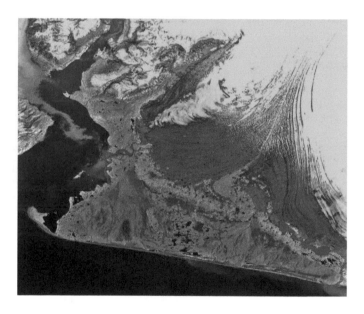

FIGURE 3.33

ASTER image of the Malaspina glacier area, Alaska. Vegetation is yellow, orange, and green. Gravel shows up as orange to red. Ice is cyan; water is black. (Courtesy of NASA and UCSB. http://www.geog.ucsb.edu/~jeff/wallpaper2/alaska_ice_bay_malaspina_glacier_veg_is_yellow-orange_and_green_gravel_appears_orange_jun8_2001_aster_wall.jpg.)

Center. Cryosphere data products (snow and sea-ice cover) are available from the National Snow and Ice Data Center (NSIDC) in Boulder, Colorado. Ocean color products and sea-surface temperature products along with information about these products are obtainable at the Ocean Color Data Processing System (OCDPS) at Goddard SFC.

The Japanese-built Advanced Spaceborne Thermal Emission and Reflection Radiometer (ASTER) consists of three instruments designed to obtain detailed surface temperature, emissivity, reflectance, and elevation data (NASA y, ASTER, online; Figure 3.33). ASTER acquires data in three visible bands with 15 m (50 ft.) resolution (520–860 nm), an additional

TABLE 3.7

ASTER Instrument Parameters

Band	Label	Wavelength (μm)	Resolution (m)	Description
B1	VNIR_Band1	0.520–0.600	15	Green/yellow
B2	VNIR_Band2	0.630–0.690	15	Visible red
B3	VNIR_Band3N	0.760–0.860	15	Near-IR
B4	VNIR_Band3B	0.760–0.860	15	Near-IR
B5	SWIR_Band4	1.600–1.700	30	Short-wave IR
B6	SWIR_Band5	2.145–2.185	30	Short-wave IR
B7	SWIR_Band6	2.185–2.225	30	Short-wave IR
B8	SWIR_Band7	2.235–2.285	30	Short-wave IR
B9	SWIR_Band8	2.295–2.365	30	Short-wave IR
B10	SWIR_Band9	2.360–2.430	30	Short-wave IR
B11	TIR_Band10	8.125–8.475	90	Thermal IR
B12	TIR_Band11	8.475–8.825	90	Thermal IR
B13	TIR_Band12	8.925–9.275	90	Thermal IR
B14	TIR_Band13	10.250–10.950	90	Thermal IR
B15	TIR_Band14	10.950–11.650	90	Thermal IR

Source: U.S.G.S. https://lpdaac.usgs.gov/products/aster_overview.

visible band (760–860 nm) for stereo, six NIR bands with 30 m (100 ft.) resolution (from 1.6 to 2.43 μm), and five TIR channels with 90 m (300 ft.) resolution (ranging from 8.125 to 11.65 μm; Table 3.7). It has a 60 km (36 mi.) wide-ground swath and a 16-day repeat cycle. In addition to imagery, the ASTER instrument gathers surface elevation data to generate a global digital elevation model (GDEM) with 30 m spatial resolution. All ASTER data are available from the Land Processes Distributed Active Archive Center, USGS.

The Multi-angle Imaging SpectroRadiometer (MISR) measures solar radiation at nine different incidence angles simultaneously in each of four channels (NASA z, MISR, online). The purpose is to study how atmospheric aerosols, clouds, and the earth's surface reflect and scatter light. One digital camera points to nadir (vertical), while the others fore-and-aft at 26.1°, 45.6°, 60.0°, and 70.5°. Repeat coverage is every nine days with a 360 km (220 mi.) swath width. Resolution is 250 m (820 ft., nadir) and 275 m (900 ft.) for all other angles. The four channels are centered on blue (443 nm), green (555 nm), red (670 nm), and NIR (865nm).

MOPITT, funded by the Canadian Space Agency (CSA) and flown on TERRA, is a nadir-oriented sensor that measures IR radiation at 4.7 μm and at 2.2–2.4 μm. The instrument uses correlation spectroscopy to create profiles of carbon monoxide, ozone, methane, and other trace gases and particulates in the lower atmosphere (NASA aa, MOPITT, online). MOPITT data are available from the Atmospheric Science Data Center at NASA LARC, Langley, Virginia.

Active Cavity Radiometer Irradiance Monitor III

NASA's Active Cavity Radiometer Irradiance Monitor III (ACRIM III) instrument is designed to study total solar irradiance. The ACRIM III package, launched on December 20, 1999, is the third in a series of long-term solar-monitoring tools built for NASA Goddard SFC by JPL. It extends the database first created by ACRIM I, launched in 1980 on the Solar Maximum Mission spacecraft. ACRIM II followed on the Upper Atmosphere Research Satellite (UARS) in 1991. ACRIM III is flown on ACRIMSAT and the project is managed by Columbia University (New York) and JPL.

ACRIMSAT data are being used to understand possible global warming, ice cap shrinkage, and ozone layer depletion. ACRIM measures incoming solar radiation. Climatologists hope to improve predictions of climate and global warming by combining ACRIM data with measurements of ocean and atmospheric currents and temperatures, as well as surface temperatures. Small solar oscillations were detected in the ACRIM I total irradiance data. ACRIM III data products are available through the Langley EOS Data Analysis and Archive Center (NASA bb, cc, ACRIMSAT, online).

Quick Scatterometer

The Quick Scatterometer (QuikSCAT) mission, launched by NASA in 1999, records sea-surface wind speed and direction under all weather and cloud conditions. QuikSCAT was created to reduce the ocean wind data gap caused by the loss of the NASA Scatterometer (NSCAT) on the Japanese Advanced Earth Observing Satellite (ADEOS), which failed in 1997. QuikSCAT flies in a near polar circular orbit at an altitude of ~800 km (500 mi.).

SeaWinds is the main instrument on QuikSCAT. SeaWinds is a radar scatterometer that transmits high-frequency microwave pulses to the ocean surface and measures the echoed radar signal. The scatterometer estimates wind speed and direction over oceans ~10 m (33 ft.) above the water surface. The instrument collects data over oceans, land, and ice in a 1800 km (1100 mi.) wide swath, covering 90% of Earth's surface every day. This information is essential for global climate research, weather forecasting, and storm warnings.

QuikSCAT was fully operational until November 2009 when the SeaWinds antenna stopped rotating. Though SeaWinds performance was not affected by the failure, the scatterometer now tracks a significantly reduced swath. These data continue to provide reliable calibration of other ocean wind sensors (NASA dd, QUICKSCAT, online).

New Millenium Program Earth Observing-1 Satellite

The New Millenium Program Earth Observing-1 (EO-1) satellite was launched in November 2000 as a technology validation mission (NASA ee, ff, Earth Observing-1, online). It carried two Earth observation instruments, the Hyperion Hyperspectral Imager and the Advanced Land Imager (ALI) multispectral instrument (Figure 3.34a, b). The mission was extended in November 2001. During June 2009, this imagery became available to the public at no cost from the USGS: EO-1 archived data are available through Internet access to the Earth Explorer (http://earthexplorer.usgs.gov) or Global Visualization Viewer (http://glovis.usgs.gov) websites. New EO-1 scenes can be obtained by submitting a Data Acquisition Request to http://eo1.usgs.gov.

The Hyperion instrument is NASA's first satellite hyperspectral imager. Hyperion contains 220 spectral bands in the range 357–2576 nm, each having a 10 nm bandwidth and 30 m (100 ft.) ground resolution (Table 3.8). The push-broom instrument provides images that cover an area 7.5 × 100 km (4.5 × 60 mi.). The Linear Etalon Imaging Spectral Array/Atmospheric Corrector (LEISA/AC) is an IR camera used to remove the effects of the atmosphere from images obtained by the ALI. A follow-on to the Landsat 7 MSS, ALI acquires data in 10 channels using a push-broom array spectrometer.

Scientific Application Satellite-C

The Scientific Application Satellite-C (SAC-C) mission is a joint project between NASA and the Argentine Commission on Space Activities (*Comisión Nacional de Actividades Espaciales*; CONAE). The satellite was launched in 2000 into a sun-synchronous 702 km (420 mi.)

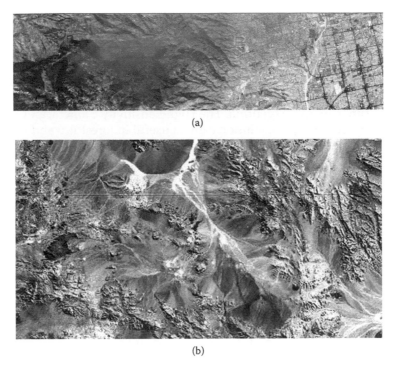

(a)

(b)

FIGURE 3.34

(a) EO-1 Hyperion image of the Aspen fire north of Tucson, Arizona, during June–July 2003. North is to the left. Bands 197-107-18 displayed as RGB. (Courtesy NASA/USGS/EROS Data Center.) (b) ALI band 5-4-3 image of Cuprite, Nevada, from NASA's EO-1 satellite. North is to the left. (Courtesy of NASA/USGS/EROS Data Center. http://eo1.gsfc.nasa.gov/new/general/imagery/Imagery/EO-1%20ALI%20Sample%20Image_Geologic%20Application.jpg. http://eo1.gsfc.nasa.gov/new/general/imagery/Imagery/Aspen_Fire-Hyp-fullHyperAspen.jpg.)

TABLE 3.8

EO-1 Instrument Parameters

Parameter	Hyperion	ALI
Swath width (km)	7.7	37
Product length (km)	42 or 185	42 or 185
Spatial resolution (meters)	30	30
Spectral range (μm)	0.4–2.5	0.4–2.5
Spectral resolution (nm)	10	Variable
Spectral coverage	Continuous	Discrete
pan band resolution (meters)	N/A	10
Total number of bands	220	10

Source: NASA, http://eo1.gsfc.nasa.gov/new/Technology/eo1Technology.html.

circular orbit with the purpose of providing images of the Earth and measuring its magnetic and gravity fields. The satellite carries 11 different instruments, of which five are used for Earth observation. The GPS Occultation and Passive Reflection Experiment (GOLPE), furnished by JPL, consists of a GPS that studies the earth's gravity field by producing postprocessed decimeter-level orbital height measurements. GOLPE uses the GPS data to study weather and climate change by measuring the refraction, or bending of GPS signals by Earth's atmosphere and ionosphere. The Scalar Helium Magnetometer (SHM) provided by

NASA/JPL, is part of the Magnetic Mapping Payload (MMP). The MMP is designed to define the Earth's magnetic field. The MMP provides continuous measurements with an accuracy of one part in 50,000. The Multispectral Medium Resolution Scanner (MMRS), designed and built by CONAE, studies the terrestrial and marine environment. The MMRS is operational mainly over Argentina. The High-Resolution Technological Camera (HRTC), also built by CONAE, provides detailed images of parts of the MMRS scenes. The HRTC is operational primarily over Argentina. The High-Sensitivity Camera, a CONAE supplied instrument, is an intermediate resolution camera useful in forest fire and storm detection, among other things (NASA gg, SAC-C, online).

Jason-1

Jason-1 is a satellite mission to monitor global ocean circulation, study the ties between the ocean and the atmosphere, improve global climate forecasts, and monitor El Niño and ocean eddies (NASA hh, JASON-1, online). A joint project between NASA, CNES, and EUMETSAT, Jason-1 is the successor to TOPEX/Poseidon.

Jason-1 was designed to measure global sea level changes. Measurements of sea surface topography allow scientists to calculate the speed and direction of ocean currents and monitor global ocean circulation. Jason-1 was launched in 2001 and has a 10-day repeat cycle. There are two Earth observation instruments:

- Poseidon 2—a nadir-pointing RA using C band and Ku band for measuring height above sea surface.

- Jason Microwave Radiometer (JMR)—measures water vapor along the altimeter beam path to correct for pulse delay (Figure 3.35).

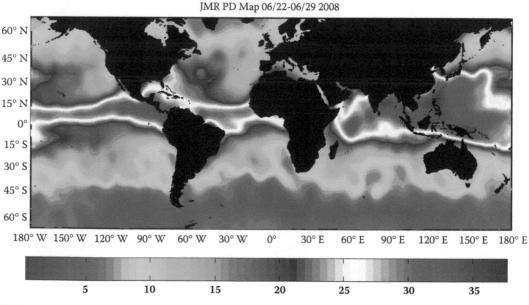

FIGURE 3.35

Atmospheric water vapor map created by the Jason-1 Microwave Radiometer. Water vapor delays the time it takes for the radar pulse from the spacecraft's altimeter to travel to the ocean surface and back. One can calculate how much water vapor is in the signal's path by knowing the time taken by the signal to return and the distance between the satellite and the ocean surface. (Image courtesy of NASA. http://photojournal.jpl.nasa .gov/catalog/PIA10954.)

Jason-2

The Ocean Surface Topography Mission (OSTM) on the Jason-2 satellite was launched by NASA and EUMETSAT in 2008 and continues the sea-surface height measurements begun by the TOPEX/Poseidon and Jason-1 missions (NASA pp, OSTM/Jason-2, online). The main Jason-2 instrument is the Poseidon-3 altimeter (Figure 3.36). This instrument sends a microwave pulse to the ocean's surface and times how long it takes to return. An MWR corrects any delay that may be caused by water vapor in the atmosphere. The altimeter is able to determine sea-surface height to within a few centimeters. The strength and shape of the signal provides information on wind speed and wave height. These data are used in ocean models to calculate the speed and direction of currents and the amount and location of heat stored in the ocean, which, in turn, is related to global climate variations. The satellite is in a 1,336 km (830 mi.) circular, non-sun-synchronous orbit that allows it to monitor 95% of Earth's ice-free ocean every 10 days. It flies in tandem with Jason-1 with ground tracks about 315 km (196 mi.) apart at the equator. This interleaved tandem mission allows detection of relatively small features such as ocean eddies.

AQUA (EOS PM 1)

Aqua is a joint project between the United States, Japan, and Brazil to monitor the earth's oceans and surface water systems. Launched in 2002 into a 691 × 708 km orbit, Aqua contains six Earth observing instruments (NASA ii, Earth Observing System AQUA, online). The Atmospheric Infrared Sounder (AIRS) along with AMSU-A measures humidity, temperature, cloud properties, and the amounts of greenhouse gases in the atmosphere.

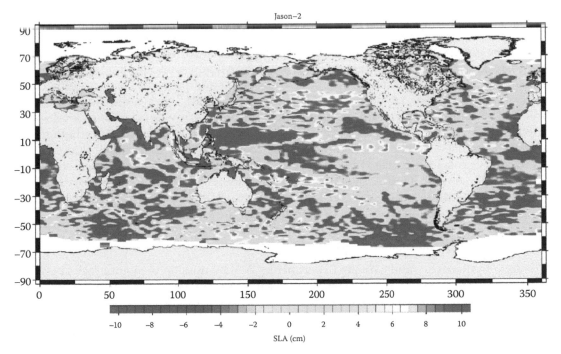

FIGURE 3.36

Jason-2 satellite's Poseidon-3 altimeter data from the Ocean Surface Topography Mission. Sea level anomaly map from July 31 and August 10, 2008. (Courtesy of CNES/CLS. http://www.aviso.oceanobs.com/fileadmin/images/news/mod_actus/SLA_J2_003.gif.)

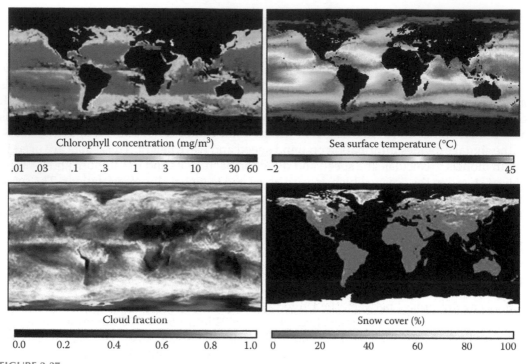

Chlorophyll concentration (mg/m³)   Sea surface temperature (°C)

.01 .03   .1   .3   1   3   10   30 60 −2   45

Cloud fraction   Snow cover (%)

0.0   0.2   0.4   0.6   0.8   1.0   0   20   40   60   80   100

FIGURE 3.37

Typical product maps derived from AQUA/EOS-PM1. (Courtesy of NASA. http://www.dailygalaxy.com/photos/uncategorized/neoglobes_oct2006_1.jpg.)

The Advanced Microwave Scanning Radiometer (AMSR-E), provided by Japan's NASDA, measures precipitation rate, cloud water, water vapor, sea-surface winds, sea-surface temperature, ice, snow, and soil moisture. The Advanced Microwave Sounding Unit (AMSU-A) obtains temperature profiles in the upper atmosphere and provides cloud-filtering capability for tropospheric temperature observations. The CERES, measures the earth's total thermal radiation budget, and, along with MODIS data, provides detailed cloud information. MODIS is a 36-band spectroradiometer measuring visible and IR radiation and obtains data used to map vegetation, land surface cover, ocean chlorophyll fluorescence, cloud and aerosol properties, fire occurrence, snow cover on land, and sea ice on oceans (Figure 3.37; see also TERRA satellite). The Humidity Sounder for Brazil (HSB), provided by Brazil's Instituto Nacional de Pesquisas Espaciais (INPE, the Brazilian Institute for Space Research), obtains humidity profiles through the atmosphere.

Gravity Recovery and Climate Experiment

The Gravity Recovery and Climate Experiment (GRACE) mission is a joint project between NASA and the German Space Agency (Deutsches Zentrum für Luft- und Raumfahrt or DLR). The twin satellites (known as GRACE-A and GRACE-B) were built by Astrium and launched together in 2002. The primary objective of GRACE is to measure the earth's gravity field. Gravity is determined by mass: by measuring gravity, GRACE shows how mass is distributed around the planet and how it varies over time (Figure 3.38). GRACE maps the earth's gravity fields by making accurate measurements of the distance between the two satellites using GPS and a microwave ranging system. The results from this mission provide information about the distribution and flow of mass within the earth (NASA jj, GRACE, online).

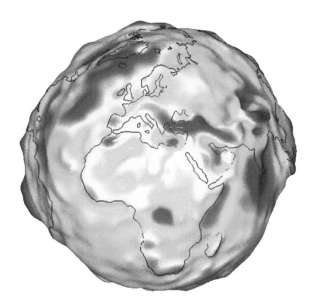

FIGURE 3.38
Geoid based on GRACE gravity measurements. (Courtesy of NASA. http://www.csr.utexas.edu/grace/gallery/gravity/ggm01_euro2_full.jpg.)

A second science objective of GRACE is to obtain approximately 150 precise, global vertical humidity and temperature profiles of the atmosphere per day using the GPS radio occultation technique. Challenging Minisatellite Payload (CHAMP) and GRACE formed a satellite configuration for precise atmospheric sounding during July 2004. Occultations were recorded aboard both satellites, providing global vertical profiles of temperature and specific humidity.

Ice, Cloud, and Land Elevation Satellite

The Ice, Cloud, and land Elevation Satellite (ICESat) is part of NASA's Earth Observing System. The mission is designed to measure changing ice sheets, cloud and aerosol heights, topography, and vegetation characteristics. ICESat was launched in 2003 into a near-circular, near-polar orbit at an altitude of 600 km (360 mi.). It operated until February 2010 (NASA kk, ICESat, online).

The sole instrument on ICESat was the Geoscience Laser Altimeter System (GLAS), a space-based Lidar. GLAS combined a precision surface Lidar with a sensitive dual-wavelength cloud and aerosol Lidar. The GLAS lasers emit IR and visible laser pulses at 1064 and 532 nm wavelengths. As ICESat orbited, GLAS gathered data from a series of ~70 m (230 ft.) diameter laser spots every 170 m (560 ft.) along the spacecraft's ground track. The satellite had a 91-day repeat cycle. A follow-on mission is planned for 2016.

Solar Radiation and Climate Experiment

The Solar Radiation and Climate Experiment (SORCE) is a NASA mission launched in 2003 that measures incoming x-ray, UV, visible, NIR, and total solar radiation (from 1 nm to 2 μm). SORCE carries four instruments including the Total Irradiance Monitor (TIM), Solar Stellar Irradiance Comparison Experiment (SOLSTICE), Spectral Irradiance Monitor (SIM), the extreme ultraviolet (XUV) Photometer System (XPS).

The SORCE satellite orbits Earth at 645 km accumulating solar data. Spectral measurements characterize the sun's energy and emissions in wavelengths that allow estimation of quantities of atmospheric molecules. Data obtained by the SORCE experiment is used to model the sun's output and to explain and predict the effect of the sun's radiation on the earth's atmosphere and climate.

The program is operated by the Laboratory for Atmospheric and Space Physics (LASP) at the University of Colorado, Boulder. SORCE measurements address long-term climate change, natural climate variability, improved climate prediction, atmospheric ozone, and UV-B radiation. These measurements are critical to studies of the sun's effects on Earth and humans (NASA ll, SORCE; University of Colorado, SORCE, online).

Aura

The Aura satellite (formerly "EOS/Chem-1") is a joint mission of NASA and the ESA with the objective of studying the chemistry and dynamics of Earth's atmosphere. In particular, it monitors ozone and other trace gases and tropospheric pollutants (NASA mm, Aura, online). Aura was launched in 2004 into a 705 km sun-synchronous orbit with a 16-day repeat cycle. The Ozone Monitoring Instrument (OMI) is a nadir-viewing UV/Visible imaging spectrograph, which measures the solar radiation backscattered by the earth's atmosphere and surface over the wavelength range from 270 to 500 nm with a spectral resolution of 0.5 nm (Figure 3.39). Swath width is 2600 km (1560 mi.).

CBERS Series

The CBERS satellite series is operated by NASA in cooperation with Chinese Academy of Space Technology (CAST) of the People's Republic of China, and INPE of Brazil. These satellites are in a 778 km (470 mi.) sun-synchronous orbit with a 16-day repeat cycle. CBERS-1 was operational from 1999 through 2003. CBERS-2 was operational from 2003 to 2007. CBERS-2B was operational from 2007 to 2010. These satellites all carried three instruments: the High-Resolution CCD Camera (HRCC), Infrared Multispectral Scanner (IRMSS), and Wide-Field Imager (WFI). The WFI, designed to observe large areas, has two bands (630–690 nm and 770–890 nm) that image an 890 km (535 mi.) swath with 260 m (850 ft.) spatial resolution. The HRCC has five channels (510–730 nm panchromatic, 450–520 nm, 520–590 nm, 630–690 nm, and 770–890 nm) that cover a 113 km (68 mi.) swath at 20 m (66 ft.) surface resolution. The IRMSS covers a 120 km (72 mi.) swath in four bands (PAN, two SWIR, one LWIR) with 80 m (260 ft.) spatial resolution, 160 m (525 ft.) in the thermal channel. Satellite images from CBERS-2 are used to monitor deforestation and fires in the Amazon basin, for water resources monitoring, urban growth planning, for soil studies, and for education purposes (Satellite Imaging Corporation a, online).

Polar Operational Environmental Satellite N Series

The NOAA Polar Operational Environmental Satellite N Series (NOAA-N, or POES-N) was launched in 2005 and is a polar-orbiting satellite developed by NASA for the NOAA. NOAA-N collects information about Earth's atmosphere and environment to improve weather prediction and climate research. NOAA-N is the 15th in a series of polar-orbiting satellites that began in 1978. NOAA uses two satellites, a morning and an afternoon satellite, to ensure every part of the earth is observed at least twice every 12 hours. Severe weather is monitored and reported to the U.S. National Weather Service, which broadcasts

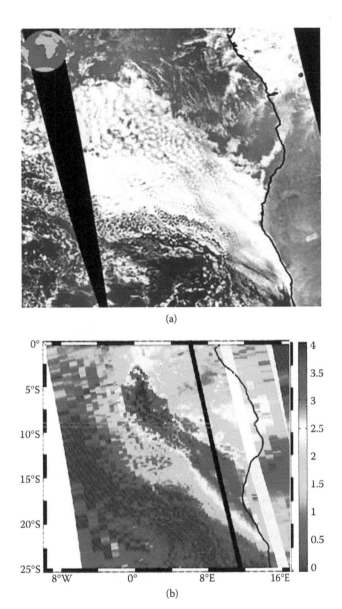

(a)

(b)

FIGURE 3.39
Aerosols as seen by two satellite sensors on August 4, 2007 over the South Atlantic off the coasts of Angola and Namibia. (a) Aqua MODIS visible image of low level clouds; (b) Aura-OMI image showing a smoke layer above clouds in terms of the UV aerosol index (a measure of aerosol scattering/absorption). Measuring aerosol–cloud interactions helps understand the radiative forcing effects of aerosols above clouds. Work by O.Torres, P. Bhartia of NASA GSFC and H. Jethva of Hampton University. (Images courtesy of NASA. http://aura.gsfc.nasa.gov/science/feature-20120305c.html.)

the findings to the global community. With early warning, the effects of catastrophic weather events can be minimized.

The spacecraft carries instruments for imaging and measuring the earth's atmosphere, its surface, and cloud cover, including Earth radiation, atmospheric ozone, aerosol distribution, sea-surface temperature, and vertical temperature and water profiles in the troposphere and stratosphere.

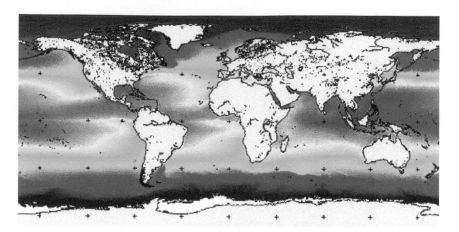

FIGURE 3.40

Global sea-surface temperatures as determined by the AVHRR on the NOAA-N satellite averaged for the period June 20–24 over the years 1985 through 1997. Temperature ranges from 0°C (violet) to 35°C (red). (Courtesy of NOAA and NASA.)

NOAA-N' (NOAA-N prime) was launched in 2009 with dedicated microwave instruments for the generation of temperature, moisture, surface, and hydrological products in cloudy regions where visible and IR instruments have decreased capability (NASA nn, NOAA-N, online).

The spacecraft carries the following five instruments: an Advanced Very High Resolution Radiometer (AVHRR/3), a High Resolution Infrared Radiation Sounder (HIRS/4), an Advanced Microwave Sounding Unit-A (AMSU-A), a Microwave Humidity Sounder (MHS), and Solar Backscatter Ultraviolet Radiometer (SBUV/2).

AVHRR/3 is a six-channel imaging radiometer that detects energy in the visible-IR portions of the spectrum. It measures reflected solar (visible and NIR) energy and radiated thermal energy from land, sea, clouds, and the intervening atmosphere (Figure 3.40).

The HIRS/4 is an atmospheric sounding instrument that provides multispectral data from 1 visible channel, 7 SWIR channels, and 12 LWIR channels using a single telescope and a rotating filter wheel containing 20 spectral filters.

The AMSU-A is a cross-track scanning radiometer. AMSU-A measures scene radiance in the microwave spectrum. The data from this instrument is used in conjunction with the HIRS and provides precipitation and surface measurements including snow cover, sea ice conditions, and soil moisture.

The MHS is a five-channel microwave instrument intended to measure profiles of atmospheric humidity. It also measures cloud liquid water content and provides estimates of precipitation rates.

The SBUV/2 is a nadir-pointing, scanning UV radiometer. This instrument measures solar irradiance and Earth radiance (backscattered solar energy) in the near-UV spectrum.

CloudSat

CloudSat is a NASA Earth observation satellite launched in 2006. The Cloud Profiling Radar (CPR) used on CloudSat is a 94 GHz (millimeter wave) nadir-looking radar that measures power backscattered by clouds as a function of distance from the radar (Figure 3.41). This instrument provides an along-track vertical profile of cloud structure with a 500 m (1500 ft.)

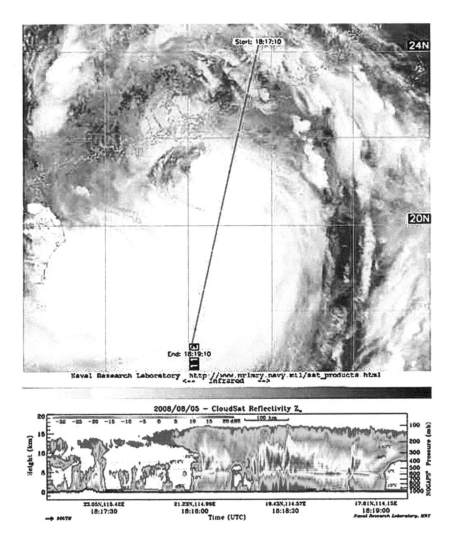

FIGURE 3.41
CloudSat satellite Cloud Profiling Radar side view through the deep clouds of Tropical Storm Kammuri as it made landfall along the southern coast of China on August 5, 2008. The Aqua satellite image (top) shows Kammuri at the same time that the CloudSat image (below) was acquired along the red line. Colors in the CloudSat image indicate the intensity of reflected radar energy, with red indicating the most intense rainfall. Blue areas along the top of the clouds indicate ice, while the wavy blue lines on the bottom of the image indicate intense surface rainfall. (Image courtesy of NASA. http://www.nasa.gov/mission_pages/cloudsat/multimedia/cloudsat-20080806.html.)

vertical resolution, 1.4 km (0.9 mi.) cross-track resolution, and 1.7 km (1 mi.) along-track resolution (Im et al., 2005). The radar instrument was developed at JPL with contributions from the Canadian Space Agency. The design and system configuration are essentially identical to the Airborne Cloud Radar, which has flown on the NASA DC-8 aircraft since 1998 (Colorado State University, CLOUDSAT, online).

CloudSat performed the first 3D study of clouds. It gathers data on cloud structure, frequency, and volume. It probes clouds to determine their thickness, altitude at base and peak, and the volume of water and ice within them. CloudSat also analyzes the way light is absorbed by the various layers of the atmosphere, and the influence of atmospheric aerosols.

CALIPSO

NASA's CALIPSO satellite was launched in 2006. The spacecraft carries three co-aligned nadir-viewing instruments for monitoring air quality and climate: the Cloud-Aerosol Lidar with Orthogonal Polarization (CALIOP), the Imaging Infrared Radiometer (IIR), and the Wide-Field Camera (WFC). CALIOP is a two-wavelength polarization-sensitive Lidar that provides high-resolution vertical profiles of aerosols and clouds. CALIOP uses three receiver channels: one measuring the 1064 nm backscatter intensity and two channels measuring orthogonally polarized components of the 532 nm backscattered signal. The WFC is a nadir-viewing imager with a single channel covering 270–620 nm. This was selected to match band 1 of the MODIS instrument on AQUA. The three-channel IIR is provided by CNES of France. IIR is a nadir-viewing imager with a 64 × 64 km (38 × 38 mi.) swath and a ground resolution of 1 km (0.6 mi.). The instrument uses a single-detector array with a rotating filter wheel that provides measurements at three channels in the TIR: 8.7, 10.5, and 12.0 µm. These wavelength bands were chosen to optimize joint CALIOP/IIR measurements of cirrus cloud emissivity and particle size. The CALIOP beam is nominally aligned with the center of the IIR image (NASA oo, CALIPSO, online).

Aquarius/SAC-D

Aquarius is a mission to measure sea-surface salinity and provide the global view of salinity variability needed for climate studies (NASA qq, AQUARIUS, online). The mission is a collaboration between NASA and CONAE, the Space Agency of Argentina (Comisión Nacional de Actividades Espaciales).

Aquarius, launched in 2011, provides monthly global measurements of how sea salinity varies at the ocean surface, that is, how the ocean responds to the combined effects of evaporation, precipitation, ice melt, and river runoff on seasonal time scales (Figure 3.42). The Aquarius/SAC-D mission streams real-time global salinity data. The instruments include a set of three radiometers that are sensitive to salinity (1.413 GHz) (L band) and a scatterometer that corrects for the ocean's surface roughness. Salinity can be measured because salt concentration changes water conductivity, which in turn affects thermal

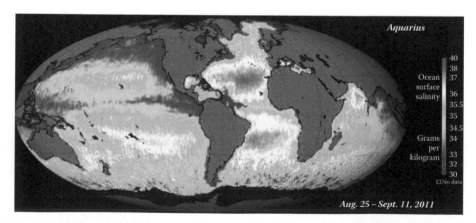

FIGURE 3.42

Aquarius/SAC-D mission global sea surface salinity data, August-September 2011. High salinity is shown as yellow and red; lower salinity is blue to violet. (Courtesy NASA/JPL. http://photojournal.jpl.nasa.gov/jpegMod/PIA14786_modest.jpg.)

emission at the water surface. At microwave frequencies this emission change can be measured. The instrument is a three-beam push-broom polarimetric radiometer and radar that operates in a nighttime sun-synchronous orbit at 675 km (405 mi.) with a 7-day revisit time. The instruments can measure salinity variations as low as 0.5 g/kg.

Suomi National Polar-Orbiting Partnership

The Suomi National Polar-orbiting Partnership (Suomi NPP), launched during October 2011, is a NOAA mission designed to monitor long-term climate trends and global biological productivity (NASA rr, SUOMI NPP, online). It extends the measurements initiated with EOS TERRA and AQUA by providing a bridge between NASA's EOS missions and the Joint Polar Satellite System (JPSS).

Measurements are taken by five different sensors: Visible Infrared Imaging Radiometer Suite (VIIRS), Cross-track Infrared Sounder (CrIS), Advanced Technology Microwave Sounder (ATMS), Ozone Mapping and Profiler Suite (OMPS) and the Clouds and the Earth's Radiant Energy System (CERES). These sensors collect data on atmospheric and sea-surface temperatures, humidity soundings, land and ocean biological productivity, cloud and aerosol properties, ozone concentration and profiles, and Earth's radiation.

The ATMS is a 22-channel passive MWR that provides input to global models of temperature and moisture profiles that meteorologists use in weather forecasting. The CrIS is a Michelson interferometer that monitors atmospheric moisture and pressure and is used to improve both short- and long-term weather forecasting. The OMPS incorporates a nadir-viewing sensor and a limb-viewing sensor to measure atmospheric ozone. The VIIRS has a 22-band radiometer similar to the MODIS instrument flown on TERRA. It collects visible and IR images of Earth's surface processes such as wildfires, land change, and ice movement. VIIRS also measures atmospheric and oceanic properties, including clouds and sea-surface temperature. CERES is a 3-channel radiometer that measures reflected solar radiation, emitted terrestrial radiation, and total radiation. CERES monitors and attempts to discern natural and anthropogenic effects on the earth's total thermal radiation (NASA rr, SUOMI NPP, online; NASA ss, SUOMI NPP, online).

Canada

The Canadian Space Agency has launched a number of satellites carrying primarily radar instruments for monitoring sea ice. Recently, however, they have participated with other space agencies to fly instruments that monitor climate change and the environment (Canadian Space Agency a, online). The CSA is the Canadian government agency responsible for Canada's space program. It was established in March 1989.

The Canada Centre for Remote Sensing (CCRS) is a branch of Natural Resources Canada's Earth Science Sector. It was created in 1970. The responsibilities of the CCRS are to provide remotely sensed geographical information to decision makers, related industries, and the general public. The Centre supports remote sensing technology development and applications.

RADARSAT-1

RADARSAT-1 is a project by the CSA to provide radar imagery, particularly in high latitudes. Launched in late 1995, the C band (5.6 cm; 5.3 GHz) Synthetic Aperture Radar (SAR) has five modes of operation: standard, wide swath, fine resolution, and two "scansar" modes (Figure 3.43). The standard beam provides 100 km (60 mi.) wide swaths with 25 × 28 m (81 × 91 ft.) resolution. The wide-beam mode provides a 150 km (90 mi.) swath, also with 25 × 28 m resolution.

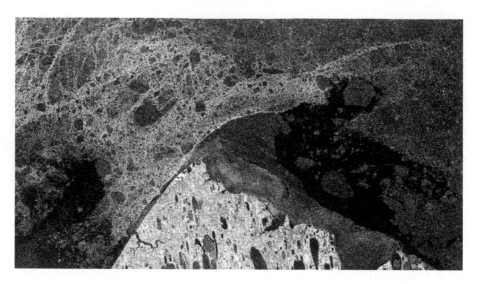

FIGURE 3.43

Annual break-up of landfast sea ice off Point Barrow, Alaska, June 20, 2009. This RadarSat-1 image clearly documents the progress of summer ice break-up. Image is from the Sea Ice Group and was received by the Alaska Satellite Facility, both located at the Geophysical Institute, University of Alaska Fairbanks. (Courtesy of Sea Ice Group, Geophysical Institute, University of Alaska Fairbanks, USGS, and RadarSat. http://seaice.alaska.edu/gi/data/barrow_breakup/brw_fc_2009.)

The fine beam provides swaths 50 km (30 mi.) wide with ~9 × 11 m (29 × 36 ft.) ground resolution. The Scansar modes provide wide swaths (up to 500 km, or 300 mi.) with moderate resolutions (50 × 50 m to 100 × 100 m). The satellite orbits at 798 km (475 mi.) in a polar sun-synchronous orbit with between three and 24-day repeat coverage. The radar is capable of both right and left look directions. The incidence angle varies from 10° to 60°. The Canadian Data Processing Facility (CDPF) in Gattineau, Quebec, produces up to 44 images per day with a turnaround time of four hours or less for rush orders.

RADARSAT-2

Launched in 2007, Canada's RADARSAT-2 carries a C band (5.4 GHz) polarimetric Synthetic Aperture Radar (SAR) in a 798 km sun-synchronous orbit with a 24-day repeat cycle. The purpose of the satellite is marine surveillance, ice monitoring, disaster management, environmental monitoring, and resource mapping. The SAR is capable of transmitting and receiving in horizontal or vertical polarizations and has left- and right-looking modes. There are seven imaging modes: "Fine," with a 50 km (30 mi.) swath and 10 × 9 m (32 × 29 ft.) ground resolution; "Standard," with a 100 km (60 mi.) swath and 25 × 28 m (82 × 92 ft.) resolution; "Low Incidence," with a 170 km (102 mi.) swath and 40 × 28 m (131 × 92 ft.) resolution; "High Incidence," with a 150 km (90 mi.) swath and 25 × 28 m resolution; "Wide," with a 100 km swath and 25 × 28 m resolution; "ScanSAR Narrow," with 300 km (180 mi.) swath and 50 m (164 ft.) resolution; and "ScanSAR Wide," with a 500 km (300 mi.) swath and 100 m (328 ft.) resolution (Canadian Space Agency b, RADARSAT-2, online).

SCISAT-1

The Canadian SCISAT-1 mission, launched in 2003, is designed to observe the atmosphere, particularly ozone (Canadian Space Agency c, SCISAT-1, online). It carries three instruments, the Atmospheric Chemistry Experiment-Fourier Transform Infrared Spectrometer (ACE-FTS), an Ultraviolet Spectrophotometer (MAESTRO), and a Visible/Near Infrared Imager (VNI). These sensors record solar radiation as it passes through the atmosphere to analyze concentrations of atmospheric gases, elements, and particulates (Figure 3.44). The ACE-FTS uses the solar occultation technique to measure how solar spectra vary as they pass through the troposphere and stratosphere. The spectrometer is a version of the Michelson interferometer. The VNI monitors aerosols using two detectors at 525 and 1020 nm to look for the extinction of solar radiation. Measurements of Aerosol Extinction in the Stratosphere and Troposphere Retrieved by Occultation (MAESTRO) measures the vertical distribution of ozone, nitrogen dioxide, and aerosols in the atmosphere using an ultraviolet-visible-near infrared (285–1030 nm) spectrophotometer.

Chile

The Chilean Space Agency (*Agencia Chilena del Espacio*) is part of the Chilean Ministry of National Defense. It facilitates the development of space-based information and technology for application in different national activities. It promotes the use of geospatial technologies for the benefit of public policy and the economic and social development of the country, including regional integration.

FASat-Bravo

Following the launch failure of FASat-Alfa a second Chilean satellite, FASat-Bravo, was launched in 1998. FASat-Bravo was put into a 650 km (390 mi.) near-polar orbit with the purpose of monitoring the ozone layer and imaging the earth. The Experimento de Monitoreo la Capa de Ozono (Ozone Layer Monitoring Experiment, OLME) contained

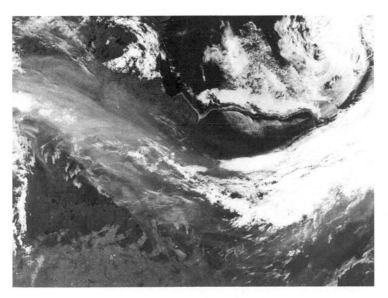

FIGURE 3.44
SCISAT-1 mission Visible/Near Infrared Imager. A plume of smoke from fires in northern Canada stretches over Manitoba on July 25, 2004. SCISAT is being used to measure the chemical content of these pollutant plumes. (From NASA, Canadian Space Agency. http://www.asc-csa.gc.ca/eng/satellites/scisat/scisat.asp. Reproduced with the permission of the Minister of Public Works and Government Services Canada, 2013.)

two instruments, the Ozone Camera (based on CCDs) and UV Photodiodes. OLME measures UV solar backscattered radiation in frequency bands around 300 nm. The Sistema de Imagenes Terrestres (Earth Imaging System, EIS) consists of two visible light cameras, a WFC with a ground resolution of 1500 m (4920 ft.; similar to the OLME system), and a narrow-field camera with 150 m (492 ft.) ground resolution. The satellite's batteries failed in 2001 after almost three years in orbit (Gunter's Space Page a, FASat Alfa/Bravo, online).

SSOT

Sistema Satelital para Observacion de la Tierra (SSOT), also known as FASat-Charlie, is a Chilean Earth observation satellite built and operated by European Aeronautic Defense and Space Company (EADS) Astrium and launched in 2011. The satellite carries an instrument capable of supplying images with a resolution of 1.45 m (6.7 ft.) in PAN mode and 5.8 m (19 ft.) in Multispectral mode (in each of blue, green, red, and NIR bands). The satellite flies in a 620 km (372 mi.) polar orbit (Gunter's Space Page b, FASat Charlie; ASTRIUM, SSOT, online).

China

The China Aerospace Science and Technology Corporation (CASC) is the main contractor for the Chinese space program. It is state-owned and has a number of subordinate entities that design, develop, and manufacture a range of spacecraft, launch vehicles, strategic and tactical missile systems, and ground equipment. The Chinese Academy of Space Technology (CAST) is a unit of CASC. CAST designs and manufactures satellites.

The Center for Earth Observation and Digital Earth (CEODE), an agency of the Chinese Academy of Science (CAS), was established in August 2007 by consolidating three CAS units: the Remote Sensing Satellite Ground Station, the Center for Airborne Remote Sensing, and the Laboratory of Digital Earth Sciences. As an institution noted for both scientific research and professional services, CEODE is committed to operating Earth observation systems and related data services; to providing cutting-edge technologies for Earth observation and their application; and to theoretical and technological research into key issues concerning Earth and integrated applications at the global, national, and regional scales.

Yaogan 1 Series

Yaogan (Yaogan Weixing or "Remote Sensing Satellite") refers to a series of Chinese satellites launched starting in April 2006 (Yaogan 1) and continuing through November 2011 (Yaogan 13). The satellites were designed by the Shanghai Academy of Space Flight Technology and CAST for the China Aerospace Science and Technology Corporation. Information on these satellites is tightly held. Yaogan 1, 3, 6, 8, 10, and 13 are believed to carry a SAR instrument; Yaogan 2, 4, 5, 7, 11, and 12 are believed to carry an electro-optical digital imaging scanner; Yaogan 9A and 9B are thought to carry both instruments. Chinese media describe the purpose of these satellites as scientific experiments, land surveying, crop assessment, and disaster monitoring (NASA tt, Long March 3C; China Defense Blog, YaoGan, online).

HaiYang-2A

China launched the HaiYang-2A, an oceanographic satellite, during August 2011. The purpose of the mission was to monitor ocean winds, sea level and temperature, waves, currents, tides, and storms to provide disaster and weather forecasting information. Instruments include a Microwave Imager (microwave brightness temperature), a dual-Ku band and C band radar altimeter to measure sea levels and wind speeds—and a Ku band Radar Scatterometer for measuring the sea-surface wind field. The satellite was built by CAST and flies in a 963 km (580 mi.) sun-synchronous orbit. The mission has two phases: the first 2 years it will have a 14-day repeat cycle, then 1 year in a geodetic orbit with a 168-day cycle (NASA Spaceflight.com a, HaiYang-2A, online).

Ziyuan Series

The Ziyuan 1-2C satellite was launched into a 780 km (470 mi.) circular orbit in December 2011 by CEODE. This civilian remote sensing satellite provides images for disaster relief, agriculture development, forestry, water conservation, urban planning, environmental monitoring, and natural resource mapping. Although details are kept tight, the spacecraft is reported by Chinese media to carry two High-Resolution Cameras with spatial resolution of 2.36 m (8 ft.) and a swath of 54 km (32 mi.) (Figure 3.45). It also features a PAN/Multispectral camera, with 5–10 m (16–33 ft.) resolution over a 60 km (36 mi.) swath (CAS, ZY1-02C; NASA Spaceflight.com b, Long March 4B ZiYuan-1, online).

China launched the ZiYuan-3 (ZY-3) satellite during January 2012. The satellite is operated jointly by the Center for Earth Observation and Digital Earth (CEODE) and the Brazil's INPE. It flies in a 506 km (304 mi.) sun-synchronous near-polar orbit with a 5-day revisit cycle. This is a civilian remote sensing satellite to be used for geologic mapping and for this reason has stereo capability. The satellite carries three PAN cameras, one front-facing, one rear-facing, and one nadir-pointing. The fore-and-aft facing instruments have spatial

FIGURE 3.45
Chinese Ziyuan 1-2C satellite infrared multispectral camera image of the Yangtze River west of Wuhu, China. Note the dark discharge near the large island. Not much information is available for these images. (Courtesy of CEODE and Dragoninspace.com. http://www.dragoninspace.com/earth-observation/ziyuan1-cbers.aspx.)

resolution of 3.5 m (11.5 ft.) over a 52.3 km (30 mi.) ground swath. The ground-facing camera has a spatial resolution of 2.1 m (7 ft.) over a 51 km (30 mi.) swath. The IRMSS has 6.0 m (20 ft.) resolution over a 51 km ground swath (NASA Spaceflight.com c, Long March 4B ZiYuan-3, online).

Denmark

Danish Meteorological Institute (DMI) is responsible for the meteorological needs of Denmark, the Faroe Islands, and Greenland including territorial waters and airspace. This involves monitoring weather, climate, and environmental conditions in the atmosphere, on the land, and at sea. The primary aim of these activities is to safeguard human life and property, as well as to provide a foundation for economic and environmental planning.

Ørsted

The Ørsted mission was launched in 1999 by DMI for the purpose of mapping the earth's magnetic field and monitoring any changes (Figure 3.46). The satellite is the result of collaboration between the DMI, NASA, ESA, CNES (France), and DLR (Germany). It flies in a 500 × 850 km (300 × 510 mi.) elliptical near-polar orbit. The satellite carries a compact spherical coil (CSC) fluxgate vector magnetometer to measure the magnetic field strength and direction, as well as an Overhauser magnetometer to measure the strength of the total magnetic field. It is accurate to less than 0.5 nT. The main purpose of the Overhauser instrument is the calibration to an absolute scale of the measurements of the CSC magnetometer. Based

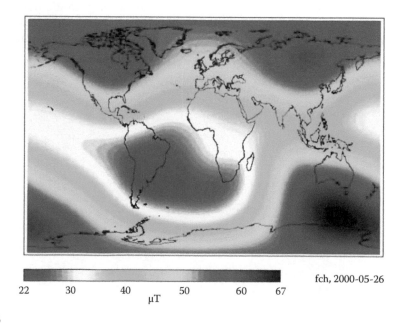

FIGURE 3.46
Ørsted magnetometer measurements were used for the latest International Geomagnetic Reference Field, the IGRF2000. This is a map of the total magnetic field strength at the earth's surface derived from the IGRF2000. Blue–black colors represent above-average field strength and reddish-yellow indicates field strength below the mean. (Image courtesy of Dr. Freddy Christiansen and Danish Meteorological Institute. http://www.science-daily.com/releases/2008/06/080619102553.htm.)

on measurements from this satellite, the Danish Space Research Institute has determined that the earth's magnetic poles are shifting, that the speed of the polar shift is increasing, and that a pole reversal may be in progress (DMI, Ørsted Satellite Project, online).

DigitalGlobe

DigitalGlobe is a private U.S. company out of Longmont, Colorado, that has launched a number of high-resolution satellites over the past two decades. DigitalGlobe was founded in 1993 under the name WorldView Imaging Corporation, became EarthWatch Incorporated in 1995, and finally DigitalGlobe in 2002.

Early Bird

Launched in 1997, this private satellite was built and operated by EarthWatch (now DigitalGlobe). Although it failed a few days after launch, Early Bird showed that private commercial satellite systems can be built and launched. Early Bird carried a PAN camera that covered the spectral range 450–800 nm (3 m resolution) and a multispectral camera with three channels in the range 450–890 nm (15 m resolution), truly revolutionary at the time.

QuickBird

QuickBird, the follow on to Early Bird, was launched by DigitalGlobe in 2001. It carries a PAN camera (450–900 nm) with 60 cm (2 ft.) resolution and a four-channel multispectral sensor with 2.4 m (8 ft.) resolution. The multispectral sensor has a blue band (450–520 nm), a green band (520–600 nm), a red band (630–690 nm), and a NIR band (760–900 nm). The swath is 22 km (13 mi.), extendable to 44 km (26 mi.), and it flies in a sun-synchronous orbit with a 3-day repeat cycle (eoPortal Directory a, Quickbird-2, online).

WorldView-1

DigitalGlobe launched the WorldView-1 satellite in 2007. It has a single, 50 cm (1.5 ft.) resolution PAN band and a 1.7-day revisit time. Swath is 17.6 km (10.6 mi.) at nadir.

WorldView-2

DigitalGlobe launched the WorldView-2 satellite in 2009 (Figure 3.47). It has a 46 cm (1.5 ft.) resolution PAN band (450–800 nm) and eight MSS bands with 1.84 m (6 ft.) ground resolution: band 1 (770–895 nm), band 2 (630–690 nm), band 3 (510–580 nm), band 4 (450–510 nm), band 5 (705–745 nm), band 6 (585–625 nm), band 7 (400–450 nm), and band 8 (860–1040 nm). The satellite flies in a 770 km (462 mi.) sun-synchronous orbit with a 1.7-day revisit time and 16.4 km (10 mi.) swath at nadir (DigitalGlobe, WorldView-2, online).

European Organization for the Exploitation of Meteorological Satellites

The European Organization for the Exploitation of Meteorological Satellites (EUMETSAT) delivers weather and climate-related satellite data and images 24 hours a day, 365 days a year. This information is supplied to European National Meteorological Services.

FIGURE 3.47
WorldView-2 image of Khone Falls on the Mekong River, Laos. (False color image courtesy of DigitalGlobe, Inc and GeoEye © 2012.)

Meteosat Second Generation

The first Meteosat Second Generation (MSG-1) satellite, known as "Meteosat-8," was launched in 2002 into a 36,000 km (22,000 mi.) geostationary orbit over Africa and Europe. Among other things it provides images every 5 minutes and supports the Indian Ocean Tsunami Warning System. It carries two main instruments, the Spinning Enhanced Visible and Infrared Imager (SEVIRI) and the Geostationary Earth Radiation Budget (GERB) instrument. Meteosat-9 (MSG-2) was launched in 2005 and Meteosat-10 (MSG-3) was launched in 2012. The purpose of these satellites is visible and IR imaging of the earth for weather monitoring and forecasting.

SEVIRI has 12 spectral channels. These provide precise data input for numerical weather prediction models. Eight of the channels are in the TIR, providing data on the temperatures of clouds, land, and sea surfaces. The High-Resolution Visible (HRV) channel has a 1 km (0.6 mi.) pixel at nadir, compared to 3 km (1.8 mi.) resolution for the other visible channels. SEVERI measures ozone, water vapor, and CO_2 absorption bands, providing information on their atmospheric concentrations (EUMETSAT b, Meteosat Second Generation, online).

The GERB instrument is a visible-infrared radiometer for Earth radiation budget studies. It makes accurate measurements of the short-wave and long-wave components of the radiation at the top of the atmosphere. GERB provides data on reflected solar radiation and thermal radiation emitted by the earth and atmosphere.

European Space Agency

ESA is an intergovernmental organization dedicated to Earth observation missions and exploration missions to other planets and the moon. Established in 1975 and headquartered in Paris, France, ESA currently has 20 member states. ESA science missions are based at ESTEC in Noordwijk, Netherlands, whereas Earth Observation missions are run out of ESRIN in Frascati, Italy.

ERS-1 and -2

ERS-1 and ERS-2 were launched in 1991 and 1995, respectively, by ESA into sun-synchronous orbits at a 777 km (466 mi.) altitude. They have repeat cycles of 3, 35, and 176 days. There are three instruments on the ERS satellites that are of interest to Earth scientists (ESA b, ERS Overview, online).

The Active Microwave Instrument (AMI) includes a SAR and a Wind Scatterometer. The SAR (Figure 3.48) operates in C band (5.6 cm). In image mode it covers a swath 80–100 km (48–60 mi.) wide with a ground resolution of 26 m (85 ft.) across track and 6–30 m (20–98 ft.) along track. It has an adjustable incidence angle of either 23° or 35° off nadir. In wave mode, the SAR produces 5 km × 5 km (3 × 3 mi.) images at regular intervals for the derivation of the length and direction of ocean waves. The Wind Scatterometer uses three antennas to observe sea-surface wind speed and direction. It operates by recording the change in radar reflectivity of the sea due to the perturbation of small ripples by wind close to the ocean surface.

The radar altimeter (RA), a Ku band (13.8 GHz) nadir-pointing sensor, provides measurements of sea-surface elevation, wave heights, various ice parameters, and an estimate of surface wind speed.

The Along-Track Scanning Radiometer (ATSR) combines an IR radiometer and a Microwave Sounder (MS). The MS measures sea-surface temperature, cloud top temperature, cloud cover and atmospheric water vapor. The Infrared Radiometer (IRR) on ERS-1 is a four-channel radiometer used for measuring sea-surface and cloud top temperatures.

In addition to the AMI, RA, and ATSR, the ERS-2 satellite carries two more instruments, GOME and MWR. The Global Ozone Monitoring Experiment is an absorption

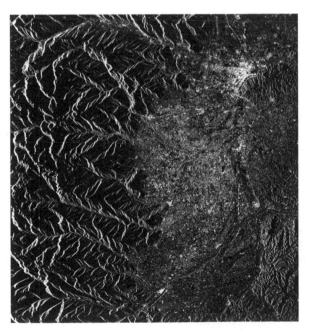

FIGURE 3.48
This ERS-2 SAR image covers the Alps and the Western part of Piemonte, Italy, with the town of Torino clearly evident at the top right of the image. Some flooded areas are evident along the Po river. (Courtesy of ESA. http://www.esa.int/esa-mmg/mmg.pl?b=b&type=I&mission=ERS-2&start=3.)

spectrometer that measures the presence of ozone, nitrogen dioxide and other trace gases and aerosols in the stratosphere and troposphere. GOME is a nadir-scanning UV and visible spectrometer that also provides cloud information. The Microwave Radiometer records the integrated atmospheric water vapor column and cloud liquid water content in order to correct the RA signal.

Cluster II Series

The Cluster II mission was launched by ESA in collaboration with NASA in 2000 and was scheduled to end in December 2012. The purpose is to study Earth's magnetosphere during an entire solar cycle. It consists of four identical satellites in highly elliptical 19,000 × 119,000 km (11,400 × 71,400 mi.) orbits. Each satellite carries eleven instruments to measure solar wind, the magnetopause, and other plasma effects. The one Earth observation instrument is a Fluxgate Magnetometer used to measure magnetic field magnitude and direction (Escoubet et al., 2001; ESA c, CLUSTER, online).

Project for On-Board Autonomy

The Project for On-Board Autonomy (Proba), an ESA technology demonstration mission was launched in 2001 (ESA d, Proba, online). Proba-1 continues to operate as an Earth Observation Third Party Mission. It is in a 681 × 561 km (408 × 337 mi.) sun-synchronous orbit and carries three Earth observation instruments. The Compact High Resolution Imaging Spectrometer (CHRIS) has 19 bands in the range 415 to 1050 nm with 5 to 12 nm spectral resolution and 15 to 20 m (49 to 66 ft.) spatial resolution over a 14 km (8.4 mi) ground swath. The High-Resolution Camera (HRC) is a PAN instrument with a miniaturized Cassegrain telescope that provides 5 m (16.4 ft.) ground resolution. The Wide-Angle Camera (WAC) is a miniaturized B/W camera used for Earth observation as well as for public relations and educational purposes.

ODIN

The ODIN satellite was launched by ESA into a 600 km (360 mi) sun-synchronous orbit in 2001. This Swedish satellite was co-funded by Finland, Canada, and France. Its mission is to monitor ozone layer depletion. The Canadian-built OSIRIS instrument measures vertical profiles of dispersed and scattered sunlight in the atmosphere and uses these to estimate concentrations of ozone, nitrous oxide, and aerosols along the profiles. The OSIRIS spectrograph acquires light in the range from 274 to 810 nm and has a vertical resolution of 1.5–2.0 km (1–1.2 mi.). OSIRIS is limited to sunlit observations in the northern hemisphere from May to August, and in the southern hemisphere from November to February (Canadian Space Agency d, Odin OSIRIS, online). A second instrument on ODIN is the Sub-Millimeter Radiometer (SMR), used to measure atmospheric profiles of other gases.

Environmental Satellite

ENVISAT, launched into a 791 × 785 km (475 × 470 mi.) sun-synchronous polar orbit by the ESA in 2002, stopped transmitting data during April 2012. The satellite carries eight Earth observing instruments. The MEdium Resolution Imaging Spectrometer (MERIS) collects

FIGURE 3.49
ENVISAT MERIS image showing distribution of plankton blooms in the Baltic Sea during July 2005. (Courtesy of ESA. http://phys.org/news69420332.html.)

useful data in 15 spectral bands which are programmable in position, width and gain (Figure 3.49). The spatial resolution of the detectors in near nadir mode allows sampling every 300 m (1000 ft.) at the earth's surface. This is the "Full Resolution" product. The more common "Reduced Resolution" product has nominal resolution of 1200 m (3950 ft.). The total FOV of MERIS is 68.5° at nadir, providing an 1150 km (690 mi.) swath sufficient to collect data for the entire planet every 3 days in equatorial regions. Polar regions are visited more frequently due to the convergence of orbits (ESA e, ENVISAT, online).

The Advanced Along-Track Scanning Radiometer (AATSR), funded by the U.K. and flown on ENVISAT, collects sea-surface temperatures with an accuracy to 0.3° K or better. The objective is to assist climate research. AATSR has three channels in the TIR that are used to derive surface temperatures. AATSR also has four VNIR channels used to identify clouds and to measure atmospheric aerosols and solar radiation scattered and reflected from the earth's surface and atmosphere. Two measurements of each point on the surface through different atmospheric path lengths allow separating surface and atmospheric components of the signal. This permits the effects of atmospheric scattering by aerosols to be removed from surface reflectance (ESA f, AATSR, online).

The Advanced Synthetic Aperture Radar (ASAR) is a C band SAR with five polarization modes (VV, HH, VV/HH, HV/HH, and VH/VV). The instrument has variable spatial resolution: "Image," "Wave," and "Alternating Polarization" modes have ~30 m (100 ft.) resolution; "Wide Swath" mode resolution is roughly 150 m (492 ft.); "Global Monitoring" mode resolution is 1000 m (3280 ft.). Swath width is up to 100 km (60 mi.) in Image and Alternating Polarization modes; it is 5 km (3 mi.) in Wave mode, and is 400 km (240 mi.) or more in Wide-Swath and Global Monitoring modes. Vertical resolution is sub-millimeter (ESA g, ASAR, online).

The MWR on ENVISAT measures the integrated atmospheric water vapor column and cloud liquid water content, surface emissivity and soil moisture over land, and helps characterize ice (ESA h, MWR, online). MWR is a nadir-pointing dual-channel radiometer operating at 23.8 and 36.5 GHz. Temperature accuracy is 2.6°K over a 20 km (12 mi.) footprint.

The Global Ozone Monitoring by Occultation of Stars (GOMOS) instrument on ENVISAT measures atmospheric constituents through analysis of three spectral bands: from 250 to 675 nm, from 756 to 773 nm, and from 926to 952 nm. By detecting the absorption of starlight in UV through NIR wavelengths GOMOS is able to measure concentrations of ozone, nitrogen dioxide, oxygen, water vapor, and aerosols. Vertical sampling resolution is on the order of 1.7 km (1 mi.). GOMOS also carries two photometers that operate in the range from 470 to 520 nm and 650 to 700 nm (ESA i, GOMOS, online).

The Michelson Interferometer for Passive Atmospheric Sounding (MIPAS) operates in the near- to mid-infrared and measures the emission spectra of trace gases. MIPAS is a Fourier Transforming IR spectrometer that provides pressure, temperature, and trace gas profiles. Gases that can be measured include nitrogen dioxide, nitrous oxide, methane, nitric acid, ozone, and water. Vertical resolution along profiles is 3 to 5 km (1.8 to 3 mi.) (ESA j, MIPAS, online).

The SCanning Imaging Absorption spectroMeter for Atmospheric CartograpHY (SCIAMACHY) is an imaging spectrometer on ENVISAT that uses different viewing geometries (nadir, limb, and sun/moon occultations) to measure total column values as well as distribution profiles in the stratosphere for trace gases and aerosols. Limb vertical resolution is 3×132 km (1.8×80 mi.); nadir horizontal resolution is 32×215 km (19×130 mi.), and swath width is 1000 km (600 mi.). Solar radiation transmitted, backscattered and reflected from the atmosphere is recorded at 0.2 nm to 0.5 nm spectral resolution. This allows detection of aerosols and trace gases (ozone, oxygen, nitrous oxide, nitrogen dioxide, carbon monoxide, CO_2, methane, and water, among others) at low concentrations. Channels include 240–314 nm, 309–405 nm, 394–620 nm, 604–805 nm, 785–1050 nm, 1000–1750 nm, 1940–2040 nm and 2265–2380nm (ESA k, l, SCIAMACHY, online).

The radar altimeter 2 (RA-2) on ENVISAT is a SAR that operates in both S band and Ku band. It is a nadir-pointing radar used to measure changes in mean sea level, map ocean surface topography, and monitor sea ice.

FORMOSAT-2

"FORMOSAT-2" is a joint ESA-Taiwan (National Space Program Office) imaging satellite launched in 2004 with the objective to collect high-resolution (2 m; 6.5 ft.) panchromatic and multispectral (8 m; 26 ft.) imagery over a 24 km (14.4 mi.) swath. FORMOSAT-2 is in an 888 km (533 mi.) sun-synchronous orbit. The instrument package includes the Imager of Sprites and Upper Atmospheric Lightning (ISUAL) and the Remote Sensing Instrument (RSI). The RSI has five channels in the blue (450–520 nm), green (520–600 nm), red (630–690 nm), NIR (760–900 nm) and PAN (450–900 nm) ranges. Besides providing imagery for Taiwan's domestic needs, the spacecraft is used to obtain high-resolution imagery for disaster monitoring (such as coverage of the Indian Ocean Tsunami of 26 December 2004, coverage of Hurricane Katrina in August 2005, coverage of typhoons in the Pacific, and coverage of earthquake regions).

The ISUAL instrument consists of a limb viewing low light level imager with a set of bore-sighted limb-viewing photometers. ISUAL observes lightning, sprites, elves, blue jets, aurora, and airglow. ISUAL attempts to determine the mechanism producing the flashes. The ISUAL imaging spectrophotometer and array photometer measure the spectral content and the spatial and temporal intensity distribution of the optical flashes to determine whether these events depend on factors other than the magnitude of storms (NSPO, FORMOSAT-2, online).

COSMO-SkyMed

Constellation Of small Satellites for Mediterranean basin Observation (COSMO-SkyMed) is a four-spacecraft constellation, launched between 2007 and 2010, conceived by ASI (Agenzia Spaziale Italiana). The satellites orbit at 617 km (370 mi.) in a sun-synchronous orbit with a 16-day repeat cycle. Each of the four satellites is equipped with the "SAR-2000," an X band SAR with imaging, interferometric, and polarimetric capabilities. The SAR-2000 has a 10 km (6 mi.; Spotlight mode) to 200 km (120 mi.; Wide-field mode) swath with ground resolution from 1 m to 100 m (3.3–328 ft.), respectively. The objective is to provide global Earth observations for the military and civilian communities (e-geos, COSMO-SkyMed, online).

Soil Moisture Ocean Salinity

The Soil Moisture Ocean Salinity satellite (SMOS) was launched by ESA in 2009. The satellite, part of the Living Planet program, is in a 758 km (455 mi.) orbit designed to monitor global soil moisture with an accuracy of 4% at 35–50 km (21–30 mi.) spatial resolution, and to monitor ocean-surface salinity with an accuracy of 0.1 psu (practical salinity units; 10- to 30-day average) and spatial resolution of 200 km (120 mi.). The satellite carries the Microwave Imaging Radiometer with Aperture Synthesis (MIRAS), based on interferometric L band radiometry (1.4 GHz) that measures changes in terrestrial moisture content and ocean salinity by observing variations in microwaves emitted by the earth's surface (Figure 3.50; ESA m, SMOS, online).

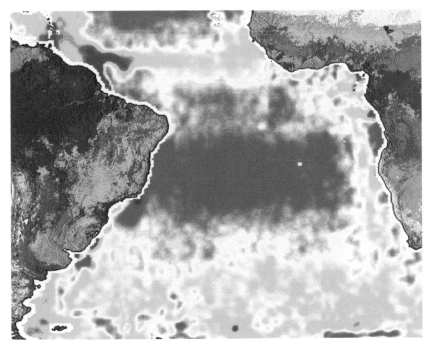

FIGURE 3.50
SMOS image of South Atlantic salinity, October 2012. Red is higher than mean, whereas blue is lower. Satellite relative salinity is tied to measurements at buoys. (Courtesy of ESA. http://www.esa.int/esa-mmg/mmg.pl?b=b&type=I&mission=SMOS&single=y&start=2.)

Gravity Field and Ocean Circulation Explorer

ESA's GOCE is in a 260 km (156 mi.) low Earth orbit to maximize its sensitivity to variations in Earth's gravity. Launched in 2009, GOCE maps the earth's geoid using the Electrostatic Gravity Gradiometer (EGG). EGG measures three components of the gravity-gradient to within 1 mGal (10^{-5} m/s^2) using a three-axis gradiometer. Because of their different positions in the gravitational field, the accelerometers undergo slightly different gravitational accelerations. The objective is to determine the Geoid to within 2 cm (0.8 in.) over a ground footprint of 100 km (60 mi.; Figure 3.51; ESA n, GOCE, online).

CryoSat-2

CryoSat-2 was launched by ESA in April 2010. It provides data about the polar ice caps and tracks changes in the thickness of the ice sheets with a vertical resolution of 1.3 cm (0.5 in.) and 250 m (820 ft.) along-track spatial resolution (Figure 3.52). This information may be useful for monitoring climate change. CryoSat-2 is in a 720 km × 732 km (432 × 439 mi.) orbit. The primary instrument package is SIRAL-2, a pair of SAR/Interferometric Radar Altimeters used to determine and monitor the spacecraft's altitude in order to measure the elevation of the ice (ESA o, CryoSat, online).

Laser Relativity Satellite

A joint venture between the ASI and ESA, the LAser RElativity Satellite (LARES) was launched in 2012 into a 1450 km (870 mi.) orbit. The satellite, completely passive, is made of tungsten alloy and houses 92 cube corner reflectors used to track the satellite via Earth-based lasers. The main scientific objective of the LARES mission is measurement of the Lense–Thirring effect (rotational frame-dragging effect) with an accuracy of about 1%. This effect derives

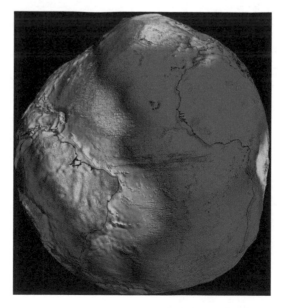

FIGURE 3.51
GOCE gravity measurements provide the best yet representation of the Earth's geoid. Colors represent deviations in meters from an ideal geoid. From +100 m (yellow–orange) to –100 m (dark blue). (Copyright of ESA. http://www.esa.int/esa-mmg/mmg.pl?b=b&type=I&mission=SMOS&single=y&start=2.)

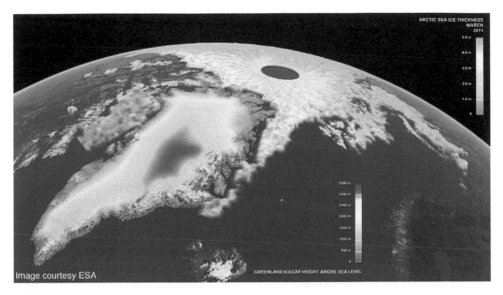

FIGURE 3.52

CryoSat-2 radar-derived map of Arctic sea ice thickness and elevation of the Greenland ice sheet during March, 2011. Green is thin ice, whereas yellow to orange indicates thicker ice. (Courtesy of NSIDC, CPOM/UCL/Leeds and ESA/PVL. http://nsidc.org/arcticseaicenews/category/analysis/page/2/.)

from the theory of relativity and states that rotation of a massive object should distort space–time, making the orbit of a nearby particle precess. A side benefit for Earth scientists is that, together with the LAGEOS-1, LAGEOS-2, and GRACE satellites, it will help accurately determine the earth's gravitational field (International Laser Ranging Service, LARES, online).

France

The National Center for Space Studies (*Centre National d'Etudes Spatiales*, CNES) is the French government space agency. Established in 1961, it is under the supervision of the French Ministries of Defense and Research with headquarters in Paris. It operates out of the Toulouse Space Center and Guiana Space Center, but also has payloads launched from space centers operated by other countries.

SPOT Image, a public limited company created in 1982 by a consortium of CNES, the French *Institut Géographique National*, and various space manufacturers, is presently a subsidiary of EADS Astrium. SPOT Image is the commercial operator for the SPOT Earth observation satellites. Astrium, headquartered in Paris, is an aerospace subsidiary of EADS that provides civil and military space systems and services.

STARLETTE and STELLA

Starlette (*Satellite de Taille Adaptée avec Réflecteurs Laser pour les Etudes de la Terre*), launched by CNES in 1975 into an 800 km (480 mi.) circular orbit, was the world's first "passive laser satellite" for solid Earth research. It is covered with 60 laser retroreflectors for geodetic ranging. Satellite Laser Reflector (SLR) tracking data are used to study the long-wavelength gravity field and its temporal changes. SLR is also used to determine the tidal responses of the solid Earth and to improve knowledge of the long wavelengths of the ocean tides. Stella, launched in 1993, is almost identical and has the same purpose. Their small size compared to their mass gives them a much larger sensitivity to gravitational attraction than to surface forces due either to the residual atmosphere at the satellite or to radiation pressure (NASA vv, Starlette and Stella; Janes Space Systems a, Starlette and Stella, online).

SPOT Series

Beginning in 1986, the French Space Agency launched the *System Probatoire d'Observation de la Terre* (SPOT 1) satellites to gather Earth resources imagery (ESA p, SPOT, online). The four SPOT satellites orbit at 832 km (500 mi.) in a sun-synchronous orbit with a 26-day repeat cycle. Like Landsat, they cross mid-latitudes at mid-morning (around 10 AM). An innovation is the ability to tilt or point the sensor up to 27° from vertical in a cross-track direction, so that the same area can be imaged on successive orbits with the resulting overlap used for stereo viewing. The area covered in vertical mode is 60 × 60 km (36 × 36 mi.). Two sensors are available: SPOT P (PAN), and SPOT XS (multispectral).

SPOT P acquires a single image in the range 510–730 nm, roughly green and red visible wavelengths (Figure 3.53). Resolution is 10 m (33 ft.). SPOT P trades spectral resolution for spatial resolution. SPOT XS records three channels of data (500–590 nm, or green light; 610–680 nm, or red light; and 790–890 nm, or NIR. These are usually combined to create a true color or CIR images (Figure 3.54). The resolution of the XS system is 20 m (65 ft.).

SPOT-6

Launched in September of 2012, this follow-on to the SPOT series is part of a constellation of Earth-imaging satellites designed to provide continuity of high-resolution, wide-swath data till 2023. EADS Astrium decided to build this constellation in 2009 based on a perceived need for these kinds of data. SPOT Image owns the satellites and ground segments. SPOT 6 is in a 694 km (430 mi.) orbit. Image resolution is 1.5 m (5 ft.) PAN and 8 m (26 ft.)

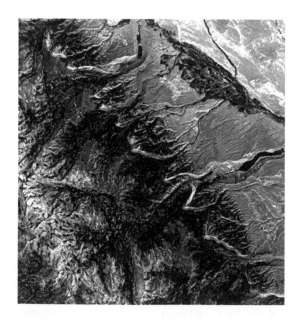

FIGURE 3.53

The Wind River Range in Wyoming as captured by the SPOT 5 PAN instrument on August 29, 2007. Copyright© Astrium Services 2007. (Courtesy of Astrium Services–copyright Astrium Services 2013. http://www.imaging-notes.com/go/article_freeJ.php?mp_id=235.)

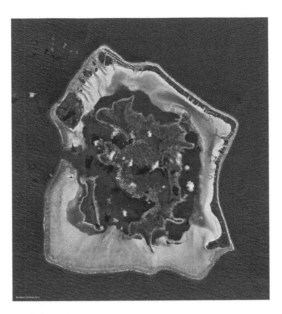

FIGURE 3.54
A 1.5 m resolution true-color image collected by SPOT-6 of Bora Bora, French Polynesia. (Courtesy of Astrium Services—Copyright © Astrium Services 2013.)

multispectral. Bands include PAN (450–745 nm), blue (450–525 nm), green (530–590 nm), red (625–695 nm), and NIR (760–890 nm). Swath width is 60 km (37 mi.) (eoPortal k, SPOT-6 and SPOT-7, online).

Detection of Electro-Magnetic Emissions Transmitted from Earthquake Regions

The Detection of Electro-Magnetic Emissions Transmitted from Earthquake Regions (DEMETER) mission is the first of the Myriade satellite series operated by CNES (CNES a, DEMETER, online). Its mission is to investigate ionospheric disturbances caused by seismic and volcanic activity. It was launched in 2004 and placed in a quasi-sun-synchronous circular orbit at an altitude of about 710 km (426 mi.). The altitude was changed to about 660 km (396 mi.) in December, 2005. Demeter carries five instruments:

- IMSC: three magnetic sensors from a few Hertz up to 18 kHz
- ICE: three electric sensors from DC up to 3.5 MHz
- IAP: an ion analyzer
- ISL: a Langmuir probe (used to determine the electron temperature, electron density, and electric potential of a plasma)
- IDP: an energetic particle detector

Polarization and Anisotropy of Reflectances for Atmospheric Science Coupled with Observations from Lidar

Launched into a 705 km (423 mi.) orbit in 2004, PARASOL is the second microsatellite in the Myriade series developed by CNES. It carries a wide-field imaging radiometer/polarimeter called POLDER (Polarization and Directionality of the Earth's Reflectances). POLDER measures the directional characteristics and polarization of light reflected by the earth and atmosphere to understand the radiative and physical properties of clouds and aerosols. This instrument should be able to distinguish natural from man-made aerosols. Flying in formation with AQUA and AURA (NASA), CALIPSO (NASA/CNES) and CloudSat (NASA/CSA) as part of the "A-Train," the five satellites cross the equator one at a time, a few minutes apart, at around 1:30 PM local time. These satellites, to be joined by NASA's Orbiting Carbon Observatory (OCO-2) in 2013, will for the first time combine a full suite of instruments for observing clouds and aerosols (CNES b, CNES c, PARASOL, online).

Pléiades Series

Launched by CNES in 2003 and 2011, the twin Pléiades satellites (Pléiades 1 and 2) have a "PAN" band with 50 cm (1.5 ft.) ground resolution, and multispectral bands with 2 m (7 ft.) resolution (Figure 3.55). The instruments provide a 20 km (12 mi.) swath width at nadir and up to 100 km (60 mi.) width as a mosaic. The 694 km (416 mi.) near-circular sun-synchronous orbit has a 26-day repeat cycle (CNES d, Pléiades, online). Pléiades radiometer features are provided in Table 3.9.

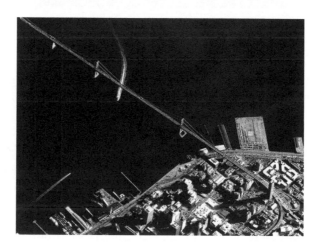

FIGURE 3.55
Pléiades PAN image of part of the Oakland Bay Bridge, San Francisco, California, acquired December 2011. Copyright© CNES 2012—Distribution Astrium Services. (Courtesy of Astrium Services—copyright Astrium Services 2013. *http://www.e2v.com/news/pleiades—the-cnes-earth-observation-satellite—captures-its-first-images-using-e2v-sensors/*.)

TABLE 3.9

Pléiades Instrument Specifications

Mode	Channel	Spectral Band	Resolution (m)
Multispectral	1	430–550 nm (blue)	2
	2	490–610 nm (green)	2
	3	600–720 nm (red)	2
	4	750–950 nm (near-IR)	2
PAN	P	480–830 nm (B/W)	0.50

Source: eoPortal, https://directory.eoportal.org/web/eoportal/satellite-missions/p/pleiades.

Pléiades 1B

Pléiades 1B was launched in December, 2012. The satellite is operated by CNES. The system contains the same five instruments as in Pléiades 1: it has a PAN band with 50 cm (1.6 ft.) ground resolution, and multispectral bands with 2 m (7 ft.) resolution. The instruments have a 20 km (12 mi.) swath at nadir and up to a 100 km (62 mi.) swath as a mosaic. The 694 km (430 mi.) sun-synchronous orbit has a 26-day repeat cycle (eoPortal l, Pléiades, online).

Germany

The German Aerospace Center (*Deutsches Zentrum für Luft- und Raumfahrt e.V.*), abbreviated DLR, is the national center for aerospace, energy, and transportation research of the Federal Republic of Germany. Previously known as the German Aerospace Research Establishment (DFVLR), its headquarters is in Cologne, Germany. The DLR is engaged in research and development projects in national and international partnerships. In addition to conducting its own projects, DLR also acts as the German space agency. Data processing, archiving of products and their distribution to users is made by the German Remote Data Center (DFD) in Oberpfaffenhofen, FRG.

Challenging Minisatellite Payload

CHAMP is a German satellite built and managed by DLR and launched during July 2000. The mission lasted until September 2010. The purpose of this satellite was atmospheric and ionospheric research and geoscience applications. The spacecraft flew in a circular near-polar 454 km (272 mi.) sun-synchronous orbit. CHAMP was managed by German Research Center for Geosciences (GFZ) Potsdam, in cooperation with NASA and JPL. The mission generated gravity and magnetic field measurements. CHAMP carried a three-component fluxgate vector magnetometer for measuring the earth's magnetic field and a scalar Overhauser magnetometer to provide in-flight calibration. In addition, CHAMP carried an electrostatic Spatial Triaxial Accelerometer for Research (STAR) accelerometer for measuring the earth's gravity field (NASA ww, CHAMP; GFZ b, CHAMP, online).

TerraSAR-X

TerraSAR-X, a German Earth observation satellite, is a partnership between the German Aerospace Center (DLR) and EADS Astrium (NASA xx, TerraSAR-X, online). TerraSAR-X was launched in 2007. The instrument is an X-band SAR (SAR-X) that provides SAR interferometry as well as orthorectified imagery in various polarity combinations (Figure 3.56). Four resolutions are available from the 514 km (308 mi.) orbit: 1 m (5 × 10 km surface footprint), 2 m (10 × 10 km footprint), 3 m (30 × 50 km footprint), and 18.5 m (100 × 150 km footprint). The satellite also carries a Tracking, Occultation and Ranging (TOR) package. TOR consists of a two-frequency GPS receiver and a CHAMP Laser Retroreflector to carry out along-track interferometry (e.g., for measurement of ocean currents).

TanDEM-X

Launched in 2010 by DLR and in service since 2011, this radar satellite is identical to TerraSAR-X and carries the X band SAR and TOR instruments (NASA yy, TanDEM-X, online). By flying in a tandem orbit with TerraSAR-X, this satellite will help generate a highly accurate global DEM.

FIGURE 3.56
TerraSAR-X radar image of the Komo River delta, Gabon, ~60 km southeast of Libreville. Copyright© DLR. (Courtesy of Astrium Services—copyright Astrium Services 2013. http://commons.wikimedia.org/wiki/ File:TerraSAR-X_image_of_Gabon,_60_kilometres_south_east_of_the_capital_Libreville.jpg.)

GEOEYE, INC.

GeoEye, Inc. (formerly "Orbital Imaging Corporation" or "ORBIMAGE") of Herndon, Virginia, owns and operates Earth-imaging satellites including the GeoEye-1, Orbview, and IKONOS series. These satellites acquire high-resolution imagery. GeoEye, owned by Cerberus Capital Management, is the world's largest space imaging corporation. As of early 2013, they merged with DigitalGlobe.

IKONOS-1 and IKONOS-2

Space Imaging-EOSAT (now GeoEye/DigitalGlobe), launched IKONOS-1 in 1999. It failed soon after launch, so its twin, IKONOS-2, was launched later in 1999. The satellite has a PAN channel (450–900 nm) with 82 cm (2.7 ft.) resolution and four multispectral sensors with 3.2 m (10.5 ft.) resolution: blue (445–516 nm), green (506–595 nm), red (632–698 nm), and NIR (757–853 nm) (Figure 3.57). The ground swath is 11 km (6.7 km) wide, and the instrument has the ability to tilt to acquire stereo imagery. The 680 km (408 mi.) sun-synchronous orbit has 11-day repeat coverage (ESA q, Ikonos-2, online).

ORBVIEW-1

Orbital Imaging (now GeoEye/DigitalGlobe) launched Orbview-1 in 1995 to monitor the atmosphere for weather forecasting. Spatial resolution is coarse at 10 km (6 mi.), and swath width is 1300 km (780 mi.) from an altitude of 740 km (444 mi.). There is one channel centered at 777 nm. Revisit time is 2 days.

ORBVIEW-2

Orbital Imaging (now GeoEye/DigitalGlobe), in conjunction with NASA, launched Orbview-2 (also known as SeaStar) in 1997 with the purpose of multispectral Earth observation, in particular over the oceans (NASA zz, Orbview-2, online). This satellite carried the Sea-viewing Wide Field of View Sensor (SeaWiFS). The 705 km (423 mi.) sun-synchronous orbit provides 1.1 km (3450 ft.) resolution (Local Area Coverage, or LAC instrument) and

FIGURE 3.57

Ikonos 3.2 m resolution image of El Capitan, Yosemite National Park, California, acquired on October 15, 2003. (Satellite Images Courtesy of GeoEye and Spatial Energy. DigitalGlobe, Inc and GeoEye © 2012.)

4.5 km (2.7 mi.) resolution (Global Area Coverage, or GAC instrument) with a 2801 km (1681 mi.) swath width and one-day revisit time. It had eight multispectral sensors: Band 1 (402–422 nm), Band 2 (433–453 nm), Band 3 (480–500 nm), Band 4 (500–520 nm), Band 5 (545–565 nm), Band 6 (660–680 nm), Band 7 (745–785 nm), and Band 8 (845–885 nm). Orbview-2 stopped collecting data in December 2010. The instrument was specifically designed to monitor ocean characteristics such as chlorophyll-a concentration and water clarity. It was able to tilt up to 20° to avoid sunlight from the sea surface. The SeaWiFS Mission is an industry/government partnership, with NASA's Ocean Biology Processing Group at Goddard SFC having responsibility for data collection, processing, calibration, archiving, and distribution.

ORBVIEW-3

Orbview-3 was launched in 2003. It contained a 1 m (3.3 ft.) resolution PAN channel (450–900 nm) and 4 m (13 ft.) resolution MSS with blue (450–520 nm), green (520–620 nm), red (630–690 nm), and NIR (760–900 nm) channels. Swath width was 8 km (4.8 mi.) from a sun-synchronous orbit with a 3-day repeat cycle (Orbital, OrbView-3, online). The instrument failed in April 2007.

GeoEye-1

GeoEye-1 (formerly OrbView-5) was launched in 2008 and provides panchromatic images with 41 cm (1.3 ft.) resolution and multispectral imagery with 1.65 m (5.4 ft.) resolution over a 15.2 km (9 mi.) wide ground swath (Figure 3.58). The spacecraft flies in a sun-synchronous orbit at an altitude of 684 km (425 mi.) and can image up to 60° off nadir. The highest resolution is available only to the U.S. government (Satellite Imaging Corporation b, GEOEYE-1, online).

FIGURE 3.58
GeoEye-1 1.65 m resolution true color image of Towra Point nature reserve, Australia, acquired on February 19, 2012. (Satellite Images Courtesy of GeoEye and Spatial Energy. DigitalGlobe, Inc and GeoEye © 2012.)

India

The Indian Space Research Organization (ISRO), headquartered in Bangalore, is responsible for Indian remote sensing satellites, presently the largest national satellite fleet for civilian purposes (ISRO a, Earth Observation Satellites, online). Starting with IRS-1A, launched in 1988, a total of 17 satellites were put into orbit for Earth sciences purposes. Completed missions include IRS-1B (1991), IRS-P2 (1994), IRS-1C (1995), IRS-P3, IRS-1D, and IRS-P4 (also known as Oceansat-1, 1999). Satellites still in service include Technology Experiment Satellite (TES), 2001), IRS-P6 (Resourcesat-1, 2003), IRS-P5 (Cartosat-1, 2005), IRS-P7 (Cartosat-2, 2007), Cartosat 2A (2008), IMS-1 (2008), Oceansat-2 (2009), Cartosat-2B (2010), and ResourceSat-2 (2011).

IRS Series

IRS satellites contain several sensors including the Linear Imaging Self Scanning Sensor-1 (LISS-1), LISS-2, and LISS-3. The satellites are in a 904 km (542 mi.) sun-synchronous orbit with 22-day repeat coverage. LISS-I has a ground swath of 148 km (89 mi.) with 72 m (234 ft.) resolution. LISS-2 has a 74 km (44 mi.) swath and 36 m (117 ft.) ground resolution. Each sensor has four spectral bands in the range 450–860 nm. LISS-3, launched on the IRS-1C in 1995, has a 24-day repeat cycle with 23.5 m (77 ft.) resolution in the VNIR range and 70.5 m (231 ft.) resolution in the SWIR region (Figure 3.59). The VNIR swath is 142 km (85 mi.), while the SWIR swath is 148 km (89 mi.) wide. It is in a polar sun-synchronous orbit at 817 km (490 mi.; ISRO b, c, IRS-1A, IRS-1E, online).

FIGURE 3.59

IRS LISS-3 color-infrared image of Kaiserstuhl Mountain, Germany, with 23 m resolution. A synthetic blue band is available to generate pseudo-natural color images. (Courtesy of IRS and GAFAG. http://www.gaf.de/content/irs.)

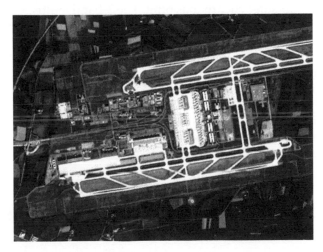

FIGURE 3.60

IRS PAN/LISS-4 Mono Mode of Munich airport. IRS PAN images have 5.8 m resolution. (Courtesy of IRS and GAFAG. http://www.gaf.de/content/irs.)

These satellites also contained the Wide-Field Sensor (WiFS) and PAN sensors. WiFS has 188 m (617 ft.) resolution over a 774 km (464 mi.) swath and 24-day nadir and 5-day off-track repeat coverage. IRS-1C has two channels, 620–680 nm and 770–860 nm. IRS-P3 had, in addition to these two, a channel from 1.55 to 1.70 μm. The PAN sensor records a single channel from 500 to 750 nm with 5 m (16.4 ft.) resolution over a 70 km (42 mi.) swath (Figure 3.60).

IRS-P4 (Oceansat-1)

Oceansat-1, launched in 1999, carried the Ocean Color Monitor (OCM) pushbroom scanner with eight spectral bands for the measurements of physical and biological oceanographic parameters: Band 1: 402–422 nm, Band 2: 433–453 nm, Band 3: 480–500 nm,

Band 4: 500–520 nm, Band 5: 545–565 nm, Band 6: 660–680 nm, Band 7: 745–785 nm, and Band 8: 845–885 nm. IRS-P4 also carried the Multifrequency Scanning Microwave Radiometers, operating at 6.6, 10.65, 18.0, and 21 GHz frequencies with H and V polarizations and spatial resolution of 150 km (90 mi.), 75 km (45 mi.), 50 km (30 mi.), and 50 km, respectively. The mission was completed in August 2010 (ISRO d, IRS-P4, online).

Technology Experiment Satellite

TES showed the viability of technologies that would be used in future ISRO satellites. TES was placed in a 568 km (340 mi.) sun-synchronous orbit in October 2001. TES has a PAN camera capable of producing images of 1 m (3 ft.) resolution. This camera made India the second country in the world after the United States to offer commercial images with 1 m resolution. It is used for civilian and military remote sensing, mapping, industrial applications, and for geographical information services (ISRO e, TES, online).

IRS-P6

ResourceSat-1 (IRS-P6) was launched in 2003 and carries three Earth observing systems in an 817 km (490 mi.) polar sun-synchronous orbit. The 3-band multispectral Linear Imaging Self Scanning sensor-4 (LISS-4) has spatial resolution better than 5.86 m (19 ft.) and a 25 km (15 mi.) swath as well as across track steerability for selected area monitoring (Figure 3.61). IRS-P6 carries an improved LISS-3 with 4 bands (red, green, NIR, and SWIR), all at 23.5 m (77 ft.) resolution and a 140 km (84 mi.) ground swath. The Advanced Wide Field Sensor (AWiFS) operates in three spectral bands in the visible-NIR and one band in SWIR. This sensor has 56–80 m (184–263 ft.) resolution and a 1400 km (840 mi.) swath. The satellite provides data useful for geologic and vegetation-related applications and allows crop- and species-level discrimination. The satellite is helping with integrated land and water resources planning and related applications (ISRO f, IRS-P6, online).

FIGURE 3.61

The Grand Canyon is displayed in this false-color satellite image from India's Resourcesat-1 collected on June 19, 2004. (Copyright DigitalGlobe, Inc and GeoEye © 2012. http://earthobservatory.nasa.gov/IOTD/view.php?id=6929.)

IRS-P5

Cartosat-1 (IRS-P5) was launched in May 2005 into a sun-synchronous orbit carrying two PAN cameras with 2.5 m (8.2 ft.) resolution and with fore—and-aft-looking capability for stereo viewing over a 30 km (18 mi.) wide swath. This system is primarily for cartography and terrain modeling (ISRO g, CARTOSAT-1, online).

IRS-P7

Cartosat-2 (IRS-P7) was launched in 2007 carrying a single PAN camera with 0.80 m (2.6 ft.) resolution (ISRO h, CARTOSAT-2, online). The swath is 9.6 km (5.7 mi.). This mission has the ability to steer sensors along and across the track up to 45°. It flies in a sun-synchronous polar orbit at an altitude of 635 km (380 mi.) with a revisit period of 4 days, which can be improved to one day with suitable orbit maneuvers. The purpose of the satellite, as suggested by its name, is cadastral-level cartographic applications (comprehensive mapping of real property, commonly including details of the ownership, tenure, and precise location).

Indian Micro Satellite-1

The Indian Micro Satellite (IMS-1) was launched into a sun-synchronous orbit during 2008. It carries two instruments: the four band Multispectral CCD Camera (MxT) and Hyper Spectral Imager (HySi-T). The MxT bands cover the VNIR (blue, green, red, and NIR bands) and have a ground resolution of 37 m (120 ft.) over a swath of 151 km (94 mi.). This system is being used for natural resources management including agriculture, forest coverage, land use, and disaster management. The HySi-T is primarily for ocean and atmospheric studies. The instrument has 64 bands in the spectral region from 400 to 950 nm (SSRO i, IMS-1, online).

Oceansat-2

Oceansat-2, launched in 2009, is an Indian satellite designed to study surface winds and the ocean surface, observe chlorophyll concentrations, monitor phytoplankton blooms and suspended sediments, and study atmospheric aerosols. Oceansat-2 carries three instruments. The Ocean Color Monitor (OCM) is an 8-band multispectral camera operating in the visible-NIR range. This camera has an IFOV of 360 m and a swath of 1420 km. OCM can be tilted up to +20° along track. The Scanning Scatterometer (SCAT) is an active microwave device used to determine ocean-surface-level wind direction by evaluation of radar backscatter. The scatterometer generates two pencil beams and scans at a rate of 20.5 rpm to cover the entire swath. The Ku band pencil beam scatterometer is an active microwave radar operating at 13.515 GHz, providing a ground resolution of 50 km (31 mi.). It covers a continuous swath from 1400 km (868 mi.) inner beam to 1840 km (1140 mi.) outer beam. The sensor provides global ocean coverage and wind vector retrieval with a revisit time of 2 days. The Radio Occultation Sounder for Atmospheric Studies (ROSA) is a GPS occultation receiver provided by ASI. The objective of this instrument is to characterize the lower atmosphere and the ionosphere using radio occultation data (ISRO j, Oceansat-2, online).

RISAT 2

Radar Imaging Satellite-2 (RISAT-2) was the first of the RISAT series put into a 550 km (340 mi.) orbit. It was launched successfully in April, 2009. No technical specifications of RISAT-2 were published. However, it contains a C band SAR and is likely to have a spatial resolution of about 1 m. The purpose is border surveillance, to deter insurgent infiltration, and to support anti-terrorist operations. For geosciences purposes, the satellite has applications in the area of disaster management and agriculture-related activities (ISRO k, RISAT-2, online).

CARTOSAT 2B

CARTOSAT 2B, launched in July 2010, carries a visible range PAN camera capable of being steered up to 26° along as well as across the direction of its movement to facilitate imaging of any area more frequently (ISRO l, CARTOSAT-2B, online).

ResourceSat-2

ResourceSat-2, launched in 2011, carries the same instruments as ResourceSat-1. It does, however, have a 70 km (43 mi.) swath instead of 25 km, and improved radiometric accuracy (ISRO m, RESOURCESAT-2, online).

RISAT-1

This C band (5.35 GHz) radar satellite was launched during April 2012 into a 536 km (332 mi.) sun-synchronous polar orbit with a 25-day repeat cycle. RISAT-1 includes a dual-polarization phased array antenna. (ISRO n, RISAT-1, online).

Indonesia

National Institute of Aeronautics and Space (*Lembaga Penerbangan dan Antariksa Nasional*, LAPAN) is the Indonesian government space agency. It was established in 1963 and is responsible for civilian and military aerospace research.

Lapan-TUBsat

Launched in 2007 by LAPAN, Lapan-TUBsat carries a three CCD color video camera with 6 m (20 ft.) ground resolution and a second color video camera with 120 m (394 ft.) resolution over an 80 km (50 mi.) swath. It flies in a 630 km (390 mi.) sun-synchronous polar orbit. The purpose is to gather imagery for environmental monitoring and disaster management (Technische Universität Berlin, TUBSAT, online).

ImageSat International

ImageSat International N.V. is a commercial provider of high-resolution, Earth imagery collected by its Earth Resources Observation Satellites. ImageSat's principal business is operating high-resolution satellites and providing exclusive, autonomous satellite imaging services to governments as well as providing civilian and commercial geospatial information applications. ImageSat is a Netherlands Antilles company with offices in Limassol, Cyprus and Tel Aviv, Israel. The company's offices in Tel Aviv supervise the construction of the EROS family of satellites and the operation of ImageSat's main ground control station.

Earth Resources Observation Satellite

The EROS A commercial Earth Resources Observation Satellite was launched by ImageSat International in 2000 into a 480 km (298 mi.) sun-synchronous orbit. The PAN sensor (500–900 nm) provides imagery, including stereo pairs, with a ground resolution of 1.8 m (6 ft.) over

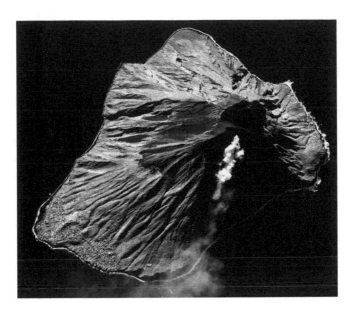

FIGURE 3.62
EROS-A 1.8 m resolution panchromatic image of Stromboli Volcano, Italy. (Copyrights by ImageSat International N.V. © 2013. http://www.sarracenia.com/astronomy/remotesensing/primer0430.html.)

a 14 km (8.7 mi.) swath (Figure 3.62). By oversampling the instrument is capable of 1.0 m (3.3 ft.) resolution over a 10 km (6 mi.) swath (ImageSat International a, EROS A, online).

Earth Resources Observation Satellite B

The EROS B satellite was launched by ImageSat International in 2006 into a 480 km sun-synchronous orbit. Its PAN (500–900 nm) pushbroom scanner provides imagery with a ground resolution of 0.7 m (2.3 ft.; ImageSat International b, EROS B; eoPortal Directory b, EROS-B, online).

Japan

The Japanese aerospace program is run by the Japan Aerospace Exploration Agency (JAXA) and National Space Development Agency of Japan (NASDA), both headquartered in Tokyo. The Japan Meteorological Agency (JMA), Tokyo, has as its goals the prevention and mitigation of natural disasters, transportation safety, industrial development, and improvement of the public welfare. To meet these goals, JMA monitors the earth's environment and attempts to forecast natural phenomena related to the atmosphere, the oceans, and the earth. JMA also conducts research and technical development in related fields.

Geostationary Meteorological Satellite Series

The Geostationary Meteorological Satellite (GMS) 1 to 5 series (also known as Himawari 1 to 5) was a joint NASA-JAXA mission consisting of five weather satellites launched between 1977 and 1995 into a low Earth geostationary orbit. These satellites contained IR radiometers with ground resolution of 1.25–5 km (0.75–3 mi.). The mission is no longer operational. Data acquired by this mission was used for cloud-type and cloud-motion detection, and also measured sea-surface temperatures. Spectral bands covered the visible, NIR, SWIR, MWIR, and TIR (400 nm to 15 μm) over a few tens of channels. Spatial resolution ranged from 30 m to 3 km (100 ft. to 1.8 mi.). The satellites were replaced by the Multifunctional Transport Satellite series (JAXA a, Himawari, online).

Marine Observation Satellite-1 and -1b

The Marine Observation Satellite (MOS, known in Japan as "Momo") was designed for remote sensing of oceans and coastal zones. Japan launched MOS-1 in 1987 into a 909 km (564 mi.) sun-synchronous orbit with a 17-day repeat cycle. MOS-1 contained three instruments that recorded in the visible, NIR, TIR, and in the microwave regions. The Multispectral Electronic Self-Scanning Radiometer (MESSR) had 4 bands between 510 nm and 1.1 μm. Each band had 50 m (164 ft.) resolution and a ground swath that varied from 100 to 185 km (60–111 mi.). The Visible and Thermal Infrared Radiometer (VTIR) had one visible and three IR bands with a 1500 km swath. The Micro Scanning Radiometer (MSR) measured microwave emission at 23 and 31 GHz. The MOS-1 mission ended in 1995. The MOS-1b satellite had the same functions as MOS-1 and ended operations during 1996 (JAXA b, MOS-1, online).

Advanced Earth Observing Satellite Series

The Japanese Advanced Earth Observing Satellite 1 (ADEOS-1) was launched in August 1996 and carried a number of instruments of interest to the earth observation community: the Ocean Color and Temperature Scanner (OCTS) had 12 bands (eoPortal Directory c, ADEOS/Midori, online). The Advanced Visible and Near Infrared Radiometer (AVNIR) had five bands, each with ground resolution of better than 20 m (65 ft.). The NASA Scatterometer (NSCAT) used a fan-beam Doppler signal to measure wind speed over bodies of water. The Total Ozone Mapping Spectrometer (TOMS) was built to study changes in Earth's ozone layer. The Polarization and Directionality of the Earth's Reflectance (POLDER) device was an optical imaging radiometer and polarimeter. The device scanned between 443 and 910 nm, depending on the objective: shorter wavelengths (443–565 nm) typically measured ocean color, whereas longer wavelengths (670–910 nm) were used to study vegetation and water vapor content. The Improved Limb Atmospheric Spectrometer (ILAS) used a grating spectrometer to measure the properties of trace gases by solar occultation. The Retroreflector in Space (RIS) and Interferometric Monitor for Greenhouse Gases (IMG) studied atmospheric trace gases and greenhouse gases. This satellite was cosponsored by France and the United States, and was being used primarily for global environmental change monitoring. After almost a year in orbit, ADEOS-1 failed during June, 1997.

ADEOS-2 (also known as Midori 2 and SeaWinds) was an Earth observation satellite launched by NASDA, NASA, and CNES in December 2002 (JAXA c, Midori II, online). The three primary mission objectives were the following:

1. Monitor water and energy cycles as a part of the global climate system
2. Quantitatively estimate biomass and productivity of the carbon cycle
3. Detect long-term climate change by continuing the observations of ADEOS-1

The satellite was equipped with five primary instruments—an Advanced Microwave Scanning Radiometer (AMSR) Global Imager (GLI), ILAS-II, POLDER, and SeaWinds (Figure 3.63).

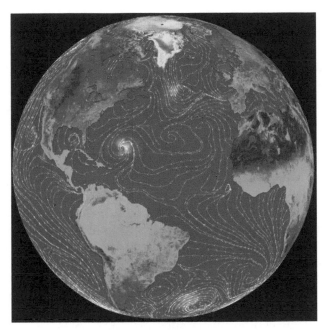

FIGURE 3.63

Midori-2 SeaWinds scatterometer image from the Japanese ADEOS-2 satellite. Map of wind patterns over the Atlantic acquired on May 8, 2000. (Courtesy NASA/JPL ; METI/Japan Space Systems image. http://archive.org/details/PIA02458.)

The AMSR was a multifrequency dual-polarized passive MWR that observed water-related microwave emissions from the earth's surface and atmosphere. Bands were centered between 6.9 and 52.8 GHz, with IFOV ranging from 6×10 km to 40×70 km (3.6×6 to 25×43 mi.) over a 1600 km (992 mi.) swath. Temperature sensitivity ranged from 0.34°K to 1.8°K (Kawanishi et al., 2003).

GLI had 250–1000 m (820–3280 ft.) resolution over a 1600 km (1000 mi.) swath (Borengasser, 2003; NASDA, Global Imager, online). It was a 36 channel multispectral scanner covering the visible and NIR (375-12,500 nm).

ILAS was a spectrometer that observed the atmospheric limb absorption spectrum from the upper troposphere to the stratosphere using the solar occultation technique. It measured the TIR region (8.5–16.1 μm) and the near visible region (753–784 nm). ILAS could measure the vertical profile of ozone (O_3), nitrogen dioxide (NO_2), aerosols, water vapor, CFC_{11}, methane (CH_4), and nitrous oxide (N_2O) as well as temperature and pressure. IFOV was 2 km (1.2 mi.) vertical and 13 km (8 mi.) horizontal (Earth Observation Research Center a, ILAS, online).

POLDER was a passive optical imaging radiometer and polarimeter with a wide FOV lens and pushbroom CCD array that observed an area from 14 different directions to help understand angular characteristics of Earth's reflectance. POLDER could observe multiple polarizations in 16 bands by rotating interference filters and polarizers. The device scanned between 443 and 910 nm. The shorter wavelengths (443–565 nm) typically measured ocean color, whereas the longer wavelengths (670–910 nm) were used to study vegetation and water vapor content. The wide FOV enabled the entire Earth surface to be scanned four times in five days (Earth Observation Research Center b, POLDER, online).

The SeaWinds scatterometer measured near-surface wind (speed and direction) over Earth's oceans. SeaWinds used a rotating dish antenna with two beams. The antenna radiated microwave pulses at a frequency of 13.4 GHz across a continuous 1800 km (1116 mi.) swath. Wind vector resolution is 50 km (31 mi.); wind speed accuracy is 2 m/s (7 ft./s); directional accuracy is 20° (NASA aaa, SeaWinds, online).

On October 23, 2003, 10 months after launch, the mission ended because of failure of the satellite's power systems (NASA bbb, ADEOS-2, online).

JERS-1

The JERS-1 satellite was launched into a sun-synchronous 568 km (340 mi.) orbit in February 1992. It has a 44-day repeat cycle and contains two onboard systems that can be co-registered: the Optical Sensor (OPS) and a SAR. The OPS has eight bands between 0.52 and 2.40 µm. Ground resolution is 18.3 m (59.5 ft.) cross track and 24.2 m (78.7 ft.) along track, with a 75 km (46 mi.) wide swath (Figure 3.64). This instrument can produce stereoscopic images using Band 3 (views nadir) and Band 4 (looks 15.3° forward in the orbit plane). Both of these bands image the 760–860 nm, or NIR region (Earth Observation Research Center c, JERS, online).

The L band (23.5 cm) SAR also has a 75 km (46.5 mi.) swath and 18 m (58.5 ft.) ground resolution. The SAR has H–H polarization and a viewing angle 35° off nadir (Figure 3.65).

Data are available to the public from the national archives in Tokyo at the Remote Sensing Technology Center (RESTEC), an agency sponsored by NASDA.

Multifunctional Transport Satellites Series

Multifunctional Transport Satellites (MTSAT 1 and 2) are geostationary weather satellites owned and operated by the Japanese Ministry of Land, Infrastructure, and Transport and the JMA. Launched in 2005 and 2006, they provide coverage for the hemisphere centered on 140° East; this includes Japan and Australia, the principal users of the imagery. The satellites provide imagery in five wavelength bands— one visible and four IR, including

FIGURE 3.64
JERS-1 OPS image of part of the Neuquen Basin, Argentina. A north-south breached anticline can be seen in the west-central part of the image due to exposure of lighter, older units. (Courtesy of ERSDAC; METI/Japan Space Systems.)

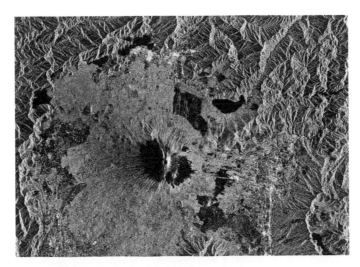

FIGURE 3.65
JERS-1 radar image of Mt. Fuji, Japan. (METI/Japan Space Systems image).

TABLE 3.10

ALOS Sensor Characteristics

Resolution	2.5 m Panchromatic; 10 m Multispectral
Orbit	Sun-Synchronous Sub-Recurrent Orbit
Repeat cycle	46 days
Altitude	~692 km

Source: eoPortal, https://directory.eoportal.org/web/eoportal/satellite-missions/a/alos.

a water vapor channel. The visible light camera has a resolution of 1 km (0.6 mi.); the IR cameras have 4 km (2.4 mi.) resolution. Resolution is lower away from the equator. The spacecraft have a planned lifespan of 5 years (JAXA d, MTSAT-1R, online).

Advanced Land Observing Satellite

The Advanced Land Observing Satellite (ALOS, also called Daichi) was launched on January 24, 2006 (Table 3.10). The satellite contained three sensors used to map terrain in Asia and the Pacific: the Panchromatic Remote-sensing Instrument for Stereo Mapping (PRISM) for digital elevation mapping, the AVNIR-2 for land cover observation, and the Phased Array-type L band Synthetic Aperture Radar (PALSAR) for day–night and all-weather land observation (Figure 3.66; Tables 3.11 through 3.13). Mission objectives included cartography, disaster monitoring, natural resource surveys, and technology development.

Full polarimetry (multipolarization), off-nadir pointing, and other functions of the SAR improved the ability to analyze geological structure and the distribution of rocks. At the same time, multipolarization was useful for acquiring vegetation information.

PALSAR had three observation modes: "Fine" mode was the most commonly used. Ground resolution could be as good as 7 m (23 ft.). "ScanSAR" mode enabled off-nadir viewing to cover a swath from 250 to 350 km (150–210 mi.). Resolution was degraded to

FIGURE 3.66

Japanese ALOS (Daichi) true color image of the Betsiboka River delta, Madagascar. (Courtesy of NASDA; METI/Japan Space Systems image. http://newsflick.net/post/13777878781/madagascars-monster-madagascars-largest-river.)

TABLE 3.11

AVNIR-2 Sensor Characteristics

Band	Wavelength Region (μm)	Resolution (m)
1	0.42–0.50 (blue)	10
2	0.52–0.60 (green)	10
3	0.61–0.69 (red)	10
4	0.76–0.89 (near-IR)	10

Source: eoPortal, http://www.eoportal.org/directory/pres_ADEOS AdvancedEarthObservingSatelliteMidori.html.

TABLE 3.12

PRISM Sensor Characteristics

Band	Wavelength Region (μm)	Resolution (m)
PAN	0.52–0.77	2.5

Source: eoPortal, https://directory.eoportal.org/web/ eoportal/satellite-missions/p/prism

about 100 m, however. In "Polarimetric" mode, PALSAR was capable of both horizontal and vertical polarization. PALSAR could also simultaneously receive horizontal and vertical polarization for each polarized transmission.

The ALOS satellite failed on April 21, 2011 (Satellite Imaging Corporation c, ALOS, online).

TABLE 3.13

Main Characteristics of the PALSAR Instrument

Mode	Fine	ScanSAR	Polarimetric	
Center Frequency	1270 MHz (L band)			
Bandwidth	28 MHz	14 MHz	14, 28 MHz	14 MHz
Polarization	HH or VV	HH + HV or VV + VH	HH or VV	HH + HV + VH + VV
Incidence angle	8°–60°	8°–60°	18°–43°	8°–30°
Range Resolution	7–44 m	14–88 m	100 m	24–89 m
Swath	40–70 km	40–70 km	250–350 km	20–65 km

Source: Satellite Imaging Corp., http://www.satimagingcorp.com/satellite-sensors/alos.html and Belgian Earth Observing Platform, http://eo.belspo.be/directory/SensorDetail.aspx?senID=15.

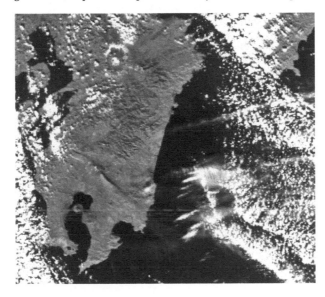

FIGURE 3.67

Japanese Greenhouse Gases Observing Satellite (Ibuki) Cloud and Aerosol Imager image shows the volcanic plume of Mt. Shinmoedake (red triangle) in the Kirishima mountain range, Japan, acquired on January 26, 2011. The volcanic plume can be seen spreading southeastward. This false color image was produced using Bands 2-3-1 as RGB. (Courtesy of JAXA/NIES/MOE ; METI/Japan Space Systems image. http://www.gosat.nies.go.jp/eng/related/2011/201101.htm.)

Greenhouse Gases Observing Satellite

The Greenhouse Gases Observing Satellite (GOSAT, also known as Ibuki) is dedicated to monitoring the greenhouse gases carbon dioxide and methane (Figure 3.67). Developed by JAXA, GOSAT was launched in 2009 and carries a gas observation sensor (TANSO-FTS) and a cloud-aerosol sensor (TANSO-CAI). TANSO-FTS records NIR through TIR wavelengths and uses a spectrometer to measure the response of elements and compounds to these wavelengths (JAXA e, Greenhouse gases Observing Satellite, online).

Korea

The Korea Aerospace Research Institute (KARI), founded in 1989, is the aeronautics and space agency of South Korea. Its main laboratories are located in Daejeon, in the Daedeok Science Town.

Korea Multipurpose Satellite-1

The Korea Multi-Purpose Satellite-1 (KOMPSAT 1, also known as Arirang-1) was launched in December 1999 by KARI into a 685 km (425 mi.) sun-synchronous orbit with a 26-day repeat cycle. The main instrument is the Electro-Optical Camera (EOC), which has 6 m (20 ft.) spatial resolution and a 24 km (15 mi.) wide swath. Other instruments performed ionosphere and magnetic field measurements. The purpose of the satellite was to measure features of the ocean and land as well as upper atmosphere. Contact with the satellite was lost in December of 2007.

Korea Multipurpose Satellite-2

The Korea Multi-Purpose Satellite-2 (KOMPSAT 2, or Arirang-2) was launched in July of 2006 into a 685 km sun-synchronous orbit with a 14-day repeat cycle. The main mission objectives are to provide surveillance of large-scale disasters, acquisition of independent high-resolution images for geographic information systems, generating printed and digitized maps for domestic and overseas areas, development of Korean territories, and natural resource surveys. The primary instruments include the Advanced EOC, which has 1 m (3.3 ft.) resolution, and the 4 m (13 ft.) resolution Multispectral Camera (MSC). Image distortion problems were announced late in 2006.

Korea Multipurpose Satellite-3

The Korea Multi-Purpose Satellite-3 (KOMPSAT 3) was successfully launched in May 2012. It acquires 1 m panchromatic imagery and 4 m multispectral imagery in the blue, green, red, and NIR regions. The objective is to continue the missions of Kompsat 1 and 2 for a wide range of applications including agriculture, environment, and ocean monitoring (eoPortal d, KOMPSAT-3, online).

Soviet Union/Russian Federation

Soyuzkarta was established in 1987 to market high-resolution (2 m) aerial/space-based photographic, cartographic, and geodetic products. In 1990, it undertook agreements with several western companies for the distribution of high-resolution images. In 1992, it became a joint stock company; Kosmokarta is the space division. This imagery is now available in digitized form direct from Priroda and its WorldMap consortium, which has one of the world's largest remote sensing archives (Janes Space Systems b, Soyuzkarta, online). Russia's Research Center for Earth Operative Monitoring (NTs OMZ), headquartered in Moscow, carries out acquisition, processing, archiving, and dissemination of space data from domestic and foreign remote sensing spacecraft. NTs OMZ offers remote sensing data from Russian spacecraft— RESURS-DK, MONITOR-E, METEOR-3M No. 1, METEOR-M, ELEKTRO-L, RESURS-O1, and OKEAN-O, as well as foreign spacecraft— QUICKBIRD, IKONOS, EROS, SPOT, IRS, LANDSAT, RADARSAT, TERRA (MODIS, ASTER radiometers), NOAA (AVHRR radiometer), and ERS, as well as processing and value-added product generation.

Satellite photography is available from several sensors including the KFA-1000, MK-4, KATE-200, DD-5, KVR-1000, and TK-350 camera systems (Figures 3.68 and 3.69). The KFA-1000 camera has a resolution of 5 m (16 ft.) and each frame covers 100 × 100 km (60 × 60 mi.). Both PAN and dual-emulsion films (sensitive to 570–670 and 670–810 nm) are used.

FIGURE 3.68
Russian KFA-3000 2 m resolution panchromatic image of part of Vienna, Austria. (Courtesy of Russian Federal Space Agency Roscosmos ©2012; © Official Site of Research Center for Earth Operative Monitoring [NTS OMZ]. http://gdsc.nlr.nl/gdsc/en/information/earth_observation/sensor_examples/panchromatic.)

FIGURE 3.69
Russian KFA-1000 high-resolution color image of Rome, Italy. (Courtesy of Russian Federal Space Agency Roscosmos ©2012; © Official Site of Research Center for Earth Operative Monitoring [NTS OMZ]. http://www.mentallandscape.com/c_catalogearth.htm.)

The MK-4 multispectral camera has 6 m (19.5 ft.) resolution and records six channels. It can achieve 60% stereo overlap using two available cameras, and covers an area 126 × 126 km (75 × 75 mi.). The KATE-200 system has 15–30 m (49–98 ft.) resolution and covers an area 180 × 180 km (108 × 108 mi.). Photographs are acquired from an altitude of 200 km (120 mi.) with a 200 mm (8 in) lens in three channels: 500–600 nm (blue), 600–700 nm (green), and

700–850 nm (red-NIR). Stereo overlap from 10%–80% is available. The DD-5 system was originally high-resolution photography, but has been degraded to 2 m data by digitizing. These data are available on computer-compatible tapes and each scene varies in area between 104 and 200 km². The KVR-1000 camera, available since 1984, acquires B/W photographs with 2–3 m (7–10 ft.) resolution. It is flown on the Kosmos spacecraft at an approximate orbital altitude of 220 km (136 mi.). Image size is about 40 × 50 km (25 × 31 mi.), depending on altitude. There is one band in the range 510–760 nm. The TK-350 camera, available since 1982, produces panchromatic stereo images (510–760 nm) with 60%–80% overlap and 5–10 m (16–33 ft.) resolution. Images cover roughly 200 × 300 km (124–186 mi.), depending on orbital altitude (Jacobsen, 1997, online).

KOSMOS Series

Digital products are available from the Resurs instrument on the Russian Kosmos series of satellites flown between 1984 and 1986. The principle sensor is the MSU-SK, which contains five channels in the visible-NIR and SWIR: Band 1 from 500 to 600 nm, Band 2 from 600 to 700 nm, Band 3 from 700 to 800 nm, Band 4 from 800 to 1100 nm, and a thermal band (Band 5) from 10.4 to 12.6 µm. Resolution in the visible-NIR is 70 m (230 ft.), although this is routinely resampled to 160 m (525 ft.); in channel 5 it is 600 m (2000 ft.). Swath width is 600 km (372 mi.), and revisit time is 4 days (eoPortal e, Resurs-01, online).

ETALON Series

Etalon-1 and -2 satellites are Russian passive geodetic satellites launched in 1989 into a circular 19,120 km (11,855 mi.) orbit. They consist of retroreflectors that, in conjunction with ground-based lasers, determine surface locations and help better define Earth's geoid (NASA ccc, Etalon-1 and -2, online).

ALMAZ-1

The Russian Almaz radar satellite was launched during March 1992 into a 300 km (180 mi.) orbit. This mission carried an S band radar with 13–30 m (40–98 ft.) resolution and a 1- to 3-day repeat cycle. The swath width is variable from 200 to 350 km (120–210 mi.) on either side of the orbital path. Incidence angle varied from 30° to 60° off nadir (NASA ddd, Almaz 1, online).

Monitor-E

The Monitor-E Earth observation satellite was launched in 2005 into a 540 km (335 mi.) sun-synchronous orbit. The purpose of this mission is to assist geologic mapping and environmental monitoring. The satellite carries an 8 m (26 ft.) resolution PAN camera that has a 90 km (56 mi.) ground swath. Monitor-E also has an MSS with three channels (540–590 nm, 630–680 nm, and 790–900 nm; Figure 3.70). The MSS has 20–40 m (66–131 ft.) resolution and a 160 km (99 mi.) ground swath (eoPortal f, Monitor-E, online).

Resurs-DK1

Resurs-DK1, launched in 2006, was originally in an elliptical 355 × 573 km altitude orbit, but the orbit was modified to near-circular 567 × 573 km (352 × 355 mi.) in 2010. It has a swath up to 28.3 km (17.5 mi.) at nadir and 40 km (25 mi.) at 30° off nadir. Revisit

FIGURE 3.70
Russian Monitor-E true-color image of the Danube River cutting through Derdap National Park near Drobeta-Turnu Severin, Romania. (Courtesy of Russian Federal Space Agency Roscosmos ©2012; © Official Site of Research Center for Earth Operative Monitoring [NTS OMZ]. http://www.hazard.maks.net/blog/index.php?op=Default&postCategoryId=2&blogId=1.)

FIGURE 3.71
Russian RESURS-DK1 panchromatic image of Naples, Italy. (Courtesy of Russian Federal Space Agency Roscosmos ©2012; © Official Site of Research Center for Earth Operative Monitoring [NTS OMZ]. http://eng.ntsomz.ru/news/news_center/newdk_090712.)

time is 5–7 days off nadir. The purpose of the satellite is remote sensing of the earth's surface in near real time. Resurs carries a PAN Sensor (580–800 nm) with 0.9–1.0 m (3–3.3 ft.) ground resolution (Figure 3.71), and a three channel MSS (500–600 nm, 600–700 nm, and 700–800 nm) with 1.5–2.0 m (5–7 ft.) resolution (eoPortal g, Resurs-DK1, online).

Meteor-M-1

Meteor-M-1 (also referred to as Meteor-M N1) is a polar-orbiting meteorological and Earth-observing mission. NTs OMZ is responsible for the acquisition, recording, processing, archiving, cataloging, and dissemination of Meteor-M-1 data. The objective of the Meteor mission is to provide operational meteorological services, to acquire multispectral and radar imagery of the earth's surface, and to obtain measurements of the earth–atmosphere system.

Meteor-M-1 was launched in September 2009 into a sun-synchronous 832 km (516 mi.) near-circular orbit crossing the local equator at 12:00 hours. Earth-observing systems include the MSU-MR, OBRC, KMSS, and MTVZA-GY. The MSU-MR (Low-resolution Multispectral Scanner) has six bands: Band 1 visible (500–700 nm), Band 2 VNIR (700–1100 nm), Band 3 SWIR (1.6–1.8 µm), Band 4 MWIR (3.5–4.1 µm), Band 5 TIR (10.5–11.5 µm), and Band 6 TIR (11.5–12.5 µm). Swath width is 2800 km (1736 mi.) and spatial resolution is 1 km (0.6 mi.) at nadir.

OBRC (Onboard Radar Complex, also called "Severyanin") is a vertical polarized X band SAR operating at 9.623 GHz. Incidence angle is 25°–48°, defining a 450–600 km (279–372 mi.) swath. Spatial resolution is 400–500 m (1300–1650 ft.) in "moderate resolution" mode or 700–1000 m (2300–3280 ft.) in "low resolution" mode. The main objective is ice monitoring in polar regions.

The KMSS (Multispectral Imaging System) is designed for Earth-surface monitoring. The instrument comprises three pushbroom cameras in the visible-NIR range; two cameras (MSU-100) have a focal length of 100 mm; the third one (MSU-50), has a 50 mm focal length. The two MSU-100 cameras are tilted ±14° cross-track; together they cover a swath of 960 km (595 mi.). Spatial resolution is 60 m (200 ft.) for the MSU-100 and 120 m (400 ft.) for the MSU-50. The MSU-100 has three spectral bands: Band 1 blue (535–575 nm), Band 2 green (630–680 nm), and Band 3 red-NIR (760–900 nm). The MSU-50 also has three bands: Band 1 UV (370–450 nm), Band 2 UV-blue (450–510 nm), and Band 3 blue-green (580–690 nm).

The MTVZA-GY (Microwave Imaging/Sounding Radiometer) is a passive 29-channel MWR similar to NOAA's AMSU-A and -B radiometers. The operating frequencies are located in the atmospheric windows at 10.6, 18.7, 23.8, 31.5, 36.7, 42, 48, and 91.65 GHz, as well as at the oxygen absorption lines at 52–57 GHz and water vapor absorption at 183.31 GHz. Swath width is 1500 km (930 mi.) and spatial resolution is 16–198 km (10–123 mi.) horizontal and 1.5–7 km (0.9–4.2 mi.) vertical. (eoPortal h, Meteor-M-1, online).

Electro-L

Elektro-L No. 1 (also known as Geostationary Operational Meteorological Satellite 2 or GOMS 2), is a Russian meteorology satellite launched into a geostationary orbit over the Indian Ocean in 2011. It produces visible and IR images of the earth's hemisphere at half hour intervals. Visible light imagery has a 1 km (0.6 mi.) ground resolution, whereas the IR frequencies have 4 km (2.4 mi.) resolution. The purpose is to gather information for weather forecasting (eoPortal i, Electro-L, online).

Singapore

Defense Science Organization (DSO) is the defense research agency of Singapore. It was established in 1972; with expansion of its R&D scope, it was renamed the DSO in 1977. In 1997, the DSO was incorporated as a not-for-profit company and became known as DSO National Laboratories.

X-Sat

Developed by the Nanyang Technological University in conjunction with DSO, X-Sat was launched in 2011 for the purpose of Earth observation, environmental monitoring, and disaster monitoring. It carries the IRIS imaging system (named after the Greek goddess of the rainbow), a three-channel pushbroom scanner. Channels include green (520–600 nm), red (630–690 nm), and NIR (760–890 nm), each with 10 m (33 ft.) spatial resolution from a 685 km (425 mi.) orbit. Swath width is 50 km (31 mi.) (eoPortal j, XSat, online).

Surrey Satellite Technology Limited

Surrey Satellite Technology Limited (SSTL) is an independent British company within the EADS Astrium group. SSTL builds and launches satellite payloads under 1000 kg. Although SSTL supplies both the satellite and payload, they also integrate customer supplied payloads with SSTL-built platforms. In-house capabilities include producing satellites for Earth observation and imaging, research, military/defense, and testing instruments in space. SSTL is located in Surrey Research Park, Guildford, United Kingdom.

Disaster Monitoring Constellation

The Disaster Monitoring Constellation (DMC) is an international program led by SSTL and operated by DMC International Imaging (DMCii), a wholly-owned subsidiary of SSTL. The purpose is to construct a network of affordable low-Earth-orbit microsatellites that can provide daily global imaging capability at medium resolution (30–40 m), in 3–4 spectral bands, for rapid-response disaster monitoring and mitigation (Figure 3.72). This effort has brought together organizations from Algeria, China, Nigeria, Thailand, Turkey,

FIGURE 3.72
Pan-sharpened 2.5 m resolution NigeriaSat-2 MSS image of the Salt Lake City, Utah, airport. NigeriaSat-2 is one of the Disaster Monitoring Constellation that include Alsat, Bilsat, NigeriaSat-X, U.K. DMCSat-1, Beijing-1, Deimos, and Formosat. (Image courtesy of NASRDA. http://eomag.eu/articles/1662/dmcii-news.)

the United Kingdom, and Spain. Beginning in 2003 with Alsat -1, the DMC Consortium is the first microsatellite constellation bringing Earth observation capabilities to individual satellite owners and to international humanitarian aid efforts (Surrey Satellite Technology Ltd a, Disaster Monitoring Constellation, online).

Alsat-1 and Alsat-2

Alsat-1, a collaboration between the Algerian Space Agency (ASAL) and SSTL, is the first Algerian satellite and part of the DMC. It was launched in November 2002 carrying the "Earth Imaging Camera" that provided 32 m (105 ft.) resolution in three spectral bands (NIR, red, green) over a ground swath of 600 km (372 mi.). The satellite completed its mission in August 2010 (Surrey Satellite Technology Ltd b, Alsat-1, online).

Alsat-2, launched in 2010 into a 670 km (415 mi.) sun-synchronous orbit, allows Algeria to obtain images for use in cartography, management of agriculture, forestry, water, mineral and oil resources, crop protection, land use planning, and management of natural disasters (EADS Astrium a, Alsat-2 Program, online). Alsat-2 carries two sensors as part of NAOMI (the New AstroSat Optical Modular Instrument): a PAN scanner with 2.5 m (8 ft.) resolution and a pushbroom multispectral scanner (known as MS) with 10 m (33 ft.) ground resolution in each of four color bands (blue, green red, NIR).

Bilsat-1

Launched in 2003 by Turkey's TUBITAK-ODTU-BILTEN (Space Technology Research Institute) as part of the DMC, the BILSAT-1 Earth observation sensor is ÇOBAN (the name is an abbreviation for "ÇOk-BANtlı Kamera," or Multi-Band Camera). COBAN has five bands: PAN with 4 m (13 ft.) resolution and four multispectral bands (blue, green, red, IR) with 26 m (85 ft.) resolution over a 600 km (372 mi.) swath. Satellite operations ended in August 2006 (NASA uu, Bilsat 1, online).

NigeriaSat-1 and NigeriaSat-2

Launched in 2003, NigeriaSat-1 is a collaboration between the National Space Research and Development Agency (NASRDA) of Nigeria and SSTL in the United Kingdom. The satellite, part of the DMC, is in a 686 km (425 mi.) sun-synchronous orbit and carries a three-band multispectral scanner (SLIM) with 32 m (105 ft.) ground resolution and a 640 km (397 mi.) wide-ground swath. Band 1 covers the NIR (770–900 nm); Band 2 (red) is from 630 to 690 nm; Band 3 (green) covers the spectrum from 520 to 600 nm (Ogunbadewa, 2008).

NigeriaSat-2 was launched in August, 2011 into a 700 km (434 mi.) sun-synchronous orbit. NigeriaSat-2 carries a Very High Resolution Imager (VHRI) with a 2.5 m (8 ft.) ground resolution PAN instrument and a 5 m (16.4 ft.) resolution MSS over a 20 km (12 mi.) swath. A second instrument, the Medium Resolution Imager (MRI), is a four-band MSS with 32 m resolution, and a 300 km (186 mi.) swath that provides data continuity with Nigeria's previous NigeriaSat-1. Applications for NigeriaSat include mapping and planning of population surveys, mapping land use, mapping rural and urban growth, and to give advance warnings of natural disasters like floods, earthquakes, and volcanic eruptions. It is also being used to manage man-made disasters such as oil pollution, desertification, erosion, forest fires, and deforestation. In agriculture, it is used for management of sustainable grazing, logging, planning reforestation programs, crop inventory, and yield forecasts.

NigeriaSat-X

NigeriaSat-X was launched in August, 2011, at the same time as NigeriaSat-2, into a 700 km sun-synchronous orbit. The Surrey Linear Imager Multispectral 6 (SLIM6) instrument provides three-band multispectral (red, green, NIR) imagery with 22 m (72 ft.) ground resolution across a 600 km (373 mi.) swath (NASRDA, NigeriaSat-X, online).

U.K. DMCSat-1

The U.K. DMCSat-1, launched in 2003 as part of the DMC, carried a visible and IR imaging radiometer, the Extended Swath Imaging System (ESIS) into a 686 km (425 mi.) sun-synchronous orbit. ESIS has four channels (520–620 nm, 630–690 nm, 760–900 nm, and 750–1300 nm) with 32 m (105 ft.) ground resolution over a 600 km (372 mi.) swath. The objective is to provide medium-resolution multispectral imagery for disaster monitoring over large areas (BEOP a, UK-DMCSat-1, online).

Beijing-1

Beijing-1 was launched by the People's Republic of China in 2005 for the purpose of disaster monitoring and mitigation. The satellite is in a 686 km (425 mi.) sun-synchronous orbit. It carries two instruments: the High-Resolution Optical Imager (CMT) and the Imaging Multispectral Radiometer (ESIS). The CMT is a single-channel instrument that acquires data in the range 400–750 nm with 4 m (13 ft.) resolution over a 24 km (14.4 mi.) swath. ESIS records reflected light in three visible bands (520–620 nm, 630–690 nm, and 760–900 nm) and the NIR (750–1300 nm) with 32 m (105 ft.) ground resolution across a 600 km swath (BEOP b, Beijing-1, online).

Deimos-1

Deimos-1 is a Spanish Earth imaging satellite operated by Deimos Imaging in conjunction with ESA. Part of the DMC, Deimos-1 was launched in 2009 into a 686 km (426 mi.) sun-synchronous orbit. The satellite carries a multispectral imager covering the green (520–600 nm), red (630–690 nm), and NIR (770–900 nm) with a resolution of 22 m (72 ft.) across a 600 km (373 mi.) ground swath (EADS Astrium b, DEIMOS-1, online).

THEOS

THEOS (Thailand Earth Observation System) is an Earth observation mission of Thailand, developed and operated by EADS Astrium, and launched in 2008. THEOS flies in an 822 km (510 mi.) near-circular sun-synchronous orbit with a 26-day repeat cycle. The primary objective of THEOS is to provide Thailand with a state-of-the-art Earth observation satellite to promote the country's capabilities and infrastructure for the development of future space missions. The main instruments are the PAN camera, which provides B/W imagery with 2 m (7 ft.) ground resolution over a 22 km (14 mi.) swath, and a multispectral camera (known as the MSC), with 15 m (49 ft.) resolution across a 90 km (56 mi.) swath. These images are used to generate georeferenced image products for cartography, land use, agricultural monitoring, forestry management, coastal zone monitoring, and flood risk management (EADS Astrium c, THEOS, online).

Future Satellite Systems

There are several systems being prepared for future missions. A partial list of proposed missions, their purpose, and primary Earth observation instruments is presented here in the order of planned launch. For an up-to-date list of NASA missions, current and future, see http://science.nasa.gov/missions/. Other excellent sites for current and future missions can be found at eoPortal, https://directory.eoportal.org/web/eoportal/satellite-missions and the ITC database maintained by the Faculty of Geo-Information Science and Earth Observation of the University of Twente, http://www.itc.nl/research/products/sensordb/searchsat.aspx.

Swarm Series

The Swarm mission, to be launched by ESA in spring 2013 as part of its Living Planet program, will provide a complete and detailed global geomagnetic field showing temporal variations over hours and days. Each of the three Swarm satellites will carry a Vector Field (fluxgate) Magnetometer (VFM) for measuring the strength and direction of the earth's magnetic field, and an Absolute Scalar Magnetometer (ASM) for calibration of the VFM. Two of the Swarm satellites will be launched into 450 km (280 mi.) orbits and one satellite into a 530 km (330 mi.) polar orbit (ESA r, SWARM, online).

LDCM (Landsat-8)

The Landsat Data Continuity Mission (LDCM, also known as "Landsat 8") will continue acquiring Earth observation data consistent with previous Landsats. Data will be available from the USGS, who will be responsible for mission operations as well as data collection, archiving, processing, and distribution. LDCM will carry two instruments. The Operational Land Imager (OLI) is a push-broom sensor with 12-bit quantization. OLI will collect data in the visible, NIR, and SWIR spectral regions, and will also have a PAN band. The Thermal Infrared Sensor (TIRS) will provide thermal imaging and support climate change and water management applications. TIRS is being built by NASA GSFC. The 100 m (328 ft.) ground resolution TIRS data will be registered to OLI data to create radiometrically, geometrically, and terrain-corrected 12-bit imagery. Scheduled for a mid-2013 launch, the satellite will be installed in a 705 km (440 mi.) circular orbit. OLI instrument parameters are provided in Table 3.14.

TABLE 3.14

OLI Instrument Parameters

Band	Bandwidth (nm)	Resolution (m)	Swath (km)
1, Coastal/aerosol	433–453	30	185
2, Blue	450–515	30	185
3, Green	525–600	30	185
4, Red	630–680	30	185
5, NIR	845–885	30	185
6, SWIR 1	1560–1660	30	185
7, SWIR 2	2100–2300	30	185
8, PAN	500–680	15	185
9, Cirrus	1360–1390	30	185

Source: Geovar, LDCM; Gunter's Space Page c, Landsat 8, online.

ATTREX

NASA's Airborne Tropical Tropopause Experiment (ATTREX) mission is scheduled to begin in late 2013. The instruments will provide measurements that trace the movement of reactive halogen-containing compounds and other chemical species, the size and shape of cirrus cloud particles, water vapor, and winds in three dimensions through the tropical tropopause layer (TTL). TTL is the part of the tropical atmosphere between the top of the cumulus outflow layer (~12 km) and the thermal tropopause (~16 km) that is crucial to understanding dehydration of air entering the stratosphere. In particular, bromine-containing gases will be measured to improve our understanding of stratospheric ozone. ATTREX will consist of four NASA Global Hawk Uninhabited Aerial System campaigns deployed from NASA's Dryden Flight Research Center (DFRC), Guam, Hawaii, and Darwin, Australia. Flights will occur in Boreal summer, winter, fall, and northern summer. The project is being managed by NASA Ames Research Center and Langley Research Center (NASA eee, ATTREX, online).

GeoEye-2

DigitalGlobe's GeoEye-2 is planned for launch in 2013. This follow-on to GeoEye-1 will have a 50 cm resolution PAN band and eight multispectral bands with 1.8 m resolution and a 16.4 km swath width. Flying at 770 km, this satellite will have an average 1.1-day revisit time and, of particular interest to geologists, off-nadir stereo capability (Satellite Imaging Corporation d, GeoEye-2, online).

CBERS Series

The CBERS satellite series, operated by NASA, is a joint project between CAST (China) and INPE (Brazil). These satellites will be launched into in a 778 km (482 mi.) sun-synchronous orbit with a 16-day repeat cycle. CBERS-3 is scheduled for a February 2013 launch, and CBERS-4 is scheduled for later in 2013. These satellites will carry a PAN (a Panchromatic Camera) with 5 m (16.4 ft.) resolution and a MUX (Multispectral Camera) with blue, green, red, and NIR bands having 10 m (33 ft.) resolution over a 60 km (37 mi.) swath. In addition, there will be an improved Wide Field Imager (WFI) having four bands (blue, green, red, NIR) with 70 m (230 ft.) spatial resolution at nadir over an 860 km (533 mi.) swath. The IRS (Infrared System) will have four bands (PAN two SWIR, one TIR) with 40 m (80 m TIR) spatial resolution (Imaging Notes, CBERS, online).

Sentinel Series

ESA is planning to launch and operate two satellites in this group, Sentinel-1A scheduled for 2013, and Sentinel-1B scheduled for 2015. The objective is to monitor forests, soil, and agriculture as well as the marine environment and sea ice. Both satellites will carry the C-SAR sensor, a C band SAR, in a 693 km (430 mi.) sun-synchronous orbit with a 12-day repeat cycle. "Strip Map" mode will have an 80 km (50 mi.) swath and 5 m (16.4 ft.) spatial resolution; "Interferometric" mode will have a 250 km (155 mi.) swath with 5 × 20 m (16.4 × 66 ft.) resolution; "Extra-Wide Swath" mode will have a 400 km (248 mi.) swath and 25 × 100 m (82 × 328 ft.) resolution; and "Wave" mode will have a 20 km (12.4 mi.) footprint with 5 × 20 m spatial resolution.

ESA's Sentinel-2 program also has plans for two satellites, Sentinel-2A, planned for 2014, and Sentinel-2B, to be launched in 2014 or 2015. The objective is to cover the earth's surface every 15–30 days using a Multispectral Imager (MSI) from a 786 km (476 mi.) sun-synchronous orbit with a 5-day repeat cycle. The MSI will have a 290 km (180 mi.) swath with 10 m, 20 m, and 60 m (3.3, 65, and 197 ft.) resolution. This land observation satellite will monitor land use, vegetation, soil and water cover, inland waterways, and coastal areas.

ESA plans to launch Sentinel-3A in 2013, Sentinel-3B in 2015, and Sentinel-3C before 2020. The satellites will be in an 814 km (505 mi.) sun-synchronous orbit with a 27-day repeat cycle. Mission objectives are to measure sea-surface topography, ocean and land temperatures, ocean and land surface color, and monitor sea ice, pollution, and land use change. Sentinel-3 will carry four Earth-observation instruments. The Ocean and Land Color Instrument (OLCI) is a medium-resolution imaging spectrometer with a 1270 km (787 mi.) swath. The Sea and Land Surface Temperature Radiometer (SLSTR) is a nine-channel visible and IR radiometer with dual-viewing directions for atmospheric corrections. SLSTR has a cross-nadir swath of 1675 km (1040 mi.) and a forward viewing swath of 750 km (465 mi.) with 0.5 km (1640 ft.) surface resolution for visible and NIR and 1.0 km (0.6 mi.) for the thermal channels. Visible bands will be 20 nm wide and centered at 555 nm, 659 nm, and 865 nm. NIR bands will be centered at 1375 nm (15 nm bandwidth), 1610 nm (60 nm bandwidth), 2250 nm (50 nm bandwidth), and 3740 nm (380 nm bandwidth). A thermal band centered at 10.85 μm will be 900 nm wide, and another centered at 12.0 μm will be 1000 nm wide (World Meteorological Association a, SLSTR, online). The SAR Altimeter (SRAL) will be a dual-frequency radar altimeter (5.3 and 13.58 GHz) with 20 km (12 mi.) IFOV and possible SAR capability (300 m ground resolution). Global coverage will be every 10–30 days, depending on spacing (World Meteorological Association b, SRAL, online). The MWR will measure water vapor, cloud water content, and thermal radiation emitted by the earth to correct the wet tropospheric path delay of the altimeter data. The radiometer will record brightness temperature at 23.8 GHz and 36.5 GHz (Bergadà et al., 2010).

Atmospheric Dynamics Mission (ADM-Aeolus)

The ESA ADM-Aeolus mission, part of the Living Planet Program, is scheduled for launch in mid-2014 and will provide global wind profiles to improve weather forecasts and improve our understanding of atmospheric dynamics. The primary instrument carried will be ALADIN, an Atmospheric Laser Doppler Instrument using the active Doppler Wind Lidar (DWL) method to provide information on wind strength, direction, cloud top heights, and the distribution of clouds and aerosols as inputs for climate modeling. The satellite will fly in a 400 km sun-synchronous orbit and have a 285 km cross-track by 90 km along-track footprint. A magnetometer will also fly on ADM-Aeolus (ESA t, ADM-Aeolus, online).

SPOT-7

Scheduled launch for SPOT-7 is in 2014. SPOT-7 will carry the same instrument package as SPOT-6 (EADS Astrium d, SPOT 6 and SPOT 7, online).

Global Precipitation Measurement

Global Precipitation Measurement (GPM) is an international satellite mission to be launched in 2014. It will provide observations of rain and snow worldwide every 3 hours. NASA and JAXA will launch a "Core" satellite carrying instruments that will provide data used to unify precipitation measurements made by an international network of partner satellites and will quantify when, where, and how much it rains or snows around the world. The GPM mission will help advance our understanding of Earth's water and energy cycles and improve the forecasting of extreme events (NASA fff, GPM, online).

The Core observatory will carry a conically scanning radiometer and a cross-track scanning radar:

- GPM Microwave Imager (GMI)
- Dual-frequency Precipitation Radar (DPR)

The GMI measurements and images obtained from the DPR will combine to provide a reference against which to calibrate other MWRs in the GPM constellation when overlapping measurements of the same area are made. The DPR will be able to make detailed 3D measurements of cloud structure, rainfall, and rain rates (NASA ggg, GPM, online).

Worldview-3

DigitalGlobe has plans to launch Worldview-3 in 2014. It is expected to have 1.2 m (4.7 ft.) ground resolution for its eight-band multispectral sensor and 0.31 m (1 ft.) resolution for the PAN instrument. An eight-band SWIR instrument will cover the range 1.195–2.362 µm with a ground resolution of 3.7 m (12 ft.) (SpaceNews, Worldview-3, online).

National Polar-Orbiting Operational Environmental Satellite System

The National Polar-orbiting Operational Environmental Satellite System (NPOESS), to be launched by NASA in early 2014, will monitor global environmental conditions and collect data related to weather, atmosphere, oceans, and land environments. Polar-orbiting satellites such as NPOESS are able to monitor the entire planet and provide data for long-range weather and climate forecasts. The program is managed by the U.S. Department of Commerce, Department of Defense, and NASA (NASA iii, NPOESS, online).

SAOCOM Series

Argentina's space agency CONAE plans to launch the SAOCOM 1A and 1B satellites in 2014 and 2015, respectively. These satellites will carry L band polarimetric SARs with a ground resolution between 7 and 100 m (23–328 ft.) and a swath width between 50 and 400 km (31–248 mi.). The satellites will be launched into sun-synchronous orbits where they will be used for disaster management and natural resources monitoring (Gunter's Space Page d, SAOCOM; Astronomy, SAOCOM 1A and 1B, online).

OceanSat-3

Indian Remote Sensing's OceanSat-3 will carry a Thermal Infrared Sensor, 12 channel Ocean Color Monitor (OCM), Scatterometer, and Passive Microwave Radiometer. The IR Sensor and OCM will be used in the analysis of potential fishing zones. The satellite is mainly for Ocean biology and sea state applications. It is slated to the launched in the 2014 time frame (ESA w, OCEANSAT-3, online).

International Space Station (ISS)

SAGE III-ISS (Stratospheric Aerosol and Gas Experiment III – International Space Station), is a NASA instrument expected to be launched into orbit during 2014 to study Earth's ozone layer. More than 25 years ago scientists realized Earth's ozone was thinning. The SAGE instruments were designed to make accurate measurements of the amount of ozone in Earth's atmosphere.

SAGE III will be mounted to the ISS. SAGE III – ISS will provide global, long-term measurements of the vertical distribution of aerosols and ozone from the upper troposphere through the stratosphere. SAGE III also provides unique measurements of temperature in the stratosphere and mesosphere and profiles of trace gases such as water vapor and nitrogen dioxide.

The SAGE III sensor contains a UV/visible spectrometer. The spectrometer uses a CCD linear array to provide continuous spectral coverage between 290 and 1030 nm. Additional aerosol information is provided by a discrete photodiode recording at 1550 nm. This configuration enables SAGE III to make multiple measurements of target gas absorption features and multi-wavelength measurements of broadband extinction by aerosols. The SAGE III - ISS instruments are being developed and managed by NASA's Langley Research Center (NASA lll, SAGE III-ISS; NASA mmm, SAGE III-ISS, online).

Jason-3

Jason-3 is planned for launch in 2014. A joint project between NASA, CNES, and EUMETSAT, Jason-3 is the successor to the Jason-1 and -2 missions (ESA x, Jason-3, online). The purpose will be to monitor global ocean circulation, sea level changes, and currents, to study the ties between the ocean and the atmosphere, improve global climate forecasts, and monitor El Niño and ocean eddies. It will likely have the following instruments:

- A nadir pointing RA using C band and Ku band for measuring height above sea surface
- An MWR that measures water vapor

Orbiting Carbon Observatory-2

The Orbiting Carbon Observatory-2 (OCO-2) is meant to replace OCO-1 which was launched in 2009 but failed to reach orbit. Scheduled for launch in 2015, its mission is to make time-dependent global measurements of atmospheric CO_2. The NASA/JPL satellite will fly in a near-polar orbit with a 16-day repeat cycle. OCO will carry a single instrument that will provide the most precise measurements of atmospheric CO_2 yet made from space. The instrument consists of three parallel, high-resolution spectrometers connected to a common telescope. The spectrometers will make simultaneous measurements of the CO_2 and molecular oxygen absorption of sunlight reflected off the same location on Earth's surface when viewed in the NIR. Since different gases absorb different wavelengths, the pattern of absorption lines will provide a spectral "fingerprint" for that molecule. The amount of light absorbed at each spectral line increases with the number of molecules along the optical path. OCO's spectrometers will measure the fraction of the light absorbed in each of these lines. This data can be used to estimate the number of molecules along the path between the top of the atmosphere and the surface (NASA, hhh, online).

EarthCARE

EarthCARE (Earth Clouds, Aerosols and Radiation Explorer) is the sixth Earth Explorer mission to be launched by ESA in conjunction with the Japanese Space Agency (JAXA) as part of its "Living Planet" program (ESA s, EarthCARE, online). EarthCARE will focus on clouds and aerosols and their influence on atmospheric radiation. EarthCARE will generate vertical profiles of natural and man-made aerosols, determine the distribution of cloud water and ice, and investigate the relationship between clouds and precipitation. Profiles of atmospheric heating and cooling by clouds will be determined by examining a combination of measured aerosols and "cloud elements." EarthCARE will carry four instruments: a Backscatter Lidar (ATLID) operating at 355 nm to produce vertical profiles of atmospheric aerosols and clouds; a 94 GHz Cloud Profiling Radar (CPR) that will provide Doppler measurements of winds in clouds and vertical cloud profiles; an MSI with seven channels covering the visible, NIR, SWIR and TIR with 500 m (1640 ft.) resolution and with a 150 km (93 mi.) ground swath; and a Broadband Radiometer (BBR) with two channels viewing nadir, fore, and aft to provide information on top-of-atmosphere radiance and radiant flux using one short-wave and one long-wave channel. The EarthCARE satellite is scheduled to be launched late in 2015. EarthCARE will be in a sun-synchronous orbit at an altitude of 393 km (244 mi.) with a 25-day repeat cycle.

GOES NEXT Series

The fourth generation GOES satellites, known as GOES-NEXT or GOES-R and GOES-S, are planned for launch starting in 2015. The joint NASA/NOAA program has two satellites on order, with options for two more (GOES-T and –U). The primary instruments will be the Imager and the Sounder. The Imager is a multichannel instrument that senses IR energy and visible reflected light from the Earth's surface and atmosphere. The Sounder provides data for vertical atmospheric temperature and moisture profiles, surface and cloud top temperature, and ozone distribution (NASA jjj, GOES-R, online).

HyspIRI

The HyspIRI mission includes two instruments mounted on a satellite in Low Earth Orbit. The Hyperspectral Infrared Imager (HyspIRI) measures visible to short-wave infrared (380 nm–2.5 µm) in 10 nm contiguous bands; a TIR multispectral imager measures from 3 to 12 µm in the mid- and thermal IR. Both instruments have spatial resolution of 60 m (197 ft.) at nadir. HyspIRI will have a revisit of 19 days and the TIR will have a revisit of 5 days. The HyspIRI mission (NASA-JPL, launch 2013-16), will study the world's ecosystems and provide critical information on natural disasters such as volcanoes, wildfires and drought. HyspIRI will be able to identify vegetation types and determine whether the vegetation is healthy or stressed (NASA kkk, HyspIRI; National Academy of Sciences, HYSPIRI, online).

ResourceSat-3

A follow on to the Indian Remote Sensing ResourceSat-2, ResourceSat-3 will carry a higher resolution LISS-3-WS (Wide Swath) sensor having similar swath width and revisit capability as the AWiFS. The satellite will also carry the Atmospheric Correction Sensor (ACS) for quantitative interpretation and geophysical parameter retrieval. It is slated to be launched in 2015 (ESA u, RESOURCESAT-3, online).

CartoSat-3

Cartosat-3 is a continuation of the Indian Remote Sensing Cartosat series. This satellite will have a 30 cm (1 ft.) resolution PAN instrument and a < 1 m (3.3 ft.) resolution multi-spectral (500–750 nm) instrument with a 16 km (10 mi.) swath suitable for infrastructure mapping and analysis. CartoSat-3 will provide disaster monitoring and damage assessment imagery. It is scheduled to be launched during 2015 (ESA v, CartoSat-3, online).

Soil Moisture Active-Passive

The NASA SMAP mission is planned for launch in 2015. SMAP will use a combined radiometer and high-resolution radar to measure surface soil moisture and freeze-thaw state. Direct measurements of soil moisture and freeze/thaw state are needed to improve our understanding of regional water cycles, ecosystem productivity, and processes that link the water, energy, and carbon cycles. Soil moisture information at high resolution enables improvements in flood and drought forecasts and predictions of agricultural productivity. Combined radar-radiometer and L band mapping of global soil moisture will allow SMAP to exceed the NPOESS soil moisture minimum performance requirements for sensing depth and spatial resolution (NASA nnn, SMAP, online).

High-resolution radar data will be created within the real-aperture footprint by synthetic aperture processing in range and azimuth. The spatial resolution of the data product is better than 3 km (1.8 mi.) over the outer 70% of the 1000-km (620 mi.) swath. The like-polarized (HH and VV) data have uncertainty from all sources (excluding rain) of 1.0 dB or less defined at 3 km spatial resolution. The cross-polarized (HV) data have uncertainty from all sources (excluding rain) of 1.5 dB or less defined at 3 km spatial resolution (NASA ooo, SMAP, online).

Geosat Follow-On-2

The successor to the U.S. Navy's Geosat Follow-On (GFO) mission is "GFO-2." GFO-2 is planned for launch in 2016, and will feature a "dual-band radar altimeter," instead of the single-band altimeter carried by the previous spacecraft. The purpose is to measure ocean height and thermal properties. Until it is operational the Navy has arranged with NASA and CNES to use Jason-1 and Jason-2 data (Gunter's Space Page e, GFO-2, online).

Ice, Cloud, and Land Elevation Satellite-2

ICESat-2 (Ice, Cloud, and Land Elevation Satellite) is the NASA EOS mission for measuring ice sheets, cloud and aerosol heights, land topography, and vegetation characteristics. Planned for 2016, the ICESat-2 mission will provide multi-year elevation data needed to determine ice sheet mass balance as well as cloud property information for stratospheric clouds that are common over polar areas. It will also provide topography and vegetation data around the globe, in addition to the polar-specific coverage over the Greenland and Antarctic ice sheets. The ICESat-2 mission is a follow-on to ICESat-1 and will continue the assessment of polar ice changes. ICESat-2 is also expected to measure vegetation canopy heights, allowing estimates of biomass and carbon in vegetation, and allow measurements of solid Earth properties (NASA ppp, ICESat-2, online).

Joint Polar Satellite System

The JPSS is a polar-orbiting operational environmental satellite system, requested by NOAA through NASA, and planned for launch in 2016 (NOAA d, JPSS, online). JPSS will provide continuity of critical observations for accurate weather forecasting, severe storm outlooks, global measurements of atmospheric and oceanic conditions such as sea-surface temperatures and ozone concentrations. JPSS's primary user, NOAA's National Weather Service, will use the JPSS data in models for medium- and long-term weather forecasting.

JPSS will provide operational continuity of satellite-based observations and products for NOAA Polar-orbiting Operational Environmental Satellites (POES) and the Suomi NPP mission. The first JPSS spacecraft, JPSS-1, will take advantage of technologies developed through the Suomi NPP satellite that was launched in October 2011. The JPSS payload will include

- The Visible Infrared Imaging Radiometer Suite (VIIRS), which takes global visible and IR observations of land, ocean, and atmospheric parameters at high temporal resolution.
- The Cross-track Infrared Sounder (CrIS) will produce high-resolution, 3D temperature, pressure, and moisture profiles. These profiles will be used to enhance weather forecasting models, and will assist both short- and long-term weather forecasting. Over longer timescales, they will help improve understanding of climate phenomena such as El Niño and La Niña.
- The Advanced Technology Microwave Sounder (ATMS) is a cross-track scanner with 22 channels, provides sounding observations needed to retrieve profiles of atmospheric temperature and moisture for civilian operational weather forecasting as well as continuity of these measurements for climate monitoring purposes.
- Ozone Mapping and Profiler Suite (OMPS) is an advanced suite of three hyperspectral instruments. The improved vertical resolution of OMPS data products will allow for better testing and monitoring of the complex chemistry involved in ozone destruction near the troposphere. OMPS products, when combined with cloud predictions, will help produce better UV index forecasts.
- The Clouds and Earth's Radiant Energy System (CERES) will sense both solar-reflected and Earth-emitted radiation from the top of the atmosphere to the Earth's surface. Cloud properties will be determined using simultaneous measurements by other JPSS instruments such as the VIIRS and will lead to a better understanding of the role of clouds and the energy cycle in global climate change.

RADARSAT-3, -4, and -5

The CSA announced on January 9, 2013, that they will build three new RADARSAT satellites to be launched starting in 2018. The project will build on technologies developed during the RADARSAT-1 and -2 programs.

ASCENDS

The ASCENDS mission is a NASA project scheduled for 2018–2020. Instruments will fly in a low-Earth sun-synchronous orbit. The purpose is to detect CO_2 concentrations during the day and night and at all latitudes and during all seasons, and to determine whether they are natural or man made. The mission is still working on which of three laser absorption spectrometers will fly, but the choices include (1) a LaRC/ITT 1.6 µm laser at two wavelengths to infer CO_2 columns, (2) a JPL/Lockheed 2.0 µm laser using two wavelengths to detect CO_2, and (3) a Goddard SFC pulsed 1.6 µm laser at six wavelengths to measure CO_2. There will also be lasers tuned to the O_2A and/or O_2B bands. Spatial resolution will be from 1 km to 500 km (0.6–310 mi.), with errors less than 5 ppm (Jucks, 2009; National Research Council, 2008).

Deformation, Ecosystem Structure, and Dynamics of Ice

Deformation, Ecosystem Structure, and Dynamics of Ice (DESDynI), a NASA program to map surface deformation linked to earthquakes, volcanic eruptions, and landslides, is planned for a 2021 launch. Observations of surface deformation are used to forecast the likelihood of earthquakes as well as predicting both the place and time that volcanic eruptions and landslides might occur. Monitoring surface deformation is also important for improving the safety and efficiency of extraction of hydrocarbons, for managing ground water resources, and providing information for managing CO_2 sequestration.

DESDynI mission objectives include

- Determine the likelihood of earthquakes, volcanic eruptions, and landslides.
- Predict the response of ice sheets to climate change and the subsequent impact on sea level.
- Characterize the effects of changing climate and land use on habitats and the carbon cycle.
- Monitor the migration of fluids associated with hydrocarbon and groundwater production.

This mission combines two sensors that provide observations important for surface deformation, ecosystems (terrestrial biomass structure), and climate (ice dynamics):

1. An L band Interferometric SAR (InSAR) system with multiple polarization
2. A multiple beam Lidar operating in the IR (~1064 nm) with ~25 m (82 ft.) spatial resolution and 1 m (3.3 ft.) vertical accuracy

The mission using InSAR requires a satellite in a 700–800 km (434–496 mi.) sun-synchronous orbit to maximize available power from the solar arrays. An 8-day revisit frequency balances temporal decorrelation with required coverage. Onboard GPS achieves centimeter-level orbit resolution. The mission should have a 5-year life (NASA qqq, DESDYNL; NASA rrr, DESDYNI, online).

References

Andreoli, G., B. Bulgarelli, B. Hosgood, D. Tarchi. 2007. *Hyperspectral analysis of oil and oil-impacted soils for remote sensing purposes*. Institute for the Protection and Security of the Citizen. European Commission Joint Research Centre. Luxembourg: European Communities: 36 p.

ASTRIUM, *SSOT*. http://www.astrium.eads.net/en/press_centre/astrium-successfully-completes-in-orbit-delivery-of-the-ssot-satellite-system.html (accessed September 16, 2012).

Astronomy, *SAOCOM 1A and 1B*. http://astronomy.activeboard.com/t27071053/saocom-1a-and-1b-satellites/ (accessed September 22, 2012).

Battelle Labs. 1983. *Geophysical Survey Capabilities*: Richland, WA: Pacific Northwest Labs: 17 p.

BEOP a, *UK-DMCSat-1*. http://eo.belspo.be/directory/SatelliteDetail.aspx?satID=21 (accessed September 17, 2012).

BEOP b, *Beijing-1*. http://eo.belspo.be/directory/SatelliteDetail.aspx?satID=22 (accessed September 17, 2012).

Bergadà, M., P. Brotons, Y. Camacho, L. Díez, A. Gamonal, J.L. García, R. González et al. 2010. Design and development of the Sentinel-3 microwave radiometer. In R. Meynart, S.P. Neeck, H. Shimoda (ed.), *Sensors, Systems, and Next-Generation Satellites XIV. Proceedings of the SPIE*. Bellingham, WA: International Society for Optics and Photonics. 7826: 78260M-78260M-13.

Borengasser, M. 2003. *Hyperspectral Remote Sensing*. AccessScience, McGraw-Hill Companies. http://www.accessscience.com (accessed September 22, 2012).

Borstad ASL, *Compact airborne spectrographic Imager (CASI)*. http://www.borstad.com/casi.html (accessed September 15, 2012).

Canadian Space Agency a. http://www.asc-csa.gc.ca/eng/default.asp (accessed September 16, 2012).

Canadian Space Agency b, *RADARSAT-2*. http://www.asc-csa.gc.ca/eng/satellites/radarsat2/ (accessed September 16, 2012).

Canadian Space Agency c, *SCISAT-1*. http://www.asc-csa.gc.ca/eng/educators/resources/scisat/grade6-factsheets-facts.asp (accessed September 16, 2012).

Canadian Space Agency d, *Odin OSIRIS*. http://www.asc-csa.gc.ca/eng/satellites/osiris.asp (accessed September 16, 2012).

Carter, W.E., R.L. Shrestha, K.C. Slatton. 2007. Geodetic laser scanning. *Physics Today*. 60(12): 41–47.

Chavez, P.S. 1986. Processing techniques for digital sonar images from GLORIA. *Photogram. Engr. and Rem. Sens.* 52: 1133–1145.

China Defense Blog, *YaoGan*. http://china-defense.blogspot.ca/2011/11/number-of-yaogan-military.html (accessed September 16, 2012).

Chinese Academy of Science, *ZY1-02C*. http://english.ceode.cas.cn/ns/icn/201201/t20120104_80566.html (accessed September 16, 2012).

Clausner, J.E., J. Pope. 1988. *Application of side-scan sonar for inspection of coastal structures*. Houston: Offshore Technology Conference: 329–336.

CNES a, *DEMETER*. http://smsc.cnes.fr/DEMETER/ (accessed September 18, 2012).

CNES b, *PARASOL*. http://smsc.cnes.fr/PARASOL/ (accessed September 18, 2012).

CNES c, *PARASOL*. http://smsc.cnes.fr/PARASOL/dossier_presse_parasol.pdf (accessed September 18, 2012).

CNES d, *PLÉIADES*. http://smsc.cnes.fr/PLEIADES/ (accessed September 18, 2012).

Colorado State University, *CLOUDSAT*. http://cloudsat.atmos.colostate.edu/instrument (accessed September 16, 2012).

Colwell, R.N. 1983. *Manual of Remote Sensing, 2nd ed.* American Society of Photogrammetry. Falls Church, VA: 2440 p.

Condit, H.R. 1970. The spectral reflectance of American soils. *Photogram. Engr. and Rem. Sens.* 36: 955–966.

Considine, D.M. ed. 1988. *Van Nostrand's Scientific Encyclopedia*. New York: Van Nostrand Reinhold: 1678–1679.

Corbley, K. 1996. Remote sensing skies filling with satellite plans. *Earth Observation Mag.* 5:26–28.

Davis, C.O. *Airborne hyperspectral remote sensing*. http://www.opl.ucsb.edu/hycode/pubs/onr01/op19.pdf (accessed September 15, 2012).

Davis, J.L., A.P. Annan. 1989. Ground penetrating radar for high-resolution mapping of soil and rock stratigraphy. *Geophys. Prospect.* 37: 531–551.

Degnan, J.J. *The History and Future of Satellite Laser Ranging*: Lanham: Sigma Space Corporation. http://ilrs.gsfc.nasa.gov/docs/degnan_0603.pdf (accessed September 15, 2012).

DFVLR a. *German Research Institute for Aviation and Space*. http://www.dlr.de/dlr/en/desktopdefault.aspx/tabid-10012/ (accessed September 16, 2012).

DigitalGlobe, *WorldView-2*. http://worldview2.digitalglobe.com/ (accessed September 16, 2012).

DLR, *The Digital Airborne Imaging Spectrometer DIAS 7915*. http://www.op.dlr.de/dais/dais-scr.htm (accessed September 15, 2012).

DMI, *Ørsted Satellite Project*. http://web.dmi.dk/projects/oersted/mission/instruments.html (accessed September 16, 2012).

Drury, S.A. 1987. *Image Interpretation in Geology*. London: Allen & Unwin.: 165–194.

EADS Astrium a, *ALSAT-2 Programme*. http://seminar.spaceutm.edu.my/eoss2007/Material/Session6/Presenter/07-11-26%20GL%20Alsat%20Presentation.PDF (accessed September 17, 2012).

EADS Astrium b, *DEIMOS-1*. http://www.astrium-geo.com/en/84-deimos-1-optical-satellite-imagery (accessed September 17, 2012).

EADS Astrium c, *THEOS*. http://www.astrium.eads.net/en/programme/theos.html (accessed September 17, 2012).

EADS Astrium d, *SPOT 6 and SPOT 7*. http://www.astrium-geo.com/en/147-spot-6-7 (accessed September 22, 2012).

Earth Observation Research Center a, *ILAS*. http://suzaku.eorc.jaxa.jp/GLI2/adeos/Project/Ilas.html (accessed September 20, 2012).

Earth Observation Research Center b, *POLDER*. http://suzaku.eorc.jaxa.jp/GLI2/adeos/Project/Polder.html (accessed September 20, 2012).

Earth Observation Research Center c, *JERS*. http://www.eorc.jaxa.jp/JERS-1/en/index.html (accessed September 20, 2012).

e-geos, *COSMO-SkyMed*. http://www.eurimage.com/products/cosmo.html (accessed September 17, 2012).

Ehrismann, J., B. Armour, M. van der Kooij, H. Schwichow. 1996. Mapping a Broken Land: using repeat pass, space-based SAR interferometry, researchers measure effects of the Kobe, Japan earthquake. *Earth Observation Mag*. 5:26–29.

El-Sheimy, N., C. Valeo, A. Habib. 2005. *Digital Terrain Modeling: Acquisition, Manipulation, and Applications*. Boston: Artech House.: 257 p.

Encyclopedia Astronautica, *CryoSat*. http://www.astronautix.com/craft/cryosat.htm (accessed September 13, 2012).

eoPortal Directory a, *Quickbird-2*. https://directory.eoportal.org/web/eoportal/satellite-missions/q/quickbird-2 (accessed September 16, 2012).

eoPortal Directory b, *EROS-B*. http://www.eoportal.org/directory/pres_EROSBEarthRemoteObservationSatelliteB.html (accessed September 19, 2012).

eoPortal Directory c, *ADEOS/Midori*. http://www.eoportal.org/directory/pres_ADEOSAdvancedEarthObservingSatelliteMidori.html (accessed September 22, 2012).

eoPortal d, *KOMPSAT-3*. https://directory.eoportal.org/web/eoportal/satellite-missions/k/kompsat-3 (accessed September 20, 2012).

eoPortal e, *Resurs-01*. https://directory.eoportal.org/web/eoportal/satellite-missions/r/resurs-o1 (accessed September 22, 2012).

eoPortal f, *Monitor-E*. https://directory.eoportal.org/web/eoportal/satellite-missions/m/monitor-e (accessed September 22, 2012).

eoPortal g, *Resurs-DK1*. https://directory.eoportal.org/web/eoportal/satellite-missions/r/resurs-dk1 (accessed September 22, 2012).

eoPortal h, *Meteor-M-1*. https://directory.eoportal.org/web/eoportal/satellite-missions/m/meteor-m-1 (accessed January 5, 2013).

eoPortal i, *Electro-L*. https://directory.eoportal.org/web/eoportal/satellite-missions/e/electro-l (accessed September 22, 2012).

eoPortal j, *XSat*. https://directory.eoportal.org/web/eoportal/satellite-missions/v-w-x-y-z/xsat (accessed September 22, 2012).

eoPortal k, *SPOT-6 and SPOT-7*. https://directory.eoportal.org/web/eoportal/satellite-missions/s/spot-6-7 (accessed September 22, 2012).

eoPortal l, *Pléiades*. https://directory.eoportal.org/web/eoportal/satellite-missions/p/pleiades (accessed September 22, 2012).

ESA a, *MODIS*. http://www.uv.es/~leo/sen2flex/modis.htm (accessed September 16, 2012).

ESA b, *ERS Overview*. http://www.esa.int/esaEO/SEMGWH2VQUD_index_0_m.html (accessed September 16, 2012).

ESA c, *CLUSTER*. http://sci.esa.int/science-e/www/object/index.cfm?fobjectid=35500# (accessed September 16, 2012).

ESA d, *Proba*. http://www.esa.int/esaMI/Proba/index.html (accessed September 16, 2012).

ESA e, *ENVISAT*. https://earth.esa.int/web/guest/missions/esa-operational-eo-missions/envisat/instruments/meris (accessed September 16, 2012).

ESA f, *AATSR*. https://earth.esa.int/web/guest/missions/esa-operational-eo-missions/envisat/instruments/aatsr (accessed September 16, 2012).

ESA g, *ASAR*. https://earth.esa.int/web/guest/missions/esa-operational-eo-missions/envisat/instruments/asar (accessed September 16, 2012).

ESA h, *MWR* https://earth.esa.int/web/guest/missions/esa-operational-eo-missions/envisat/instruments/mwr (accessed September 16, 2012).

ESA i, *GOMOS*. https://earth.esa.int/web/guest/missions/esa-operational-eo-missions/envisat/instruments/gomos (accessed September 16, 2012).

ESA j, *MIPAS*. http://envisat.esa.int/instruments/mipas/index.html (accessed September 16, 2012).

ESA k, *SCIAMACHY*. http://envisat.esa.int/instruments/sciamachy/index.html (accessed September 14, 2012).

ESA l, *SCIAMACHY*. http://www.wmo.int/pages/prog/sat/Instruments_and_missions/SCIAMACHY.html (accessed September 14, 2012).

ESA m, *SMOS*. http://www.esa.int/esaLP/LPsmos.html (accessed September 17, 2012).

ESA n, *GOCE*. http://www.esa.int/esaLP/ESAYEK1VMOC_LPgoce_0.html (accessed September 17, 2012).

ESA o, *CryoSat*. http://www.esa.int/esaLP/ESA0DL1VMOC_LPcryosat_2.html (accessed September 17, 2012).

ESA p, *SPOT*. http://www.esa.int/esaMI/Eduspace_EN/SEMIW04Z2OF_0.html (accessed September 17, 2012).

ESA q, *Ikonos-2*. https://earth.esa.int/web/guest/missions/3rd-party-missions/current-missions/ikonos-2 (accessed September 18, 2012).

ESA r, *SWARM*. http://www.esa.int/esaMI/Operations/SEM27Z8L6VE_0.html (accessed September 22, 2012).

ESA s, *EarthCARE*. http://www.esa.int/esaLP/SEM75KTWLUG_LPearthcare_0.html (accessed September 22, 2012).

ESA t, *ADM-Aeolus*. http://www.esa.int/esaLP/ESAVO62VMOC_LPadmaeolus_1.html (accessed September 22, 2012).

ESA u, *RESOURCESAT-3*. http://database.eohandbook.com/database/missionsummary.aspx?missionID=573 (accessed September 22, 2012).

ESA v, *CartoSat-3*. http://database.eohandbook.com/database/missionsummary.aspx?missionID=565 (accessed September 22, 2012).

ESA w, *OCEANSAT-3*. http://database.eohandbook.com/database/missionsummary.aspx?missionID=571 (accessed September 22, 2012).

ESA x, *Jason-3*. http://earth.eo.esa.int/brat/html/missions/jason3/welcome_en.html (accessed September 22, 2012).

Escoubet, C.P., M. Fehringer, M. Goldstein. 2001. The Cluster mission. *Ann. Geophys.* 19: 1197–1200.

EUMETSAT a, *MetOp*. http://www.eumetsat.int/Home/Main/Satellites/Metop/index.htm?l=en (accessed September 16, 2012).

EUMETSAT b, *Meteosat Second Generation*. http://www.eumetsat.int/Home/Main/Satellites/MeteosatSecondGeneration/index.htm?l=en (accessed September 16, 2012).

Evans, D.E., T.G. Farr, J.P. Ford, T.W. Thompson, C.L. Werner. 1986. Multipolarization radar images for geologic mapping and vegetation discrimination. *IEEE Trans. Geosci. Rem. Sens.* GE-24: 246–257.

Ferguson, S.T., Y. Hammada. 2001. Experiences with AIRGrav: results from a new airborne gravimeter. *Proceedings of the IAG International Symposium 'Gravity, Geoid and Geodynamics 2000', v. 123 of IAG Symposia*, Banff, Canada. Berlin: Springer–Verlag: 211–216.

Gabriel, A.K., R.M. Goldstein, H.A. Zebker. 1989. Mapping small elevation changes over large areas: differential radar interferometry. *J. Geophys. Res.* 94 B7: 9183–9191.

GeoExplo, *Airborne survey companies.* http://www.geoexplo.com/Geophysics_links.html http://www.geoexplo.com/airborne_survey_workshop_rad.html (accessed September 14, 2012).

Geovar, LDCM, *Landsat 8.* http://www.geovar.com/ldcm.htm (accessed September 22, 2012).

GFZa, *Terrestrial and airborne gravimetry.* http://www.gfz-potsdam.de/portal/gfz/Struktur/Departments/Department+1/sec12/topics/terrestrial_and_airborne_gravimetry (accessed September 15, 2012).

GFZ b, *CHAMP.* http://op.gfz-potsdam.de/champ/systems/index_SYSTEMS.html (accessed September 18, 2012).

Giroux, J., L. Moreau, G. Girard, M. Soucy. 2010. Technological evolutions on the FTS instrument for follow-on missions to SCISAT atmospheric chemistry experiment. *Proceedings of SPIE.* 7826: Bellingham, WA: International Society for Optics and Photonics: 12.

Goetz, A.F.H. 1976. *Remote Sensing Geology, Landsat and Beyond.* SP43-30. Pasadena, CA: Jet Propulsion Lab: 8-1–8-8.

Graf, F.L. 1990. Using ground-penetrating radar to pinpoint pipeline leaks. *Matter Performance* 29: 27–29.

GSSI, *Service providers.* http://www.geophysical.com/serviceproviders.htm (accessed September 15, 2012).

Gunter's Space Page a, *FASat Alfa, Bravo.* http://space.skyrocket.de/doc_sdat/fasat-alfa.htm (accessed September 16, 2012).

Gunter's Space Page b, *FASat Charlie.* http://space.skyrocket.de/doc_sdat/ssot.htm (accessed September 16, 2012).

Gunter's Space Page c, *Landsat 8.* http://space.skyrocket.de/doc_sdat/ldcm.htm (accessed September 22, 2012).

Gunter's Space Page d, *SAOCOM.* http://space.skyrocket.de/doc_sdat/saocom-1.htm (accessed September 22, 2012).

Gunter's Space Page e, *GFO-2.* http://space.skyrocket.de/doc_sdat/gfo-2.htm (accessed September 22, 2012).

Gupta, R.P. 1991. *Remote Sensing Geology.* Berlin: Springer–Verlag: 149–179.

Haddad, D.E., E. Nissen, J.R. Arrowsmith. 2012. *High-Resolution Lidar Digital Topography in Active Tectonics Research.* Active Tectonics, Quantitative Structural Geology, and Geomorphology Laboratory, School of Earth and Space Exploration. Tempe, AZ: Arizona State University.

Hadden, S., C.D. Green. 1999. Remote classification of seabed types derived from interferometric sonar data (abs.). *Bull. Am. Assn. Pet. Geol.* 83: 1315.

Harris, R. 1987. *Satellite Remote Sensing—an Introduction.* London: Routledge Kegan & Paul: 220 p.

Hauff, P.L., P. Kowalczyk, M. Ehling, G. Borstad, G. Edmundo, R. Kern, R. Neville et al. 1996. The CCRS SWIR Full Spectrum Imager: Mission to Nevada. In *Proceedings of the 11th Thematic Conference on Geologic Remote Sensing.* v. 1, Ann Arbor: ERIM: I-38–I-47.

Hook, S.J., J.E. Dmochowski, K.A. Howard, L.C. Rowan, K.E. Karlstrom, J.M. Stock. 2005. Mapping variations in weight percent silica measured from multispectral thermal infrared imagery—examples from the Hiller Mountains, Nevada, USA and Tres Virgenes-La Reforma, Baja California Sur, Mexico. *Rem. Sens. Environ.* 95: 273–289.

Hook, S.J., J.J. Myers, K.J. Thome, M. Fitzgerald, A.B. Kahle. 2001. The MODIS/ASTER airborne simulator (MASTER)—an instrument for earth science studies. *Rem. Sens. Environ.* 76: 93–102.

Hunt, G.R., R.P. Ashley. 1979. Spectra of altered rocks in the visible and near infrared. *Econ. Geol.* 74: 1613–1629.

Hunt, G.R., J. Salisbury. 1976. Visible and near-infrared spectra of minerals and rocks: XI, sedimentary rocks. *Mod. Geol.* 5: 211–217.

HyperPhysics, *Michelson interferometer.* http://hyperphysics.phy-astr.gsu.edu/hbase/phyopt/michel.html (accessed September 14, 2012).

Im, E., C. Wu, S.L. Durden. 2005. Cloud Profiling Radar for the Cloudsat Mission. http://trs-new.jpl. nasa.gov/dspace/bitstream/2014/37690/1/042971.pdf (accessed September 16, 2012).

ImageSat International a, *EROS A*. http://www.imagesatintl.com/?catid={2B426A6D-F4A8-4ED5-8B11-A07F7EE8B459} (accessed September 19, 2012).

ImageSat International b, *EROS-B*. http://www.imagesatintl.com/?catid={58902695-6CAC-4189-962A-22AADAE85F85} (accessed September 19, 2012).

Imaging Notes, *CBERS*. http://www.imagingnotes.com/go/article_free.php?mp_id=134 (accessed September 22, 2012).

Inkster, D.R., J.R. Rossiter, R. Goodman, M. Galbraith, J.L. Davis. 1989. Ground penetrating radar for subsurface environmental applications. *Proceedings of the 7th Thematic Conference on Remote Sensing for Exploration Geology* Ann Arbor: ERIM: 127–140.

International Laser Ranging Service, *LARES*. http://ilrs.gsfc.nasa.gov/satellite_missions/list_of_satellites/lars_general.html (accessed September 18, 2012).

ISRO a, *Earth Observation Satellites*. http://www.isro.org/satellites/earthobservationsatellites.aspx (accessed September 18, 2012).

ISRO b, *IRS-1A*. http://www.isro.org/satellites/irs-1a.aspx (accessed September 19, 2012).

ISRO c, *IRS-1E*. http://www.isro.org/satellites/irs-1e.aspx (accessed September 19, 2012).

ISRO d, *IRS-P4*. http://www.isro.org/satellites/irs-p4_oceansat.aspx (accessed September 19, 2012).

ISRO e, *TES*. http://www.isro.org/satellites/technology_experiment_satellite_tes.aspx (accessed September 19, 2012).

ISRO f, *IRS-P6*. http://www.isro.org/satellites/irs-p6resourcesat-1.aspx (accessed September 19, 2012).

ISRO g, *CARTOSAT-1*. http://www.isro.org/satellites/cartosat-1.aspx (accessed September 19, 2012).

ISRO h, *CARTOSAT-2*. http://www.isro.org/satellites/cartosat-2.aspx (accessed September 19, 2012).

SSRO i, *IMS-1*. http://www.isro.org/satellites/ims-1.aspx (accessed September 19, 2012).

ISRO j, *Oceansat-2*. http://www.isro.org/satellites/oceansat-2.aspx (accessed September 19, 2012).

ISRO k, *RISAT-2*. http://www.isro.org/satellites/RISAT-2.aspx (accessed September 19, 2012).

ISRO l, *CARTOSAT-2B*. http://www.isro.org/satellites/cartosat-2b.aspx (accessed September 19, 2012).

ISRO m, *RESOURCESAT-2*. http://www.isro.org/satellites/cartosat-2b.aspx (accessed September 19, 2012).

ISRO n, *RISAT-1*. http://www.isro.org/satellites/RISAT-1.aspx (accessed September 19, 2012).

Jacobsen, K. 1997. *Comparison of information contents of different space images*. http://www.ipi.uni-hannover.de/uploads/tx_tkpublikationen/jac_97_infocontents.pdf (accessed September 20, 2012).

Janes Space Systems a, *Starlette and Stella*. http://articles.janes.com/articles/Janes-Space-Systems-and-Industry/Starlette-and-Stella-France.html (accessed September 18, 2012).

Janes Space Systems b, *Soyuzkarta*. http://articles.janes.com/articles/Janes-Space-Systems-and-Industry/Soyuzkarta-Russian-Federation.html (accessed September 20, 2012).

Jansen, W.T. 1992. Mineralogical and geological analysis using hyperspectral data. *Geobyte* 7:46–49.

JAXA a, *Himawari*. http://www.jaxa.jp/projects/sat/gms/index_e.html. (accessed September 20, 2012).

JAXA b, *MOS-1*. http://science.nasa.gov/missions/mos/ (accessed September 20, 2012).

JAXA c, *Midori II*. http://www.jaxa.jp/projects/sat/adeos2/index_e.html (accessed September 20, 2012).

JAXA d, *MTSAT-1R*. http://airex.tksc.jaxa.jp/pl/dr/AA0029607003/en (accessed September 20, 2012).

JAXA e, *Greenhouse gases Observing Satellite*. http://www.jaxa.jp/projects/sat/gosat/index_e.html (accessed September 20, 2012).

Jenkinson, W.D. 1977. Side-scan sonar: applications in exploration and exploitation. *Oil and Gas Journal* 75: 97–102.

JPL, *AVIRIS*. http://aviris.jpl.nasa.gov/(accessed September 15, 2012).

Jucks, K. 2009. Active Sensing of Carbon Dioxide Emissions over Nights, Days, and Seasons. NASA: 18 p.

Kawanishi, T., T. Sezai, Y. Ito, K. Imaoka, T. Takeshima, Y. Ishido, A. Shibata et al. 2003. Geoscience and remote sensing. *IEEE Trans.* 41(2): 184–194.

Khale, A.B., A.F.H. Goetz. 1983. Mineralogic information from a new thermal infrared multispectral scanner. *Science* 222: 24–27.

Khale, A.B., L.C. Rowan. 1980. Airborne geological mapping using infrared emission spectra. *Geology* 8: 234–239.

Kruse, F.A. 1999. Visible-infrared sensors and case studies. *Manual of Remote Sensing*, 3rd ed. v. 3 Chap. 11. New York: John Wiley and Sons: 567–611.

Kruse, F.A., K.S. Kierein-Young, J.W. Boardman. 1990. Mineral mapping at Cuprite, Nevada, with a 63-channel imaging spectrometer. *Photogram. Engr. and Rem. Sens.* 56: 83–92.

LaCoste & Romberg. 2004. *Instruction Manual Model G & D Gravity Meters*. Austin, TX: LaCoste & Romberg, 127 p. http://www.ifg.tu-clausthal.de/java/grav/gdmanual.pdf (accessed September 15, 2012).

Langel, R., G. Ousley, J. Berbert, J. Murphy, M. Settle. 1982. The MAGSAT Mission. *Geophys. Res. Lett.* 9(4): 243–245.

Laughton, A.S. 1981. The first decade of GLORIA. *J. Geophys. Res.* 86: 11,511–11,534.

Searle, R.C., T.P. LeBas, N.C. Mitchell, M.L. Somers, L.M. Parson, P. Patriat. 1990. GLORIA image processing: the state of the art. *Marine Geophys. Res.* 12: 21–39.

Lidar data web directory, *Airborne Lidar Directory*. http://sites.google.com/a/full-links.pl/Lidar-data/Airborne_Lidar (accessed September 14, 2012).

Lidar News, *Lidar Directory*. http://www.Lidarnews.com/component/option,com_mtree/task,listcats/cat_id,78/Itemid,193/ (accessed September 14, 2012).

MALÅ, *Ground penetrating radar (GPR) links*. http://www.malags.com/Resources/ground-penetrating-radar-link-directory/gpr-service-providers (accessed September 15, 2012).

McCauley, J.F., G.G. Schaber, C.S. Breed, M.J. Grolier, C.V. Haines, B. Issawi, C. Elachi, R. Blom. 1982. Subsurface valleys and geoarchaeology of the Eastern Sahara revealed by Shuttle radar. *Science* 218: 1004–1020.

McCormick, J. 14 June 1999. A high-tech search for a missing Navy wife. *Newsweek*: 54–55.

Miller V.C., C.F. Miller. 1961. *Photogeology*. New York: McGraw-Hill Book Co.: 8–50.

Moon, C.J., M.K.G. Whateley, A.M. Evans. 2006. *Introduction to Mineral Exploration*, 2nd ed.: Oxford: Blackwell Publishing: 481 p.

Moran, D.E. 1973. *Geology, Seismicity, and Environmental Impact*. Los Angeles, CA: University Publishers: 141–155.

Murphy, C.A. 2004. The Air-FTG™ airborne gravity gradiometer system. In R. Lane (ed.), *Airborne Gravity 2004: Abstracts from the ASEG-PESA Airborne Gravity Workshop*: Sydney, Australia: Geoscience Australia: 7–14.

NASA a, Broadband Lidar instrument successfully tested on NASA's DC-8. http://www.nasa.gov/centers/dryden/Features/broadband_Lidar_tested.html (accessed September 14, 2012).

NASA b, *Planetary Magnetospheres Laboratory*. http://science.gsfc.nasa.gov/solarsystem/magnetospheres/(accessed September 15, 2012).

NASA c, Alpha particle X-ray spectrometer. http://msl-scicorner.jpl.nasa.gov/Instruments/APXS/(accessed September 15, 2012).

NASA d, *Thematic Mapper Simulator*. http://www.nasa.gov/centers/dryden/research/AirSci/ER-2/tms.html (accessed September 15, 2012).

NASA e, *AIRSAR/Airborne Synthetic Aperture Radar*. http://airsar.jpl.nasa.gov/index_detail.html (accessed September 15, 2012).

NASA f, *Operation IceBridge*. http://science.nasa.gov/missions/operation-ice-bridge/ (accessed September 16, 2012).

NASA g. *Operation IceBridge*. http://www.espo.nasa.gov/oib/ (accessed September 16, 2012).

NASA h, *DISCOVER AQ*. http://discover-aq.larc.nasa.gov/instruments.php; (accessed September 16, 2012).

NASA i, *DISCOVER AQ*. http://science.nasa.gov/missions/discover-aq/(accessed September 16, 2012).

NASA j, *AirMOSS*. http://science.nasa.gov/missions/airmoss/ (accessed September 16, 2012).

NASA k, *UAVSAR*. http://uavsar.jpl.nasa.gov/ (accessed September 16, 2012).

NASA l, *CARVE*. http://science.nasa.gov/missions/carve/ (accessed September 16, 2012).

NASA m, *HS3*. http://science.nasa.gov/missions/hs3/ (accessed September 16, 2012).

NASA n, *OGO series*. http://heasarc.nasa.gov/docs/heasarc/missions/ogo.html (accessed September 16, 2012).

NASA o, *Skylab*. http://www.nasa.gov/mission_pages/skylab/(accessed September 16, 2012).

NASA p, *Landsat*. http://www.nasa.gov/mission_pages/landsat/main/ (accessed September 16, 2012).

NASA q, *GOES-P*. http://www.nasa.gov/mission_pages/GOES-P/main/index.html (accessed September 16, 2012).

NASA r, *LAGEOS*. http://ilrs.gsfc.nasa.gov/satellite_missions/list_of_satellites/lag1_general.html (accessed September 16, 2012).

NASA s, *NOAA-N*. http://www.nasa.gov/mission_pages/noaa-n/main/index.html (accessed September 16, 2012).

NASA t, *SeaSat*. http://www.jpl.nasa.gov/missions/details.php?id=5971 (accessed September 16, 2012).

NASA u, *TOPEX/Poseidon*. http://sealevel.jpl.nasa.gov/missions/topex/ (accessed September 16, 2012).

NASA v, *TRMM*. http://science.nasa.gov/missions/trmm/ (accessed September 16, 2012).

NASA w, *TRMM*. http://trmm.gsfc.nasa.gov/(accessed September 16, 2012).

NASA x, *GEOSAT Follow-On*. http://gcmd.nasa.gov/records/GCMD_GEOSAT_FOLLOWON.html (accessed September 16, 2012).

NASA y, *ASTER*. http://asterweb.jpl.nasa.gov/ (accessed September 16, 2012).

NASA z, *MISR*. http://www-misr.jpl.nasa.gov/ (accessed September 16, 2012).

NASA aa, *MOPITT*. http://eosweb.larc.nasa.gov/PRODOCS/mopitt/table_mopitt.html (accessed September 16, 2012).

NASA bb, *ACRIMSAT*. http://acrim.jpl.nasa.gov/ (accessed September 16, 2012).

NASA cc, *ACRIMSAT*. http://science.nasa.gov/missions/acrimsat/ (accessed September 16, 2012).

NASA dd, *QUICKSCAT*. http://science.nasa.gov/missions/quikscat/ (accessed September 16, 2012).

NASA ee, *Earth Observing-1*. http://science.nasa.gov/missions/eo-1/ (accessed September 16, 2012).

NASA ff, *Earth Observing-1*. http://nmp.jpl.nasa.gov/ (accessed September 16, 2012).

NASA gg, *SAC-C*. http://www.nasa.gov/centers/goddard/pdf/110896main_FS-2000-11-012-GSFC-SAS-C.pdf (accessed September 16, 2012).

NASA hh, *JASON-1*. http://science.nasa.gov/missions/jason-1/ (accessed September 16, 2012).

NASA ii, *Earth Observing System AQUA*. http://www.nasa.gov/pdf/151987main_Aqua_FactSheet.pdf (accessed September 16, 2012).

NASA jj, *GRACE*. http://instrumentsystems.jpl.nasa.gov/projects/grace/ (accessed September 16, 2012).

NASA kk, *ICESat*. http://icesat.gsfc.nasa.gov/ (accessed September 16, 2012).

NASA ll, *SORCE*. http://science.nasa.gov/missions/sorce/ (accessed September 16, 2012).

NASA mm, *Aura*. http://www.nasa.gov/mission_pages/aura/main/index.html (accessed September 16, 2012).

NASA nn, *NOAA-N*. http://science.nasa.gov/missions/noaa-n/ (accessed September 16, 2012).

NASA oo, *CALIPSO*. http://www.nasa.gov/mission_pages/calipso/main/ (accessed September 16, 2012).

NASA pp, *OSTM/Jason-2*. http://www.nasa.gov/mission_pages/ostm/main/index.html (accessed September 16, 2012).

NASA qq, *AQUARIUS*. http://aquarius.nasa.gov/ (accessed September 16, 2012).

NASA rr, *SUOMI NPP*. http://science.nasa.gov/missions/suomi-npp/ (accessed September 16, 2012).

NASA ss, *SUOMI NPP*. http://jointmission.gsfc.nasa.gov/spacecraft_inst.html (accessed September 16, 2012).

NASA tt, *Long March 3C*. http://www.nasaspaceflight.com/news/chinese/ (accessed September 16, 2012).

NASA uu, *Bilsat 1*. http://nssdc.gsfc.nasa.gov/nmc/masterCatalog.do?sc=2003-042E (accessed September 17, 2012).

NASA vv, *Starlette and Stella*. http://ilrs.gsfc.nasa.gov/satellite_missions/list_of_satellites/star_general.html#obj (accessed September 18, 2012).

NASA ww, *CHAMP*. http://science.nasa.gov/missions/champ/ (accessed September 18, 2012).

NASA ww, *CHAMP*; GFZ, CHAMP. http://science.nasa.gov/missions/champ/ (accessed September 18, 2012).

NASA xx, *TerraSAR-X*. http://ilrs.gsfc.nasa.gov/satellite_missions/list_of_satellites/tsar_general.html (accessed September 18, 2012).

NASA yy, *TanDEM-X*. http://ilrs.gsfc.nasa.gov/satellite_missions/list_of_satellites/tand_general.html (accessed September 18, 2012).

NASA zz, *Orbview-2*. http://eospso.gsfc.nasa.gov/eos_homepage/mission_profiles/show_mission.php?id=31&mission_cat_id=20 (accessed September 18, 2012).

NASA aaa, *SeaWinds*. http://winds.jpl.nasa.gov/missions/seawinds/ (accessed September 20, 2012).

NASA bbb, *ADEOS-2*. http://ilrs.gsfc.nasa.gov/satellite_missions/list_of_satellites/ade2_general.html (accessed September 20, 2012).

NASA ccc, *Etalon-1 and -2*. http://ilrs.gsfc.nasa.gov/satellite_missions/list_of_satellites/eta1_general.html (accessed September 22, 2012).

NASA ddd, *Almaz 1*. http://nssdc.gsfc.nasa.gov/nmc/spacecraftDisplay.do?id=1991-024A (accessed September 22, 2012).

NASA eee, *ATTREX*. http://science.nasa.gov/missions/attrex/ (accessed September 22, 2012).

NASA fff, *GPM*. http://science.nasa.gov/missions/gpm/ (accessed September 22, 2012).

NASA ggg, *GPM*. http://pmm.nasa.gov/GPM/flight-project/spacecraft-and-instruments (accessed September 22, 2012).

NASA, hhh, *OCO*. http://oco.jpl.nasa.gov/(accessed September 22, 2012).

NASA iii, *NPOESS*. http://science.nasa.gov/missions/npoes/ (accessed September 22, 2012).

NASA jjj, *GOES-R*. http://science.nasa.gov/missions/goes-r/ (accessed September 22, 2012).

NASA kkk, *HyspIRI*. http://hyspiri.jpl.nasa.gov/ (accessed September 22, 2012).

NASA lll, *SAGE III-ISS*. http://science.nasa.gov/missions/sage-3-iss/ (accessed September 22, 2012).

NASA mmm, *SAGE III-ISS*. http://sage.nasa.gov/SAGE3ISS/ (accessed September 22, 2012).

NASA nnn, *SMAP*. http://science.nasa.gov/missions/smap/ (accessed September 22, 2012).

NASA ooo, *SMAP*. http://smap.jpl.nasa.gov/instrument/ (accessed September 22, 2012).

NASA ppp, *ICESat-2*. http://science.nasa.gov/missions/icesat-ii/ (accessed September 22, 2012).

NASA qqq, *DESDYNL*. http://science.nasa.gov/missions/desdyni/ (accessed September 22, 2012).

NASA rrr, *DESDYNL*. http://desdyni.jpl.nasa.gov/ (accessed September 22, 2012).

NASA Spaceflight.com a, *HaiYang-2A*. http://www.nasaspaceflight.com/2011/08/chinas-surge-haiyang-21a-launch-long-march-4b/ (accessed September 16, 2012).

NASA Spaceflight.com b, *Long March 4B ZiYuan-1*. http://www.nasaspaceflight.com/2011/12/china-in-surprise-launch-of-long-march-4b-with-ziyuan-1/ (accessed September 16, 2012).

NASA Spaceflight.com c, *Long March 4B ZiYuan-3*. http://www.nasaspaceflight.com/2012/01/china-opens-2012-ziyuan-3-launch-long-march-4b/ (accessed September 16, 2012).

NASDA, *Global Imager*. http://www.ioccg.org/sensors/gli/tanaka.pdf (accessed September 20, 2012).

NASRDA, *NigeriaSat-X*. http://www.nasrda.gov.ng/NIGERIASATx.html (accessed September 17, 2012).

National Academy of Sciences, *HYSPIRI*. http://cce.nasa.gov/pdfs/HYSPIRI.pdf (accessed September 22, 2012).

National Research Council, 2008. *Satellite Observations to Benefit Science and Society*. Washington, DC: The National Academies Press: 32 p.

National Security Space Road Map, *Cloud profiling system*. http://www.fas.org/spp/military/program/nssrm/initiatives/cloud.htm (accessed September 13, 2012).

NOAA a, *High resolution Doppler Lidar (HRDL)*. http://www.esrl.noaa.gov/csd/groups/csd3/instruments/hrdl/(accessed September 14, 2012).

NOAA b, *GEOSAT*. http://www.nodc.noaa.gov/sog/geosat_recovery/ (accessed September 16, 2012).

NOAA c, *JPSS*. http://www.nesdis.noaa.gov/jpss/ (accessed September 22, 2012).

NOAA d, *JPSS*. http://www.nesdis.noaa.gov/jpss/(accessed September 22, 2012).

NSF, First-ever Use of Airborne Resistivity System in Antarctica Allows Researchers to Look Beneath Surface in Untapped Territories: Press Release 12–055.http://www.nsf.gov/news/news_summ.jsp?cntn_id=123620 (accessed September 15, 2012).

NSPO, *FORMOSAT-2*. http://www.nspo.org.tw/2008e/projects/project2/instrument.htm (accessed September 17, 2012).

Ogunbadewa, E.Y. 2008. The Characteristics of NigeriaSat-1 and its Potential Application for Environmental Monitoring. AFRICAN SKIES/CIEUX AFRICAINS, No. 12, October 2008: 64–70.

Open Directory Project, *Products and services*. http://www.dmoz.org/Science/Earth_Sciences/Geophysics/Products_and_Services/(accessed September 14, 2012).

Orbital, *OrbView-3*. http://www.orbital.com/SatellitesSpace/ImagingDefense/OV3/index.shtml (accessed September 18, 2012).

Pearlman, M., G. Appleby, G. Kirchner, J. Mcgarry, T. Murphy, C. Noll, E. Pavlis, F. Pierron. 2010. *Current trends in satellite laser ranging*. Poster G13C-07. http://cddis.nasa.gov/docs/slrposter_agubrazil_1008.pdf (accessed September 15, 2012).

Plevin, J. 1974. *Passive microwave radiometry and its potential application to earth resources survey*. in: Fundamentals of remote sensing; Proceedings of the First Technical Session, London, February 13, 1974. (A75-3083 13-35) Birmingham, UK: University of Aston: 87–106.

Prior, D.B., J.M. Coleman, H.H. Roberts. 1981. Mapping with side-scan sonar. *Offshore* 41(4): 151–161.

Radar Altimeter Tutorial, *Radar altimetry*. http://www.altimetry.info/ (accessed September 13, 2012).

Resmini, R.G., M.E. Kappus, W.S. Aldrich, J.C. Harsanyi, M. Anderson. 1996. Use of hyperspectral digital imagery collection experiment (HYDICE) sensor data for quantitative mineral mapping at Cuprite, Nevada. *Proceedings of the 11th Thematic Conference on Geologic Remote Sensing*. v. 1, Ann Arbor, ERIM: I-48–I-65.

Rowan, L.C., M.J. Pawlewicz, O.D. Jones. 1992. Mapping thermal maturity in the Chainman Shale, near Eureka, Nevada, with Landsat Thematic Mapper images. *Bull. Am. Assn. Pet. Geol.* 76: 1008–1023.

Rowlands, N.A., R.A. Neville. 1994. A SWIR imaging spectrometer for remote sensing. In *Proceedings of SPIE Infrared Technology XX Conference*. Bellingham, WA: International Society for Optics and Photonics: 2269.

Sabins, F. F., 1997, *Remote Sensing - Principles and Interpretation*: 3rd ed., Long Grove, IL: Waveland Press: 432 p.

Satellite Imaging Corporation a, *CBERS*. http://www.satimagingcorp.com/satellite-sensors/cbers-2.html (accessed September 16, 2012).

Satellite Imaging Corporation b, *GEOEYE-1*. http://www.satimagingcorp.com/satellite-sensors/geoeye-1.html (accessed September 18, 2012).

Satellite Imaging Corporation c, *ALOS*. http://www.satimagingcorp.com/satellite-sensors/alos.html (accessed September 20, 2012).

Satellite Imaging Corporation d, *GeoEye-2*. http://www.satimagingcorp.com/satellite-sensors/geoeye-2.html (accessed September 22, 2012).

Scholz, C. H. 2002. *The Mechanics of Earthquakes and Faulting*. Cambridge: Cambridge University Press: 496 p.

Schumacher, D. 2012. *Power Imaging—A Passive Electromagnetic Hydrocarbon Detection Method: Examples from Railroad Valley*. Nevada (abs.): AAPG Search and Discovery Article #90142: 1 p.

SeaDiscovery, Hydrographic & Coastal Surveying Directory. http://www.seadiscovery.com/mtDirectory.aspx?mnl2=l2directoryissielisting&IssueID=140 (accessed September 15, 2012).

Shan, S., M. Bevis, E. Kendrick, G.L. Mader, D. Raleigh, K. Hudnut, M. Sartori, D. Phillips. 2007. Kinematic GPS solutions for aircraft trajectories: Identifying and minimizing systematic height errors associated with atmospheric propagation delays. *Geophys. Res. Lett.* 34, no. 23.

Silliman, S.W., P. Farnsworth, K.G. Lightfoot. 2000. Magnetometer prospecting in historical archaeology: evaluating survey options at a 19th century rancho site in California. *Historical Archaeology* 34(2): 89–109.

Slater, P.N. 1983. *Manual of Remote Sensing*, 2nd ed. In R.N. Colwell (ed.), Falls Church, VA: American Society of Photogrammetry: 231–291.

Smith, R.S., J. Klein. 1996. A special circumstance of airborne induced-polarization measurements. *Geophysics* 61(1): 66–73

SpaceNews, *Worldview-3*. http://www.spacenews.com/earth_observation/120417-digitalglobe-adding-ir-worldview3.html (accessed September 22, 2012).

Spectrasyne. *What is DIAL and how does it work?* http://www.spectrasyne.ltd.uk/html/about_dial.html (accessed September 14, 2012).

Stone, C. November 2008. Technology sweeps data from basement: HRAM 'flies low' over subtleties. *AAPG Explorer*: 29:22–26.

Stove, G. 2012. Ground penetrating abilities of a LIDAR-like imaging spectrometer for finding, classifying, and monitoring subsurface hydrocarbons and minerals (abs.): AAPG Search and Discovery Article #90142: 1 p.

Surrey Satellite Technology Ltd a, *Disaster Monitoring Constellation*. http://www.sstl.co.uk/divisions/earth-observation-science/eo-constellations (accessed September 16, 2012).

Surrey Satellite Technology Ltd b, *AlSAT-1*. http://www.sstl.co.uk/missions/alsat-1—launched-2002/alsat-1/alsat-1—the-mission (accessed September 17, 2012).

Sylvester, A. G. 1999. *Rifting, Transpression, and Neotectonics in the Central Mecca Hills, Salton Trough*. Santa Barbara: University of California Santa Barbara: 52 p.

Technische Universität Berlin, *TUBSAT*. http://www.raumfahrttechnik.tu-berlin.de/tubsat/LAPAN-TUBSAT (accessed September 19, 2012).

Thomas, M. Feburary 1997. DIY remote sensing. *Mapping Awareness*: 11:32–33.

University of Colorado, *SORCE*. http://lasp.colorado.edu/sorce/ (accessed September 16, 2012).

U.S. DOT, *Locating Shallow Sand and Gravel Deposits*. http://www.cflhd.gov/resources/agm/engApplications/SubsurfaceChartacter/614LocShallowSandGravelDeposits(2).cfm (accessed September 15, 2012).

U.S. EPA. 1993. Use of Airborne, Surface, and Borehole Geophysical Techniques at Contaminated Sites–A Reference Guide; EPA/625/R-92/007: 304 p.

U.S. EPA. 2000. Meteorological Monitoring Guidance for Regulatory Modeling Applications. Research Triangle Park: U.S. EPA: 9-11 to 9-13.

Vane, G., R.O. Green, T.G. Chrien, H.T. Enmark, E.G. Hansen, W.M. Porter. 1993. The airborne visible/infrared imaging spectrometer (AVIRIS). *Remote Sensing Environ.* 44: 127–143.

Walker, S., J. Rudd. 2008. *Airborne resistivity mapping with helicopter TEM: an oil sands case study*; AEM 2008—5th International Conference on Airborne Electromagnetics. Haikko Manor, Finland: 5 p.

Watson, K., F.A. Kruse, S. Hummler-Miller. 1990. Thermal infrared exploration in the Carlin Trend, northern Nevada. *Geophysics* 55: 70–79.

Vedder, J. G., R.E. Wallace. 1970. *Recent active breaks along the San Andreas fault between Cholame Valley and Tejon Pass, California*. U.S. Geological Survey Miscellaneous Geological Investigations Map I-741, scale 1:24,000, 3 sheets.

Wilford, J. 2002. Airborne gamma-ray spectrometry. In E. Papp (ed.), *Geophysical and remote sensing methods for regolith exploration*. CRCLEME Open File Report 144: Canberra, Australia: 46–52.

World Meteorological Association a, *SLSTR*. http://www.wmo.int/pages/prog/sat/Instruments_and_missions/SLSTR.html (accessed September 22, 2012).

World Meteorological Association b, *SRAL*. http://www.wmo.int/pages/prog/sat/Instruments_ and_missions/SRAL.html (accessed September 22, 2012).

Zebker, H.A. 1991. The TOPSAR interferometric radar topographic mapping instrument. In J.J. van Zyl (ed.), *Proceedings of 3rd Airborne Synthetic Aperture Radar Workshop, JPL Pub. 91–30* Pasadena, CA: Jet Propulsion Lab: 230–233.

Additional Reading

Bianchi, R., C.M. Marino. 1997. Airborne multispectral and hyperspectral remote sensing: Examples of applications to the study of environmental and engineering problems. *Environ. Eng. Geophys. Soc. Appl. of Geophys. To Eng. & Environ. Probl. Symp. Proc.* 2: 811–816.

Fujiwara, S., P.A. Rosen, M. Tobita, M. Murakami. 1998. Crustal deformation measurements using repeat-pass JERS 1 synthetic aperture radar interferometry near the Izu Peninsula, Japan. *J. Geophys. Res.* 103 B2: 2411–2426.

Hagemann, J. 1958. *Facsimile recording of sonic values of the ocean bottom.* United States Patent Office.

Henderson, F.M, A.J. Lewis, eds. 1999. *Principles and Applications of Imaging Radar, Manual of Remote Sensing* v. 2. Falls Church, VA: American Society of Photogrammetry.

Hook, S.J., E.A. Abbott, C. Grove, A.B. Kahle, and F. Palluconi. 1999. *Use of Multispectral Thermal Infrared Data in Geological Studies. Manual of Remote Sensing* v. 3, 3rd ed. New York: John Wiley and Sons: 59–110.

Kramer, H.J. 1994. *Observation of the Earth and Its Environment. Survey of Missions and Sensors*, 2nd ed. Berlin: Springer–Verlag: 580 p.

Kruse, F.A. 1996. Identification and mapping of minerals in drill core using hyperspectral image analysis of infrared reflectance spectra. *Int. J. Remote Sensing* 17: 1623–1632.

Kruse, F.A., J.W. Boardman, A.B. Lefkoff, J.M. Young, K.S. Kierein-Young, T.D. Cocks, R. Jenssen, P.A. Cocks. 2000. HyMap: An Australian Hyperspectral Sensor Solving Global Problems—Results from USA HyMap Data Acquisitions: in Proceedings of the 10th Australasian Remote Sensing and Photogrammetry Conference, Adelaide, Australia, 21–25 August 2000, Causal Productions (www.causalproductions.com), CD-ROM.

Lawrance, C., R. Byard, and P. Beaven. 1993. *Terrain Evaluation Manual*. (Transport Research Lab, Department of Transport State of the Art Review 7. London: HMSO Publications Centre: 64–86.

Morain, S.A., A.M. Budge, eds. 1999. *Earth Observing Platforms & Sensors. Manual of Remote Sensing* v. 1. Falls Church, VA: American Society of Photogrammetry, CD-ROM.

Rosen, P.A., S. Hensley, H.A. Zebker, F.H. Webb, and ElJ. Fielding. 1996. Surface deformation and coherence measurements of Kilauea Volcano, Hawaii, from SIR-C radar interferometry. *Geophys. Res.* 101 E10: 23,109–23,125.

Wicks, C. Jr., W. Thatcher, D. Dzurisin. 1998. Migration of fluids beneath Yellowstone Caldera inferred from satellite radar interferometry. *Science* 282: 458–462.

Section II

Remote Sensing in Geologic Mapping and Resource Exploration

Introduction

For years now exploration for minerals and hydrocarbons has taken the geoscientist into increasingly remote, complex, and poorly understood areas. The days of locating obvious gossans and drilling out an economic mineral deposit, or drilling surface anticlines and discovering a new oil field for the most part are in the past. Today geophysical and geological surveys use advanced technology to map subtle indications of mineral alteration, structure, and hydrocarbon traps. Remote sensing is one high tech answer to surface mapping. Through the acquisition and analysis of airphotos, multispectral and hyperspectral aircraft and satellite imagery, thermal, radar, gravity, magnetic, radiometrics, and sonar images the explorationist is able to preview large areas quickly and economically and start mapping in places that are remote or structurally complex. One can quickly and inexpensively evaluate the potential of a region and pinpoint targets where field work is necessary to answer questions. Imagery carried into the field, especially with a global positioning system, allows rapid and accurate location of a worker's position and serves as an accurate base for plotting faults, contacts, and sample locations. This section provides a guide to the analysis of imagery to map geologic units, alteration, and geologic structures that can serve as a guide to oil and gas traps or conduits for mineralization.

The uses of remote sensing are as varied as the explorationist's imagination. From regional tectonic analysis to site-specific problems such as step out well site selection, there is imagery appropriate to every need. Generally, the problem that needs to be addressed and the detail on available maps will determine the imagery to be used.

Geologic Mapping

Geologic maps show the distribution of rock types, their age, and the deformation they have been subject to. Specialized maps for mineral exploration focus on alteration associated with mineral deposits, while many petroleum or coal exploration maps show the depositional environment of units (marine, eolian, fluvial) and structures (folds, faults, strike and dip of units). Soil scientists map the distribution of soil types for agriculture (gradations of silt, sand, clay, and loam end members) or engineering purposes (well-drained sandy, clay rich) based on grain size, constituent materials, and strength properties. Hydrologists map surface water features (lakes, ponds, streams, springs), fault zones, flood prone areas, and the potentiometric or piezometric surface (level to which water in a confined aquifer would rise if tapped by wells). At the very least, a generalized "stratigraphic column" (the vertical distribution of rocks in an area) can be generated showing which units are older and which are younger, what are their erosional properties (easily eroded vs ledge former) and color (gray, red, buff) or shade (light, intermediate, dark).

Remote sensing can contribute to each of these types of maps through analysis of imagery from one or more system or instrument. For example, true color imagery can provide information on the distribution and depth of surface water. Hyperspectral imagery can identify and show the distribution of argillic alteration associated with hydrothermal mineral deposits, clays associated with certain soil types, and hydrocarbon seeps. Predawn thermal imagery can show the location of springs and geothermal resources. Radar imagery can map geologic structure through a veneer of wind-blown sand or in thick rainforests, and can map offshore oil seeps. Radiometric surveys show the distribution of potassium feldspars (granites) or uranium, while gravity surveys indicate depth to basement, density contrasts, and structural highs. Magnetic surveys indicate type of basement, the location of magnetic mineral accumulations, fault zones, and magnetic (e.g., volcanic) versus nonmagnetic (e.g., carbonate) rocks. The power of this type of analysis comes from the integration of multiple complementary surveys.

Frontier Area Analysis

In frontier areas, remote sensing data, usually satellite imagery, are used prior to leasing to determine which areas contain favorable host rocks, intrusives, or alteration. One can identify which basins, or parts of a basin, contain structures large enough to be worth pursuing for oil and gas exploration. Types of alteration can be mapped, source rocks can be identified, and access can be determined. Areas to be leased can be delineated and prioritized.

In hydrocarbon exploration, the size and geometry of structures can be measured and their origin deduced, thus providing information regarding subsurface configuration and timing of development. When combined with subsidence modeling, this suggests which structures are pre-, syn-, or post-hydrocarbon generation (structures developed after hydrocarbons were generated and migrated would obviously be of less interest for oil and gas exploration unless they cause a trap to leak). Evidence of seals can be obtained by mapping shales or evaporites stratigraphically above reservoir units along a basin margin. Source rock distribution can be deduced from outcrops around the basin: these would be dark, organic rich, slightly radioactive, and easily eroded shales. The thickness of source units can be estimated and, together with information on the areal extent and total organic carbon from ground sampling, can be used to calculate the amount of hydrocarbons generated. The presence of intrusive or extrusive units may provide information on the thermal maturity of source rocks (and the distribution of volcanics is good to know when planning seismic programs for they often prevent energy from penetrating into deeper layers). Granitic outcrops in a dominantly volcanic setting can suggest the provenance of good hydrocarbon reservoir: volcanic rock fragments weather easily to clay-rich rock and tend to make low porosity, low-permeability reservoir, whereas granite-derived sediments tend to be quartz rich and make good-quality reservoir. Sediment distribution networks (surface drainage systems) may provide evidence pointing to the location of paleodepocenters and lateral variation of clastic mineralogy and grain size. Depocenters distant from their source areas tend to have finer grained, better rounded, and better sorted sediments with more stable minerals than proximal depocenters. Such remote sensing-based basin analysis can contribute significantly, for example, to Technical Evaluation Agreements between a company and foreign government.

In mineral exploration, one can attempt to map the extent and types of alteration, the nature and extent of a host rock, the location of highly fractured host rock, or where major conduits of mineralizing fluids exist or intersect one another to form a deposit. Imagery combined with magnetics can suggest the location of buried intrusives. Imagery merged with gravity data can suggest the depth to a buried intrusive. Paleochannels may be identified that can lead to placer deposits. An abrupt decrease in vegetation or change in vegetation type observed on imagery can help identify massive sulfides, ultramafic-associated deposits, or kimberlites because of the vegetation-stressing soils associated with these rock types. Limonitic (iron rich) or argillic (clay rich) alteration, identified on imagery and used in conjunction with a geochemical survey, should help identify mineral prospects. Identification of old workings would support the interpretation of a nearby deposit.

Images also serve a scouting or reconnaissance role: they can reveal tracks, old tailings, and abandoned workings; indicate which structures are already producing; and show precisely where recently drilled wells are located, the extent and layout of a seismic program by observing the cut lines, or the location of pipelines and oil storage tanks. Such information can be invaluable in planning a bidding strategy.

Lease Evaluation

During evaluation of a mining or petroleum lease, it is important to make the best possible geologic map of the surface in order to locate prospects, lay out geophysical or potential fields surveys, and identify surface structures (mineralized fault zones, anticlines and salt domes) to help interpret seismic and tie into geochemical surveys. At this stage in a program, it is important to achieve the proper balance of spatial resolution versus synoptic coverage. The image analysis should either replace or supplement existing geologic maps. It is important for an interpreter to understand the objective in each area in order to focus on specific evidence that may not be shown on an existing, generalized geologic map. For example, most geologic maps show the distribution of lithologic units and major structures, but leave out details of subtle structure or any mention of alteration. In some cases, units are generalized as "Cretaceous marine," for example, with no indication of the lithology present in a particular area. The image analyst, being aware that a hydrocarbon play is in fractured carbonates, can look specifically for fracture zones at the surface above carbonate units. Alternatively, surface limestone can be targeted if the objective is a Mississippi-Valley type lead–zinc–silver deposit or a manto or skarn-type mineral deposit.

In a petroleum lease, individual traps can be assessed and prioritized on the basis of size alone, or on structural complexity, or using the ratio of area under closure divided by "fetch," the area that would have sourced migrating hydrocarbons. Proximity to known production, seeps, pipelines and other infrastructure, mature source rock, and good reservoir outcrops all factor into an evaluation. In a mining concession proximity of alteration to known mines, abandoned tailings, old workings, or to mills, roads, and other infrastructure are important factors in prioritizing and evaluating a lease. Many of the parameters that define a prospect and determine its value can be derived entirely from imagery.

Finally, many concessions now require that the land be returned to its natural state after the development (mining, field production) is completed. Remediation requires knowing the natural state of an area prior to development. In order to avoid lengthy negotiations and potentially costly litigation, it is wise to establish and document an environmental baseline. Acquiring imagery is frequently the first step in understanding surface water drainage patterns, areas prone to flooding, the contour of the land, erosional patterns, sources of stream silting, landslides, and vegetation types and distribution.

Established Concessions

One might think that an area under lease has no need for remote observations and surveys. In many situations, that is not the case. When an offshore production platform notices a pressure drop during a waterflood, they will want to know if hydrocarbons are leaking to the seafloor. An oil sands Steam-Assisted Gravity Drainage development needs to know if they are causing surface heave above their steam injection pads. An Arctic oil field producer needs to know sea ice location, movement direction, and speed to plan a seismic program or plan tanker routes. A surface mine needs to determine the best location for a tailings pond. The owner of an old oil field in an urban area needs to know the extent and magnitude of surface subsidence. A field extension into a floodplain should take into account the extent of historical flooding. An underground mine is on a steep hillside: What is the likelihood of a landslide? A new road is being built to an old mine: What is the chance of avalanches in winter?

There are an infinite number of problems that can arise that would benefit from imagery to answer a question or help plan a safe development.

The Field Check

It is not possible to overemphasize the importance of field checking an image analysis. Although ground "truth" may be subjective in many ways (observational bias, sampling bias), there are certain measurements (e.g., dip of fault plane) and features (lithologic details, direction of slickensides, age of units) that are just not available on imagery, regardless how high the resolution. Most image-derived dips are, for the most part, estimates. For every feature described here, there are numerous other features that can have a similar appearance (Is it a radar shadow or a body of water? Is the dip steep or overturned?). Ideally, an image analyst will visit the field briefly prior to each project in order to determine the types of ground cover and erosional patterns, measure the lithologic section, and obtain some strikes/dips for control. In reality, it is not often that an image interpreter gets the opportunity to do field work even after a project is completed. But remember: image analysis remains a subjective interpretation until one has been in the field and confirmed the mapping.

The purpose of a thorough analysis is to create a map of identifiable structures, alteration, and units, and highlight those areas where critical questions remain. The purpose of field mapping is to focus on those problematic areas and answer the outstanding questions. The interpretation thus serves many objectives, including the following:

1. Provide a map in a short period of time that can answer key questions and direct field work where questions remain, thus saving the time and expense of a drawn-out field season.
2. Provide efficient access to remote areas.
3. Optimize the logistics of a geochemical sampling campaign, seismic survey, or drilling program.
4. Assist with the interpretation of geophysical, geochemical, and other surveys.
5. Keep an exploration program confidential prior to ground work, leasing, or bidding.

Field work not only completes the analysis, but also adds to the image interpreter's experience and knowledge, making the entire process more reliable for future projects.

4

Recognizing Rock Types

Chapter Overview

Rock types generally can be grouped according to characteristics such as resistance to erosion, color, texture, jointing, type of drainage development, cross-sectional shape of gullies and valleys, soil development, and vegetation cover, among others. Color or tonal banding may be a result of metamorphism or sedimentary layering. Both sandstone and limestone can be jointed, but only the limestone terrain will have sinkholes and deranged drainage. Although one cannot usually identify surface units, it is often possible to group them into recognizable categories such as sedimentary or igneous, or easily eroded versus resistant.

When starting to work in a new area, the analyst/image interpreter may not have a clear idea about the local stratigraphy. In this case, it is necessary to generate a "photogeologic" stratigraphic column based not on age or field criteria, but on features that can be distinguished on imagery. Photogeologic units are defined by their tone, texture, color, erosional characteristics, and stratigraphic position. It should be possible to distinguish between various lithologies such as sandstones and shales (if the vegetation cover is not too dense), but it may not be possible to identify lithologies in many cases. One can still use these photogeologic units to evaluate the structure of an area or to presume their engineering characteristics (infiltration, slope stability, resistance to erosion).

Igneous Rocks

It is useful to map the presence and distribution of igneous rocks in an area of interest. Many intrusives, such as porphyries, are host to ore deposits. Others provide the hydrothermal solutions necessary to mineralize country rock. Dikes compartmentalize a petroleum reservoir, but usually do not contribute significant heat flow beyond 1 or 2 m into the host rock. On the other hand, an intrusive stock can contribute significant heat and assist in generating hydrocarbons or worse, cause them to become over mature. Basalt flows make it difficult to interpret magnetic data and to get enough seismic energy into the ground for good reflections. Seismic programs generally try to avoid lava flows at or near the surface.

Intrusive Igneous Rocks

Intrusive rocks usually occur as rounded outcrops covering tens of square kilometers and are fairly homogeneous in texture. Intrusives tend to have uniform plant cover because of their homogenous composition. These rocks may have arcuate cooling joints, faults, or dikes along their margins (concave inward) and are characterized by multiple, often orthogonal joint systems (Figure 4.1). A buried intrusive center may be indicated at the surface only by the conjunction of three sets of fractures at 120°, similar to the pattern formed at an oceanic triple junction. The large crystals found in coarse-grained intrusive rocks tend to erode readily into thick, well-drained soils; fine-grained intrusives are generally more resistant to erosion. Granite terrain tends to weather into smooth, rounded landscapes characterized by domes (Yosemite and the Sierra Nevada) and large rounded boulders. Granitic rocks can be distinguished by radiometric surveys due to their high potassium mineral content. Felsic intrusives tend to be light in color, whereas mafic intrusives are dark colored (Figure 4.2). On gravity surveys, felsic terrains are less dense than mafic landscapes. Mafic minerals readily break down under surface conditions into clays and form gentler topography than granitic terrains. Ultramafic, carbonatite, and kimberlite units usually have anomalous or sparse vegetation due to exotic soil minerals and

FIGURE 4.1
True color image of Precambrian granites of the Sinai Peninsula near Sharm el Sheikh, Egypt. Note the multiple, orthogonal joint sets, faults, and dikes. (Courtesy of DigitalGlobe and Google Earth © 2012 Google.)

FIGURE 4.2
Light (acidic) and dark (basic) intrusives, Sinai Peninsula northwest of Mt. Sinai, Egypt. (Image courtesy of DigitalGlobe, GeoEye, and Google Earth, © 2012 Google.)

clays that cause poor soil drainage (Ray, 1960; Miller and Miller, 1961), are slightly toxic to plants, or weather into sinkholes and depressions. In fact, kimberlites are often found in slight depressions or under ponds and lakes.

Vein-type mineral deposits are found in large fracture systems (faults, cooling joints) associated with the margins of intrusives. The vein systems may be emplaced along faults concentric to the core (ring faults), form along faults that radiate outward from the intrusive center, or form at the intersection of these radial and concentric faults. Large, late-stage, dominantly quartz veins are recognized by their association with faults, their resistance to erosion, associated bleaching or iron staining (alteration), and/ or their light color.

Contact metamorphic halos and skarns around the margins of intrusive centers are often more resistant to erosion than the surrounding country rock or may be associated with hydrothermal alteration and related color and texture changes.

Extrusive Igneous Rocks

Extrusive rocks frequently have distinctive landforms. Volcanoes, cinder cones, and flow lavas are all easily recognized surface features. Volcanoes have a classic cone or shield shape, usually with a central crater or caldera. Flows spread across a landscape, flowing into low areas and burying older topography and structures under a nearly flat, usually dark layer. A simple example of using a digital elevation model (DEM) to show topography in gray shades and color serves to illustrate how landforms can be interpreted from topography alone (Figures 4.3 and 4.4). These examples show classic cone-shaped stratovolcanoes. The DEM of Mt. Elgon clearly shows the composite cone of the volcano as well as the central crater. The merged DEM and aeromagnetics image of Mt. St. Helens is an example of data fusion used to identify a geologic feature.

Calderas are often characterized by ring dikes, radial dikes, and craters filled by lakes. Sometimes lava flows can be seen extending from a cone or vent. Laccoliths, shallow folds caused by volcanic intrusion, may be indistinguishable from normal folds unless aeromagnetic surveys exist in the area and reveal a magnetic core to the fold. Most volcanic

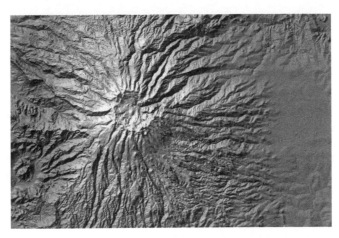

FIGURE 4.3
DEM of Mt. Elgon volcano, Kenya and Uganda, generated by the Shuttle Radar Topography Mission. Note the classic cone-shaped composite volcano with centralcaldera. (Courtesy of NASA. http://photojournal.jpl.nasa .gov/catalog/PIA04958.)

FIGURE 4.4
DEM merged with aeromagnetics over Mt. St. Helens, Oregon. Magnetic highs, shown in reds and yellows, coincide with the peak at Mt. St. Helens as well as indicate other, buried intrusives. (Courtesy of USGS.)

FIGURE 4.5
True color image of basalt flows at SP Crater near Gray Mountain, Arizona. Youngest basalts are darkest since they have the least weathering and vegetation cover. Some flows originate at cinder cones, others at faults. (Courtesy of USDA and Google Earth, © 2012 Google.)

rocks are finely crystalline, form layers highly resistant to erosion, and form steep cliffs. However, air-fall tuffs may be difficult to distinguish from easily eroded sedimentary rocks unless they are welded. Tuffs tend to be light colored and erode into a badlands topography. Sulfotaric alteration associated with extrusive volcanics is recognized by its rusty to yellow color and/or bleached rocks that are the result of breakdown of pyrite and other iron minerals and alteration of feldspars to clays.

Basalts are essentially black and highly fluid and therefore flow through topographic lows (Figure 4.5). They flow from calderas and vents and form broad, gentle shield volcanoes rather than steep composite cones. Quite often river meanders are preserved as topographic highs because of basalt flows that filled the stream channel. In other areas, flood basalts form dark, resistant layers that cover hundreds of square kilometers. The relative age of flows can be determined by noting the density of vegetation growing on them: no vegetation indicates a relatively young age (a few tens to hundreds of years); dense vegetation cover indicates an older age. Most volcanic terrains have associated reddish soils that form as a result of rapid weathering of iron minerals (Drury, 1987).

Metamorphic Rocks

Metamorphic rocks are recognized by their disharmonic, ptygmatic, or flexural flow folding, or by pervasive, parallel lineations (foliation or cleavage). Metamorphic units are crystalline and therefore generally resistant to erosion. In contrast to intrusive units, they tend to be heterogeneous, and thus do not have uniform surface texture, tone, color, or erosional characteristics (Figure 4.6). These rocks may be revealed by a pervasive, aligned fine scale, often curving texture. They may be harder or softer than the surrounding terrain, depending on composition and level of metamorphism.

It is not always possible to discriminate between banding in metamorphic rocks (due to compositional changes or cleavage) and sedimentary layering unless there is a recognizable foliation or complex metamorphic folding. Foliation may also be confused with glacial scour or striations in boreal or formerly glaciated terrain. Low-grade metamorphics may be indistinguishable from sedimentary rocks, whereas high-grade metamorphic terrain may appear nearly identical to intrusive igneous landscapes.

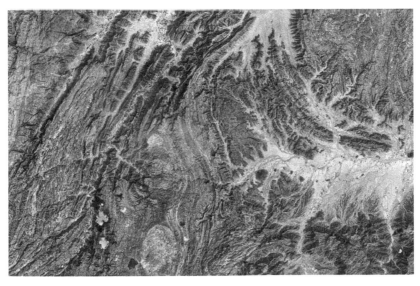

FIGURE 4.6
True color image of metamorphic outcrops near Shabwa, Yemen. Note layering, faulting, and complex folding. (Courtesy of GeoEye, DigitalGlobe, CNES/SPOT Image, and Google Earth, © 2012 Google.)

Sedimentary Rocks

The most revealing feature of sedimentary rocks is depositional layering. Sedimentary rocks are characterized by alternating weak and resistant layers of varying thickness and color. The recognition of layering depends on contrasts in grain size, color, resistance to erosion, and composition, which are in turn influenced by the weathering environment, erosion, and vegetation cover.

Sandstones

Sandstones, composed largely of quartz, feldspar, and lithic fragments, tend to form resistant ridges (Figure 4.7). This, of course, depends on the degree of cementation and type of cement. In most environments, silica cements are more resistant than calcite cement or iron oxide cement, both of which are more resistant than clay or gypsum cement. The relatively high porosity generally associated with sandstone causes high infiltration and low runoff of precipitation and results in widely spaced drainages. High infiltration also favors vegetation, like pine trees, with deep root systems: every county in the semi-arid western United States seems to have a "pine ridge" that is usually a sandstone outcrop surrounded by grassy, shale-filled valleys. Sandstones display fracturing prominently because they tend to be resistant and brittle, and the fractures weather more readily. Gullies are generally V shaped in cross section, becoming slightly U shaped in siltier soils (Way, 1978). Conglomerates and their erosion product, lag gravels, also tend to be highly resistant to erosion.

Sandstones can be thin bedded, interbedded with shales or carbonates, or massive. On high-resolution imagery, it may be possible to distinguish cross bedding, particularly in massive sandstones. Sandstones are generally ledge formers and more resistant to erosion in arid and arctic environments. In humid and tropical climates, they form more rounded outcrops.

Shales

Shale, the most common sedimentary rock, can be composed either of clay-sized grains (<3.90625 μm or <0.00015 in) or clay minerals. The small grain size and composition makes them easy to erode, and the resulting landscape is either a highly dissected badlands in an

FIGURE 4.7
The resistant Cretaceous Dakota sandstone forms a prominent east-west hogback south of Durango, Colorado (arrow). Oblique view toward the east-northeast. (True color image courtesy of TerraMetrics and Google Earth, © 2012 Google.)

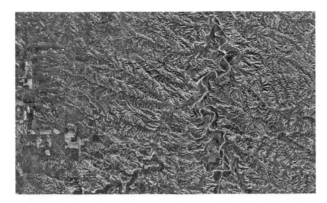

FIGURE 4.8
Badlands are developed in shale and form highly dissected terrain in arid climates. Note the closely spaced drainage in the badlands of North Dakota. (True color image courtesy of USDA, DigitalGlobe, and Google Earth, © 2012 Google.)

arid climate or flat to gently rolling valleys and rounded knobs in a humid climate. The thin bedding that may exist over large intervals also contributes to ease of erosion. Drainage is closely spaced because infiltration is low and runoff is high (Figure 4.8). Joints are not well developed because the units tend to be ductile (Lattman and Ray, 1965). Gullies range from U shaped in cross section to gently rounded with increasing clay in the soil (Way, 1978). Outcrops can be sparse. Banding is a result more of different colored layers than of varying resistance to erosion.

Shales are extremely susceptible to landslides, slumps, and downslope creep, particularly when they contain the expanding clay minerals montmorillonite, smectite, bentonite, or illite. Slopes as low as 1 to 2 degrees in shale-rich zones have been known to fail when waterlogged.

Carbonates

Carbonates are crystalline, and the interlocking crystal structure helps them resist physical weathering. In arid and arctic climates, where physical weathering dominates, they form ledges and are easily fractured by tectonics or freeze-and-thaw. In humid climates, they are still resistant, but more rounded in form. Carbonates usually weather light colored, although they can also be quite dark. The distinguishing feature of carbonate rocks in all climates is the karst or sinkhole caused by solution of the rock by slightly acidic rain and surface water. These dissolution features form a mottled, pockmarked terrain (Figures 4.9 and 4.10) and ultimately hums, or buttes, in humid climates. Drainage patterns are considered "deranged" as a result of streams suddenly ending in sinkholes. There may be no recognizable drainage pattern, since all precipitation infiltrates through joints. Undisturbed, flat lying carbonates have rhombic to orthogonal joint patterns at all scales. In arid areas, carbonates develop little soil, form steep-sided valleys, and concentrate vegetation in sinkholes where moisture accumulates and some soil is developed (Figure 4.11). In humid areas, the vegetation can be dense. Residual clay associated with carbonates in humid areas often form a "terra rosa" or red soil.

There is no effective way to distinguish between limestones and dolomites except with multispectral or hyperspectral imagery in areas with little or no vegetation cover (Bellian et al., 2006; Ouajhain et al., 2011). Dolomites tend to fracture more readily. Otherwise, if one knows that, for example, a dolomite is black and a limestone is gray in a specific area, one can discriminate them on this basis.

FIGURE 4.9
Karst topography and deranged drainage is typical of carbonate terrain. This true color image of a tropical, low-relief karst terrain near Kunche, Mexico, shows essentially flat terrain with an occasional cenote (sinkhole). There is no throughgoing drainage pattern. (Image courtesy of INEGI/CNES/SPOT Image and Google Earth, © 2012 Google.)

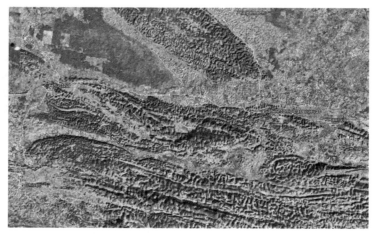

FIGURE 4.10
Karsted carbonate highlands can be recognized by their hummocky, or pockmarked texture in this folded highland area in Chiapas, northern Guatemala. (True color image courtesy of USGS, GeoEye, CNES/SPOT Image, and Google Earth, © 2012 Google.)

Evaporites

Evaporites such as halite and gypsum/anhydrite exist at the surface only in arid climates because they are water soluble. They are recognized because they are generally very bright, and can be nearly white in color. Evaporites accumulate as recent or ancient deposits in dry lake beds and evaporating, shallow seas. They occur as layered beds, diapirs, and, rarely, as sand dunes (e.g., White Sands, New Mexico). Subsurface deposits are of greater interest because they act as seals and decollements, and may generate structural traps for hydrocarbon accumulations.

Evaporites are recognized in the subsurface by the land forms they generate. Dissolution along faults can cause collapse in overlying units (Figure 4.11). Salt becomes mobile under

FIGURE 4.11

Limestones of the Miocene Upper Dam Formation on this peninsula in western Qatar form northeast-southwest linear karsts that appear to have coalesced along probable fault zones. These linear karsts form surface "synclines" with inward-dipping beds, and they appear lighter than background when filled with sand. (Image courtesy of DigitalGlobe and Google Earth, © 2012 Google.)

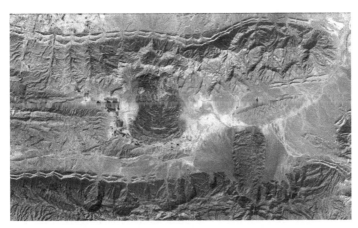

FIGURE 4.12

Salt-cored anticlines and their salt glaciers are common in this arid part of the Zagros fold belt near Gerash, Iran. Evaporites in the core of these two folds have breached the surface to form salt glaciers (dark areas, center and center right of image). (Courtesy of GeoEye, CNES/SPOT Image, and Google Earth, © 2012 Google.)

pressure and tends to form diapirs (roughly concentric domes) or anticlines in otherwise nearly flat strata (e.g., Gulf of Mexico, Persian Gulf). In folded terrain, evaporites can facilitate anticlinal folding over basement faults through ductile flow and bedding plane slip (e.g., Paradox Basin, Utah). They are found in the mobile core of thrusted folds, and occasionally reach the surface in the eroded core of folds or as salt glaciers (e.g., Zagros, Iran, or Atlas, Algeria; Figure 4.12). Along the south flank of the Salt Ranges in Pakistan, the Cambrian salt that acts as a decollement for thrusting is exposed as a bright linear zone along the mountain front. In areas where evaporite layers have been dissolved by ground water, they can create a chaotic pattern of irregular synclinal collapse structures (e.g., Qatar, Figure 4.13), long, linear synclinal collapse features along fracture zones (Paradox Valley, Utah), or domes with collapsed cores (one proposed origin for Upheaval Dome, Utah).

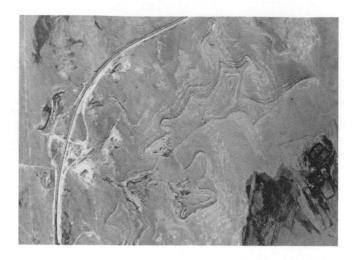

FIGURE 4.13
Chaotic collapse of Miocene Upper Dam Formation limestone can be seen where gypsum of the Lower Dam Formation has been dissolved by groundwater in the subsurface near Salwa, Qatar. (True color image courtesy of DigitalGlobe and Google Earth, © 2012 Google.)

References

Bellian, J.A., C. Kerans, R.A. Beck, Y. Price, K. Nedunuri. 2006. *Calcite-Dolomite Delineation using Airborne Hyperspectral Data for Ordovician Paleokarst Mapping*. AAPG Discovery Digest (abs.) #90052.

Drury, S.A. 1987. *Image Interpretation in Geology*, Chap. 4. London: Allen & Unwin: 72–88.

Lattman, L.H., R.G. Ray. 1965. *Aerial Photographs in Field Geology*, Chap. 5. Austin: Holt, Rinehart and Winston: 162–188.

Miller, V.C., C.F. Miller. 1961. *Photogeology*, Chap. 7. New York: McGraw-Hill Book Co.: 80–100.

Ouajhain, B., K. Labbassi, P. Launeau, R. Baissa, H. Jabour, A. Gaudin, P. Pinet. 2011. *Carbonate Mineral mapping and Geologic Outcrop Modeling using Hyperspectral in the Agadir Basin Targeting the Jurassic Sequence, Western High Atlas, Morocco* (abs.) AAPG Search and Discovery Article #90137.

Ray, R.G. 1960. *Aerial Photographs in Geologic Interpretation and Mapping*. U.S. Geol. Survey Professional Paper 373: 16–19.

Way, D.S. 1978. *Terrain Analysis: a Guide to Site Selection Using Aerial Photographic Interpretation*. New York: Van Nostrand Reinhold: 31–57.

5

Recognizing Structure

Chapter Overview

In order to recognize structures, it is necessary to begin by understanding how structures form and how they influence surface processes. Simple structures are examined first, followed by increasingly complex forms. The result of a structural analysis is a structure map or structural form line map that integrates what is seen at the surface (geomorphology) and what is known or presumed about the subsurface from well data, seismic data, and potential fields data. This map should reveal the structural fabric of the region and show individual prospect-scale features. Structure is an essential part of basic geologic maps, and it will help plan seismic, geochemical, potential fields, and geochemical sampling programs by providing the location and orientation of folds and faults. The structure analysis will also assist in the interpretation of these surveys and well data.

Most imaging remote sensing systems observe the surface and near subsurface. Digital Terrain Models (DTMs) are representations of surface topography. Potential fields data sense the deeper subsurface. All these systems are useful in understanding surface and deep structure.

Undeformed Terrain

It may be that there are no truly undeformed terrains. However, such a landscape serves as one end-member in the range of deformation from simple to severe. What landforms might be expected in an area with essentially flat-lying, unfaulted, and unfolded strata?

In this case, drainage will develop into a dendritic network on a flat plain (Figure 5.1) (Howard, 1967). As erosion progresses toward base level, terraces will develop along the margins of the flood plains. These terraces correspond to the various layers of rock being exposed (Cotton, 1968). As the terraces erode, erosional remnants are left standing as plateaus and mesas, then as buttes, and finally as small rises. An example of this is expressed in Mesozoic rock of great beauty in Monument Valley, Arizona (Figure 5.2).

In areas with good outcrops, the trace of bedding in flat-lying units appears as closed polygonal to circular tonal bands on slopes and hillsides (Figure 5.3).

It is now possible to begin to recognize geologic structure based on deviation from the undeformed condition described above.

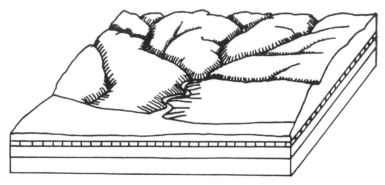

FIGURE 5.1
Diagram illustrating the development of dendritic drainage on a homogeneous, flat or gently inclined surface.

FIGURE 5.2
Buttes and mesas are erosional remnants of flat-lying strata, such as this example from the Monument uplift, northern Arizona.

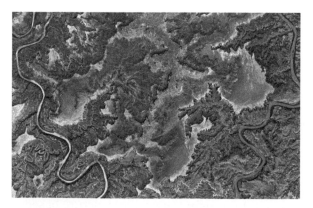

FIGURE 5.3
Irregular closed tonal bands indicate flat-lying bedding in the White Rim area, Canyonlands National Park, Utah, near the confluence of the Colorado and Green Rivers. (True color image courtesy of DigitalGlobe, GeoEye, and Google Earth, © 2012 Google.)

Recognizing Dip

"Dip" is the angle at which a rock layer is inclined from the horizontal. Dips may vary from gentle to vertical and overturned, with different topographic expressions for each range.

Gently inclined layers tend to form long, dip-slope drainages (known as "consequent" drainage) opposed by short tributaries flowing down the bedrock escarpment (known as "obsequent" drainage). Over a large area, long semi-parallel dip-slope drainage is known as a "trellis drainage" pattern (Figure 5.4). The landform commonly associated with gentle dip is the cuesta (Figure 5.5). A *cuesta* is a ridge with a gentle slope on one side (dip slope) and a steep slope on the other.

All stereo airphotos have some degree of vertical exaggeration that must be considered when estimating dip (see Chapter 1, section on Relief; Chapter 3, Instrument Systems/Cameras). Vertical exaggeration makes objects appear taller or steeper than they really are and can be helpful in estimating dips in low-relief terrain (Figures 5.6 and 5.7). Exaggeration is a function of both camera focal length and photobase, the distance between adjacent image

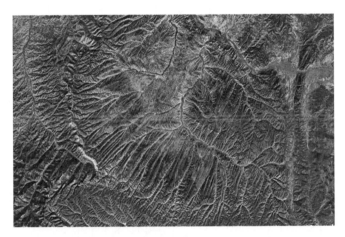

FIGURE 5.4
Trellis drainage pattern, perhaps enhanced by jointing, is developed on the Eocene Uinta Formation southwest of Meeker, Colorado, in the Piceance basin. Drainage is developed on gentle dip slopes. (True color image courtesy of TerraMetrics and Google Earth, © 2012 Google.)

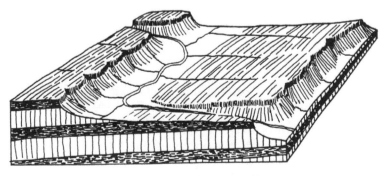

FIGURE 5.5
The cuesta, a long sloping surface opposed by an escarpment, is generally (but not always) a dip slope.

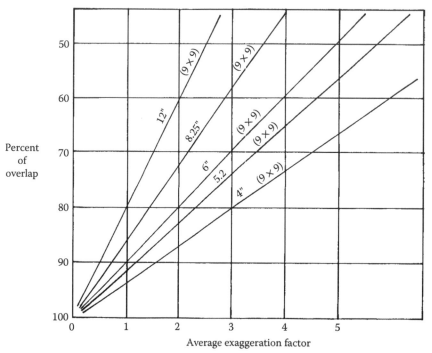

FIGURE 5.6
Chart relating the average exaggeration factor to the amount of overlap between adjacent stereopair, print size (in parentheses), and camera focal length in inches. (From Thurrell, R.F. Jr. 1953. Vertical exaggeration in stereoscopic models. *Photogram. Engr.* 19: 579–588. Reprinted with permission from the American Society for Photogrammetry & Remote Sensing, Bethesda, MD, www.asprs.org.)

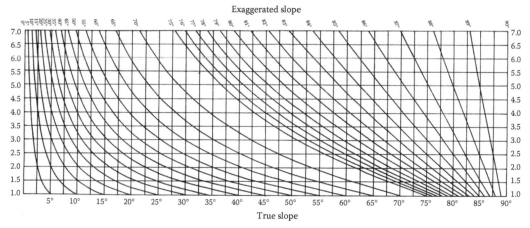

FIGURE 5.7
Chart showing the relation between true slope (or dip), exaggeration factor (vertical axis), and exaggerated slope (dip). (Modified after Ray, R.G. 1960. *U.S. Geol. Survey Prof. Paper.* 373: 64–66.)

centers. A shorter focal length or longer photo base provides greater apparent relief. On nonstereo imagery, such as Landsat, it is not too difficult to recognize wide, illuminated slopes paired with short, shadowed slopes (or vice versa) and estimate the inclination of strata. Several satellite systems (e.g., SPOT) have stereo capability and thus make it easier to estimate dip.

In moderately inclined strata (5–30°), the dip direction and relative magnitude can be estimated using the "Rule of Vs" (Figure 5.8). This rule states that the trace of bedding (either outcrop or color bands) in a valley or water gap forms a V-shape pointing in the direction of dip. This is true in all cases unless topographic relief is greater than the dip of the units (e.g., gently dipping units exposed along a steep cliff). In such cases, the bedding trace must be plotted on topographic maps and true dip could be calculated by solving a 3-point problem (Figure 5.9). When digital elevation models exist, the calculation of dip can be done manually or by using a computer (Morris, 1991). For example, Reif et al. (2011) described software that they used to calculate dip angle and the strike of bedding mapped on remote sensing imagery in Iraq using satellite images draped on digital elevation models. The authors demonstrate that, under favorable conditions (moderate dip, sparse vegetation cover, good stratigraphic contrast) computed strikes/dips are within 10° of field measurements.

As layers become increasingly inclined, the dip slope becomes narrower and the opposing escarpment becomes wider, until in the range of about 45–90° the ridge, or hogback that is formed is almost symmetrical. A "hogback" is a ridge with slopes that are almost equal in steepness. "Flatirons" are short, triangular hogbacks forming a ridge, generally on the flank of a mountain, which are characteristic of steeply dipping resistant rock.

In near-vertical beds, it is difficult to determine dip using the V-shaped bedding trace. In the case of thinly bedded units, such as sandstone and shale, however, it may be possible to see the trace of multiple beds on the escarpment, but not on the dip slope (Figure 5.10). It has been noted (Miller and Miller, 1961) that the height of hogbacks decreases with increasing dip (Figure 5.11). Thinning and thickening of an outcrop along strike may indicate changes

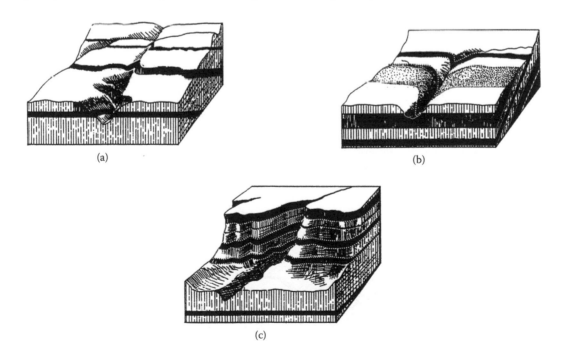

(a)   (b)

(c)

FIGURE 5.8
The "Rule of Vs". (a) The trace of beds that dip upstream form a "v" that points upstream. (b) Beds that dip downstream form a "v" pointing downstream. (c) When topographic relief is steeper than the dip of the layering, the "v" points upstream regardless of the dip. Solving a three-point problem is then required to determine dip.

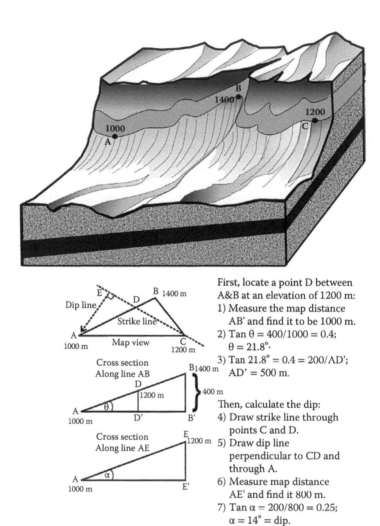

First, locate a point D between A&B at an elevation of 1200 m:

1) Measure the map distance AB' and find it to be 1000 m.
2) Tan θ = 400/1000 = 0.4; θ = 21.8°.
3) Tan 21.8° = 0.4 = 200/AD'; AD' = 500 m.

Then, calculate the dip:

4) Draw strike line through points C and D.
5) Draw dip line perpendicular to CD and through A.
6) Measure map distance AE' and find it 800 m.
7) Tan α = 200/800 = 0.25; α = 14° = dip.

FIGURE 5.9
Solving a three-point problem requires knowing the elevation of at least three points along the trace of a marker bed, and the distances between those points. This can best be obtained when the bedding is traced or transferred onto a topographic map.

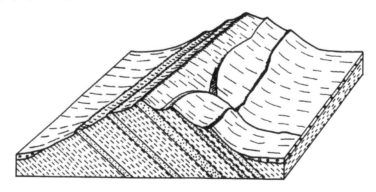

FIGURE 5.10
On nearly symmetrical ridges, the trace of bedding forms a "v" on the dip slope and multiple straight lines or tonal bands on the opposing escarpment.

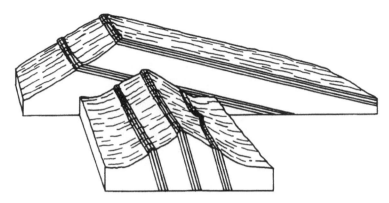

FIGURE 5.11
The relative height of a hogback decreases as the dip increases. (From Miller, V.C., C.F. Miller. 1961. *Photogeology*. New York: McGraw-Hill Book Co.: 80–100.)

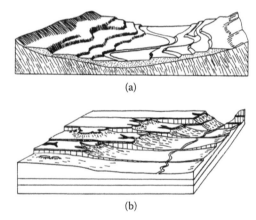

(a)

(b)

FIGURE 5.12
Erosional and stratigraphic surfaces. (a) Pediment surfaces carved by a down cutting river. (b) Stratigraphic terraces formed by successively exposed bedding surfaces. (From Cotton, C. 1968. *Geomorphology*. Christchurch: Whitcombe & Tombs Ltd.: 76–95.)

in dip (steepening or flattening, respectively) or may be due to a facies change or true thickness change. If the change occurs in several adjacent units, it is probably a structural change (change in dip or thickening across a fault or hingeline). If a unit appears to dip in one direction and, as one moves along strike, the outcrop thins and then thickens and appears to dip in the opposite direction, the bed may have become vertical and then overturned. The common flank of some anticline–syncline pairs appear to change along strike such that all bedding appears to dip in the same direction. This is caused by an overturned common flank where the fold hinge is so tight that it is not recognizable.

A gently inclined surface lacking outcrops (such as in a shale zone) is characterized by its "geomorphic dip." Geomorphic dip often coincides with true dip, especially for gentle surfaces. However, it is possible to have a gently inclined flat surface that is completely unrelated to the dip of bedding. This is common in the case of "pediment surfaces," where rivers or flash floods carve gently inclined surfaces into bedrock between a mountain front and valley bottom (Figure 5.12). Such surfaces are generally gently inclined toward the valley bottom. Gently inclined smooth surfaces can also be the

FIGURE 5.13

Examples of terraces. (a) Tertiary alluvial terrace gravels form the Meeteetse Rim along the Graybull River, Bighorn basin, Wyoming. (True color image courtesy of USDA Farm Service Agency and Google Earth, © 2012 Google.) (b) Wave-cut terrace along former shore of glacial Lake Bonneville, taken from near the Idaho-Utah border in August 2010. (Photo courtesy of Erik W. Klemetti, Denison University.)

result of "alluvial terraces" formed by fluvial processes within basins and preserved as erosional remnants because of a cap of resistant lag gravel (Figure 5.13a). "Stratigraphic terraces" are the result of successively exposed bedding surfaces. "Wave-cut terraces" form smooth, gently inclined surfaces along a coastline or around drying inland lakes (Figure 5.13b).

Anticlines, Domes, and Uplifts

Anticlines are, without question, the most important structure in the search for oil and gas. The range of shapes is large and includes, among others, circular to elliptical patterns with four-way closure, plunging noses with three-way closure, and gentle changes in the strike of bedding (Figure 5.14). The anticline is referred to as a "positive" structure or "structural high," and is commonly expressed as a topographic high at the surface. Just as commonly associated with anticlines is "inverted topography," that is, a topographic low

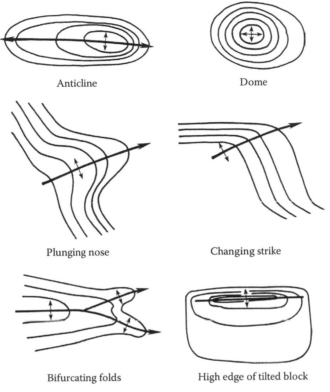

Anticline   Dome

Plunging nose   Changing strike

Bifurcating folds   High edge of tilted block

FIGURE 5.14
Some common fold forms depicted by structure formlines.

where resistant layers have been breached by erosion and the easily eroded core is exposed as a valley surrounded by outward-dipping ridges (e.g., Axial Basin anticline, Colorado; Paradox anticline, Utah).

Anticlines generally form as elongated folds in layered strata due to horizontal shortening at right angles to the fold axis. They may also form by draping of sediments over the edge of a faulted block ("forced folds"), or along the leading edges of thrust faults and over thrust ramps. Domes, a subset of anticlines that are nearly circular, generally develop as a result of density contrasts between layers, as is the case when a low-density layer (salt, gypsum, overpressured shale) lies beneath a high-density layer (sandstone, carbonate). This disequilibrium results in upward movement of the low-density layer as a "diapir," a structure where overlying rocks have been ruptured by squeezing out of the lower density ductile core material. If the overlying layers are pierced, a diapir may also be referred to as a "piercement" dome. Salt-cored anticlines have been known to form in this manner, as in the Paradox basin of Utah, or Dukhan anticline in Qatar. In these cases, the fold was probably initiated over a basement fault, thus inheriting its elongated shape. With diapirism, one would not expect folding to extend beneath the ductile layer (Figure 5.15). If it is known, for example, that the only reservoir unit in an area lies below an evaporite, and the evaporite has formed piercement domes, there would be no reason to drill the flanks or crests of these domes as the reservoir unit is unaffected by the doming. If the unit sealing this reservoir has been broken by faulting, however, one might expect hydrocarbons to leak upward and migrate into the flanks and crest of the dome.

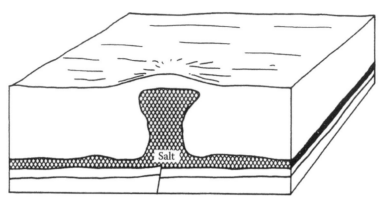

FIGURE 5.15
Most salt diapirs are expressed as domal topographic highs. Note that there need not be folding beneath the salt.

FIGURE 5.16
Drape of Upper Cretaceous Wata and Galala limestones over the southeast block corner (arrow) of Gebel Somer, western Sinai, Egypt. (True color oblique image courtesy of CNES/SPOT Image, DigitalGlobe, GeoEye, and Google Earth, © 2012 Google.)

"Horsts" are uplifted blocks bounded by faults on one or more sides. They tend to be rectangular in shape, and accumulate hydrocarbons in sediments draped over the block, at high corners, or at updip fault truncations along their flanks. Horsts are often associated with forced folds, which are recognized by their flat or gently inclined upper surfaces, near-vertical bedding around their flanks, and abrupt changes in strike at block corners (e.g., Bighorn and Beartooth uplifts, Wyoming; Gebel Somer, Egypt Figure 5.16).

Surface Folds

Domes and anticlines are expressed at the surface by opposing dips in outcrops on opposite flanks of the structure. In many cases, one can follow bedding along a circular to elliptical path, whereas in other circumstances, the bedding merely changes strike a few degrees. The intersection of bedding with the Earth's surface may be thought of as structural formlines similar to structure contours. "Closure" is a component of a structural hydrocarbon trap: four-way closure refers to dip in all directions. Ideally, bedding will

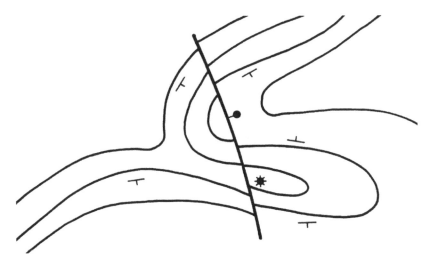

FIGURE 5.17

Open folding, shown here by structure formlines, requires a fault or other type of up-dip closure to complete a hydrocarbon trap, shown by the well symbol.

close on itself and a significant exploration target will be revealed. The structure need not be a closed fold: an updip facies change or a fault crossing the structure is then required for full closure (Figure 5.17).

Identifying dip on imagery is one way to recognize surface folding. Another is to map curvature on digital elevation models. Modern analytical methods can compute curvature if a resistant unit is exposed at the surface and the topography resembles the structure. This technique has been described for mapping folds in northern Iraq using a DEM generated by the Shuttle Radar Topography Mission (Burtscher et al., 2012). Curvature analysis is particularly useful for mapping folds, fold hinges, and potential fracture zones at the surface as guides to structural traps at depth.

Buried Folds

Buried or "blind anticlines" occur in areas lacking surface outcrops. Buried folds require geomorphic evidence in order to be recognized. There are three reasons these structures are expressed geomorphically:

1. Draping as a result of differential compaction. Differential compaction of semi-consolidated or unconsolidated sediments is a result of dewatering of shales (Mollard, 1957; DeBlieux, 1962; Foster and Soeparjadi, 1974; Conrad, 1977). The compaction causes overlying material to drape over preexisting structures. The amount of compaction depends in large part on the type of material and stratigraphic thickness of the material over the structure (e.g., shales and gypsum are more readily compressed and dewatered, and these are more prone to plastic flow than sandstones and carbonates). The surface expression is also a function of the number of unconformities between the structure and the surface. The surface expression of the buried anticline may be offset slightly from the actual position of the structure at depth, particularly if the structure is recumbent (inclined axial plane) or bedding over the buried fold is inclined (in which case, the structural crest will migrate up dip).

2. Reactivation. Folds tend to form in areas that are predisposed to deformation, that is, at points of weakness. They are, therefore, subject to reactivation with renewed tectonic activity (Weimer, 1980). Indeed, outcrops and isopachs can show repeated episodes of folding in a single structure (e.g., Marsh Creek anticline, North Slope, Alaska). In formerly glaciated terrain, old structures have been reactivated by isostatic rebound (Kupsch, 1956; Maslowski, 1985).

3. Effect on overlying materials. Buried folds can sometimes affect overlying materials by influencing soil development, vegetation patterns, and present-day deposition or erosion.

Topographic Expression of Folding

Buried structural highs can be expressed as gentle topographic highs in alluvial cover (Figure 5.18). This is a result either of compaction over the structure or reactivation of the folding. However, one must be cautious not to interpret every erosional remnant as a buried structure. It may not always be possible to distinguish between erosional remnants and buried structure. In permafrost terrain, one finds pingos, or frost heaves, that appear to be symmetrical domal uplifts. These are purely surface effects and unrelated to structure. Many erosional remnants, and even buried volcanic/intrusive features, appear similar to uplifted structures.

In arid environments, the "zero edge of alluviation" is where alluvial fans (distributary drainage) meet a pediment or bajada (tributary drainage). Subtle structure is suggested where the edge of alluviation forms an embayment due to diversion of fan deposition (Lattman, oral communication, 7/78).

Color Anomalies Indicating Folds

Color or tonal "anomalies" may be associated with uplifts. These often circular to elliptical tonal changes may be the result of older material being exposed by erosion in the core of an uplift, or by changes in soil type, soil chemistry, moisture differences, or changing vegetation patterns as one proceeds from a low area onto an uplifted area. One word of

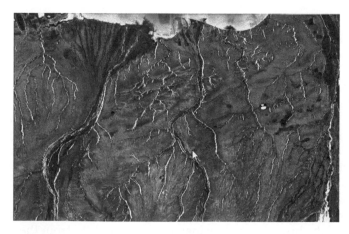

FIGURE 5.18
Buried structures can be expressed as subtle topographic rises. The Marsh Creek Anticline (arrow) in the Arctic National Wildlife Refuge, North Slope Alaska, is expressed by drainage incised into alluvial cover and stream deflection. (True color image courtesy of IBCAO, TerraMetrics, and Google Earth, © 2012 Google.)

caution: many of the "hazy" tonal anomalies associated with oil fields that were seen on some early satellite imagery turned out to be the result of surface disturbances (tracks, drill pads, mud pits) that were too small to be resolved on the imagery, yet contributed to a lighter overall tone in the area of production. This misidentification gave the concept of tonal anomalies a bad reputation.

Classic examples of soil tonal anomalies revealed on black and white airphotos were presented by Kupsch (1956) for the surface expression of Braddock Dome, Nottingham field, and other areas in the Williston Basin of Saskatchewan. These were thought to reflect a change from the background A horizon (lighter) to B soil horizon (darker) along the margins of the topographic high (buried fold), or a change from high moisture (darker) off the structure to low moisture (lighter) over the structure. Many similar examples of tonal anomalies detected from satellite altitudes have been described (see, for example, Saunders, 1980; Morgan et al., 1982).

Vegetation Anomalies Suggesting Folds

Topographic highs related to positive structures are frequently associated with changes in vegetation communities or densities. This may be a result of changing moisture conditions (as mentioned above) from wetter in the background to drier over the uplift; it may be a result of a change in subcrop or soil chemistry or infiltration characteristics that affects the type or density of plant cover; or it may simply be the result of a change from north- to south-facing slopes across the uplift. In middle latitudes of the northern hemisphere, for example, north-facing slopes are more often in shadow. Evaporation rates are therefore lower, snow melts more slowly, and soils are better developed and contain more moisture. Thus, north-facing slopes support a more robust plant community than south-facing slopes that receive more direct sunlight (Schoenherr, 1992). An example of vegetation change related to moisture conditions over subtle topographic (and structural) highs is given by DeBlieux (1962), who described cane stands growing over salt domes in the coastal marshes of Louisiana. Water over the domes is shallow enough to support rooted vegetation, whereas water in surrounding areas is too deep for vegetation to take hold (Figure 5.19).

Mapping vegetation patterns is one of several remote sensing methods used to identify salt domes. The Monte Real salt structure, onshore west-central Portugal, was mapped using a combination of optical, thermal, and radar imagery combined with gravity mapping (Lopes et al., 2012). This multifaceted analysis used outcrops, soil texture, and topography along with vegetation patterns to outline a shallow salt dome and its associated faults and

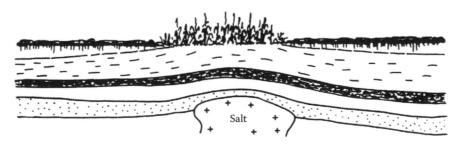

FIGURE 5.19
Cane stands grow over salt domes in coastal swamps in Louisiana. (From DeBlieux, C. 1962. *Gulf Coast Assn. Geol. Soc. Trans. 12th Ann. Mtg.*: 231–241. Republished by permission of the Gulf Coast Association of Geological Societies, whose permission is required for further publication use.)

fractures. A supervised classification was applied to Landsat 7 ETM+ imagery to create a digital geologic map of the surface. JERS horizontal transmitted, horizontal received (HH) radar imagery was used to map soil textures related to subcrop composition. Optical imagery was used to map vegetation related to the changing substrate. Plants over the salt dome have reduced chlorophyll or perturbed ectomycorrhizal fungi (organisms that support nutrient uptake through root systems). It is thought that these plants are sensitive to high soil salinity. This integrated approach allowed recognition of the underlying salt structure.

Fracture Patterns Indicating Folds

Fracture orientations and intensity have been used to locate buried folds. Many workers have mapped zones of intense fracturing in order to locate the zones of rapidly changing dip associated with the flanks of folds (Blanchet, 1957; Harris, 1959; Murray, 1968; Gorham et al., 1979). A tight fold will have a single zone of fracturing along the crest of the fold, whereas a fold with a gentle crest and steep limbs would have a halo of intense jointing above the flanks (Figures 5.20 and 5.21). These joints are a result of flexural folding, and the joint zones extend from the subsurface upward as a result of differential compaction and/or reactivation.

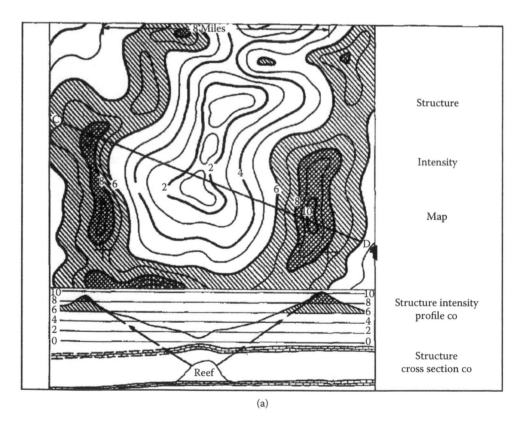

(a)

FIGURE 5.20
Examples of fields that produce out of zones of intense fracturing where the rate of change of dip is greatest.
(a) Wizard Lake reef, Alberta. (From Blanchet, P.H. 1957. *Bull. Am. Assn. Pet. Geol.* 41: 1748–1759.)

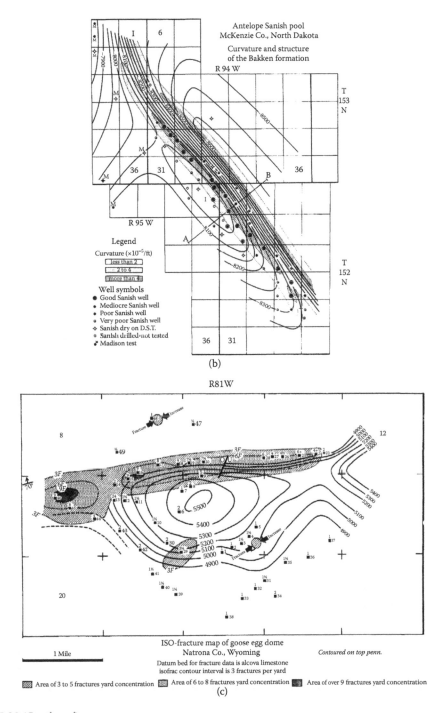

(b)

FIGURE 5.20 (*Continued*)

Examples of fields that produce out of zones of intense fracturing where the rate of change of dip is greatest. (b) Curvature on the Antelope-Sanish pool, North Dakota. (From Murray, G.H. Jr. 1968. *Bull. Am. Assn. Pet. Geol.* 52: 57–65.) (c) Iso-fracture map of Goose Egg dome, Wyoming. (From Harris, J.F. et al 1960. *Bull. Am. Assn. Pet. Geol.* 44: 1853–1873.)

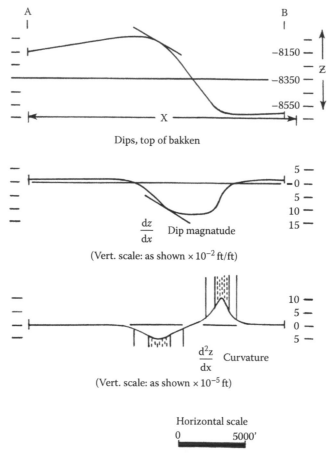

FIGURE 5.21
Comparison of structural profile, dip magnitude, and structural curvature across the Antelope Sanish field, McKenzie County, North Dakota. Line AB shown in Figure 5.20b. (From Murray, G.H. Jr. 1968. *Bull. Am. Assn. Pet. Geol.* 52: 57–65.)

A good example of fold-related jointing can be found at the Verde field, San Juan County, New Mexico (Figure 5.22). The field produces out of zones in the Cretaceous Niobrara (interbedded black shales and thin quartz siltstones) at depths around 1000 m. Matrix porosities and permeabilities are too low to produce oil unless the formation is fractured. The main structure is the Hogback monocline, which changes strike and forms both a gentle anticlinal nose and a syncline that cross the hogback. The best wells produce in the syncline at the position where maximum curvature exists rather than at the crest of the anticline (Gorham et al., 1979).

Another aspect of jointing associated with folds is joint orientation (Figure 5.23). Joint trends expected in a dome-shaped granite hill were first described by Chapman (1958) as being radial and concentric to the hill. Jointing in anticlines was described by Price (1966) as tensional, sheer, and cross-axis. Aguilera (1980) described asymmetric anticlines with intense open tension joints at the plunging ends oriented at roughly 45° to the fold axis. On the gentle flank (back limb), the open tension joints are essentially parallel to the fold axis. The steep flank (forelimb) contains joints that brecciated the rock and were cemented by secondary mineralization (Aguilera, 1980). Fold-related fracture sets were summarized

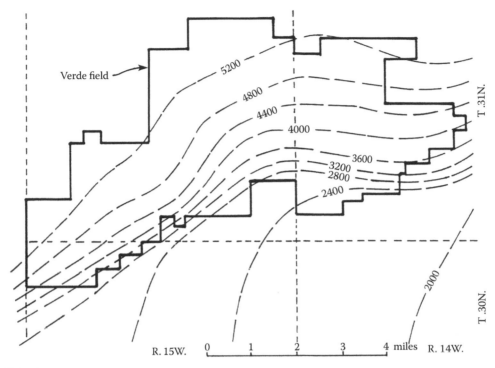

FIGURE 5.22
Structure contour map on top of the Point Lookout member of the Mesaverde Group, Verde field, San Juan County, New Mexico. The field is located in a syncline where it plunges across the Hogback monocline. (From Gorham, F.D. Jr. et al. 1979. *Bull. Am. Assn. Pet. Geol.* 63: 598–607.)

by Nelson (1985). In his classification, the position and intensity of fold-related fractures varied with shape and origin of the fold. Elaborating on the system developed by Stearns (1968), Nelson labels the sets relative to dip and bedding:

Set 1: Parallel to dip and perpendicular to bedding

Set 2: Parallel and at an angle to strike; perpendicular to bedding

Set 3a: Parallel to strike; perpendicular and at an angle to bedding

Set 3b: Parallel to strike and roughly 60° to bedding

Work at Oil Mountain, Wyoming (Hennings et al., 2000), found that "intensity of tectonically produced fractures is closely related spatially to rate of dip change and total curvature, with the former having the strongest correlation." Cooper et al. (2006) determined that the basement-cored anticline at Teapot Dome, Wyoming, contained fractures, deformation bands, and faults caused by stretching of the sedimentary cover causing normal and normal-oblique faults both parallel and perpendicular to the fold hinge.

Fracture orientations in thrust-related folds were described by Cooper (1992). Fractures were axis-normal extensional joints, axis-parallel extensional joints, or conjugate shear fractures at roughly 60° to the fold axis. Fischer and Wilkerson (2000) predicted the orientation of joints as a function of fold shape. They used modeling and curvature analysis to show that joints open parallel to the direction of maximum stretch of a rock layer, and thus form parallel to fold strike (what they call the "minimum curvature axis") of a fold.

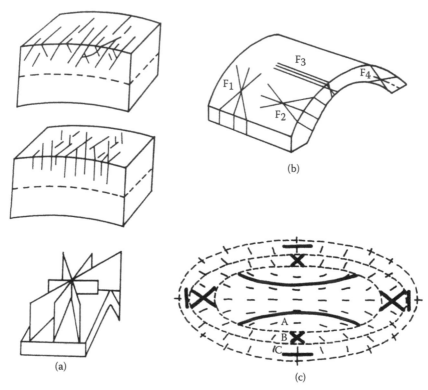

FIGURE 5.23
Joint orientations with respect to folding. (a) Joints typically developed in anticlines as described by Price (1966). (b) Joint sets defined by Stearns (1968). (c) Joints (light) and faults (heavy) observed in clay models of elliptical uplifts. (From Withjack, M.O., C. Scheiner. 1982. *Bull. Am. Assn. Pet. Geol.* 66: 302–316.)

They also note that as folds grow in amplitude and length, a given package of rock will occupy a number of structural positions and consequently will have joints with varying orientations and abutting relationships representative of the structural positions occupied by that rock package.

The Teton anticline in Montana has multiple hinges and is developed in Devonian to Mississippian carbonates (Ghosh and Mitra, 2009). The surface structure is considered a good analog for fracture patterns and fracture connectivity at depth. The two main fracture sets occur as "longitudinal" (subparallel to the fold axis) and "transverse" (normal to the fold axis). The length and density of the dominant longitudinal fractures are controlled by the position relative to the fold hinges, with longer fractures and more intense fracturing near a hinge. Fracture connectivity is also greater near fold hinges. Transverse fractures are related to changes in fold plunge. This continues to be a topic of great interest to explorers (Zaeff et al., 2010).

The foregoing discussion supports the contention that folds can sometimes be recognized by their fracture patterns and that these patterns seen on the surface can be used to predict trends such as porosity and permeability at depth.

In addition to recognizing opposing dips associated with domes, there are specific fault patterns that occur over salt domes. Yin and Groshong (2007) described active diapirs that have either two or three fault sets crossing near the crest of the dome. If these faults extend into the salt, then there are usually few if any minor faults associated with the dome. If the salt is unfaulted (the faults detach at the top of the salt), then additional faulting occurs on the flanks of the diaper and there will be rollover in the hanging wall.

Clay models reveal surprising detail regarding the orientation and location of joints and faults in uplifts and folds formed in extensional stress fields, compressive fields, and with no prevailing stress (Withjack and Scheiner, 1982; Withjack et al., 1990). These fractures are expected in domed strata, and under favorable conditions (reactivation, compaction) they might be perceived through overlying units.

In alluvial terrain, streams may show a marked angularity over buried structural highs. Fractures propagating up section due to drape or reactivation appear to control stream channel direction more than in other parts of a basin (Figures 5.24 and 5.25).

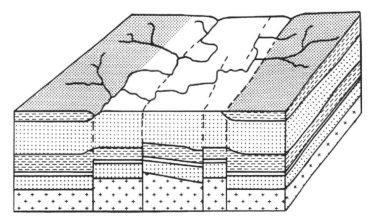

FIGURE 5.24

Jointing can control drainage over uplifts to a greater extent than in adjacent areas. This is a result of outcrops and thin soil on the uplifts that better reflect shallow jointing because of compaction and drape as compared to areas with thick soil in surrounding lowlands.

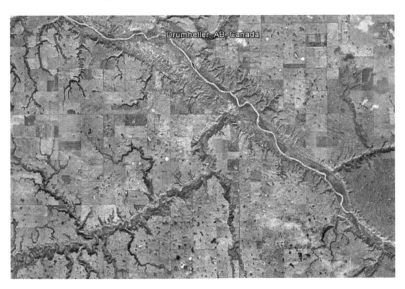

FIGURE 5.25

Red Deer River near Drumheller, Alberta. Note the orthogonal and barbed drainage that suggests northeast and northwest faulting in this post-glacial landscape. (True color image courtesy of GeoEye, CNES/SPOT Image, and Google Earth, © 2012 Google.)

Buried Folds in Glacial Terrain

In areas of continental glaciation, the abrasive action of ice sheets works to decrease the relief on the preexisting surface. Yet, as the glaciers advance, they grind down fractured bedrock faster than unbroken rock, thus deepening fault-controlled drainage. For this reason, postglacial drainages are often superimposed on preglacial drainages. Areas that remain topographically high as the ice sheet melts are characterized by "dead ice" moraines. Dead ice moraines are deposited by stagnant ice and are identified by rimmed or donut-shaped kettles, usually filled with water, which form as the ice remnants melt and deposit debris around the last ice (Mollard and Janes, 1984). The last ice melts off topographic high points since meltwater flows through and speeds melting in low areas. Dead ice moraines are developed over preglacial topographic (and in some cases structural) highs and can be seen in the Williston Basin of Saskatchewan and North Dakota along the Missouri Coteau. Kupsch (1956) found that 24 of 30 glacial (and subcrop) highs corresponded to seismic highs between Old Wives Lake and Beechy in Saskatchewan. The same correspondence of dead ice moraines to subglacial highs was reported by Mollard (1957).

The southern peninsula of Michigan is covered by as much as 330 m of glacial till that is either cultivated or forest covered. It is a challenge to map geologic structures at the surface in this area. Yet, there are reasons why surface features should reflect subsurface structure in Michigan. It has been shown (Rieck, 1976) that glacial channels develop above preexisting bedrock channels, which are in turn often eroded into faulted bedrock. In addition, a minimum of 130 m and maximum of 1000 m of uplift has been determined (Flint, 1971) as a result of postglacial isostatic rebound in Michigan. The amount of ground movement is considered enough to reactivate old fault lines and impress their trace on the postglacial surface (Drake and Vincent, 1975; Maslowski, 1985) (Figure 5.26).

Folds beneath Drifting Sand

When thin, discontinuous sand sheets cover the surface, topographic (and structural) highs tend to be windswept and fairly clear of sand cover, whereas great thicknesses of sand accumulate on the leeward side of the high (Bagnold, 1941). The high acts as a windbreak,

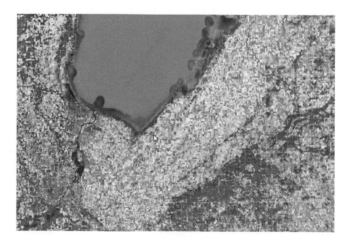

FIGURE 5.26
A fault bounding the eastern shore of the Saginaw Bay graben (arrow) is a result of glacial rebound imposing old basement faults on recent terrain. (True color image courtesy of TerraMetrics, NOAA, CNES/SPOT Image, and Google Earth, © 2012 Google.)

slowing air velocities and causing turbulence, forcing sand to drop out on the downwind side (Figure 5.27). Examples of structures expressed in such an environment include Safir dome in northern Yemen and El Borma in Algeria. These structures are expressed as wide, smooth, elliptical swells between sand waves (Figure 5.28). Thermal imagery can reveal folds either by changes in moisture and evaporative cooling or by differences in thermal inertia in folded strata that have little or no contrast in the visible range (e.g., Imler Road anticline, Imperial County, California; Sabins, 1987).

Folds Revealed by Drainage Patterns

Stream patterns are especially sensitive indicators of blind structures in flat, alluviated plains. A fractional change in regional gradient as a result of drape or reactivation can affect the course of runoff over structures too subtle to be evident during field mapping.

The following list provides some drainage indicators of folding. Drainage responds to many factors, and buried structure is one of these factors.

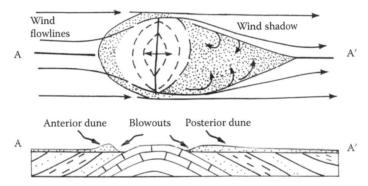

FIGURE 5.27
Wind flow and sand accumulation around topographic (and structural) highs. (Modified after Bagnold, R.A. 1941. *The Physics of Blowing Sand and Desert Dunes*. London: Metheun Publishing Co.: 265 p.)

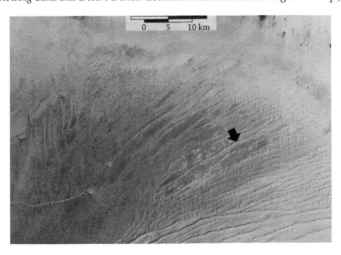

FIGURE 5.28
Large dunes are developed on the flanks of Safir dome, Yemen, where Jurassic units outcrop (arrow). The dome is windswept and appears as a color change in this false color image. (TM bands 1-4-7, processed by Earth Satellite Corporation, Rockville, MD [now MDA Federal] © MDA Information Systems LLC 2013.)

Channel Width/Depth Adjustment and Incision

The width/depth ratio of a stream channel increases as the gradient decreases upstream from the crest of a structure (Howard, 1967). As the gradient flattens, one may observe "alluvial ponding," an anomalous area of alluvium that occurs in the wider channel upstream from a structure. As it passes over the crest of the fold, the channel narrows and becomes incised between high cutbanks (Ollier, 1981). Downstream from the fold, the stream straightens as a result of increased gradient (Figure 5.29). Examples of such features are particularly well expressed in sand cover and alluvium over horst blocks along the Gulf of Suez and near Riyadh, Saudi Arabia, and at the Kekaya anticline in the Tarim Basin, China (Figure 5.30).

Compressed Meanders

Streams show local compression of meanders (progressively shortening the wavelength and increasing the amplitude), where the gradient decreases upstream from a fold crest (Figures 5.31 and 5.32). This has been noted at a fold in Kent Co., Texas (DeBlieux and Shepherd, 1951; Lattman, 1959) and in the Salawati Basin of Indonesia associated with drape over pinnacle reefs (Foster and Soeparjadi, 1974). Downstream from the fold crest the channel straightens as the gradient steepens slightly, then the stream resumes its normal course as it returns to the regional gradient.

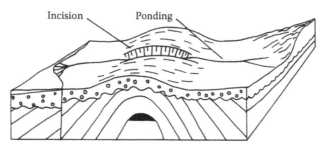

FIGURE 5.29
Stream channels become narrower and are incised over a structure. Alluvium is "ponded' upstream from the structure.

FIGURE 5.30
Stream incision (arrow) and tilted alluvium reveal the Kekeya field anticline in the southwest Tarim basin, China. (True color image courtesy of CNES/SPOT Image and Google Earth, © 2012 Google.)

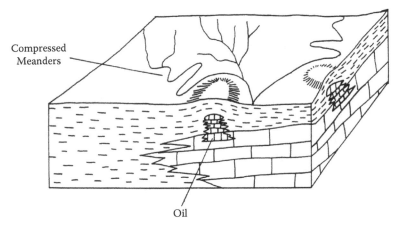

FIGURE 5.31
Compressed meanders form upstream from drape over pinnacle reefs, as in the Salawati basin, Indonesia.

FIGURE 5.32
Compressed meanders of the Washita River occur upstream from the Butterly field, Oklahoma (arrow). (True color image courtesy of Google Earth, © 2012 Google.)

Drainage Texture

Lithologic mapping can be based in part on drainage texture. Coarse clastic units tend to have better infiltration and less runoff, causing widely spaced drainage, whereas fine-grained units, particularly shales, have fine-textured and closely spaced drainage patterns. This criterion has been used in the Ucayali Basin, Peru, to map a fold in a low relief area mantled with alluvium and continuous vegetation cover (Doeringsfeld and Ivey, 1964) (Figure 5.33).

Radial and Concentric (Annular) Drainage

All topographic highs have drainage that flows away from the crest in a more or less radial pattern. If the pattern is radial in 360°, and has concentric tributaries, then it is likely to represent domed, stratified material (Figure 5.34). The concentric pattern is a result of arcuate

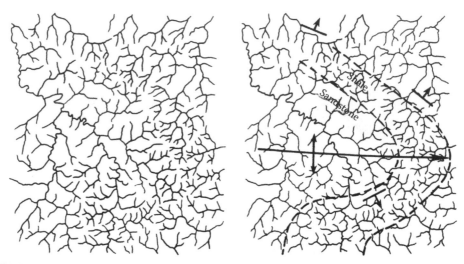

FIGURE 5.33
Folding is revealed by changes in drainage texture in the Ucayali basin, Peru. (From Doeringsfeld, W.W. Jr., J.B. Ivey. 1964. *Mountain Geol.* 183–195. Reprinted with permission of the Rocky Mountain Association of Geologists.)

FIGURE 5.34
Radial and concentric drainage are associated with anticlinal and synclinal folds.

strike valleys. Alternatively, radial inward flowing streams may represent a synclinal structure (Miller and Miller, 1961; Doeringsfeld and Ivey, 1964; Howard, 1967).

Braided Stream Segments

A stream becomes braided whenever the load exceeds the stream's capacity to carry material. One way this can happen is for a local decrease in gradient, such as might be expected over a buried structure.

Double Drainage Deflection

The deflection of two more or less parallel streams in opposite directions is considered evidence of a subtle topographic high or buried structure (Doeringsfeld and Ivey, 1964). The streams usually converge on the downstream side of the feature (Figure 5.35).

Change in Strike of Bedding

Patterns of long drainages opposed by short tributaries indicate the dip and strike of bedding at cuestas. Changes in strike can be a subtle indicator of folding (Ray, 1960).

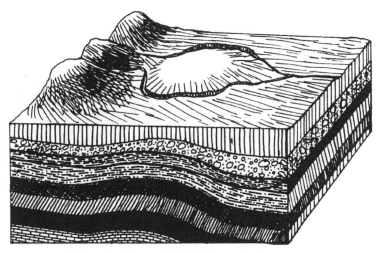

FIGURE 5.35
A double drainage deflection occurs when streams adjust their courses to account for intervening structure.

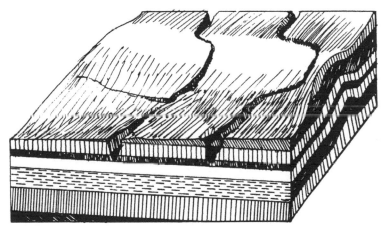

FIGURE 5.36
Parallel drainage deflections occur when streams shift their courses to accommodate the plunging nose of a fold.

Parallel Drainage Deflection

The deflection of two or more parallel streams in the same direction can be evidence for a plunging fold (Figure 5.36). The streams bend around the structural nose and become increasingly incised as erosion progresses (Doeringsfeld and Ivey, 1964).

Levee Preservation

In an actively subsiding delta plain, there may be areas where short sections of paired levees remain above water over a fold or dome. An example of such "flying levees" is given by DeBlieux (1962) at Cutoff field, La Fourche Parish, Louisiana (Figure 5.37).

More examples of drainage indicating deeply buried structures are given by Melton (1959), Trollinger (1968), and Penny (1975).

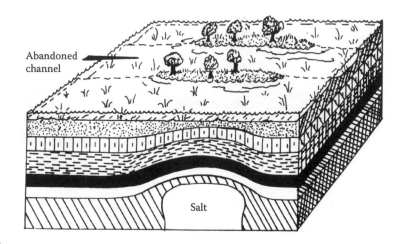

FIGURE 5.37
Abandoned, subsiding levees in a swamp area are submerged except where they form "flying levees" over a structure. (From DeBlieux, C. 1962. *Gulf Coast Assn. Geol. Soc. Trans. 12th Ann. Mtg.*: 231–241. Republished by permission of the Gulf Coast Association of Geological Societies, whose permission is required for further publication use.)

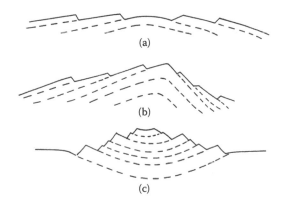

FIGURE 5.38
Topographic profiles over folds. (a) Profile over a symmetric, breached anticline. (b) Profile over an asymmetric anticline. (c) Profile over a syncline with inverted topography, that is, the synclinal axis is topographically high because the youngest units are resistant to erosion.

Topographic Profiles

Drawing topographic profiles across an area of interest may reveal structures that are not otherwise obvious (Figure 5.38). The appearance of asymmetric slopes, suggesting cuestas, arranged on opposite sides of a presumed structure, may be an indication of folded bedding, especially in vegetation-covered areas where outcrops are sparse (Doeringsfeld and Ivey, 1964). The ready availability of DEMs makes it easy to generate topographic profiles.

Trend Surface Analyses

Structures as deep as 1830 m have been mapped at the surface through one or more intervening unconformities. For example, the Devonian Leduc reef in the Western Canadian (Alberta) basin has been mapped by using trend surface analysis on the pre-Cretaceous

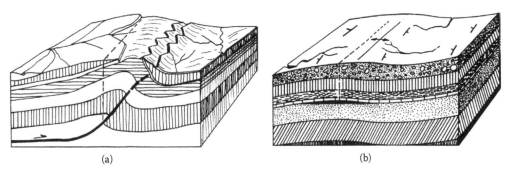

FIGURE 5.39

Asymmetric folds. (a) Thrusting causes an inclined axial plane, so the surface trace of the fold axis is offset from the fold crest at depth (dashed line). (b) In gently dipping strata, the surface trace is offset updip from the subsurface fold crest, as at the McLouth field, Kansas.

unconformity (Glass, 1981). The analysis removes the effect of regional dip, allowing smaller irregularities due to drape over the reef to become evident. This same type of analysis can be done on surface topography: the smoothing of irregularities may reveal broad, regional highs and lows.

Surface Features Offset from Subsurface Structure

The surface trace of a fold axis occurs above the subsurface high only when the fold is symmetrical. Many petroleum explorers have learned to their dismay that a structural high at depth can be offset a significant amount from the surface crest. The best evidence for an inclined axial plane is an asymmetric fold, that is, where one limb is steeper than the other. The axial plane often (but not always) bisects the interlimb angle, and thus dips away from the steeper limb. In many cases, an asymmetric fold is cored by reverse or thrust faults, which further complicate predicting the structural high at depth (Figure 5.39a).

When gently inclined strata overlie a fold (or reef, etc.) at depth, the surface trace of the high will be offset updip, as demonstrated at the McLouth oil field, Kansas (Lee and Payne, 1944) (Figure 5.39b).

Circular Features

Some image analysts consider "circular features" of unknown origin important to a petroleum or mineral exploration program. Unless these features can be attributed to some process, they are meaningless. Natural circular features of interest in petroleum exploration are usually related to salt diapirs or drape folds over pinnacle reefs (Figures 5.15 and 5.31). Gentle drape folds may appear to be no more than circular soil tone changes (Kupsch, 1956; Saunders, 1980; Morgan et al., 1982). Other common circular geologic features that may be of interest in petroleum or mineral exploration include meteor impact craters, volcanic craters, circular karst sinkholes, diatremes, and granite exfoliation domes. Oil has been produced from meteor impact craters (Ames Hole, Oklahoma; Little Knife, North Dakota; Steen River, Canada), diatremes (south Texas), and volcanic stocks (Dineh b'kiyah field, Arizona). Uranium has been produced from karst collapse breccias in northern Arizona. Diamonds are produced from kimberlite diatremes (South Africa; Russia, Australia, Canada).

Fractures and Lineaments

"Fractures" are breaks in rocks, and include both joints and faults. "Joints" are fractures that have no discernable movement parallel to the fracture surface. Joints can be open, closed, or filled with minerals (veins). "Faults" have measurable offset in a dip-slip, strike-slip, or combination mode. Faults can be further subdivided into normal, reverse, rotational, strike-slip (or wrench), and thrust based on their displacement (Figure 5.40).

"Lineaments," much discussed in the remote sensing literature, are defined here as any linear feature on the surface of the Earth or on potential fields or geochemical maps that may be evidence of geological processes such as fracturing or folding. Lineaments may indicate zones of increased porosity, the boundaries of uplifted blocks, fault traps, drape folds, fold hinges, petroleum reservoir boundaries, mineralized veins and shear zones, dikes, linear sinkholes, or springs aligned along a fault. Lineaments may represent dense fracturing along the zone of maximum curvature on a fold (Figure 5.20). Although the trace of bedding may be linear, it is generally not considered to be a geologic lineament.

Propagation Mechanisms

Some discussion of the mechanisms involved in fracture propagation is required because the relationship between surface structures and those at depth is often controversial and

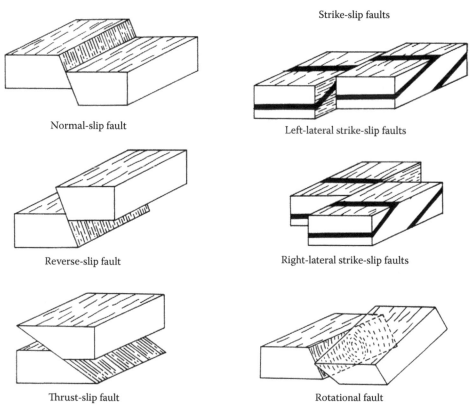

FIGURE 5.40
Characteristics of different fault types.

not always evident. Mechanisms invoked for fracture propagation include reactivation, settling, minor readjustments associated with seismicity, Earth tides, thermal contraction, uplift, and regional extension.

It has been demonstrated that fractures can extend upward from faults in the basement. The theoretical basis for the upward propagation of faults into flexures and fracture zones is found in the pioneering work of Sanford (1959) who used displacement field and stress distribution diagrams to predict the surface expression of vertical displacement at depth. Structures such as the Bright Angel fault and Butte fault-East Kaibab monocline in the Grand Canyon (Sears, 1973; Huntoon and Sears, 1975; Timmons et al., 2003) are classic examples of multiple periods of reactivation that have caused a fault to display varying displacements in successively younger units from Precambrian through Permian time. Jamison (1979) and Heyman (1983) have shown examples of faults extending upward into joints and folds along the Redlands/Monument fault system on the flank of the Uncompahgre Uplift, Colorado. Stein and Wickham (1980, p. 225) used a viscosity-based finite element model to demonstrate that "failure begins near the basement deflection and propagates ... upward" from a vertical step in the basement.

Hodgson (1961) and Alpay (1969) believed that a possible mechanism for fracture propagation without fault rejuvenation was high frequency, low amplitude vibrations related to seismic activity. They cited an analogous situation in cracked pavement, where old fractures reappeared in new surfacing after being subjected to traffic vibrations (Roberts, 1954). Free oscillations in the Earth caused by large earthquakes have "been compared to the ringing of a bell" (Stacy, 1969, p. 108). Oscillations have been measured with periods from a few minutes to almost an hour. On the other hand, microseisms are continuous vibrations with periods of 5–10 seconds and maximum displacement amplitudes exceeding 10 µm. Barosh (1968, p. 216) related the pattern of fractures observed in alluvium at Yucca Flats, Nevada, to the joint patterns in the adjacent bedrock and proposed that the mechanism "is that of upward propagation of fractures from differential movement between joint-bounded blocks during [underground nuclear] explosions." Brown and Hudson (1974) showed that it should be possible to cause fatigue failure by cycling almost any load. The implication is that even loads as small as those caused by Earth tides may cause jointing over time (Prost, 1988).

Another means of propagating fractures is by cooling a large region. Coefficients of thermal expansion for minerals and rocks were catalogued by Skinner (1966). The volume of quartz, for example, increases 3.8% as temperature increases from 20 to 570°C; the volume of calcite increases 1.8% from 20 to 600°C. Sandstones have a coefficient of thermal expansion of 0.00001 per degree Centigrade in the range 20 to 100°C, and granites have an expansion coefficient of 0.000008 per degree Centigrade in the same temperature range. This is equivalent to a change in length (contraction) of 20 m (65 ft) for a granite layer, and 25 m (81.3 ft) for a sandstone layer 10 km (6.1 miles) long uplifted and cooled from a depth of 5 km (16,250 ft) with a higher than normal thermal gradient of 5° C/100 m (as measured at Cisco, Utah, by Hallin, 1973). English (2012) makes a case for thermal contraction as a cause for regional tensile fractures during major exhumation events. The extent of fracturing is dependent on the magnitude of cooling (amount of uplift and unroofing) and the mechanical properties (e.g., Young's Modulus) of the rock. This contractional strain would be accommodated by jointing. In sedimentary rock, there is no preferred orientation to thermal contraction joints, but preexisting planar flaws may control fracture trends. Continental scale crustal extension joints (regional joint sets) due to thermal cooling were proposed by Turcotte and Oxburgh (1973).

Physical modeling is an important method for understanding the control or influence of basement features on overlying structure (Sanford, 1959; Cloos, 1968; Withjack and Scheiner,

1982; Withjack and Callaway, 2000; Schlische et al., 2002; Neely and Erslev, 2009; Bose and Mitra, 2010; Henza et al., 2010; Miller and Mitra, 2011; Paul and Mitra, 2012). By observing the surface patterns generated in physical models and reverse engineering, we can interpret basement structure from the geometry and distribution of deformation at the Earth's surface.

Regional extension may affect large areas of the Earth's surface (e.g., the Basin-Range province). Strain takes the form of normal faults and near-vertical joint patterns parallel to the maximum horizontal compressive stress. In areas with a strong preexisting structural fabric, joints may form above and parallel to or rotated slightly from the preexisting fabric.

Fractures can propagate downward from the surface due to local uplift or volumetric changes due to dewatering of sediments. Price (1959) and Nur (1978) proposed that joints could form in uplifted strata because of a volume increase (Figure 5.41). Such joints tend to form parallel to the margins of an uplift, and decrease in frequency toward the center of the uplift. Nur (1982) suggested that the depth of penetration was related to the length of the fracture and spacing between parallel fractures.

Arcuate fractures similar to glacial crevasses (including fissures and clastic dikes) have been related to down slope creep on a decollement or to collapse of the overlying section as an evaporite layer is slowly dissolved. The surface geometry suggests a listric nature at depth. Large volumes of brittle rock are capable of moving down slope on ductile layers and forming fissures. In the Canyonlands of Utah, the Colorado River has cut through 460 m (1500 ft) of strata into the Pennsylvanian Paradox evaporites (McGill and Stromquist, 1979; Melosh and McGill, 1982). The rocks dip toward the river and move slowly down slope forming grabens that are arcuate in plan view (Figures 5.42 and 5.43).

Fractures that originate as the result of overpressuring within a specific horizon are generally confined to that horizon and its overlaying or underlying seals. Once a seal is

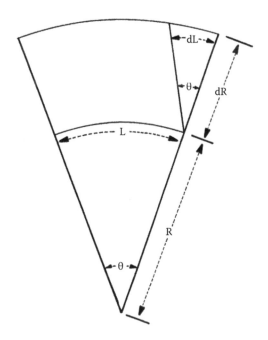

FIGURE 5.41
The volume increases in a slice of rock that has been uplifted. This volume increase is caused by distributed jointing or normal faults parallel to the edge of the uplift. (From Price, N.J. 1959. *Geol. Mag.* 46: 149–167. Copyright © 1959 Cambridge University Press.)

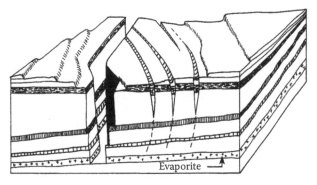

FIGURE 5.42
Arcuate extension fractures probably go listric into a decollement in the Pennsylvanian salts, Canyonlands, Utah. The section is laterally unconstrained due to erosion by the Colorado River.

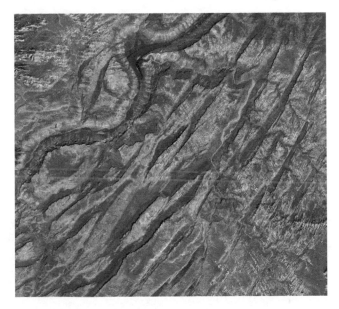

FIGURE 5.43
The Grabens area, Canyonlands, Utah. The Colorado River (top and left) has carved a space that allows the section to detach on Pennsylvanian age salt and slowly slide down dip on arcuate, listric normal faults. (True color image courtesy of Digital Globe and Google Earth, © 2012 Google.)

breached by fracturing, the pressure dissipates and there is no need for fractures to continue to propagate. One would not expect, then, and one generally does not see fracturing at the surface associated with overpressured, fractured reservoirs (e.g., in the Uinta Basin, Utah). However, one may surmise that joint orientations in an overpressured reservoir are broadly similar to those in the underlying, and possibly overlying units, because they were formed in a similar regional stress environment.

Joints

Joints at the surface can be valuable indicators of folding, as seen in the previous section (Figure 5.20), and of faulting or joint zones at depth. In particular, fractured reservoirs may be revealed at the surface as zones of intense jointing if the fracturing is tectonic (as opposed to overpressured) in origin. Tight sand plays, coalbed methane extraction, shale oil plays, and shale gas plays all depend on fractures, natural or induced, for permeability. For all these reasons, it may be important to identify and map joints at the surface as a guide to the regional stress regime and to fracture and permeability trends in the subsurface.

Joints can be represented by linear ridges or valleys, vertical rock fins, tonal and vegetation alignments, and orthogonal or parallel streams or lakes (Figure 5.44). In some basins, fracture systems have been injected by dikes, and the dikes clearly reveal the joint trends as well as the stress field at the time of injection. Oligocene dikes in the northeast San Juan basin, New Mexico, are an example (Huffman and Taylor, 1998) (Figure 5.45). Dikes intruded extension joints and are thus aligned parallel to the paleostress direction.

FIGURE 5.44
Jointing may be recognized by vegetation alignments, because plants obtain moisture from the fractured , water-bearing bedrock.

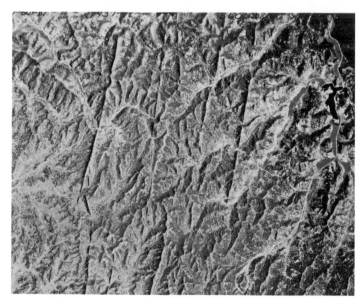

FIGURE 5.45
Miocene north-south-trending basaltic dikes intrude the Tertiary San Jose Fm. in the northeastern San Juan basin, New Mexico. Intrusion was contemporaneous with jointing, and thus reveals paleostress directions. (Color infrared airphoto courtesy of USGS EROS Data Center.)

Where carbonates lie at the surface, alignments of sinkholes indicate fracture zones. In Saudi Arabia and Qatar, the Shamal winds have etched joints such that outcrops are serrated and some have been reduced to fins of rock oriented parallel to jointing. Similar rock fins caused by erosion along joint zones are seen in Arches National Park, Utah (Figure 5.46).

Joints in intrusive rocks form orthogonal networks, reflecting the relatively homogenous nature of the rock. The margins of the intrusion may contain arcuate and concentric cooling joints. Metamorphic basement is characterized by a strong parallelism reflecting foliation and its influence on weathering. These features are unrelated to jointing.

Carbonates have characteristic rhombic and orthogonal jointing, a fractal pattern that occurs on a local to regional scale (Utah Geological Survey, 2010; Eby et al., 2010; Miller and Nelson, 2002). A convincing explanation has not been proposed for the rhombic fracture pattern: they may be two completely separate superimposed regional fracture systems, may be an upscaled fractal version of the rhombic crystal structure in carbonates, or may represent the influence of underlying structures such as those in overlapping synthetic transfer zones (Miller et al., 2007).

Joints in dipping sedimentary units are most easily recognizable on imagery when the jointing is oriented perpendicular to strike. Eroded joint zones that parallel strike generally have the same appearance as strike valleys. This can introduce a bias when mapping fracture densities if strike valleys are counted as fracture zones.

The shape and distribution of subsurface fractures are predictable with a knowledge of local stratigraphy. Several geologic factors control fracture spacing and will assist with prediction of fracture density in the subsurface. These include composition, grain size, porosity, bed thickness, and structural position (Nelson, 1985). Any factor that increases rock strength or makes rock more brittle (higher silica content, lower porosity, decreasing grain size, better sorting, thinner bedding) will increase fracture density. Joints in

FIGURE 5.46
Oblique view northeast at yardangs in the Jurassic Kayenta and Wingate sandstones of the San Juan Arm of Lake Powell, Utah. Joints are etched by eolian abrasion. (True color image courtesy of USDA Farm Service Agency, TerraMetrics, and Google Earth, © 2012 Google.)

massive or crossbedded sandstones tend to be relatively long, linear, wavy, or curved along bedding, whereas joints in siltstones and shales are shorter, more discontinuous, and more irregular (Figures 5.47 and 5.48). Fractured reservoirs in essentially undeformed rocks will be best developed in brittle, well-cemented and fine-grained rock, that is, in brittle units with the lowest tensile strength (Prost, 1986). Thin brittle units (e.g., sandstone, marl) interbedded with ductile units (shale, anhydrite) will generally be highly fractured.

It has been observed that primary joint sets are frequently at right angles in adjacent sandstones and shales. This may be a result of stress reorientation in the unfractured

FIGURE 5.47
Jointing in a 200 m high cliff of the Triassic Wingate Formation sandstone, Ute Canyon, Uncompahgre uplift, Colorado. The longest vertically continuous joints are ~42 m long.

FIGURE 5.48
Irregular joints in the Triassic Chinle Formation shales, Unaweep Canyon, Uncompahgre uplift, Colorado.

stronger units following failure in the weakest brittle strata. First, the weakest layer fails (forms joints) parallel to the largest horizontal stress (σ_{h1}), then the stronger unit fails at right angles to σ_{h1}. This may allow prediction from surface patterns to the subsurface if the mechanical stratigraphy is known.

There is some debate regarding joint orientations near the edges of cliffs. Are these the result of expansion perpendicular to the unconstrained cliff face, or are they joints that already exist in the area and are obvious only because the rock is allowed to expand along the cliff face? The orientation of cliffs is related to regional or local joint trends if the cliff has a sawtooth pattern reflecting preexisting joints; nontectonic unloading joints should be parallel to the cliff face.

Regional shale gas and shale oil resource plays try to locate subsurface joints and faults to optimize horizontal well orientation and take advantage of existing permeability. Locating subsurface fracture zones using surface fractures is inherently risky. Not only can the joint surfaces be inclined, but field measurements have shown that near surface joints rarely exceed 40 m of vertical extent in sedimentary rock (Prost, 1986). You do not want to find fractures that are continuous with the subsurface or hydrocarbons would have migrated to the surface and dissipated. What you want to find is the location and orientation of discontinuous vertical fractures at depth.

Contour maps of fracture intensity are one way to define fracture zones. Many fracture density maps ignore the bias caused by mapping in areas with soil cover (low fracture density) versus outcrops. Meaningful mapping of fracture density requires continuous, homogeneous surface units to avoid the necessity of normalizing units by applying lithologic factors. Even then, drilling in a high-density fracture zone is no guarantee of encountering fractures at depth; it only increases the chance of encountering fractures. In producing areas, the large variation in initial production within such zones implies that hitting an open joint is not guaranteed. The critical factor in a successful well is tapping into a fracture plumbing system. Horizontal drilling perpendicular to the primary open fracture trend, or at some oblique angle to two open fracture trends, improves the chance of a successful water or oil well.

Surface mapping provides information necessary to predict fracture trend and location at depth. Fracture orientation data are most reliable in basins without complex deformation or several tectonic episodes, where surface units are about the same age as reservoir units, and where there are few or no unconformities. Although surface fracture sets are usually the same as those in the subsurface, the dominant set at a given depth cannot be predicted with confidence.

A study using high-resolution satellite imagery and a DEM over surface folds in Iraq was able to characterize fracture sets (orientation, intensity, location on folds) and fold curvature and use this information to build a full-field geomodel for adjacent producing fields (Zaeff et al, 2010). The geomodel was able to incorporate multiple fracture sets identified on imagery, place them in their proper structural context, and use them to generate a permeability multiplier for dynamic reservoir simulation.

Fracturing also has an indirect effect on reservoir porosity. A study of porosity distribution in the Mississippian Sun River Dolomite in northwestern Montana indicates that production typically comes from moldic porosity localized along basement-controlled fault or flexure zones on structural closures. The moldic porosity occurs where fractures have enhanced groundwater dissolution of bioclastic debris (Pasternack, 1988). The basement zones were detected using seismic, gravity, aeromagnetics, satellite images and airphotos, and field mapping.

Faults

Faults are characterized by topographic or lithologic offsets, offset streams, streams flowing around acute angles, or flowing against a regional gradient. Within a flood plain, meander loops that show reverse stream flow 180° from the regional gradient indicate that the valley floor has been locally tilted slightly upstream by a fault crossing the valley (Figure 5.49). A fault trace may be as subtle as an alignment of springs, or as obvious as the juxtaposition of two dramatically different rock types. In glaciated terrain, they may simply be a scoured linear depression. In rainforests, they may be an alignment of vegetation. In deltas, a linear stretch of river can indicate faulting.

Faults are classified by their offset or slip. "Slip" is the three-dimensional (3D) displacement of a point on the fault surface. Unless the offset is measured on a key bed or marker that pierces the fault (e.g., a dike), the interpreted displacement is apparent and is based on the topographic expression of the fault, which can be misleading (Figure 5.50).

Movement on inclined fault planes is described in terms of the "hanging wall" (rock above the fault) and "footwall" (rock beneath the fault). "Normal faults" are those in which the hanging wall moves down the fault plane. A "listric normal" (or "growth") fault is a normal fault that flattens at depth, generally dying out in bedding. In a "reverse fault," the hanging wall moves up with respect to the footwall and places older over younger rock. "Thrust faults" are reverse faults inclined less than 45° and are often nearly flat.

"Strike-slip" (or "wrench") faults are near vertical at depth, and blocks on either side of the fault move horizontally in opposite directions more or less parallel to the fault. They are considered right-lateral or left-lateral strike-slip depending on the direction moved by the block across the fault from an observer (Figure 5.51). Transpressional strike-slip faults separate blocks that are converging; transtensional wrench faults separate blocks that are moving apart. A special category of strike-slip fault, called a "tear fault," is confined to the

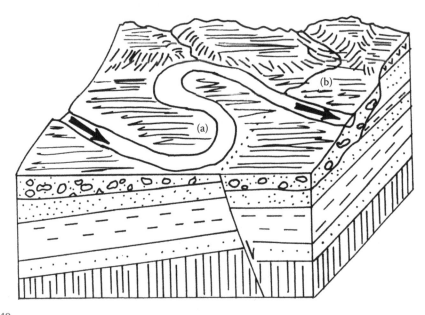

FIGURE 5.49
Streams flowing against the regional gradient suggest fault control. A meander loop that reverses the direction of flow 180 degrees (a) suggests that a fault has tilted the alluvium slightly. A barbed tributary (b) also shows fault control.

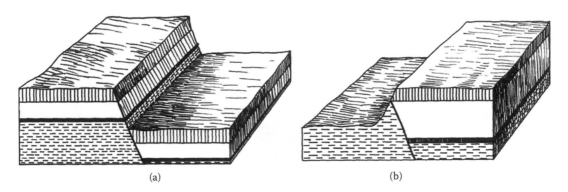

(a)   (b)

FIGURE 5.50
Topographic relief along a fault. (a) Fault scarps give a direct indication of dip slip. (b) Fault line scarps often give a misleading indication of dip slip. As the block shown in (a) erodes, the resistant layer on the downthrown side becomes topographically high (b).

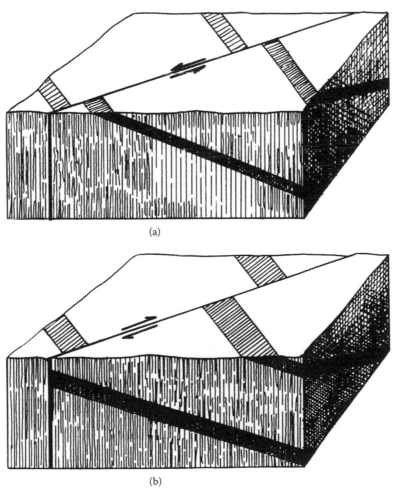

(a)

(b)

FIGURE 5.51
Strike-slip, or "wrench" faults. (a) Left-lateral offset. (b) Right-lateral offset.

hanging wall of a thrust fault, also called the thrust sheet. Tear faults allow deformation in the thrust sheet to vary in intensity across the fault (Figure 5.52).

Most faults have a combination of dip-slip and strike-slip displacement, and they are classified on the basis of the dominant component. Some are rotational, changing from normal to reverse displacement along strike. Some strike-slip faults split and curve upward and outward from a single fault plane at depth, creating "flower" or "palm tree" geometries that are essentially thrusts in cross section (Figure 5.53). Each category is examined in turn for those characteristics that make the faulting recognizable at the surface.

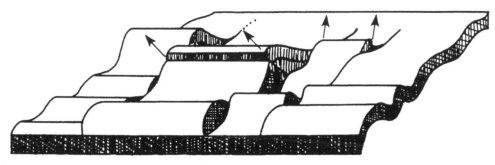

FIGURE 5.52
Tear faults in thrust sheets allow differential movement across the fault, often resulting in changes in fold vergence. (From Davis, G.H. 1984. *Structural Geology of Rocks and Regions*. New York: John Wiley and Sons: 492 p.)

FIGURE 5.53
"Flower" structures are caused by convergent strike-slip faulting, or transpression. (From Lowell, J.D. 1985. *Structural Styles in Petroleum Exploration*. Tulsa: Oil and Gas Consultants International: 477 p.)

Normal, Reverse, and Listric Faults

Normal and reverse faults are often near-vertical at the surface, and thus are characterized by a linear fault trace and are nearly or completely indistinguishable from each other. Many of these faults are marked by topographic offset, or an escarpment, with the upthrown side ranging from an abrupt mountain front to a structural terrace with only a few meters relief (Figure 5.54). Uplifted basement blocks are frequently characterized by triangular facets along the scarp, whereas drag along the fault causes sedimentary layers to dip sub-parallel to the fault plane and form pronounced flatirons (Figures 5.55 and 5.56). Faults frequently

FIGURE 5.54
A normal fault along the east side of the Teton Range, Wyoming, abruptly raises granitic basement several thousand meters along this escarpment.

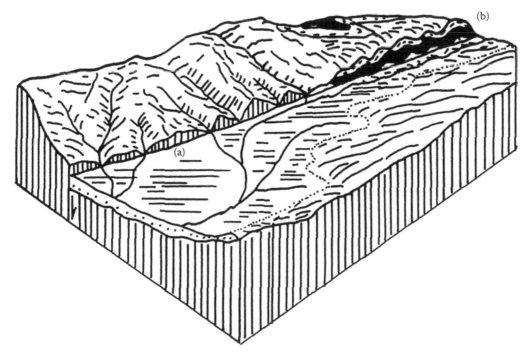

FIGURE 5.55
A fault scarp forms triangular facets (a) where ridges are cut off, and flatirons (b) where bedding dips steeply into the fault. (From Cotton, C. 1968. *Geomorphology*. Christchurch: Whitcombe & Tombs Ltd.: 76–95.)

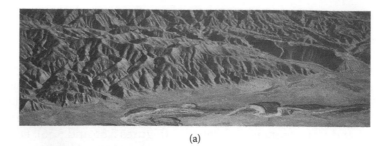

(a)

(b)

FIGURE 5.56
(a) Triangular facets where ridges are cut off along the San Andreas fault in the Carrizo plain, California. (b) Flatirons in the Pennsylvanian Fountain Formation near Boulder, Colorado, where bedding dips steeply into the basin-bounding fault.

offset landforms such as streams, coastlines, or the crest of a mountain range. Commonly two different and incongruous lithologies are abruptly juxtaposed by a fault, such as crystalline basement and sandstone, carbonate and volcanics, or outcrop and alluvium. Faults are more difficult to recognize if the same unit occurs on both sides: in such a case, an abrupt change in bedding orientation would indicate faulting. Growth faults along the Gulf coastal plain in Texas are noticeable only as slight dips in the surface (Figure 5.57a). If displacement is significant, then there can be an abrupt change in strike or dip or both across a fault trace (Figure 5.57b). High-angle normal faults have been associated with tilted crustal blocks or horst and graben topography (e.g., Al Khatieb and Norman, 1986). Intrusion of basaltic dikes along the trace of a fault suggests that these faults penetrate into the lower crust (e.g., the Hurricane fault, northwestern Arizona). Mineralized faults may appear either as resistant ridges and knobs or linear depressions, depending on the material in the fault zone. Silicified fractures resist erosion; the presence of iron oxides will make them linear zones of reddish hues. Faults filled with clay gouge will weather rapidly and may form sags filled with poorly drained soils. They may be evident due to a change in vegetation from surrounding areas.

It should be possible to make predictions about the offset on normal faults by measuring their length. Many workers have shown that the maximum displacement over length ratios (D_{max}/L) for total displacement across a fault zone (what you are most likely to be seeing on imagery) is in the range between 10^0 and 10^{-3}, and most likely between 10^{-1} and 10^{-2} (Davison, 1994; Clark and Cox, 1996; Morley, 1999; Schultz et al., 2006). This means that if an analyst maps a fault zone of length 10 km, the most likely maximum normal displacement will be between 1 km and 100 m. Ferrill et al. (2008) showed that there is a relationship between fault length and offset for normal faults (Figure 5.58). They used radar interferometry (InSAR) to map ground ruptures and fault displacements, which they

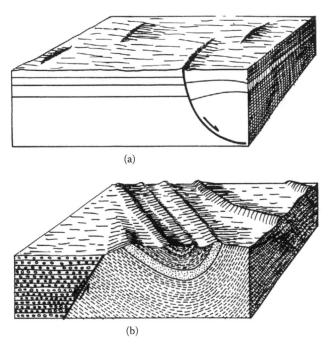

(a)

(b)

FIGURE 5.57
Fault scarps. (a) Growth faults such as those found along the Gulf coast generally have relief from 1 to 3 m. (b) Many high-angle normal faults have abrupt changes in structure and topography across the fault.

compared with 1950s photogeologic mapping. They concluded that preexisting individual fault segments tend to be reactivated and that the fault damage zone width is established early. With continued slip, the fault zone width remains stable but the active fault zone narrows as offset accumulates on a throughgoing surface.

More faults are needed in thin layers than in thick layers for total displacement to balance if brittle strain is equal in layers of different thickness (Benedicto et al., 2003). A fault zone consisting of many short segments (thin layers) leads to irregular fault shapes, whereas a fault zone consisting of few long segments (thick layers) suggests more planar faults.

When there are no outcrops, or they are covered by vegetation, near-vertical faults may appear as vegetation or soil tone alignments (Figure 5.59). Water well drillers in North Carolina would climb the hills in the spring looking for the green streaks of early grass in the fields. They knew from experience that these were the areas with the best chance of finding water (Fritz Johnson, oral communication, 1985). We can suspect that these are areas where faults brought groundwater near the surface (Lattman and Parizek, 1964). Linear stream segments or linear valleys extending for several kilometers suggest faulting, as the broken rock is more readily weathered and eroded into valleys (Lattman, 1959). Tributaries that are aligned on both sides of a major stream or across drainage divides strongly suggest structural control, as do barbed drainages and offset streams (Figures 5.25 and 5.60). Springs can indicate groundwater ponding on one side of a fault, and aligned seeps indicate a fault zone acting as a groundwater conduit. On thermal images, these faults appear cooler than adjacent rock and soil due to evaporative cooling both during the day and night (Sabins, 1987).

Normal faults can provide updip fault closure for hydrocarbon traps, and help form the traps at the high corner of tilted blocks. Accommodation zones (depocenters)

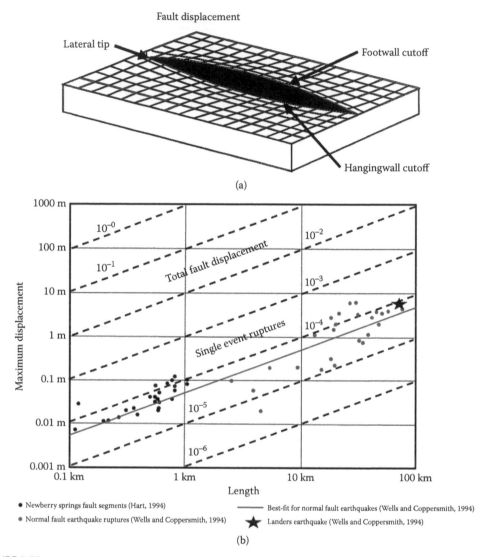

FIGURE 5.58

(a) Normal faults die out along strike as displacement decreases. As displacement is lost, the hanging wall and footwall cutoffs approach one another and merge at the fault tip. (b) Relationship between fault length and offset for normal faults. Comparison of maximum measured displacement versus rupture length for faults in the Newberry Springs Fault Zone and other single event faults. The Landers strike-slip fault is plotted for comparison. (From Ferrill, D.A. et al. 2008. Displacement-length scaling for single-event ruptures: insights from Newberry Springs Fault Zone and implications for fault zone structure. In C.A.J. Wibberley et al. (eds.), *The Internal Structure of Fault Zones: Implications for Mechanical and Fluid-Flow Properties* v. 299. The Geological Society of London: 113–122.)

and transfer zones (ramps where hydrocarbons migrate from the downthrown to the upthrown block) are formed by the overlap of normal faults in rifted settings (Figure 5.61). Known as stepover zones or relay ramps, these are areas where offset normal faults appear on imagery and maps as offsets in grabens (Fossen et al., 2010) (Figure 8.13). Overlapping normal faults in the Devil's Lane area of Canyonlands National Park, Utah, were scanned using Lidar to create a 3D model of faulting and fracturing (Rotevatn et al., 2009). Ground-based Lidar was combined with digital imagery

FIGURE 5.59
Vegetation (red) indicates springs, which are aligned along faults bounding the west side of Spring Valley, Nevada. Note the playa (dry lake) and paleo-shorelines at the north end of the valley. (This Landsat MSS color infrared image was processed by Earth Satellite Corporation, Rockville, MD [now MDA Federal]. © MDA Information Systems LLC 2013.)

FIGURE 5.60
Drainage indicators of faulting. Streams aligned across drainage divides (a) or on both sides of a major drainage (b) suggest fault control. Barbed drainage (c) and long, linear valleys (d) also suggest faulting.

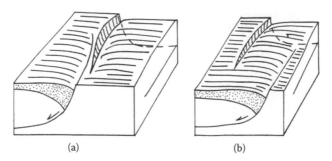

(a)   (b)

FIGURE 5.61
Antithetic *en echelon* transfer zones. (a) Back to back. (b) Overlapping. Note that the downthrown side is not only a depocenter, but the rollover associated with the faulting causes a structural high that may be a site for the accumulation of hydrocarbons. (From Morley, C.K. et al. 1990. *Bull. Am. Assn. Pet. Geol.* 74: 1234–1253.)

and a Global Positioning System to create a virtual outcrop of overlapping faults in the Needles area, where extensional grabens and relay ramps are well expressed. Together, these remote sensing data were used to generate 3D geomodels that, after assigning facies and petrophysical properties (porosity, permeability, net-to-gross, and so forth), were used in flow simulation work.

In addition to facilitating fluid migration, fault zones focus sedimentation by juxtaposing structural highs and depocenters. Provided they have updip seals, these zones can be the location of prolific oil production (e.g., Morgan Field, Gulf of Suez). Faults serve as conduits for hydrothermal and epithermal mineralizing solutions and as the locus of mineral and metal deposition, forming vein deposits along the fault, or manto deposits where acidic solutions impregnate a carbonate or reactive clastic unit. Faults commonly localize hot springs and hydrothermal sources, volcanic vents, karst collapse features (sinkholes), and even pingos in permafrost terrain.

"Listric faults" flatten with depth and die out in bedding. These faults tend to be arcuate or scoop shaped in plan view. For purposes of interpretation, faults that are concave toward the downthrown side are more likely to be listric, whereas straight faults are more likely to be planar and high angle. This relationship of curving trace to dips that flatten with depth is based on slope stability work by engineering geologists (Krynine and Judd, 1957). The ideal shape of gravity-slide blocks (slumps or landslides) is concave-up in cross section, and concave toward the downthrown side in plan view (Figure 5.62) (Ollier, 1981). Moore (1960, p. 411) observed that, in the Basin and Range of Nevada

> curvature of the fault block ranges reflects the curvature of the main bounding fault in plan; hence, the fault itself is believed to be convex toward the direction of the tilt of the range, or concave toward the downthrown side of the fault.

Recent work expands on this theme. Roberts (2007) showed that not only are listric normal faults curved and concave toward the downthrown side, but slip vectors indicate oblique slip near the fault terminations, and subseismic faults (offset below the resolution of seismic data) mimic and extend the fault patterns mapped in the field "…in map view from the footwall looking toward the hanging wall, right-lateral slip occurs at the left end of the faults, with left-lateral slip at the right end."

Rotation of bedding and subsequent continued, nested listric faulting lead to a sequence of steep to shallow dips as one proceeds into the basin (Figure 5.63).

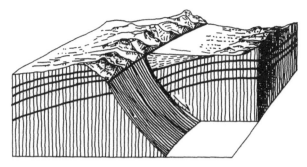

FIGURE 5.62

Gravity slide and listric normal faults are concave toward the downthrown side in both map view and cross section. Note triangular facets along the fault. (From Moore, J.G. 1960. *U.S. Geol. Survey Prof. Paper.* 400–B: 409–411.)

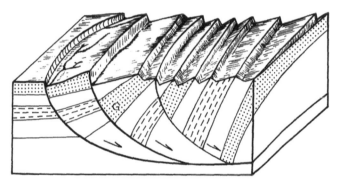

FIGURE 5.63

Block diagram illustrating the effect of successive listric faulting. Note how the oldest fault block has been rotated so that bedding is almost vertical.

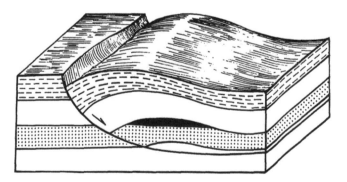

FIGURE 5.64

Oil can accumulate in a rollover anticline generated by listric growth faulting.

One type of listric normal fault, the growth fault, is characterized by thickening of units (sedimentary growth) and dip reversal into the fault, called "rollover" or "reverse drag" (Shelton, 1984) (Figure 5.64). This is, in effect, an anticline formed by collapse of the hanging wall strata as it pulls away from the footwall. Clay models bear out the curved nature of listric normal faults in map view and cross section: Bose and Mitra (2009, 2010) have shown how these curved faults form on the upthrown block set back slightly from the underlying master fault. Rollover folds are economically important in areas of rapid sediment accumulation such as the Gulf of Mexico coastal plain and river deltas.

On occasion, a normal fault will steepen with depth. These can be recognized on imagery because they are expressed at the surface by a master normal fault with nested, geometrically similar (but upward steepening) small displacement normal faults in the hanging wall (Patton, 2005). These small faults develop as the shallow part of the master fault becomes inactive.

Some uplifts, such as those in the Colorado Plateau-Southern Rocky Mountains provinces of the United States, are characterized by basement overhangs and zones of normal faults in the uplifted basement rock where the fractures are more or less parallel to the faulted margins. The uplifted basement and bounding faults form updip seals for hydrocarbons migrating out of the basin. Fracture zones in the uplifted basement rock serve as conduits for mineralization, such as at the Schwartzwalder uranium mine in the Rampart Range, Colorado. Overhang plays have been studied and observed on satellite imagery in many areas including the Rampart Range and Uinta mountains, Utah, and Wind River mountains, Wyoming. Faulting and jointing parallel to the mountain front in the uplifted block occurs in each area. Although the Uintas are composed primarily of bedded Precambrian sediments and are fundamentally different in their mechanical response to uplift and faulting than the granite-cored Wind River and Rampart Ranges, the fracture pattern parallel to the mountain-bounding faults mapped in the Uintas by Ritzma (1971) is essentially the same as that mapped in granites of the Wind River Range by Love et al. (1955, 1979a,b) and by Bryant et al. (1981) in the Rampart Range. This coincidence goes beyond the tendency for granites to form orthogonal fracture patterns. The patterns mapped on Landsat and airphotos are essentially parallel to the mountain front (Figure 5.65).

Faults in the uplifted block that are parallel to the mountain front are diagnostic of overhangs. These normal faults are extensional features that developed following uplift along reverse or thrust faults (Figure 5.66) (Price, 1959). They are likely to penetrate the overhang and extend into the sediments, and possibly hydrocarbon reservoirs below. These fractures have been targets of geochemical surveys targeting subbasement overhang oil and gas accumulations.

FIGURE 5.65
Oblique true color image of the Rampart Range, Colorado, looking north from Pikes Peak toward Castle Rock. Arrows indicate faults parallel to the mountain front. (Image courtesy of TerraMetrics, GeoEye, and Google Earth, © 2012 Google.)

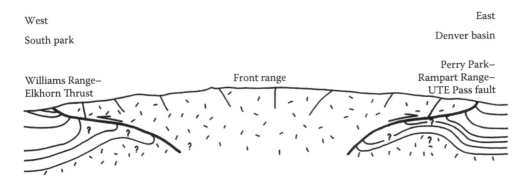

West

South park

Williams Range–
Elkhorn Thrust

Front range

East

Denver basin

Perry Park–
Rampart Range–
UTE Pass fault

FIGURE 5.66
Schematic east-west cross section of the Rampart Range, Colorado, roughly coincident with the center of Figure 5.65. Note the extension faults parallel to the margins of the uplift. (From Jacob, A.F. 1983. *Rocky Mountain Foreland Basins and Uplifts.* Denver: Rocky Mountain Association of Geologists: 229–244. Reprinted with permission of the Rocky Mountain Association of Geologists.)

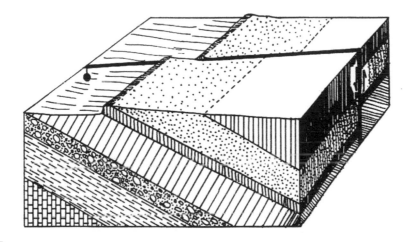

FIGURE 5.67
A normal fault can offset the contact of inclined units and give the appearance of strike-slip faulting. One should look for several supporting indicators of displacement, such as consistent offsets or *en echelon* folds, in order to make this determination.

Whenever a normal fault cuts inclined strata, there will be an "apparent strike-slip" offset, with a key bed on the downthrown side being displaced so that the trace is opposite older strata (Figure 5.67). It is not always possible to distinguish between such a normal fault and strike-slip faults. Strike-slip faults, however, often cause bedding to turn into the fault as a result of drag. Strike-slip faults are also associated with numerous structural and geomorphic features discussed in the next section.

"Structural inversion" is where a normal fault has been reactivated with an opposite sense of slip due to a change in stress state from extension to shortening. The reversal of slip causes the former depocenter to become uplifted. Inversion can be identified by recognizing thickening of sedimentary units into the fault. This may require the stratigraphic section to be inclined (dipping) such that the exposure appears as a cross section. Inversion can create a structure that looks like a thrust or reverse faulted fold, that is, the former downthrown side of the fault now appears to be an anticline bounded by a thrust fault (Grimaldi and Dorobek, 2011).

Strike-Slip Faults

The interaction of discrete segments of a strike-slip or wrench fault system generates regional and local deformation that can be detected on imagery and gravity data (Campagna and Levandowski, 1991). Strike-slip faults are recognized by consistent lateral offset of lithologic units, facies, streams, ridges, alluvial fans, or any feature crossed by the fault (ore bodies, roads, fences, homes). Strike-slip faults are generally near-vertical at the surface, and thus have long, linear exposures (Figure 5.68). Drag along the fault can cause features that cross the fault to bend or rotate into the fault. "Faceted spurs" are ridges that have been abruptly cut off by the fault. A "headless valley" occurs when these ridges block an old stream valley.

There are a host of accessory geomorphic features that are developed with these faults and help to identify them. These include *en echelon* folds oriented 30–45° to the fault and perpendicular to the direction of maximum horizontal compression (Figure 5.69). *En echelon* folds begin to develop prior to faulting and are most intense (highest amplitude) adjacent to the fault. As the strike-slip fault develops, the original folds are cut and offset, and new *en echelon* folds form along the flanks of the fault zone (Wilcox et al., 1973). The Newport-Inglewood trend in the Los Angeles basin has produced much oil from *en echelon* folds (Harding, 1974). *En echelon* normal faults (e.g., Lake Basin and Cat Cr. fault zones, Montana) have also been associated with wrench faulting.

Changes in strike of the fault trace cause other characteristic features (Crowell, 1974). When a fault bends, one will find either pull-apart zones (oblique rifting, sag ponds; Figure 5.70) or areas of compression (buckle folds, flower structures; Figure 5.71). Examples of releasing bends (pull-apart zones) include the Salton Trough, California, and the Dead Sea rift, Israel-Jordan. Examples of restraining bends (compression zones) include the Mecca Hills, California and Ocatillo badlands, California (Sylvester and Smith, 1976; Segal and Pollard, 1980). Pull-apart basins and restraining bend uplifts can be identified using gravity maps (gravity lows indicate basins; gravity highs indicate uplifts) along with imagery.

Much has been made of using fracture patterns to determine both the fault trend and stress field (and thus sense of offset) around strike-slip faults. There are a plethora of wrench-associated fractures. Those fractures parallel to the master fault are "Y" shears; "R" (Reidell) shears are *en echelon* fractures oriented 15° to the master fault; "R-prime" shears

FIGURE 5.68
Scarp of the San Andreas fault on the east side of the Carrizo Plain, California. Oblique view north. Note the linear fault trace and right-lateral offset of streams crossing the fault. (True color image courtesy of GeoEye and Google Earth, © 2012 Google.)

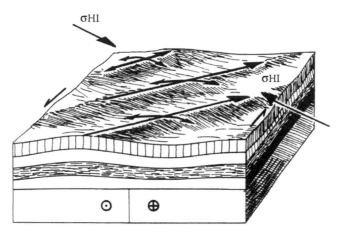

FIGURE 5.69
En echelon folds develop perpendicular to the maximum horizontal compressive stress, and at 30° to 45° to the master wrench fault.

FIGURE 5.70
San Andreas Lake and Crystal Springs Lake are sag ponds developed along the San Andreas fault just south of San Francisco, California. (True color oblique view north courtesy of TerraMetrics and Google Earth, © 2012 Google.)

intersect the master wrench at 75°; the "P" shear is symmetric to the R shear and intersects the master fault at 15° (Figure 5.72). One should take care when interpreting strike-slip offset using fault patterns: if each set of shear-related fractures is given a range of ±15°, then most fractures seen on imagery can be used to justify a wrench fault interpretation.

Another common practice is to categorize fractures into two primary sets, consider them to be conjugate shears, and interpret the principal horizontal compressive stress as bisecting the acute angle between the shear fracture sets. The problem with this approach is that most areas have at least two principal joint sets, but they are not necessarily caused by strike-slip faulting.

Under conditions of "pure shear" (homogeneous nonrotational strain), conjugate fractures form with an acute angle of 60°, bisected by the direction of maximum compression (Figure 5.73). Under these circumstances, each of the conjugate shear fractures accommodates strike-slip displacement (shear) with an opposing sense of offset (one right-lateral,

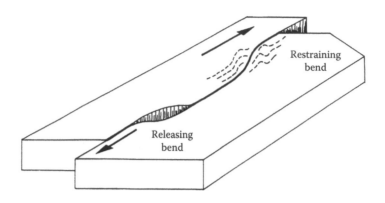

FIGURE 5.71
Releasing bends tend to form basins; restraining bends cause uplifts along strike-slip faults. (From Crowell, J.C. 1974. Origin of late Cenozoic basins in southern California. In W.R. Dickenson [ed.], *Tectonics and Sedimentation*. SEPM Spec. Pub. 22: 190–204. Published with permission of the Society for Sedimentary Geology [SEPM].)

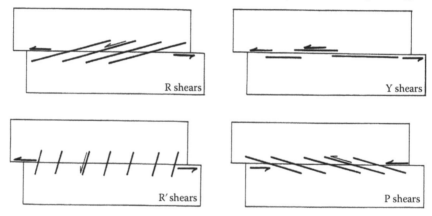

FIGURE 5.72
Subsidiary faults that have been described associated with strike-slip faulting include R, Y, R′, and P shears. These are usually not all present in any given area along the master fault.

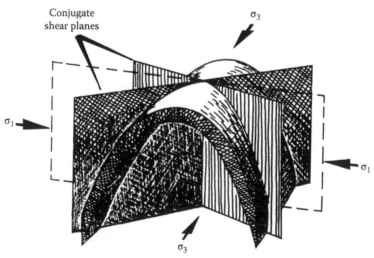

FIGURE 5.73
Conjugate shears form at 30° on either side of the maximum horizontal compression (σ_1).

one left-lateral). Wrench deformation usually involves "simple shear" (constant volume homogeneous rotational strain) within a shear zone.

Shear faults die out along strike by transferring their displacement. This can be done through a series of "horsetail faults," which disperse displacement among many segments. *En echelon* horsetail faults are sites of local extension and often contain veins and mineral deposits. Whereas a major fault frequently contains gouge, the small auxiliary faults may contain relatively clean breccias with good permeability and serve as excellent sites for mineral deposition. Likewise, clean fault breccias assist in vertical migration of hydrocarbons, and for that reason do not make good hydrocarbon traps. Shear faults can also transfer displacement by folding strata (shortening) on one side and extending (downdropping) strata on the other side of the termination.

Thrust Faults

Most faults are recognized by abrupt changes in strike, dip, or rock type, often associated with an escarpment. The trace of a high-angle or near-vertical fault tends to be linear, hence the term "lineament." In contrast, the trace of a thrust is generally irregular, following topographic contours, and making identification of these faults challenging. However, there are both direct and indirect criteria for recognizing thrusts on imagery and maps.

Direct indications of thrusting include abrupt changes in strike or dip, representing juxtaposition of different structures in the hanging wall and footwall (Figures 5.74 and 5.75). The hanging wall of a thrust is usually topographically higher than the footwall, in which case the fault trace commonly runs along a break in slope at the base of the thrust sheet. When there is little or no topographic relief, the hanging wall can be recognized because the strike of bedding in folds carried by the thrust (fault bend folds and fault propagation folds) is usually parallel to the thrust front. Back limb dips of leading edge thrusted folds are more or less parallel to the fault surface. Often the fault trace in plan view is convex in the direction of transport (Figure 5.76). Part of the irregular fault trace may be a result of erosional outliers, or klippen, in front of the main sheet.

"Leading edge anticlines," as their name implies, develop on a thrust sheet at the thrust front. They are typical of thrusts that terminate by ramping up section; otherwise, the hanging wall units generally dip the same direction as the thrust fault. "Imbricate fans," where thrust splays cut the hanging wall into stacked slivers, appear as zones of

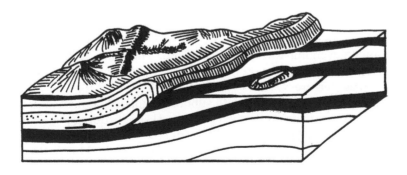

FIGURE 5.74
Leading-edge anticlines along thrust fronts tend to have irregular to arcuate fault traces and may have erosional outliers (klippen). They can be recognized because of the abrupt change in strike and dip of units on either side of the fault. Beds in the hanging wall generally strike parallel to the thrust trace and dip parallel to the fault. (From Prost, G.L. 1990. *World Oil*. 211: 39–45.)

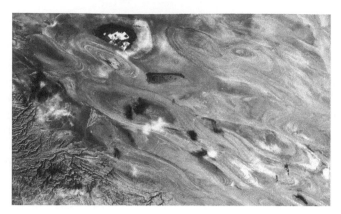

FIGURE 5.75
True color image of part of the Tsaidam basin, Tibetan Plateau, China, illustrating multiple northeast-directed thrust sheets with leading-edge anticlines. (Image courtesy of CNES/SPOT Image and Google Earth, © 2012 Google.)

FIGURE 5.76
The southern Sulaiman Range of Pakistan is a south-directed series of thrusted folds. This illustrates how a thrust front is often convex in the direction of transport. (True color image courtesy of CNES/SPOT Image and Google Earth, © 2012 Google.)

parallel ridges, and are recognized as thrusts by observing repeated section. Imbricate fans develop over footwall faults or thrust ramps. Folds carried on a thrust tend to have common asymmetry, verging toward the foreland. They may form imbricated, stacked anticlines or synclines, depending on the level of erosion, with each fold separated from the others by a thrust (Figure 5.77).

Shortening of sedimentary layers above a subhorizontal detachment (decollement) causes "detachment folds." These folds are formed by flexural slip (bedding plane slip) and flexural flow (ductile flow of bedding) above a weak or ductile detachment surface (salt, gypsum, overpressured shale). Detachment folds tend to be concentric and have steep to overturned limbs, generally have no preferred vergence and may have no obvious thrust faults at the surface (Figures 5.78 through 5.80). The axial length of these folds is generally several times greater than the wavelength, and these folds are frequently arcuate or sinuous (Vendeville, 2005; Gaullier and Vendeville, 2005). Abrupt changes in structural style, as from tight to open folding, or imbricate ridges to folds, can be observed at the transition from thrust to foreland,

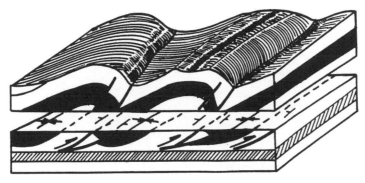

FIGURE 5.77
Stacked anticlines (or synclines, depending on the level of erosion) imply thrusting. (From Prost, G.L. 1990. *World Oil.* 211: 39–45.)

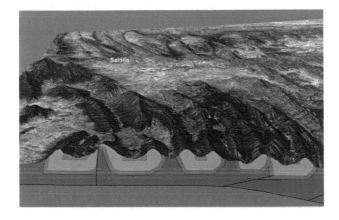

FIGURE 5.78
Landsat TM perspective image of the Monterrey Salient of the Sierra Madre Oriental near Monterrey, Mexico. View west toward Saltillo. Folds developed in Cretaceous and Jurassic strata are believed to be detached over the upper Jurassic Olvido Fm. evaporite decollement. (Color infrared image courtesy of Amoco Production Company and PEMEX.)

as in the Brooks Range, Alaska, or the Sawtooth Range, Montana. One should expect abrupt changes in structural style going from the thrust sheet into the foreland (Figure 5.81).

Indirect clues to thrusting include tear faults, lateral ramp anticlines or monoclines, and relaxation faults. "Tear faults" are linear shear zones that look like strike-slip faults, but can also have a component of normal offset or monoclines developed along strike. They end at the thrust front, either abruptly or by curving into the thrust fault (Figures 5.82 and 5.83). Folds on either side of a tear fault may have opposite vergence or be at different stages of structural development. Monoclines form in the thrust sheet where a tear fault lies above a faulted footwall. "Lateral ramp anticlines" also form over footwall faults that deform the thrust sheet (Figures 5.82 and 5.84). "Relaxation faults" develop at the end of the shortening episode that formed the thrusts. They occur on the back limb of hanging wall folds or where the thrust ramps up section. These listric normal faults are scoop-shaped, concave toward the hinterland, and sole in bedding or a thrust.

Thrusts terminate along strike by transferring their displacement to folds or overlapping thrust faults. Folds that merge into faults along strike suggest displacement transfer related to thrusting (Prost, 1990).

FIGURE 5.79
The northeast flank of the San Julian uplift near Caopas, Mexico. In this natural down-plunge projection, one can see a cross section of Laramide-age detachment folds (one shown at arrow) that were rotated to vertical by later uplift. These folds clearly show a detachment at the upper Jurassic evaporite level. North is to the left of the image. (Courtesy of GeoEye, CNES/SPOT Image, and Google Earth, © 2012 Google.)

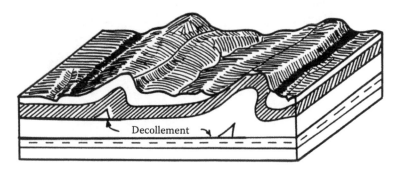

FIGURE 5.80
Folds with steep limbs that may be overturned on both sides, with no preferred vergence, and no obvious faulting at the surface imply detachment thrusting. (From Prost, G.L. 1990. *World Oil*. 211: 39–45.)

FIGURE 5.81
The transition from imbricate ridges to folds (at arrow) marks the frontal Sawtooth thrust at Crab Butte, Montana. Thrusting here is from west to east. True color oblique view north-northwest. (Courtesy of USDA Farm Service Agency, TerraMetrics, and Google Earth, © 2012 Google.)

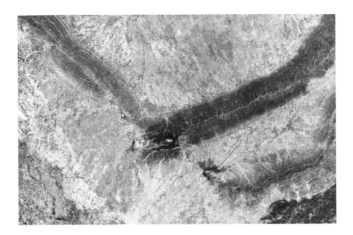

FIGURE 5.82

A major tear fault (arrow) along a lateral ramp on the west edge of the Bannu depression, Pakistan. The Pezu-Bhittani Range (including a faulted lateral ramp anticline) is formed along the right-lateral tear fault; the Marwat-Khisor Range is the leading edge anticline on the southeast-directed thrust. North is at the top of the image. (True color image courtesy of CNES/SPOT Image and Google Earth, © 2012 Google.)

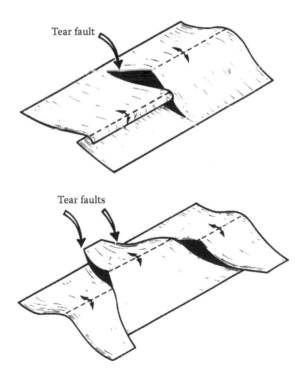

FIGURE 5.83

Tear faults separate thrust sheets with different amounts of displacement. They often curve into and offset the thrust front. Displacement can appear normal, strike-slip, or have components of both. These faults may be caused by subthrust structures or form at facies changes in the thrust sheet. Folds may have different vergence or amplitudes, or die out across these faults. (From Prost, G.L. 1990. *World Oil*. 211: 39–45.)

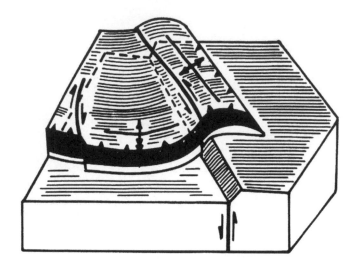

FIGURE 5.84
Lateral ramp anticlines can form over footwall faults by oblique convergence. (From Prost, G.L. 1990. *World Oil*. 211: 39–45.)

It is difficult to predict the depth to the thrust fault or detachment based on image interpretation alone. Knowledge of stratigraphic thicknesses and the units at the surface can help. Detachments usually occur on mechanically weak or overpressured horizons such as shale, gypsum, or salt. High-amplitude (tight), high-frequency folds suggest a detachment closer to the surface than broad, open folds (Morley, 1987).

Thrusts mapped along the forelimb of an asymmetric fold suggest that these are leading-edge folds, also called "truncation anticlines" or fault bend folds (Jamison, 1987). "Fault bend folds" develop over thrust ramps. These folds suggest relatively shallow thrust faults. All factors being equal, the amplitude and frequency of thrust-carried folds decrease as the thrust sheet thickens. Lithology and stratigraphic thickness also plays a part: thick, competent units form broader folds than thin, ductile units. Fold length should have no relationship to depth of detachment. On the other hand, the "excess area" of a fold, that area on a cross section that is uplifted by deformation to a position above its original regional datum level, is related to depth to detachment in compressional folding (Epard and Groshong, 1993). Imagery can be used to generate a cross section that accurately depicts a fold and the amount it has been uplifted above its regional.

Factors other than fold geometry are used to estimate thrust sheet thickness (Figure 5.85). If the section contains sequences of shales or evaporites (ductile layers) overlain by rigid sandstones or carbonates, the tendency is to form large amplitude detached buckle folds above the decollement (Jamison, 1987). Carbonates (rigid) are more apt to form imbricate thrusts (Morley, 1987). Interbedded sandstones and siltstones (varying stiffness) tend toward fault propagation folding with multiple, leading edge thrust-cored folds (Jamison, 1987). "Fault propagation folds" form at the thrust tip where the decollement has ceased propagating but displacement on the thrust continues. That displacement is accommodated by asymmetric folding. "In shale sequences, faults tend to be blind and die out in complex, hanging-wall folds" (Morley, 1987, p. 339).

"Blind thrusts" are thrusts that terminate before reaching the surface. Blind thrusts may tip out (lose all displacement) along bedding planes or as the fault approaches the surface. They typically die out upward into folds, upward branching fault splays, and lateral

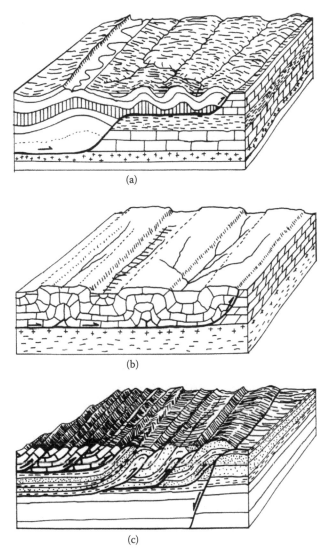

FIGURE 5.85
Clues to thickness and lithology of the thrust sheet. (a) Tight folds imply a thinner thrust sheet than broad, open folds. (b) Depth to the detachment is approximately equal to the thickness of a fold flank for box (detachment) folds. (c) Imbricate thrusts are more common in carbonates, whereas fault propagation folds are more common in clastic sequences. (Data from Morley, C.K., *J. Struct. Geol.* 9: 331–343, 1987.)

overlapping fault splays (Hart et al., 2007). Shortening due to folding at the surface is always less than shortening along the blind fault at depth. Folding and uplifted topography are formed above the area of slip on a blind thrust, with maximum uplift above the blind thrust plane. An anticlinal ridge and backthrusts indicate a blind fault at depth (Schultz, 2000). In the case of multiple stacked thrusts, deformation above a blind thrust may be either more or less intense than that in the lower thrust sheet. There are only indirect clues to lower sheet deformation. If subthrust folding has longer wavelengths than upper plate folding, the vergence of surface fold axes changes from foreland to hinterland depending on position with respect to underlying folds (Figure 5.86). If subthrust folding is higher frequency than thrust-carried folds, surface highs may develop over ramps or subthrust folds.

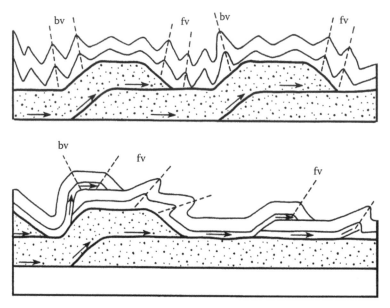

FIGURE 5.86
Fold vergence can change depending on the position of the fold with respect to subthrust structures. fv = forward verging; bv = backward verging. (From Dunne, W.M., D.A. Ferrill. 1988. *Geology.* 16: 33–36.)

Angular unconformities have some of the same characteristics as thrusts. They are characterized by abrupt changes in strike and/or dip across the unconformity. Folding that exists above an unconformity also exists below the unconformity (but not necessarily vice versa). Bedding above an unconformity is, by definition, less deformed than below, whereas a thrust sheet is usually more deformed than underlying units. Thrusts repeat the section; unconformities remove section.

Reverse faults superimpose older rocks on younger, as do thrusts, (except in special circumstances, as might occur when thrusting is out of sequence). The high angle of the reverse fault plane causes a linear fault trace, distinguishing this class of faults from thrusts.

Subthrust hydrocarbon plays are difficult to recognize. Folding that occurred postthrusting will be expressed at the surface and will extend below the hanging wall thrust sheet, but the age of folding generally cannot be determined with certainty from remote sensing data alone. Subthrust highs (which can be detected by gravity or magnetic surveys) can act as a buttress, uplifting the thrust and causing deeper erosion into the thrust sheet and along tear faults or lateral ramps at the margins of the uplift. Tear faults can sometimes be detected by potential fields data such as High Resolution AeroMagnetics (HRAM) particularly if they are located over subthrust faults that offset underlying basement rocks.

The depth to a detachment can be estimated on the basis of the geometry of detached folds. A series of curves developed by Jamison (1987) for detachment folds relate fold interlimb angle and back limb dip to the ratio of fold amplitude (from elevation data) divided by thrust sheet thickness (Figure 5.87). Ramsay and Huber (1987) showed how one can calculate the depth to a decollement for fault bend folds by measuring the uplifted area, original line length, and length after deformation (Figure 5.88). Hanging wall thickness changes can be hung from surface elevations to reveal the shape of the thrust surface. Fold vergence of minor structures can be used to speculate whether the thrust sheet is riding up onto a structure (hinterland verging) or moving over a subthrust structure (foreland verging; Figure 5.86) (Dunne and Ferrill, 1988).

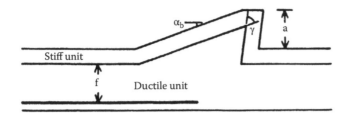

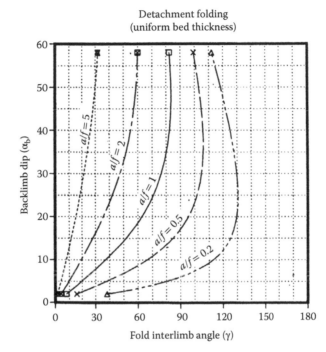

FIGURE 5.87
Graph for estimating the depth (*f*) to the detachment surface for detachments folds when the interlimb angle, back limb dip, and fold amplitude are known. (From Jamison, W.R. 1987. *J. Struct. Geol.* 9: 207–219.)

In general, the thrust fault zone itself is thin and gouge filled and does not serve as a conduit for mineralizing fluids or hydrocarbon migration. Extension faults in the upper plate, however, may be excellent conduits for vertical migration of fluids. They may also allow fluids to breach a potential trap, so an evaluation of fault offsets and geometries is critical to hydrocarbon trap evaluation.

Case History: Structure and Geomechanics of Active Fault Zones

Standard imagery can identify and map faults, but cannot measure small-scale surface displacements. Lidar can do this. This case study, contributed by David E. Haddad, Edwin Nissen, and J. Ramón Arrowsmith of the Active Tectonics, Quantitative Structural Geology, and Geomorphology Laboratory, Arizona State University (Tempe), demonstrates that Lidar is a promising technology that provides a framework on which the efficient and accurate characterization of earthquake processes may be constructed over a range of

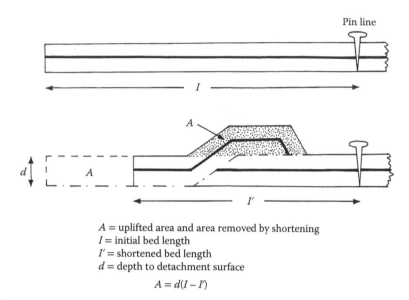

A = uplifted area and area removed by shortening
I = initial bed length
I' = shortened bed length
d = depth to detachment surface

$$A = d(I - I')$$

FIGURE 5.88
Estimates of depth to the detachment for fault bend folding can be obtained by the uplifted area technique. (Data from Ramsay, J.G., M.I. Huber. *Techniques of Modern Structural Geology 2: Folds and Fractures.* London: Academic Press, 1987.)

spatio-temporal scales. Airborne Lidar datasets presented here can be downloaded from http://www.opentopography.org.

At centennial to millennial rates, earthquakes are the primary driver for topographic deformation. Therefore, the topography of active fault zones holds a wealth of information about past earthquakes and active faulting. From a societal standpoint, fault zone topography provides crucial information about the recurrence of past earthquakes that may help forecast the likelihood of future earthquakes and prepare seismically sensitive infrastructure such as schools, hospitals, and nuclear power plants for strong ground motions (e.g., WGCEP, 2008). Thus, quantitative documentation and characterization of fault zone topography is important. Until the late 1990s, measurement of tectonically displaced features such as offset stream channels, terraces, and topographic ridges was made using total station surveys (Arrowsmith and Rhodes, 1994; Arrowsmith et al., 1998). In these surveys, thousands of measurements were made over multiple days in order to depict the topography of displacement markers. Since around 2000, a significant expansion in the development of Lidar instruments has provided an opportunity to survey topography with unprecedented speed, accuracy, and resolution. Airborne Lidar campaigns along active plate boundaries were quickly recognized as necessary to document the record of earthquake-generated deformation at appropriate spatial and temporal scales. Such campaigns have provided digital representations of topography at resolutions sufficient to make measurements of earthquake-related vertical and lateral topographic displacements (Hudnut et al., 2002; Bevis et al., 2005; Oskin et al., 2007, 2010a,b, 2012; Prentice et al., 2009; DeLong et al., 2010; Hilley et al., 2010). Fault trace geometries and stream channels that were offset by past earthquakes are clearly illuminated by Lidar datasets (Arrowsmith and Zielke, 2009; DeLong et al., 2010; Haddad et al., 2012). Systematic analysis of these data reveals geomorphic features that are barely perceivable in the field but may change the fundamentals of inferring earthquake recurrence and fault segmentation (Zielke et al., 2010).

An added benefit from Lidar datasets in active tectonics research is their educational value. With the increase in web-based 3D topographic visualization, such as the ubiquitous Google Earth, Lidar datasets can provide important teaching aids in undergraduate- and graduate-level geoscience courses. It will become important to the future of geoscience education to integrate Lidar datasets as components in undergraduate- and graduate-level curricula by bringing virtual outcrops of faults and fault-related deformation to the classroom.

Earthquake ruptures along crustal faults originate in the middle to lower depths of the seismogenic layer and transmit deformation to the Earth's surface by driving slip along faults (Figure 3.14; Scholz, 2002; Titus et al., 2011). This process is expressed as fault scarps and fractures or off-fault folding and warping (Oskin et al., 2012; Quigley et al., 2012). The extent to which these surface manifestations represent earthquake processes at depth is controlled by the geometric complexity of faults, faulting mechanisms, and the strength variations within the upper lithosphere (Sibson, 1986; Scholz, 2002). With the exception of very limited direct observations of active faults at depth (e.g., the San Andreas Fault Observatory at Depth), there is no direct access to faults in the seismogenic layer. However, Lidar allows high-resolution analyses of coseismic deformation of fault zone topography within tens to a few thousands of meters from fault ruptures. These Lidar datasets help us interpret fracture patterns observed in the seismic record and enable surface rupture patterns to be examined to see how well they represent coseismic ruptures and seismic moment released at depth (Oskin et al., 2012). Furthermore, these datasets provide important controls for measuring coseismic slip in the most recent event and slip accumulated over multiple earthquakes (Arrowsmith et al., 2011; Haddad et al., 2011; Madden et al., 2011). In this way, the Lidar data improve our understanding of fault system behavior and interactions through space and time. This case study examines the application of airborne Lidar data to understand past surface ruptures along the Garlock fault, California (Figure 5.89).

Fault Zone Mapping

The surface traces of past earthquake ruptures (faults) are mapped along the Garlock fault, California, using high-resolution aerial photographs and Lidar-derived DEMs. The resulting maps revealed that along-strike fault trace patterns differ significantly between the western and eastern sections of the Garlock fault. The topographic expression of the western section is not well defined when compared with the ubiquitous fault scarps of the central and eastern sections. Similarly, few lateral displacements in stream channels and ridges are preserved along the western section in comparison with the central and eastern sections. This may be due to intense mass wasting that has obliterated fault scarps and surface manifestations of the last few earthquakes along the western section of the Garlock fault. This mass wasting is attributed to an along-strike climate gradient where the wetter conditions of the western section are not conducive to preserving fault scarps when compared with the central and eastern sections that reside in the Mojave Desert.

Fault Zone Complexity

Lidar-derived fault trace maps for the Garlock fault were used to calculate fault complexity parameters such as segmentation and length. These metrics are correlated with fault zone geology to explore for lithologic controls on surface ruptures. Figure 5.90 presents

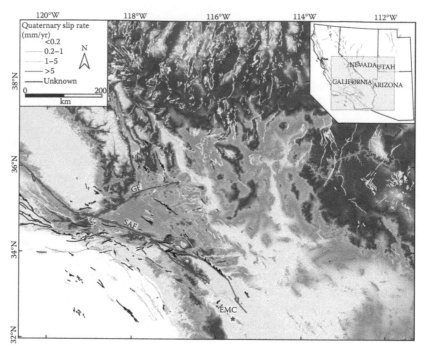

FIGURE 5.89

Seismotectonic settings of the case studies presented in this chapter. The first case study uses airborne light detection and ranging (Lidar) datasets to document earthquake-related slip distributions along the Garlock fault (GF). The second case study uses airborne Lidar data to compute three-dimensional topography displacements created by the April 4, 2010, El Mayor-Cucupah (EMC) earthquake rupture. Digital topographic data in this map were accessed from the U.S. Geological Survey (USGS) Seamless Data Warehouse (http://seamless.usgs.gov). Fault data were acquired and modified from the USGS Quaternary Fault and Fold Database (http://earthquake.usgs.gov/hazards/qfaults/). (From Haddad, D.E. et al. 2012. *Geosphere*. 8(4): 771–786.)

the results of an analysis of a spatial correlation between fault segmentation, length, and rock type for a section of the central Garlock fault. Fault complexity was calculated using 300-m-wide bins that moved along strike of the fault zone and show that, in general, fault segmentation is greater in bedrock breaks than alluvial breaks. This is counter to the general perception that unconsolidated media (e.g., Qal) distribute brittle deformation across broad fracture belts. The relatively simple Qal rupture patterns observed in our fault complexity analysis are attributed to partial masking by the thin alluvial cover that overlies the shallow bedrock and associated low confining stresses. As a result, the transmission of strain through the Qal cover to the topography of the Garlock fault zone has generated fewer fractures than are seen in the bedrock. This analysis also showed that fault segment length appears to be controlled by rock type; segments are generally longer in granodiorite (Tg) than those breaking through alluvium or quartz-monzonite (Qal, Tgm in Figure 5.90), indicating that rock type controls the local continuity and mechanics of earthquake ruptures.

Lateral Displacement of Geomorphic Markers

The Lidar datasets for the Garlock fault provide important insights into the along-strike distribution of slip created by past earthquakes. Laterally displaced geomorphic markers

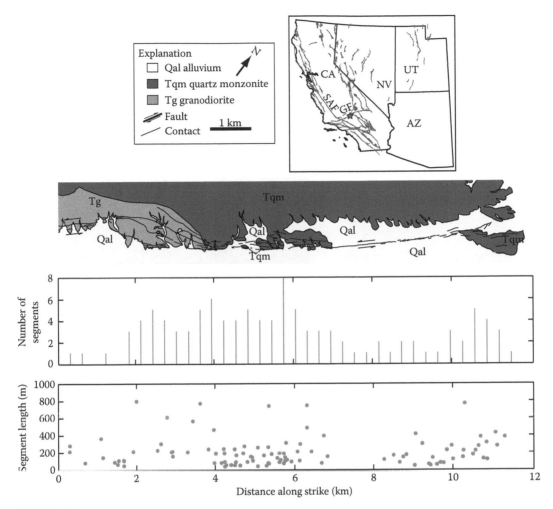

FIGURE 5.90

Results from the fault complexity analyses using Lidar-derived fault trace maps. The analyses were performed in the central Garlock fault (GF), California, where past coseismic breaks ruptured alluvium (Qal), quartz monzonite (Tqm), and granodiorite (Tg) rock units (for the area shown). Fault segment length and number of segments were computed in 300 m bins along strike. SAF – San Andreas fault. (Geology from Ludington, S. et al. 2007. Preliminary integrated geologic map databases for the United States – Western States: California, Nevada, Arizona, Washington, Oregon, Idaho, and Utah. U.S. Geological Survey Open File Rept. 2005–1305.)

such as offset ridges, stream channels, and terraces along the entire length of the Garlock fault are measured using a lateral offset calculator and submeter-resolution DEMs (Zielke and Arrowsmith, 2012). Figure 5.91 shows results from the displacement analysis of 431 offset features along the Garlock fault. To validate the Lidar-derived measurements with those made in the field, 129 offsets measured in the field by McGill and Sieh (1991) using the Lidar offset calculator are reviewed. The Lidar-derived measurements compare well with those made in the field for the same offset features, with correlation coefficients R^2 of 0.9. This validation demonstrates that the Lidar-derived offset measurements are reliable indicators of coseismic slip in the last few earthquakes. Lidar-derived offsets thus provide accurate representations of slip distributions for fault zones, especially in sections like the western Garlock fault where no field measurements of earthquake offsets are available.

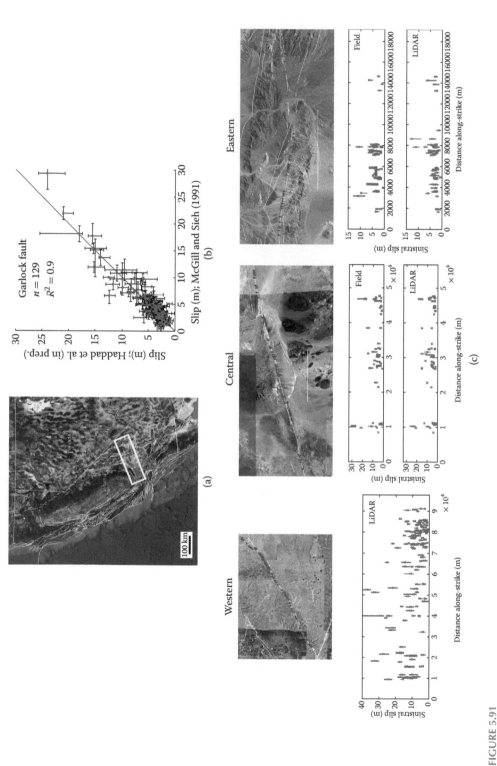

FIGURE 5.91

Results from Lidar-derived offset measurements. (a) Overview map of offset measurements compiled for major faults in California. (b) Lidar versus field-derived offset measurements for the Garlock fault (GF). Garlock field measurements were made by McGill and Sieh (1991). (c) Slip distribution plots for the GF made from Lidar- and field-derived offset measurements. (Figure courtesy of Haddad, Nissen, and Arrowsmith.)

For the western, central, and eastern sections of the Garlock fault, average surface slip in the last earthquake is calculated as 3.6 m ± 1.1 m, 3.8 m ± 0.8 m, and 3.3 m ± 0.9 m, respectively. McGill and Sieh (1991) reported the average slip from field-derived displacement measurements for the central Garlock as nearly double these Lidar-derived average slips. Such inconsistencies between Lidar- and field-derived slip measurements have important implications for how slip distributions are interpreted (Zielke et al., 2010).

Geomechanical Modeling

Lidar-derived fault trace maps constrain parameters for geomechanical models of the faulting processes that operate in the upper few kilometers of the Garlock fault. 3D geomechanical models were used to understand the subsurface structures responsible for fault segments that were mapped using Lidar. The goal is to illuminate the factors that control the fidelity of fault patterns interpreted along the Garlock fault zone.

In classical mechanics, the stability along an interface between two solids is controlled by the magnitude of shear (τ) and normal (σ_n) tractions, and the coefficients of static (μ_s) and dynamic (μ_d) friction acting on the interface. Slip along this interface occurs when

$$\tau > \mu_s \sigma_n$$

When this condition is met and sliding initiates, the value of μ_s decreases to μ_d such that the shear stress drop associated with the sliding motion is

$$\Delta\tau = (\mu_s - \mu_d)\sigma_n .$$

This approach can be applied to natural fault systems where coseismic stresses and strains due to displacements along source faults drive slip along receiver faults. The potential for receiver faults to fail is expressed in terms of Coulomb failure stress (CFS)

$$\Delta CFS = \Delta\tau + \mu'\Delta\sigma_n$$

where $\Delta\tau$ and $\Delta\sigma_n$ are the changes in shear and normal stresses along the receiver faults, respectively, and μ' is the effective friction coefficient after accounting for change in pore fluid pressure. Failure along the receiver fault occurs when the ratio of τ to σ_n exceeds the coefficient of static friction μ_s; an increase in normal stresses relative to shear stresses acting on receiver faults will reduce this ratio and inhibit conditions for slip. In terms of ΔCFS, failure along receiver faults is encouraged if ΔCFS is positive.

The above approach is applied to fault interaction modeling built on descriptions of internal deformation due to slip along rectangular dislocations in an elastic half-space (Okada, 1992; Zielke and Arrowsmith, 2008). Lidar-derived fault trace maps and offset measurements are used to guide fault models by providing kinematic parameters such as slip vectors, magnitudes, and the attitudes of receiver faults. Two models were run, one for the central and one for the eastern parts of the Garlock fault (Figure 5.92). Figure 5.92a shows a fault configuration where a driving sinistral master fault is flanked by an obliquely oriented horst and graben system and a major southeast-dipping driving normal fault. The orientations and rakes of receiver faults are identical to those of the horst and graben faults. Local positive ΔCFS lobes are present along the master strike-slip fault and are likely created by local segment irregularities along strike (e.g., small

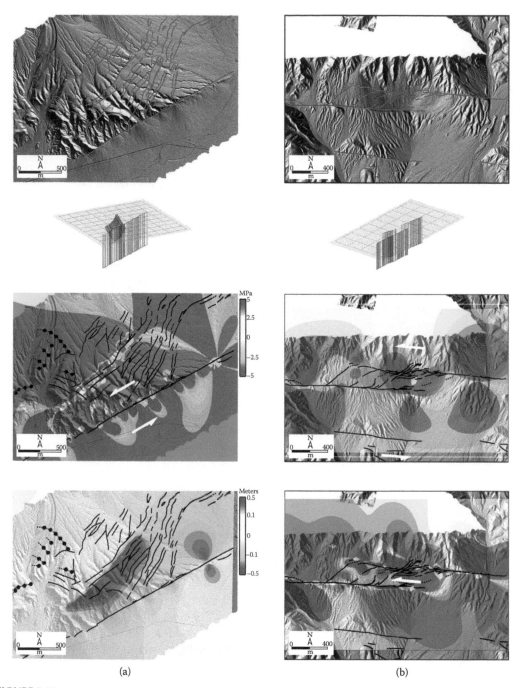

FIGURE 5.92

Examples of fault traces mapped using Lidar-derived digital elevation models. Shown are their three-dimensional representation in an elastic halfspace (top), Coulomb failure stress (CFS, middle), and vertical displacements (bottom). (a) In the central portion of the Garlock fault, a sinistral master fault is flanked by a horst and graben system, exhibiting distributed surface deformation >1.5 km across the main fault trace. (b) In the eastern section of the Garlock fault, three left-stepping and overlapping sinistral faults form at least two releasing steps where extension is accommodated by a system of normal faults. CFS and vertical displacement calculations were performed using the elastic dislocation model by Zielke and Arrowsmith (2008). (Figure courtesy of Haddad, Nissen, and Arrowsmith.)

stepovers and fault bends). The region between the driving strike-slip and normal faults exhibits reduced ΔCFS along receiver faults, which is consistent with the presence of the horst and graben system as accommodating the deformation induced by the master strike-slip fault. This is corroborated by the vertical displacement calculations where sinistral slip on the master fault and down-to-the-southeast slip on the normal fault are consistent with Lidar-derived fault mapping and topography. In Figure 5.92b, a system of three left-stepping and overlapping sinistral faults that form at least two releasing steps in an accommodation zone are shown. The ΔCFS calculations show that the stepovers and releasing steps experience enhanced ΔCFS on northeast-striking receiver faults. This is consistent with the presence of northeast-striking normal faults that bound the rhomboid-shaped pull-apart basins, and it is evident in Lidar-derived fault maps and vertical displacement calculations. These models provide viable explanations for the pattern of faulting seen in this area.

Major support for this work was provided by the National Science Foundation (NSF): Tectonics Program (EAR 0405900, EAR 0711518), Instrumentation and Facilities Program (INTERFACE; EAR 0651098), and OpenTopography (EAR 0930731, EAR 0930643). Additional support was provided by the U. S. Geological Survey (USGS) grant 07HQGR0092 and the Southern California Earthquake Center (SCEC). The SCEC is funded by NSF Cooperative Agreement EAR-0529922 and USGS Cooperative Agreement 07HQAG0008. This is SCEC contribution number 1679. Haddad was supported by a grant that was awarded jointly by the Arizona State University (ASU) Graduate and Professional Student Association, the ASU Office of the Vice President of Research and Economic Affairs, and the ASU Graduate College. San Andreas fault airborne Lidar data were collected by the NSF-funded National Center for Airborne Laser Mapping (NCALM) for the B4 Project (Bevis et al., 2005).

Earthquakes and Tectonics

Numerous tectonic studies include, usually at an early stage, the use of remote sensing imagery to generate regional tectonic maps, provide base maps for field programs, or to put structure into a tectonic framework (see, for example, Molnar and Tapponier, 1977; Prost, 2004; Jin and Cunningham, 2012). Structural style is often easier to recognize on imagery (i.e., in map view) than on the ground. Reconnaissance maps may be entirely the result of image analysis.

Interferometric radar (InSAR) has been compared to standard seismic inversion of P and S waves used to determine inferred fault parameters such as epicenter depth, location, and focal mechanism solutions/fault style (Devlin et al., 2012). C band InSAR data from ERS-1, ERS-2, and Envisat were used to obtain observations of surface deformation (uplift and subsidence on the order of several centimeters) ranging from several weeks to months before and after an earthquake. They analyzed InSAR for four subduction zone earthquakes and two crustal earthquakes in the central Andes and found that their inversion solutions for epicenter depth were within acceptable uncertainty ranges (±5 km) when earthquakes had magnitudes greater than Moment Magnitude (Mw) 5.5. The differences between the InSAR and standard solutions were greater for earthquakes with magnitudes less than Mw 5.5. Epicenters for five of the six earthquakes with Mw > 5.5 were found within 10 km of global earthquake catalog (International Seismological Centre; USGS National Earthquake Information Center) locations. Of the nine earthquakes with InSAR data, all InSAR epicenter locations were within 50 km of global catalog estimates. Centroid Moment Tensor (CMT) solutions and InSAR results were broadly in agreement regarding fault style, that is, normal, thrust, or strike-slip. This work by Devlin et al. (2012) showed that InSAR measurements of surface deformation related to earthquakes can help determine the structural style in a region.

Case History: Measuring Earthquake Deformation Using Differential Lidar

Research in active tectonics and earthquake geology is significantly enhanced by the use of airborne Lidar datasets. These datasets help refine our understanding of the geologic and geomorphic processes that act along fault zones by allowing us to study these processes at spatial and temporal scales that are relevant to surface and deformational processes. Differential analyses that span repeat Lidar datasets provide a wealth of near-fault displacement data to complement existing geodetic or field-based observations. Such displacements help constrain the slip distribution and rheology of the shallow parts of fault zones. Slip and rheology are crucial for interpreting paleoseismic and geomorphic offsets, and for studies of long-term earthquake behavior. When coupled with satellite-based measurements such as InSAR, differential Lidar analyses offer the means to explore relations between surface rupturing and deeper fault zone processes. This example is contributed by David Haddad, Edwin Nissen, and J. Ramón Arrowsmith of the Active Tectonics, Quantitative Structural Geology, and Geomorphology Laboratory, Arizona State University. Support for this work was provided by the NSF: Tectonics Program, Instrumentation and Facilities Program, and OpenTopography. Additional support was provided by the USGS and the SCEC. The SCEC is funded by the NSF and USGS. Lidar datasets presented here are available from http://www.opentopography.org.

In addition to advancing our understanding of the geomorphology and paleoseismology of active faults, Lidar datasets offer a topographic baseline against which to compare repeat Lidar surveys taken in the aftermath of a future earthquake. The typically submeter Lidar point spacing is finer than the scale of displacements caused by large earthquakes, making differential Lidar analyses well suited for capturing 3D near fault ground displacements. The development of these methods, some of which are described here, provides impetus to expand the range of active faults mapped with Lidar. In the future, differential Lidar analyses will complement satellite-based techniques such as interferometric synthetic aperture radar (InSAR) and subpixel optical matching, which map only certain components of the deformation field and which are often hindered by variable coherence close to surface faulting and in areas of dense vegetation.

El Mayor-Cucapah Earthquake Deformation

Earth surface processes can be understood from remotely sensed DTMs (Tarolli et al., 2009). The Mw 7.2 El Mayor-Cucapah (EMC), Mexico, earthquake of April 4, 2010, is currently the only complete rupture with both pre- and post-event Lidar coverage, although the pre-event point density of ~0.013 m^{-2} is orders of magnitude sparser than most modern datasets. By differencing pre- and post-event Lidar DEMs, Oskin et al. (2012)

revealed a complex pattern of surface elevation changes that included slip on numerous fault strands and tilting and warping of the ground between these segments during the EMC earthquake. However, the measured elevation changes do not correspond directly to the actual surface displacements, and a large horizontal displacement component results in apparent vertical motions that are governed by the local slope facing direction. Leprince et al. (2011) overcame this limitation by using image co-registration and subpixel correlation techniques to measure the horizontal offsets, which were then back slipped and differenced to reveal the vertical deformation caused by the EMC earthquake. However, this two-step procedure still relies on gridding the original pre- and post-event point clouds into DEMs, which introduces biases and artifacts in resulting displacement calculations.

A pair of recent studies by Borsa and Minster (2012) and Nissen et al. (2012) outlined methods for capturing 3D earthquake displacements (such as occurred during the EMC event) more directly by computing the translations that best align square windows of the pre- and post-earthquake topography. Both studies used simulated Lidar datasets to test their methods. These datasets were generated by adding synthetic earthquakes with known displacements to real Lidar point clouds (Bevis et al., 2005). This approach enables a full exploration of displacement resolutions and accuracies at a range of input point cloud densities, but does not take into account the effects of processes such as ground shaking, erosion and deposition, vegetation growth, or infrastructure development. However, as long as these processes affect shorter length scales than the window size, they are unlikely to affect the results.

In Borsa and Minster's (2012) approach, a set of harmonic basis functions was used to produce a smoothed surface model of the pre-earthquake topography onto which square subsets of the post-earthquake points were translated using a least-squares minimization scheme. This method also incorporated Lidar intensity data as an additional, independent constraint on horizontal displacements. Nissen et al. (2012) instead used an adaptation of the iterative closest point (ICP) algorithm to align pre-event (source) and post-event (target) point clouds. The ICP method works by iterating three steps: (1) identify the closest point in the target point cloud for each point in the source point cloud, (2) calculate the rigid body translation and rotation between all paired points from step 1 to minimize the mean square error in the points' 3D locations, and (3) apply the transformations in step 2 and update the mean square error between the source and target point clouds. These steps are iterated until a local minimum in closest point distances is reached, which is determined when the reduction in the mean square error falls below some threshold. In past geological applications, the ICP method was used to detect landslide displacements using repeat terrestrial Lidar datasets (Terza et al., 2007). Its main advantage over other Lidar differencing techniques is that it alleviates the need for gridding or smoothing of either dataset. The method also works well when there are large mismatches in the density of the two point clouds, eliminating the need to down-sample the denser dataset. A final, unique aspect of ICP is that it can measure rotations directly, thus providing important new kinematic data in areas of distributed faulting where block rotations may be important.

Figure 5.93 shows an example of an ICP analysis of simulated pre- and post-earthquake point clouds derived from B4 data (Bevis et al., 2005) on part of the southern San Andreas fault. The "B4" Lidar dataset was the first Lidar campaign flown specifically to collect high-resolution topographic data for the active southern San Andreas and San Jacinto faults (Carter et al., 2005). The B4 Lidar campaign was done on a segment of the San

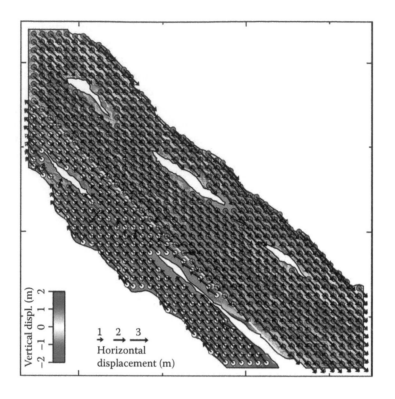

FIGURE 5.93

Results for a simulated earthquake experiment at Painted Canyon on the southern San Andreas fault. Shaded topography is a 1-m-resolution digital elevation model constructed from the B4 Lidar dataset (Bevis et al., 2005) and illuminated from the northeast. White patches show areas in which earthquake point cloud coverage is unavailable. The iterative closest point (ICP; see text) window size is 50 m. White and black arrows show input and output horizontal displacements, respectively, and colored circles show output vertical displacements. The synthetic fault is plotted in yellow. (Adapted from Nissen, E. et al. 2012. *Geophys. Res. Lett.* 39(L16301), doi: 10.1029/2012GL052460.)

Andreas fault *before* the next large earthquake. *After* the next large earthquake, Lidar data will immediately be collected again along the San Andreas fault so that ground deformation can be monitored and documented.

A synthetic fault was added to the post-event dataset to simulate a large earthquake with right-lateral slip. Points southwest of the fault were moved 2 m to the northwest, and points northeast of the fault were moved 2 m to the southeast and also raised by 1 m. The datasets were split into 50 m × 50 m windows, and the ICP algorithm was applied separately to each window. The input displacements were reproduced with horizontal and vertical accuracies of ~20 cm and ~4 cm, respectively, to mimic errors in the original point height measurements. As expected, accuracies are highest in windows containing rugged topography, but the method is fairly successful even in low-relief areas. Improved accuracies and finer resolutions should be achievable using higher point cloud densities and with further advances in survey georeferencing during an airborne Lidar campaign. This work shows that Lidar surveys of earthquake deformation before and after multiple events like the EMC earthquake can reveal horizontal offsets and elevation changes resulting from these events.

Remote Sensing and Earthquake Forecasting

Forecasting and predicting are often used synonymously. In the context of earthquake science, however, prediction implies a higher probability than a forecast (Marzocchi and Zechar, 2011), that is, a prediction is more definite than a forecast. Forecasting earthquakes is important both for public safety (to mitigate the impact) and to prove the science. There are multiple earthquake forecasting models. The Uniform California Earthquake Rupture Forecast (UCERF) approach links rupture forecasting to ground shaking (Field et al., 2009). Earthquake ruptures and deformation of the Earth's surface along active faults and tectonic zones manifest themselves as distinct geomorphic and topographic features (e.g., Kirby and Whipple, 2012). Examples of these features include fault scarps, shutter ridges, incised channel systems, and displaced geomorphic markers. Displaced geomorphic markers such as offset stream channels and terraces are of particular interest to seismic hazard studies that aim to forecast future earthquakes. Meter-scale measurements of these offsets help build surface slip histories associated with past large earthquakes, that is, they provide surface slip magnitudes and distributions in time and space. Past efforts to identify and measure earthquake-related displaced markers involved extensive fieldwork where active faults were mapped and offset markers were manually surveyed. Lidar systems make it possible to rapidly, accurately, and more completely survey the topography of active fault zones from which displaced geomorphic markers can be identified and measured. When combined with the appropriate earthquake recurrence models and geochronologic data collected from paleoseismic trenches, Lidar-derived offset measurements provide a complete picture of earthquake histories that may tell us something about the potential magnitude of future earthquakes along these faults.

It has been proposed that some faults display characteristic behavior in that rupture repeatedly produces similar slip magnitudes and distributions along strike during each successive earthquake (Schwartz and Coppersmith, 1984). One would expect to see landforms in the topography of those fault zones that developed between earthquakes and were displaced *en masse* during successive earthquakes. The smallest observable offsets should represent surface displacement created by the most recent earthquake along a fault. One can group these into offsets of similar magnitudes along that fault. For example, suppose Lidar is used to measure a group of offset stream channels that are displaced ~5 m, and another group of channels offset by ~10 m, and yet another group that was offset ~15 m. If we assume the "characteristic earthquake" model for slip accumulation along that fault, we can posit that each earthquake that ruptures this fault produces ~5 m of surface offset. Using the Wells and Coppersmith (1994) earthquake magnitude-displacement scaling relationships, we can estimate that this fault is capable of producing a ~Mw 7 earthquake. If those displaced features can be dated (e.g., from paleoseismic trenches or dating of the stream channel terraces), then we can establish an average recurrence interval for a characteristic earthquake that would rupture along that fault.

A word of caution: the above discussion is based on the "characteristic earthquake" model. That is, it assumes that a fault produces a constant displacement per earthquake at each point along strike. This may not be applicable if one were to use other earthquake recurrence models such as the uniform-slip or variable-slip models (Schwartz and Coppersmith, 1984). Using the characteristic earthquake model to forecast future earthquakes based on offset geomorphic landforms is a first-order approach.

Timing of Structural Development

Structural timing is important in mineral exploration because mineralization may occur pre- or post-folding and faulting. Early mineralization may be linked to stratigraphic changes, whereas late mineralization may be related to fractures (veins) or folds that controlled the flow of hydrothermal fluids. Likewise, hydrocarbons may have been generated before folding, in which case they would have migrated through the area, or after folding, when they could have accumulated in structural traps. Timing of structural development interpreted from imagery is relative, for example, a fold may have developed "after thrust faulting" or "contemporaneous with normal faulting." For absolute time of development, it is necessary to obtain radiometric ages, fossil ages, or some other technique that involves sampling the rocks themselves. The relative timing of structures can sometimes be interpreted from geometric relationships seen on imagery (Figure 5.94).

If one can map a thickness change across a fault, as when a unit is dipping and is exposed in cross section, it is likely that the fault was contemporaneous with sedimentation and controlled deposition either by allowing growth on the downthrown side of a fault or by condensing the section on the upthrown side (Figure 5.94a). A change in dip across a fault can cause an apparent thickening (with decreasing dip) or thinning (with increasing dip) of a section of uniform thickness.

If one can identify an angular unconformity surface on imagery, the units parallel to the unconformity are younger than those that terminate against the surface at an

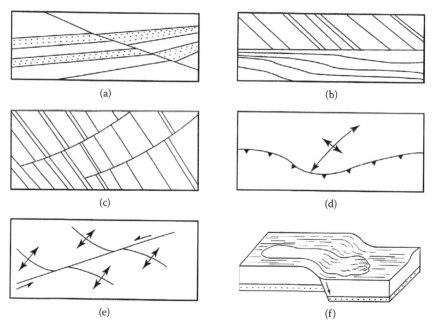

FIGURE 5.94
Some indicators of relative structural timing. (a) Unit increases in thickness across a fault indicating the fault was active during deposition. (b) Units parallel to an unconformity are younger than those terminated by it. (c) Faults die out in younger rock and coalesce in older rock. (d) A thrust embayment may indicate a preexisting buttress. (e) Folds cut by a fault are older than the fault. (f) Undeformed lava flows covering a fault scarp are younger than the scarp.

angle (Figure 5.94b). One should be aware, however, that some downlapping units may appear to form angular unconformities when in fact they are younger than the surface they are lapping onto. Clinoform bedding is the best clue to downlapping or onlapping units.

When a section is inclined such that faulted section is visible, normal faults will die out (lose displacement) upward in younger units and coalesce downward, with increasing displacement, in older units (Figure 5.94c). If the fault is listric, movement must have occurred between the age of the decollement horizon and the horizon where the fault tips out (displacement goes to zero). The same occurs in thrust faults, where displacement decreases in the direction of transport (unless a fault has broken through to the surface, in which case the displacement is constant along the fault). The age of thrusting is younger than the age of the decollement and older than the youngest deformed unit.

Folds and faults are younger than the units they deform. Cross-cutting relationships tell us that a continuous fault is younger than the unit, fold, or fault that it displaces. A folded fault, folded fold, or folded unit indicates a period of deformation that post-dates the original fault, fold, or unit (Figure 5.94d). A fault that terminates against another fault is probably younger than the throughgoing fault (unless it is offset by the throughgoing fault). The same applies to joints.

Multiple structural events can sometimes be unraveled by careful attention to regional kinematic indicators. For example, if a thrust is clearly offset by extensional faulting, one can infer that the thrust is older than normal faulting in that area. If a thrust front forms an embayment around a large foreland bulge, the bulge must have existed prior to thrusting and acted as a buttress to forward movement of the thrust sheet. If *en echelon* folds are cut and offset by a strike-slip fault, the folds were formed early in the same stress field that later caused a fault to break through and displace them (Figure 5.94e).

A lava flow, alluvial fan, or lake bed deposit that covers a fold or fault scarp with no indication of deformation or displacement is younger than the structure it covers (Figure 5.94f). Careful attention to juxtaposition and cross-cutting relationships can reveal the relative age of structures.

References

Aguilera, R. 1980. *Naturally Fractured Reservoirs*. Tulsa: PennWell Books: 720 p.

Al Khatieb, S.O., J.W. Norman. 1986. A tilted crustal block bounded by lineaments, shown by remote sensing. *J. Pet. Geol.* 9: 463–468.

Alpay, A.O. 1969. Application of aerial photographic interpretation to the study of reservoir natural fracture systems. *Soc. Pet. Eng. Ann. Fall Mtg.* SPE No. 2567: 1–11.

Arrowsmith, J.R., D.D. Rhodes. 1994. Original forms and initial modifications of the Galway Lake Road scarp formed along the Emerson Fault during the 28 June 1992 Landers, California, Earthquake. *Bull. Seism. Soc. Am.* 84(3): 511–527.

Arrowsmith, J.R., D.D. Rhodes, D.D. Pollard. 1998. Morphologic dating of scarps formed by repeated slip events along the San Andreas fault, Carrizo Plain, California. *J Geophys. Res.* 103(B5): 10,141–110,160.

Arrowsmith, J.R., O. Zielke. 2009. Tectonic geomorphology of the San Andreas Fault zone from high resolution topography: An example from the Cholame segment. *Geomorphology*. 113: 70–81.

Arrowsmith, J.R., C.M. Madden, D.E. Haddad, J.B. Salisbury, R.J. Weldon. 2011. Compilation of slip in the last event data for high slip rate faults in California for input into slip dependent rupture forecast. *American Geophysical Union, Fall Meeting 2011*, Abstract #S13B–06.

Bagnold, R.A. 1941. *The Physics of Blowing Sand and Desert Dunes*. London: Metheun Publishing Co.: 265 p.

Barosh, P.J. 1968. Relationships of explosion-produced fracture patterns to geologic structure in Yucca Flat, Nevada test site. *Geol. Soc. Am. Memoir*. 110: 199–217.

Benedicto, A., R.A. Schultz, R. Soliva. 2003. Layer thickness and the shape of faults. *Geophys. Res. Lett*. 30: 13–1 to 13–4.

Bevis, M., K. Hudnut, R. Sanchez, C. Toth, D. Grejner-Brzezinska, E. Kendrick, D. Caccamise et al. Fall Meeting 2005. The B4 Project: scanning the San Andreas and San Jacinto fault zones. *Eos Trans. AGU*. 86(52). Suppl., Abstract H34B-01.

Blanchet, P.H. 1957. Fracture analysis as an exploration method. *Bull. Am. Assn. Pet. Geol*. 41: 1748–1759.

Borsa, A., J.B. Minster. 2012. Rapid determination of near-fault earthquake deformation using differential Lidar. *Bull. Seism. Soc. Am*. 102: 1335–1347.

Bose, S., S. Mitra. 2009. Deformation along oblique and lateral ramps in listric normal faults: insights from experimental models. *Bull. Am. Assn. Pet. Geol*. 93: 431–451.

Bose, S., S. Mitra. 2010. Analog modeling of divergent and convergent transfer zones in listric normal fault systems. *Bull. Am. Assn. Pet. Geol*. 94: 1425–1452.

Brown, E.T., J.A. Hudson. 1974. Fatigue failure characteristics of some models of jointed rock. *Earthquake Engr. Struct. Dynam*. 2: 379–386.

Bryant, B., L.W. McGrew, R.A. Wobus. 1981. *Geologic map of the Denver 1 degree by 2 degrees Quadrangle, north-central Colorado*. U.S. Geological Survey Map I-1163: 1:250,000.

Burtscher, A., M. Frehner, B. Grasemann. 2012. Tectonic geomorphological investigations of antiforms using differential geometry: Permam anticline, northern Iraq. *Bull. Am. Assn. Pet. Geol*. 96: 301–314.

Campagna, D.J., D.W. Levandowski. 1991. The recognition of strike-slip fault systems using imagery, gravity, and topographic data sets. *Photogramm. Eng. Rem. S*. 57: 1195–1201.

Carter, B., M. Sartori, D. Phillips, F. Coloma, K. Stark. Fall Meeting 2005. The B4 Project: scanning the San Andreas and San Jacinto fault zones. *Eos Trans. AGU*. 86(52). Suppl., Abstract H34B-01.

Chapman, C.A. 1958. Control of jointing by topography. *J. Geol*. 66: 552–558.

Clark, R.M., S.J.D. Cox. 1996. A modern regression approach to fault displacement-length scaling relationships. *J. Struct. Geol*. 18: 147–152.

Cloos, E. 1968. Experimental analysis of Gulf Coast fracture patterns. *Bull. Am. Assn. Pet. Geol*. 52: 420–444.

Conrad, J.F. 1977. Significance of surface structure in Tertiary strata for part of the Idaho-Wyoming thrust belt. *Wyoming Geological Association—29th Annual Field Conference Guidebook*: 391–396.

Cooper, M. 1992. The analysis of fracture systems in subsurface thrust structures from the foothills of the Canadian Rockies. In K.R. McClay (ed.), *Thrust Tectonics*. London: Chapman and Hall: 391–405.

Cooper, S.P., L.B. Goodwin, J.C. Lorenz. 2006. Fracture and fault patterns associated with basement-cored anticlines: the example of Teapot Dome, Wyoming. *Bull. Am. Assn. Pet. Geol*. 90: 1903–1920.

Cotton, C. 1968. *Geomorphology*. Christchurch: Whitcombe & Tombs Ltd.: 76–95.

Crowell, J.C. 1974. Origin of late Cenozoic basins in southern California. In W.R. Dickenson (ed.), *Tectonics and Sedimentation*. SEPM Spec. Pub. 22: Tulsa, OK: SEPM: 190–204.

Davis, G.H. 1984. *Structural Geology of Rocks and Regions*. New York: John Wiley and Sons: 492 p.

Davison, I. 1994. Linked fault systems; extensional, strike-slip and contractional. In P.L. Hancock (ed.), *Continental Deformation*. New York: Pergamon: 121–142.

DeBlieux, C. 1962. Photogeology in Louisiana coastal marsh and swamp. *Gulf Coast Assn. Geol. Soc. Trans. 12th Ann. Mtg.*: Austin, TX: GCAGS: 12: 231–241.

DeBlieux, C., G.F. Shepherd. 1951. Photogeologic study in Kent County, Texas. *Oil Gas J.* 50: 86–100.

DeLong, S.B., G.E. Hilley, M.J. Rymer, C. Prentice. 2010. Fault zone structure from topography: Signatures of *en echelon* fault slip at Mustang Ridge on the San Andreas Fault, Monterey County, California. *Tectonics.* 29(TC5003), doi: 10.1029/2010Tc052673.

Devlin, S., B.L. Isacks, M.E. Pritchard, W.D. Barnhart, R.B. Lohman. 2012. Depths and focal mechanisms of crustal earthquakes in the central Andes determined from teleseismic waveform analysis and InSAR. *Tectonics.* 31(TC2002): 33 p.

Doeringsfeld, W.W. Jr., J.B. Ivey. 1964. Use of photogeology and geomorphic criteria to locate subsurface structure. *Mountain Geol.* 1: 183–195.

Drake, B., R.K. Vincent. 1975. Geologic interpretation of Landsat-1 imagery of the greater part of the Michigan Basin. *Proc. 10th Intl. Symp. Rem. Sens. Environ.* Ann Arbor: ERIM: 933–947.

Dunne, W.M., D.A. Ferrill. 1988. Blind thrust systems. *Geology.* 16: 33–36.

Eby, D.E., T.C. Chidsey, Jr., D.A. Sprinkel. 2010. *Carbonate heterogeneity based on lithofacies and petrography of the Jurassic Twin Creek Limestone in Pineview field, northern Utah.* AAPG Search & Discovery poster. New Orleans, LA: AAPG Annual Convention and Exhibition.

English, J.M. 2012. Thermomechanical origin of regional fracture systems. *Bull. Am. Assn. Pet. Geol.* 96: 1597–1625.

Epard, J.L., R.H. Groshong. 1993. Excess area and depth to detachment. *Bull. Am. Assn. Pet. Geol.* 77: 1291–1302.

Ferrill, D.A., K.J. Smart, M. Necsoiu. 2008. Displacement-length scaling for single-event ruptures: insights from Newberry Springs Fault Zone and implications for fault zone structure. In C.A.J. Wibberley, W. Kurz, J. Imber, R.E. Holdsworth, C. Collettini (eds.), *The Internal Structure of Fault Zones: Implications for Mechanical and Fluid-Flow Properties* v. 299. London: The Geological Society of London: 113–122.

Field, E.H., T.E. Dawson, K.R. Felzer, A.D. Frankel, V. Gupta, T.H. Jordan, T. Parsons et al. 2009. Uniform California earthquake rupture forecast, version 2 (UCERF 2). *Bull. Seism. Soc. Am.* 99: 2053–2107.

Fischer, M.P., M.S. Wilkerson. 2000. Predicting the orientation of joints from fold shape: results of pseudo-three-dimensional modeling and curvature analysis. *Geology.* 28: 15–18.

Flint, R.F. 1971. *Glacial and Quaternary Geology.* New York: John Wiley and Sons: 892 p.

Fossen, H., R.A. Schultz, E. Rundhovde, A. Totevatn, S.J. Buckley. 2010. Fault linkage and graben stepovers in the Canyonlands (Utah) and the North Sea Viking Graben, with implications for hydrocarbon migration and accumulation. *Bull. Am. Assn. Pet. Geol.* 94: 597–613.

Foster, N.H., R.A. Soeparjadi. 1974. Geomorphic expression of pinnacle reefs in Salawati basin, Irian Jaya, Indonesia. *Am. Assn. Pet. Geol. Soc. Econ. Paleo. Min. Ann. Mtg.* Abs. v. 1: 35.

Gaullier, V., B.C. Vendeville. 2005. Salt tectonics driven by sediment progradation: Part II—Radial spreading of sedimentary lobes prograding above salt. *Bull. Am. Assn. Pet. Geol.* 89: 1081–1089.

Ghosh, K., S. Mitra. 2009. Structural controls of fracture orientations, intensity, and connectivity, Teton anticline, Sawtooth Range, Montana. *Bull. Am. Assn. Pet. Geol.* 93: 995–1014.

Glass, D.E. 1981. Reflection of topography on pre-cretaceous unconformity through overlying section in central Alberta (abs.). *Bull. Am. Assn. Pet. Geol. Assn.* Round table: 930.

Gorham, F.D. Jr., L.A. Woodward, J.F. Callender, A.R. Greer. 1979. Fractures in Cretaceous Rocks from selected areas of San Juan Basin, New Mexico—exploration implications. *Bull. Am. Assn. Pet. Geol.* 63: 598–607.

Grimaldi, G.O., S.L. Dorobek. 2011. Fault framework and kinematic evolution of inversion structures: natural examples from the Neuquen Basin, Argentina. *Bull. Am. Assn. Pet. Geol.* 95: 27–60.

Haddad, D.E., C.M. Madden, J.B. Salisbury, J.R. Arrowsmith, R.J. Weldon. 2011. Lidar-derived measurements of slip in the most recent ground-rupturing earthquakes along elements of the San Andreas fault system. SCEC Proceedings and Abstracts, v. 21.

Haddad, D.E., S.O. Akciz, J.R. Arrowsmith, D.D. Rhodes, J.S. Oldow, O. Zielke, N.A. Toké, A.G. Haddad, J. Mauer, P. Shilpakar. 2012. Applications of airborne and terrestrial laser scanning to paleoseismology. *Geosphere*. 8(4): 771–786.

Hallin, J.S. 1973. Heat flow and radioactivity studies in Colorado and Utah, 1971–1972. MSc. Thesis. Laramie: University Wyoming: 108 p.

Harding, T.P. 1974. Petroleum traps associated with wrench faults. *Bull. Am. Assn. Pet. Geol.* 58: 1290–1304.

Harris, J.F., G.L. Taylor, J.L. Walper. 1960. Relation of deformational fractures in sedimentary rock to regional and local structure. *Bull. Am. Assn. Pet. Geol.* 44: 1853–1873.

Hart, B.S., B.L. Varban, K.J. Marfurt, A.G. Plint. 2007. Blind thrusts and fault-related folds in the Upper Cretaceous Alberta Group, deep basin, west-central Alberta: implications for fractured reservoirs. *Bull. Can. Petrol. Geol.* 55: 125–137.

Hart, E. 1994. *Fault-rupture hazard zones near Newberry Springs, San Bernardino County*. California Division of Mines and Geology Fault Evaluation Report 238. California Geological Survey CD 2002–02.

Hennings, P.H., J.E. Olson, L.B. Thompson. 2000. Combining Outcrop Data and Three-Dimensional Structural Models to Characterize Fractured Reservoirs: An Example from Wyoming. *AAPG Bull.* 84: 830–849.

Henza, A.A., M.O. Withjack, R.W. Schlische. 2010. Pre-existing zones of weakness: an experimental study of their influence on the development of extensional faults. *AAPG Search Discov. Art.* #40631: 1 p.

Heyman, O.G. 1983. Regional compression as the cause for Laramide deformation of the north-western Uncompahgre Plateau, western Colorado and eastern Utah. MSc Thesis. Laramie: University Wyoming: 133 p.

Hilley, G.E., S. DeLong, C. Prentice, K. Blisniuk, J.R. Arrowsmith. 2010. Morphologic dating of fault scarps using airborne laser swath mapping (ALSM) data. *Geophys. Res. Lett.* 37(L04301), doi: 10.1029/2009GL042044.

Hodgson, R.A. 1961. Regional study of jointing in Comb Ridge-Navajo Mountain area, Arizona and Utah. *Bull. Am. Assn. Pet. Geol.* 45: 1–38.

Howard, A.D. 1967. Drainage analysis in geologic interpretation: a summation. *Bull. Am. Assn. Pet. Geol.* 51: 2246–2259.

Hudnut, K.W., A. Borsa, C. Glennie, J.B. Minster. 2002. High-resolution topography along surface rupture of the 16 October 1999 Hector Mine, California, earthquake (Mw 7.1) from airborne laser swath mapping. *Bull. Seism. Soc. Am.* 92(4): 1570–1576.

Huffman, A.C., D.J. Taylor. 1998. Relationship of basement faulting to laccolithic centers of south-eastern Utah and vicinity. In J.D. Friedman, A.C. Huffman (eds.), *Laccolith complexes of south-eastern Utah: time of emplacement and tectonic setting—Workshop Proceedings*. U.S.G.S. Bull. 2158: 41–43.

Huntoon, P.W., J.W. Sears. 1975. Bright Angel and Eminence faults, eastern Grand Canyon, Arizona. *Bull. Geol. Soc. Am.* 86: 465–472.

Jacob, A.F. 1983. *Rocky Mountain Foreland Basins and Uplifts*. Denver: Rocky Mountain Association of Geologists: 229–244.

Jamison, W.R. 1979. Laramide deformation of the Wingate Sandstone, Colorado National Monument-A study of cataclastic flow. PhD thesis. College Station: Texas A & M University: 168 p.

Jamison, W.R. 1987. Geometric analysis of fold development in overthrust terranes. *J. Struct. Geol.* 9: 207–219.

Jin, Z., D. Cunningham. 2012. Kilometer-scale refolded folds caused by strike-slip reversal and intra-plate shortening in the Beishan region, China. *Tectonics*. 31: 19 p.

Kirby, E., K.X. Whipple. 2012. Expression of active tectonics in erosional landscapes. *J. Struct. Geol.* 44: 54–75.

Krynine, D.P., W.R. Judd. 1957. *Principles of Engineering Geology and Geotechnics*. New York: McGraw-Hill Book Co.: 730 p.

Kupsch, W.O. 1956. Submask geology in Saskatchewan. Williston Basin Symposium. 66–75.

Lattman, L.H. 1959. Geomorphology applied to oil exploration. *Mineral Industries*. 28: 1–8.

Lattman, L.H., R.R. Parizek. 1964. Relationship between fracture traces and the occurrence of ground water in carbonate rocks. *J. Hydrol.* 2: 73–91.

Lee, W., T.J. Payne. 1944. McLouth gas and oil field, Jefferson and Leavenworth Counties, Kansas. *Kans. Geol. Survey Bull.* 53: 79.

Leprince, S., K.W. Hudnut, S. Akciz, A. Hinojosa-Corona, J.M. Fletcher. 2011. Surface rupture and slip variation induced by the 2010 El Mayor-Cucupah earthquake, Baja California, quantified using COSI-Corr analysis on pre- and post-earthquake Lidar acquisitions. AGU Fall Meeting Abstracts, EP41A-0596.

Lopes, F.C., A.J. Pereira, V.M. Mantas. 2012. Mapping of salt structures and related fault lineaments based on remote-sensing and gravimetric data: The case of the Monte Real salt wall (onshore west-central Portugal). *Bull. Am. Assn. Pet. Geol.* 96: 615–634.

Love, J.D., J.L. Weitz, R.K. Hose. 1955. *Geologic map of Wyoming*. Geol. Survey Wyo. Map Series 7A: 1:500,000.

Love, J.D., A.C. Christiansen, R.W. Jones. 1979a. Preliminary geologic map of the Lander 1 degree by 2 degree Quadrangle, central Wyoming. U.S. Geological Survey Map OF 79–1301: 1:250,000.

Love, J.D., A.C. Christiansen, T.M. Brown, J.L. Earle. 1979b. Preliminary geologic map of the Thermopolis 1 degree by 2 degree Quadrangle, central Wyoming. U.S. Geological Survey Map OF 79–962: 1:250,000.

Lowell, J.D. 1985. *Structural Styles in Petroleum Exploration*. Tulsa, OK: Oil and Gas Consultants International: 477 p.

Ludington, S., B.C. Moring, R.J. Miller, P.A. Stone, A.A. Bookstrom, D.R. Bedford, J.G. Evans et al. 2007. Preliminary integrated geologic map databases for the United States—Western States: California, Nevada, Arizona, Washington, Oregon, Idaho, and Utah. *U.S. Geological Survey Open-File Report 2005–1305*.

Madden, C. M., J.R. Arrowsmith, D.E. Haddad, J.B. Salisbury, R.J. Weldon. 2011. Compilation of slip in the last earthquake data for high slip rate faults in California for input into slip-dependent rupture forecast. SCEC Proceedings and Abstracts, v. 21.

Marzocchi, W., J.D. Zechar, T.H. Jordan. 2012. Bayesian forecast evaluation and ensemble earthquake forecasting. *Bulletin of the Seismological Society of America*. 102: 2574–2584.

Maslowski, A. September 1985. Deeper section receives new interest. *Northeast Oil World*: 5: 19–21.

McGill, G.E., A.W. Stromquist. 1979. The grabens of Canyonlands National park, Utah. *J. Geophys. Res.* 84: 4547–4563.

McGill, S.F., K. Sieh. 1991. Surficial offsets on the central and eastern Garlock fault associated with prehistoric earthquakes. *J. Geophys. Res.* 96(B13): 21,597–521,621.

Melosh, H.J., G.E. McGill. 1982. Mechanical analysis of the Canyonlands grabens, Utah. *EOS*. 63 (abs): 1 p.

Melton, F.A. 1959. Aerial photographs and structural geomorphology. *J. Geol.* 67: 351–370.

Miller, J. McL., E. P. Nelson. April 2002. Three-dimensional strain during basin formation— orthorhombic fault patterns and associated MVT mineralization, Lennard shelf, Western Australia (abs.). Society of Economic Geologists Global Exploration 2002: *Integrated Methods for Discovery*, Denver, Colorado.

Miller, J. McL., E.P. Nelson, M. Hitzman, P. Muccilli, W.D.M. Hall. 2007. Orthorhombic fault-fracture patterns and non-plane strain in a synthetic transfer zone during rifting: Lennard shelf, Canning basin, Western Australia. *J. Struct. Geol.* 29: 1002–1021.

Miller, V.C., C.F. Miller. 1961. *Photogeology*. New York: McGraw-Hill Book Co.: 80–100.

Miller, J.F., S. Mitra. 2011. Deformation and secondary faulting associated with basement-involved compressional and extensional structures. *Bull. Am. Assn. Pet. Geol.* 95: 675–689.

Mollard, J.D. 1957. Aerial photographs aid petroleum search in Saskatchewan. *Can. Oil Gas Ind. J.* 10: 1–8.

Mollard, J.D., J.R. Janes. 1984. *Airphoto Interpretation and the Canadian Landscape*. Hull, Quebec, Canada: Canadian Department of Energy, Mines, and Resources: 415 p.

Molnar, P., P. Tapponnier. 1977. The Collision between India and Eurasia. *Scientific American*. 236: 30–41.

Moore, J.G. 1960. Curvature of normal faults in the Basin and Range province of the Western United States. *U.S. Geol. Survey Prof. Paper*. 400–B: 409–411.

Morgan, K.M., D.R. Morris-Jones, D.G. Koger. 1982. Applying Landsat data to oil and gas exploration along the Texas Gulf Coast. *Oil Gas J*. 80: 326–327.

Morley, C.K. 1987. Lateral and vertical changes of deformation style in the Osen-Roa thrust sheet, Oslo region. *J. Struct. Geol.* 9: 331–343.

Morley, C.K. 1999. Patterns of displacement along large normal faults: implications for basin evolution and fault propagation, based on examples from East Africa. *Bull. Am. Assn. Pet. Geol.* 83: 613–634.

Morley, C.K., R.A. Nelson, T.L. Patton, S.G. Munn. 1990. Transfer zones in the East African Rift system and their relevance to hydrocarbon exploration in rifts. *Bull. Am. Assn. Pet. Geol.* 74: 1234–1253.

Morris, K. 1991. Using knowledge-base rules to map the three-dimensional nature of geological features. *Photogramm. Eng. Rem. Sens.* 57: 1209–1216.

Murray, G.H. Jr. 1968. Quantitative fracture study—Sanish pool, Mckenzie County, North Dakota. *Bull. Am. Assn. Pet. Geol.* 52: 57–65.

Neely, T.G., E.A. Erslev. 2009. The interplay of fold mechanisms and basement weaknesses at the transition between Laramide basement-involved arches, northcentral Wyoming, U.S.A. *J. Struct. Geol.* 31: 1012–1027.

Nelson, R.A. 1985. *Geologic Analysis of Naturally Fractured Reservoirs*. Houston: Gulf Publishing Co.: 320 p.

Nissen, E., A.K. Krishnan, J.R. Arrowsmith. 2012. Three-dimensional surface displacements and rotations from differencing pre- and post-earthquake Lidar point clouds. *Geophys. Res. Lett.* 39(L16301), doi: 10.1029/2012GL052460.

Nur, A. 1978. The tensile origin of fracture-controlled lineaments. *Int. Basement Tectonics Symp. Contrib.* 22: 155–167.

Nur, A. 1982. The origin of tensile fracture lineaments. *J. Struct. Geol.* 4: 31–40.

Okada, Y. 1992. Internal deformation due to shear and tensile faults in a half-space. *Bull. Seism. Soc. Am.* 82: 1018–1040.

Ollier, C.D. 1981. *Tectonics and Landforms*. London: Longman Group Ltd.: 324 p.

Oskin, M.E., K. Le, M.D. Strane. 2007. Quantifying fault-zone activity in arid environments with high-resolution topography. *Geophys. Res. Lett.* 34(L23S05), doi: 10.1029/2007GL031295.

Oskin, M., R. Arrowsmith, A. Hinojosa, J. Gonzalez, A. Gonzalez, M. Sartori, J. Fernandez et al. 2010a. *Airborne Lidar survey of the 4 April 2010 El Mayor-Cucupah earthquake rupture* (proceedings and abstracts). Southern California Earthquake Center Annual Meeting, Palm Springs, California, September 12–16, 2009: 19 p.

Oskin, M.E., P.O. Gold, A. Hinojosa, J.R. Arrowsmith, A.J. Elliott, M.H. Taylor, A.J. Herrs et al. 2010b. *Airborne and terrestrial Lidar imaging and analysis of the 4 April 2010 El Mayor-Cucupah earthquake rupture*. Abstract T35B-2135, 2010 Fall Meeting, AGU, San Francisco, 13–17 Dec.

Oskin, M.E., J.R. Arrowsmith, A.C. Hinojosa, A.J. Elliott, J.M. Fletcher, E.J. Fielding, P.O. Gold et al. 2012. Near-field deformation from the El Mayor-Cucupah earthquake revealed by differential Lidar. *Science*. 335: 702.

Pasternack, I. 1988. *Nature and distribution of Mississippian Sun River Dolomite porosity, west flank of the Sweetgrass Arch, Northwestern Montana*. Rocky Mountain Assn. Geologists 1988 Carbonate Symposium: 129–138, 439–441.

Patton, T.L. 2005. Sandbox models of downward-steepening normal faults. *Bull. Am. Assn. Pet. Geol.* 89: 781–797.

Paul, D., S. Mitra. 2012. Controls of basement faults on the geometry and evolution of compressional basement-involved structures. *Bull. Am. Assn. Pet. Geol.* 96: 1899–1930.

Penny, F.A. 1975. *The local surface expression of some deep discoveries in the central Rocky Mountains*. Rocky Mountain Association of Geologists Symposium: 55–61.

Prentice, C.S., C.J. Crosby, C.S. Whitehill, J.R. Arrowsmith, K.P. Furlong, D.A. Phillips. 2009. Illuminating northern California's active faults. *Eos*. 90(7): 55–56.

Price, N.J. 1959. Mechanics of jointing in rocks. *Geol. Mag.* 46: 149–167.

Price, N.J. 1966. *Fault and joint development in brittle and semi-brittle rock*. Oxford: Pergamon Press, 176 p.

Prost, G.L. 1986. Jointing in relatively undeformed strata: relation to basement and exploration implications. PhD thesis. Golden: Colo. School Mines: 293 p.

Prost, G.L. 1988. Jointing at rock contacts in cyclic loading. *Int. J. Rock Mech. Min. Sci. Geomech. Abstr.* 25: 263–272.

Prost, G.L. 1990. Recognizing thrust faults on remote sensing images. *World Oil*. 211: 39–45.

Prost, G.L. 2004. Tectonics and hydrocarbon systems of the East Gobi basin, Mongolia. *Bull. Am. Assn. Pet. Geol.* 88: 483–513.

Quigley, M., R. Van Dissen, N. Litchfield, B. Duffy, D. Barrell, K. Furlong, T. Stahl, E. Bilderback, D. Noble. 2012. Surface rupture during the 2010 Mw 7.1 Darfield (Canterbury) earthquake: implications for fault rupture dynamics and seismic-hazard analysis. *Geology*. 40: 55–58.

Ramsay, J.G., M.I. Huber. 1987. *Techniques of Modern Structural Geology 2: Folds and Fractures*. London: Academic Press: 548 p.

Ray, R.G. 1960. Aerial photographs in geologic interpretation and mapping. *U.S. Geol. Surv. Prof. Paper*. 373: 64–66.

Reif, D., B. Grasemann, R.H. Faber. 2011. Quantitative structural analysis using remote sensing data: Kurdistan, northeast Iraq. *Bull. Am. Assn. Pet. Geol.* 95: 941–956.

Rieck, R.L. 1976. The glacial geomorphology of an interlobate area in southeast Michigan: relationships between landforms, sediments, and bedrock. PhD dissertation, East Lansing: Michigan State University: 216 p.

Ritzma, H.R. 1971. *Faulting on the north flank of the Uinta Mountains, Utah and Colorado*. Wyo. Geol. Assn. 23rd Ann. Field Conf. Guidebook: 145–150.

Roberts, G.P. 2007. Fault orientation variations along the strike of active normal fault systems in Italy and Greece: Implications for predicting the orientations of subseismic-resolution faults in hydrocarbon reservoirs. *Bull. Am. Assn. Pet. Geol.* 91: 1–20.

Roberts, S.E. 1954. Cracks in asphalt resurfacing affected by cracks in rigid bases. National Research Council, Highway Research Board, 33rd Annual Meeting: Washington, D.C.: 341–345.

Rotevatn, A., S.S. Buckley, J.A. Howell, H. Fossen. 2009. Overlapping faults and their effect on fluid flow in different reservoir types: a Lidar-based outcrop modeling and flow simulation study. *Bull. Am. Assn. Pet. Geol.* 93: 407–427.

Sabins, F.F. Jr. 1987. *Remote Sensing Principles and Interpretation*. New York: W.H. Freeman and Co.: 449 p.

Sanford, A.R. 1959. Analytical and experimental study of simple geologic structures. *Bull. Geol. Soc. Am.* 70: 19–52.

Saunders, D.F. 1980. *Geomorphic and tonal anomalies in petroleum prospecting*. Unconventional Methods in Exploration Symposium II: 63–82.

Schlische, R.W., M.O. Withjack, G. Eisenstadt. 2002. An experimental study of the secondary deformation produced by oblique-slip normal faulting. *Bull. Am. Assn. Pet. Geol.* 86: 885–906.

Schoenherr, A.A. 1992. *A Natural History of California*. Berkeley: University of California Press: 9 p.

Scholz, C.H. 2002. *The Mechanics of Earthquakes and Faulting*. Cambridge: Cambridge University Press: 496 p.

Schultz, R.A. 2000. Localization of bedding plane slip and backthrust faults above blind thrust faults: Keys to wrinkle ridge structure. *J. Geophys. Res.* 105(E5): 12,035–12,052.

Schultz, R.A., C.H. Okubo, S.J. Wilkins. 2006. Displacement-length scaling relations for faults on the terrestrial planets. *J. Struct. Geol.* 28: 2182–2193.

Schwartz, D. P., J. Coppersmith. 1984. Fault behaviour and characteristic earthquakes: examples from Wasatch and San Andreas faults. *J. Geophys. Res.* 89: 5681–5698.

Sears, J.W. 1973. Structural geology of the Precambrian Grand Canyon Series. MSc thesis. Laramie: University of Wyoming: 100 p.

Segal, F., D.D. Pollard. 1980. Mechanics of discontinuous faults. *J. Geophys. Res.* 85: 4337–4350.

Shelton, J.W. 1984. Listric normal faults: an illustrated summary. *Bull. Am. Assn. Pet. Geol.* 68: 801–815.

Sibson, R.H. 1986. Earthquakes and rock deformation in crustal fault zones. *Annu. Rev. Earth Pl. Sc.* 14: 149–175.

Skinner, B.J. 1966. Thermal expansion. *Geol. Soc. Am. Memoir.* 97: 76–96.

Stacy, F.D. 1969. *Physics of the Earth*. New York: John Wiley and Sons: 108–117.

Stearns, D.W. 1968. *Certain aspects of fracture in naturally deformed rocks*. NSF Advanced Science Seminar in Rock Mechanics, Special Rept. Bedford: Air Force Cambridge Res. Lab.: 97–118, 215–229.

Stein, R.J., J. Wickham. 1980. Viscosity-based numerical model for fault zone development in drape faulting. *Tectonophys.* 66: 225–251.

Sylvester, A.G., R.R. Smith. 1976. Tectonic Transpression and Basement-Controlled Deformation in San Andreas Fault Zone, Salton Trough, California. *Bull. Am. Assn. Pet. Geol.* 60: 2081–2102.

Tarolli, P., J.R. Arrowsmith, E.R. Vivoni. 2009. Understanding earth surface processes from remotely sensed digital terrain models. *Geomorphology.* 113: 1–3.

Terza, G., A. Galgaro, N. Zaltron, R. Genevois. 2007. Terrestrial laser scanner to detect landslide displacement fields: a new approach. *Int. J. Remote Sens.* 28: 3425–3446.

Thurrell, R.F. Jr. 1953. Vertical exaggeration in stereoscopic models. *Photogram. Engr.* 19: 579–588.

Timmons, J.M., K.E. Karlstrom, J.W. Sears. 2003. Geologic structure of the Grand Canyon Supergroup. In S.S. Beus, M. Morales (eds.), *Grand Canyon Geology*. New York: Oxford University Press: 76–89.

Titus, S.J., M. Dyson, C. DeMets, B. Tikoff, F. Rolandone, R. Buergmann. 2011. Geologic versus geodetic deformation adjacent to the San Andreas fault, central California. *Geol. Soc. Am. Bull.* 123: 794–820.

Trollinger, W.V. 1968. Surface evidence of deep structure in the Anadarko Basin. *Shale Shaker:* 18: 162–171.

Turcotte, D.L., E.R. Oxburgh. 1973. Mid-plate tectonics. *Nature.* 244: 337–339.

Utah Geological Survey. 2010. *Jurassic Twin Creek Limestone Reservoir, Thrust Belt—Outcrop Analogs Near Major Fields*. Poster, http://geology.utah.gov/emp/pump/pdf/pump_aapg_2010_poster.pdf (accessed 29 Sept. 2012).

Vendeville, B.C. 2005. Salt tectonics driven by sediment progradation: Part I—Mechanics and kinematics. *Bull. Am. Assn. Pet. Geol.* 89: 1071–1079.

Weimer, R.J. 1980. Recurrent movement on basement faults, a tectonic style for Colorado and adjacent areas. In Kent, H.C., and Porter, K.W. (eds.) *Colorado Geology*. Denver: Rocky Mountain Association of Geologists: 301–313.

Wells, D.L., K.J. Coppersmith. 1994. New empirical relationships among magnitude, rupture length, rupture width, rupture area, and surface displacement. *Bull. Seism. Soc. Am.* 84: 974–1002.

WGCEP (Working Group on California Earthquake Probabilities). 2008. *The uniform California earthquake rupture forecast, Version 2 (UCERF 2)*. U.S. Geological Survey Open File Report 2007–1437: 104 p.

Wilcox, R.E., T.P. Harding, D.R. Seely. 1973. Basic wrench tectonics. *Bull. Am. Assn. Pet. Geol.* 57: 74–96.

Withjack, M.O., S. Callaway. 2000. Active normal faulting beneath a salt layer: an experimental study of deformation patterns in the cover sequence. *Bull. Am. Assn. Pet. Geol.* 84: 627–651.

Withjack, M.O., J. Olson, E. Peterson. 1990. Experimental models of extensional forced folds. *Bull. Am. Assn. Pet. Geol.* 74: 1038—1054.

Withjack, M.O., C. Scheiner. 1982. Fault patterns associated with domes—an experimental and analytical study. *Bull. Am. Assn. Pet. Geol.* 66: 302–316.

Yin, H., R.H. Groshong. 2009. A three-dimensional kinematic model for the deformation above an active diaper. *Bull. Am. Assn. Pet. Geol.* 91: 343–363.

Zaeff, G.D., C. Liu, K.A. Soofi, T. Hassan, P.H. Hennings, A.P. Morris, D.A. Ferrill, R.N. McGinnis. 2010. Characterizing fracture sets at outcrop exposures using high resolution remote sensing data; developing a fracture model as input into a static geomodel (abs.). *AAPG Search Discov.* #90104: 1 p.

Zielke, O., J.R. Arrowsmith. 2008. Depth variation of coseismic stress drop explains bimodal earthquake magnitude-frequency distribution. *Geophys. Res. Lett.* 35(24): 1–5.

Zielke, O., J.R. Arrowsmith. 2012. LaDiCaoz and Lidarimager—MATLAB GUIs for Lidar data handling and lateral displacement measurement. *Geosphere.* 8: 206–221.

Zielke, O., J.R. Arrowsmith, L.G. Ludwig, S.O. Akciz. 2010. Slip in the 1857 and earlier large earthquakes along the Carrizo Plain, San Andreas Fault. *Science.* 327: 1119–1122.

Additional Reading

Harris, J.F. 1959. Relationships of deformational fractures in sedimentary rocks to regional and local structures. MSc. thesis, University of Tulsa: 56 p.

McKinstry, H.E. 1948. *Mining Geology*. Englewood Cliffs: Prentice-Hall: 290–327.

Park, C.F. Jr., R.A. MacDiarmid. 1970. *Ore Deposits*. San Francisco: W.H. Freeman and Co.: 64–100.

Powell, D. 1992. *Interpretation of Geological Structures through Maps*. New York: Longman Scientific & Technical/John Wiley & Sons: 176 p.

6

Stratigraphic and Compositional Mapping

Chapter Overview

In Chapter 4, we discussed recognizing rock types using their erosional characteristics, tone, or color on black-and-white and color images. Here, we look at mineralogical and stratigraphic remote sensing: mapping lithologic sequences, stratigraphic architecture, and mineral assemblages using instruments such as film cameras, hyperspectral imagery, and ground-penetrating radar (GPR). We will examine surface indicators of mineralization and petroleum using characteristics of their reflectance curves. Multispectral instruments offer the possibility of automated geologic mapping: an instrument records the reflectance spectra of every object on the ground and matches the spectral curves to those in a reference library to determine the composition of the surface material.

Remote sensing offers a unique opportunity to the geologist who wants to map the stratigraphy in a previously unmapped area. Mapping the sequence architecture can tell us not only about near surface units but can serve as analogs for deeper units. From the perspective of the petroleum geologist evaluating a frontier basin, it would be useful to map the distribution of source rock and its maturity, and map the distribution of outcropping reservoir and seal units around the basin margins. We would determine, directly or indirectly, if any seeps exist within the basin. These needs are being addressed using increasingly sophisticated sensors and by integrating data from multiple remote sensing systems. The minerals geologist is interested in identifying surface units (e.g., quartz monzonite porphyry versus granite, Carlin-type black shale versus shale, kimberlite versus carbonatite) and wants to identify alteration types and their mineral associations. All geologists try to understand the lithologies of an area and their geologic relationships by building a stratigraphic column (with units younger, older, laterally equivalent, or faulted out) when entering an unmapped area for the first time.

Rock Spectra

In its simplest form, spectral stratigraphy means using the unique combination of spectral reflectance, brightness, and erosional texture associated with lithologic units to make photogeologic maps. The units that are mapped are "photogeologic units" because they represent volumes of rock bounded by horizons (generally, but not always time surfaces) recognizable on the imagery. In this sense, they may differ from standard geologic units, which are bounded by horizons recognizable in the field. Photogeologic units have few genetic connotations, nor is it always possible to identify the lithology involved, particularly if there is heavy vegetation or soil cover. The definition of the units is highly dependent on the resolution of the imagery in that high-resolution airphotos, for example, allow more and better discrimination of photogeologic units than low-resolution satellite images. Photogeologic units can be followed along strike until they are covered, faulted out, reach an unconformity, or change facies. Correlation of these units across areas without outcrops is tenuous, but should be based on several supporting criteria. Besides having a similar expression, the unit that one feels is correlative should fall within the same sequence of photogeologic units, that is, it should fit in the same photostratigraphic framework. If one cannot follow an individual unit for great distances along strike, it may still be possible to follow a thicker sequence of units (Sgavetti, 1992).

It is a simple matter to compile reference curves of single minerals, and this has been done by Hunt and Salisbury (1970, 1971), among others. It is more difficult to identify mixtures of minerals in varying amounts, what we call lithologies or rock types (Hunt and Salisbury, 1976). In addition to the effect of atmospheric absorption and scattering, the mixing of rock types (e.g., a limy shale), weathering products on the surface of an outcrop, soil cover, and plant cover of varying density and composition, we have a formidable task ahead of us before we can generate geologic maps automatically. As an aside, many of these constraints (chemical weathering, plant cover) do not exist on other planets and moons, where spectral stratigraphy really comes into its own.

These maps are not "classical" geologic maps consisting of formations and members based on features mapped in the field. However, it is possible to generate a fairly accurate mineralogic or lithologic distribution map using imaging spectrometer data (Hook et al., 2005). These instruments can quite simply "see" more than the human eye because of the extended wavelength range available to high-resolution imaging sensors. The product is an excellent first step in generating a reconnaissance map showing mineralogic composition and assemblages at the surface. Although not geologic maps, they are potentially just as meaningful, if not more so, since the information provided would not be available otherwise without detailed laboratory work such as x-ray diffraction, gas chromatography–mass spectrometry (GCMS), or petrographic study. Spectral remote sensing can assist standard photogeologic mapping by making images easier for the geologist to interpret. Among other things, it can assist in (1) the remote identification of specific lithologies (e.g., limestones versus dolomites) that might not be readily separated on standard photos or imagery, (2) extending known lithologies into unmapped territory by means of their spectral characteristics, (3) making it easier to identify surface alteration (e.g., dolomitization, sericitization), and (4) making possible the identification of organic-rich source rock, and perhaps its maturity.

Mineral Spectra

In the early 1970s, workers at the Air Force Cambridge Research Laboratory began measuring the spectral reflectance of suites of minerals grouped by anions, for example, silicates, carbonates, oxides. The curves are characterized by reflectance peaks and absorption bands caused by electronic and vibrational processes within the mineral crystal lattice. It was determined, for example, that iron oxides have a reflectance peak in the visible red (730 nm) and absorption features near 400 and 900 nm. These minerals, which include hematite, goethite, limonite, and jarosite, have a diagnostic spectral reflectance that was used by Rowan et al. (1974) to process the earliest Landsat multispectral scanner (MSS) images to identify ferric iron associated with mineralized gossans in the Goldfield mining district, Nevada. This involved a processing technique called ratioing (defined in the section on image processing) where brightness (reflectance) values in one channel are divided by values for the same pixel in another channel (Figure 6.1). The ratio technique can generate images showing the distribution of specific minerals based on their unique reflectance highs and absorption features (Figure 6.2). The mining industry was an early user of this technique because there are a number of hydrothermal alteration minerals, including iron oxides, clay minerals, and carbonates that are readily mapped using band ratios (Abrams et al., 1977; Krohn et al., 1978; Prost, 1980).

One can not only enhance, but can also identify many surface materials on the basis of their unique spectral curves. These curves are characterized by "absorption bands" (reflectance or emissivity minima) caused by the presence of OH, H_2O, CO_3, SO_4, CH, and SiO_2 (Figure 0.1). Vegetation has an absorption band due to chlorophyll at wavelengths shorter than 750 nm. Clay minerals have a characteristic absorption centered near 2.2 μm. This absorption minimum shifts slightly depending on the clay type (Figure 6.3). Calcite has absorption features at 2.32 and 2.51 μm in the short-wave infrared (SWIR) region (e.g., ASTER bands 8, 2.295–2.365 μm and band 9, 2.360–2.430 μm)

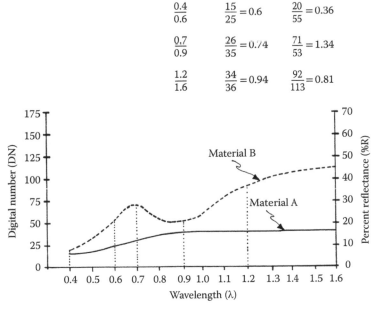

Ratio (λ)	Material A	Material B
$\dfrac{0.4}{0.6}$	$\dfrac{15}{25} = 0.6$	$\dfrac{20}{55} = 0.36$
$\dfrac{0.7}{0.9}$	$\dfrac{26}{35} = 0.74$	$\dfrac{71}{53} = 1.34$
$\dfrac{1.2}{1.6}$	$\dfrac{34}{36} = 0.94$	$\dfrac{92}{113} = 0.81$

FIGURE 6.1

Hypothetical reflectance curves showing how ratios enhance minor reflectance variations. (From Prost, G.L. 1980. *Econ. Geol.* 75: 894–906. Copyright SEG.)

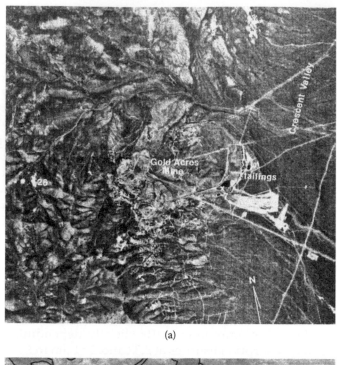

(a)

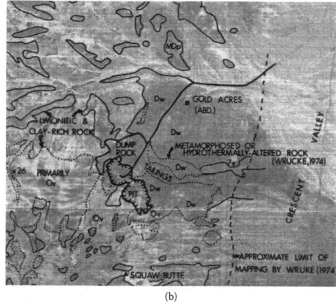

(b)

FIGURE 6.2

(a) True color airphoto over the Gold Acres mining district, Nevada. (b) Airborne Thematic Mapper Simulator ratio image over the same area as in (a). The ratio consists of TM bands 7/4 color-coded magenta, 6/10 coded yellow, and 12/1 coded cyan. Magenta colors indicate iron oxide and clay minerals; unaltered greenstone is dark green; sparse vegetation is yellow and green. Ov = Valmy Greenstone; Dw = Wenban Limestone; Se = Elder Sandstone; MDp = Pilot Shale; Qa = alluvium. (From Prost, G.L. 1980. *Econ. Geol.* 75: 894–906. Copyright SEG.)

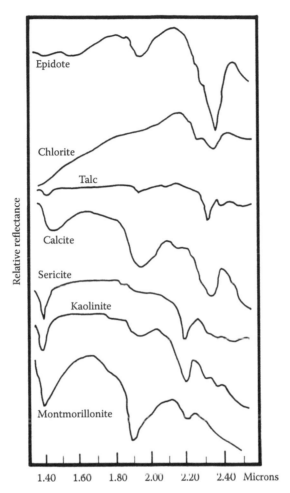

FIGURE 6.3
Near infrared reflectances for some common minerals. (From Henderson, F.B. III., B.N. Rock. eds. 1983. *Frontiers for Geological Remote Sensing from Space*. Fourth Geosat Workshop. Falls Church: Am. Soc. Photogram. Reprinted with permission from the American Society for Photogrammetry & Remote Sensing, Bethesda, MD, www.asprs.org.)

and at 11.45–11.75 µm and 13.92–14.0 µm in the thermal region. These bands are useful for mapping the distribution of limestone. Dolomite has absorption bands at 2.32138 µm and 2.51485 µm and 11.42–11.67 µm and 13.44–13.65 µm (Zaini, 2009). Gypsum has an absorption at 2.4 µm. Silicate minerals have "reststrahlen bands" (emissivity minima) between 8 and 11 µm, depending on the dominant mineral (Figure 6.4). The minimum shifts to longer wavelengths as the quartz content decreases. Organic matter (including oil and tar) has a characteristic minimum at 1.75 µm. Minerals can be characterized by their emissivity curves and dielectric constants in the microwave region (Jinhai, 1990).

Many workers have compiled libraries of spectral curves (Gaffey, 1985; Krohn, 1986; Clark et al., 1990). They have demonstrated the ability to identify the differences between calcite, dolomite, and aragonite (Gaffey, 1985). In some cases, they have shown that the fine detail in the spectra vary with elemental abundance, for example, the 2.2 µm absorption minimum in montmorillonites shifts to longer wavelengths with increasing calcium content (Clark et al., 1990).

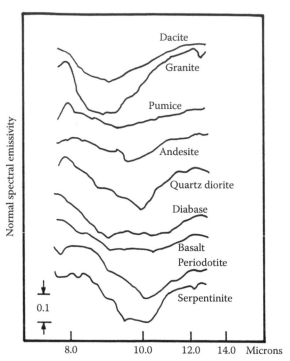

FIGURE 6.4

Thermal infrared emittance for some silicate minerals and rocks. (From Henderson, F.B. III., B.N. Rock. eds. 1983. *Frontiers for Geological Remote Sensing from Space*. Fourth Geosat Workshop. Falls Church: Am. Soc. Photogram. Reprinted with permission from the American Society for Photogrammetry & Remote Sensing, Bethesda, MD, www.asprs.org.)

Hyperspectral Imagery

The shape of an object can be used to recognize what it is; its spectra can be used to identify what it is made of. Imagery is available with tens or even hundreds of closely spaced narrow bandwidth channels (Chapter 3). The acquisition, processing, and analysis of these data are known as imaging spectrometry or hyperspectral remote sensing (Goetz et al., 1985). Hyperspectral imagery can be thought as a cube of information: the top of the cube consists of a map view of the Earth's surface in a given channel, whereas the depth of the cube consists of multiple layers, each representing surface reflectance in a different wavelength or band (Figures 6.5 and 6.6). For any given point on the surface of this data cube, it is possible to generate a reflectance curve by plotting the reflectance values through the depth of the cube. This technique has been useful for the identification of surface materials such as rock types, soils, and vegetation. Hyperspectral imagery is used during environmental site characterization to map the distribution of vegetation communities and contaminants, to identify and map the distribution of minerals and alteration in mineral exploration programs, and to map the distribution of rock types such as organic-rich shales for petroleum basin evaluations.

People first began working with hyperspectral data sets in the mid 1980s. Goetting and Lyon (1987) developed an expert system that identified minerals by examining 16 spectral windows in the range of 1.36 to 2.50 μm. This system characterized the minerals by the position of their absorption minima and the strength of the minimum relative to other

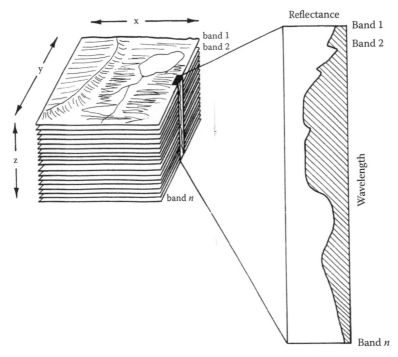

FIGURE 6.5
Diagram of a hyperspectral image cube consisting of "*n*" layers of images in "*n*" wavelengths. One may extract a reflectance curve for any given pixel in the image.

absorptions. Kruse et al. (1990) used the 63-channel Geophysical Environmental Research Imaging Spectrometer (GERIS) to identify minerals based on their field and laboratory spectra (Figure 3.11), and made mineral distribution maps. Since that time workers have been developing expert systems to interactively process imaging spectrometer data sets to extract spectra from each surface pixel, characterize the absorptions, match them to curves in a spectral library, and generate mineral distribution maps (Kruse and Lefkoff, 1993; Kruse et al., 1993a,b). These knowledge-based systems generate continuum-removed (normalized) spectra and characterize them using attributes of the absorption bands. A human expert decides interactively which spectral attributes are characteristic of a material and assigns a weight to their importance. The system then analyzes unknown spectra and generates images showing the distribution of materials and the measure of certainty of occurrence of a given mineral at each pixel (Kruse and Lefkoff, 1993).

Mineral Mixing and Unmixing

The problems of mineral mixing, vegetation cover, weathering, soil cover, instrument drift or calibration artifacts, and atmospheric effects do not allow consistent, unambiguous mapping. We cannot do much about weathering, soil cover, and plant cover, but mineral unmixing and atmospheric correction are areas of active research. Hyperspectral unmixing algorithms may help resolve some of the uncertainty in surface material identification.

Most surface pixels contain brightness information from a combination of several materials as well as from the atmosphere between the surface and sensor. The pixel can consist of several discrete materials, or a homogeneous blend of many materials. Multispectral or hyperspectral "unmixing algorithms" reverse the mixing process by decomposing mixed

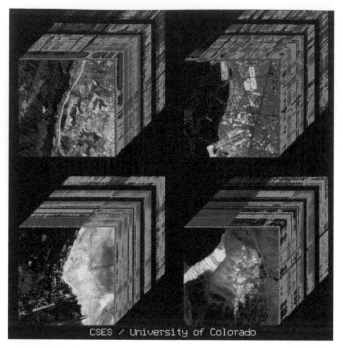

FIGURE 6.6
AVIRIS hyperspectral cubes. The front face is a color infrared image of four different areas. The sides represent the color-coded radiance of the edge pixels in 224 bands from 400 nm in front to 2.45 μm in the back. Reds are high reflectance values; blues are low values; and black zones are atmospheric absorption bands. Courtesy of CSES, University of Colorado, Boulder. (From Goetz, A.F.H. 1992. *Episodes*. 15: 7–14.)

pixels into their constituent substances. Each pixel is modeled as the sum of all reflected or radiated energy curves of materials making up the pixel. "Linear mixing" models assume a well-defined proportional mixture of discrete materials with a single reflection of the incoming solar radiation. "Nonlinear mixing" models assume a randomly distributed, homogeneous mixture of materials, with multiple reflections of the illuminating radiation (Keshava, 2003). These algorithms use reflectance curves taken from a reference library and combined in varying proportions to approximate the measured brightness in order to estimate the proportion of surface materials in a pixel.

Earth scientists have developed a number of unmixing algorithms to solve particular problems. Geologists have approached unmixing from the perspective of precise physical models that carefully capture the interactions of light with mixed matter. Despite their accuracy, these models are unable to convey the statistical variability inherent in remote observations, and the unmixing results, while accurate for the situation, lack robustness. In contrast, engineers and statisticians tend to favor simpler descriptions that use robust statistical models. Unfortunately, statistical modeling fails to reflect the high degree of physical detail that guarantees precision and physically plausible answers for individual pixels (Keshava, 2003; Parente and Plaza, 2010). All this is to say that, despite several decades of working the problem, spectral unmixing remains as much art as science.

Mustard and Pieters (1987a,b) proposed an unmixing technique that works well in parts of the spectrum that do not contain strong absorptions. The spatial distribution and abundance of primary surface components are calculated using a nonlinear mixing model that requires one to input the probable unmixing end-members and particle sizes. They deconvolved (unmixed) Airborne Imaging Scanner 128 band spectrometer data over Moses Rock dike, Utah, using six spectral end-members (ultramafic breccia, two sandstones, gypsiferous soil, clay-rich soil, and desert varnish). Their calculations of surface composition and abundance are similar to field observations. Bierwirth (1990) used a linear unmixing algorithm on eight-band NS001 aircraft imagery at a test site in Queensland, Australia. Six images were generated that show the spatial abundance of green vegetation, dry vegetation, kaolinite, hematite, goethite, and quartz. An inverse linear unmixing model was developed and applied by Boardman (1991) and Boardman and Kruse (1994). They unmixed 224-channel AVIRIS data by using a human expert to pick the most likely spectral end-members (minerals, rock or soil units, vegetation), then solve for the unknown abundances of these materials. The end product is a set of images showing the distribution of single end-members. These images can be combined in color to make mineral or lithologic distribution maps.

A review of unmixing algorithms by Parente and Plaza (2010) breaks them into geometric and statistical algorithms. "Geometric end-member determination" is based on the relationship between mixing models and the geometry of hyperspectral data in multidimensional space. Linear spectral unmixing assumes that the spectra recorded by the hyperspectral scanner can be expressed as a linear combination of end-members weighted by their abundance. Linear unmixing algorithms attempt to find the minimum volume that encloses the data cloud. These algorithms determine end-members by searching for a geometric form within the data cloud. Some code runs autonomously and finds "pure" pixels (e.g., a pixel with spectra that perfectly matches a single mineral or plant) that can then be used as end-members to describe the mixed pixels in the scene. Abundance maps are generated and updated during each iteration of the process.

"Statistical spectral unmixing" code processes a mixed pixel using statistics. Parametric statistical representations are analytical expressions that represent probability density functions. An example of this is the stochastic mixing model, in which each end-member distribution is Gaussian. Each pixel of the hyperspectral image is decomposed as a linear combination of pure end-member spectra. The estimation of end-members and abundances is conducted by generating the posterior distribution of abundances and end member parameters using a Bayesian model. Another Bayesian model for unmixing uses Bayesian self-organizing maps combined with Gaussian mixture models to generate spectral mixtures. A number of nonparametric statistical unmixing approaches have been used, among them neural network models. Some methods use spatial statistics to improve the selection of end-members. The spatial-spectral end-member extraction algorithm is an unmixing method that analyzes a scene in sections such that the spectral contrast of low-contrast end-members is increased, thus improving the potential for these end-members to be selected.

Crosta et al. (1998) examined publicly available software programs to determine the effectiveness of hyperspectral mineral classification where no prior ground spectra or atmospheric information was available. AVIRIS data were used to examine hydrothermal alteration at the Bodie and Paramount mining districts north of Mono Lake, California. Bodie is a lode quartz vein gold deposit with stockworks developed in a zone of hydrothermal alteration along fault zones in dacitic, andesitic, and rhyolitic lavas and tuffs. Silicification near the center of the deposit is surrounded by zones of potassic, argillic, sericitic, and propylitic alteration as one moves farther from the center. Clay mineral distribution maps were made for SWIR bands (2.0–2.45 μm), and iron oxide mineral maps were made for the visible-near infrared (0.42–1.34 μm). They concluded that the algorithms were capable of recognizing and mapping a number of hydrothermal minerals without independent ground information. This is important since exploration activity often does not allow for collection of ground data during the reconnaissance phase of a program.

Atmospheric Corrections

Atmospheric gases and aerosols cause scattering and absorption that affect imaging spectrometer data. Atmospheric effects should be removed in order to use hyperspectral data to properly identify surface materials. Atmospheric absorption bands for water vapor are centered at 0.94, 1.14, 1.38, and 1.88 µm. An oxygen absorption occurs at 760 nm, and the carbon dioxide absorption band is at 2.08 µm. Wavelengths shorter than 1 µm are affected by molecular and aerosol scattering.

Gao et al. (1993) developed some of the original atmospheric correction algorithms in the 1980s. These algorithms gather "scaled surface reflectance" spectra by making the assumption that horizontal surfaces have Lambertian reflectance properties. A "Lambertian surface" reflector is a perfectly diffuse reflector, that is, it has the same reflectance regardless of the observer's perspective. Water vapor effects are derived from the 940 nm and the 1.14 µm water vapor absorption features. The transmission spectrum of water vapor, carbon dioxide, ozone, nitrous oxide, carbon monoxide, methane, and oxygen in the 0.4–2.5 µm region is simulated based on the derived water vapor value, consideration of the location of the sun and observer, and by use of spectral models. Scattering due to atmospheric molecules and aerosols is also modeled. Scaled surface reflectances are derived from the apparent reflectances using the simulated gas transmittances and the simulated molecular and aerosol scattering. A number of atmospheric correction algorithms exist for hyperspectral image data (Gao et al., 2006). Some of these algorithms include spectral smoothing, topographic corrections, and adjacency effect corrections.

One empirical approach requires field measurements of reflectance spectra for at least one bright and one dark target (Conel et al., 1987). The imaging spectrometer data over the surface targets are compared to field-measured reflectance spectra to derive gain and offset corrections. The gain and offset curves are then applied to the entire image. This process generates spectra that are comparable to reflectance spectra measured in the field (Aspinall et al., 2002).

Ocean surfaces are darker than land surfaces, so accurate modeling of atmospheric absorption and scattering effects and specular reflection effects is necessary to derive correct water reflectances from hyperspectral data. Reasonable results have been achieved by applying these algorithms to hyperspectral data acquired by the AVIRIS instrument on an ER-2 aircraft and the Hyperion instrument on the EO-1 satellite platform (Gao et al., 2006).

Ground-Penetrating Radar

Mapping outcrops is inherently limiting: most do not have good three-dimensional (3D) exposures. Adding core holes can complete the picture, but the information derived from shallow core is also limited by the small volume of rock encountered. Another technique is needed to fill in the gaps, especially when complex facies relationships are involved.

Ground-penetrating radar (GPR) imagery can be acquired, processed, and interpreted much like 3D seismic. In dry, consolidated clastic units, the depth of penetration is on the order of 15 to 50 m and the vertical resolution is on the order of 0.25 to 0.5 m, depending on the electrical properties of the rock (Szerbiak et al., 2001). As always, there is a tradeoff between depth of penetration and resolution: greater penetration is available with longer wavelengths, but finer resolution is a function of shorter wavelengths (higher frequencies). Flat, vegetation-free surfaces are ideal for GPR surveys, as are arid or semi-arid environments with deep water tables.

A project to map the stratigraphic architecture of the Ferron Sandstone Member of the Upper Cretaceous Mancos Shale in east-central Utah achieved adequate resolution using a 100 MHz GPR system with antennas oriented parallel to each other and perpendicular to the survey lines and offset by 3 m (Corbeanu et al., 2001, 2004). Both 2D and 3D surveys were conducted. The 2D lines had 10 and 75 m spacing between lines with 0.25 and 0.5 m spacing between traces, respectively. The 3D survey had interline and intertrace spacing of 0.5 m. Vertical and cross-hole GPR data were collected at three locations to control the vertical velocity profile of the data. The GPR data were processed and interpreted similar to seismic data: it was examined both as cross-sectional lines and horizontal depth images. The radar reflections were generated by layers with high clay content in a sandstone background, that is, mudstone drapes and mudstone conglomerates in point bar sandstones of a tidally influenced delta-plain distributary channel system (Figure 6.7). The spatial distribution of the point bar mudstones was determined by a combination of outcrop mapping, drill-hole data, and interpretation of the 2D and 3D GPR imagery.

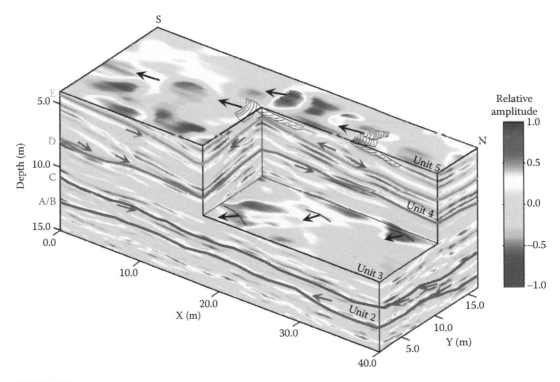

FIGURE 6.7
Three-dimensional (3D) ground penetrating radar (GPR) image cube of Ferron Sandstone Member of the Upper Cretaceous Mancos Shale, east-central Utah. Red, blue, orange, and green labels on the left side of the cube mark the interpreted bounding surfaces. Inside the vertical GPR profiles, purple arrows mark downlap, onlap, and truncation of GPR reflections against major bounding surfaces. The relation between high-GPR-amplitude zones on the horizontal slice and the inclined reflections on the vertical profiles in unit 5 is illustrated using thin black lines portraying the climbing cross-beds in the vertical plane and their shape on the surface. In unit 5, the black arrows show paleoflow direction, and in unit 3 they show the dip direction of the mudstone and mudstone intraclast conglomerate layers. (From Corbeanu, R.M., K. Soegaard, R.B. Szerbiak, J.B. Thurmond, G.A. McMechan, D. Wang, S. Snelgrove, C.B. Forster, A. Menitove. 2001. *Bull. Am. Assn. Pet. Geol.* 85: 1583–1608.)

Lithologic Mapping

It is important for the Earth scientist to map lithologic units, such as sandstones or limestones, rather than to map individual minerals. Again, pioneering work was done by Hunt and Salisbury (1976). They found that similar lithologies often have similar reflectance curves (Figures 6.8 and 6.9). This was then taken as the basis for work by others, such as Kahle and Rowan (1980) and Conel et al. (1985), who set about making remote sensing

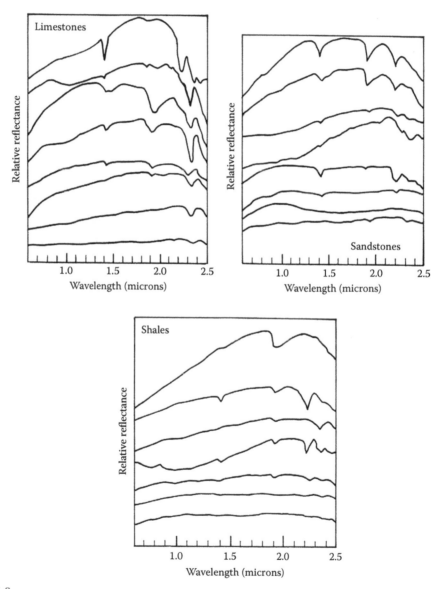

FIGURE 6.8

Visible and near-infrared reflectance curves for rock families, including sandstones, shales, and limestones. Spectra are displaced vertically. (From Hunt, G.R., J.W. Salisbury. 1970. *Mod. Geol.* 1: 283–300.)

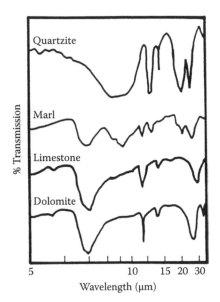

FIGURE 6.9

Thermal infrared transmission spectra of some sedimentary rocks from the East Tintic mountains, Utah. (From Kahle, A.B., L.C. Rowan. 1980. *Geology* 8: 234 239. Reproduced with permission of the Geological Society of America.)

geologic maps using spectral units that usually, but not always corresponded to lithologic units (Figure 6.10). The work by Jet Propulsion Laboratory (Lang et al., 1987, 1990) in the Wind River and Bighorn basins, Wyoming, showed that a combination of photogeologic and spectral interpretation of multispectral data was useful for characterizing the attitude, thickness, and lithology of various strata. They were able to map the distribution of quartz, calcite, dolomite, smectite clays, and gypsum in the stratigraphic section using visible, near infrared, and thermal infrared multispectral data. When digital elevation data were co-registered to the imagery, the contacts of various units could be traced, three-point problems could be solved automatically, and strike and dip information was generated (Lang et al., 1987). This type of map allows a petroleum geologist, for example, to rapidly locate not only structures, but also all outcrops of sandstone, organic-rich shale, and evaporites in an area of interest, and thus map the distribution of potential reservoir, source, and seal units around the margin of a basin.

Sgavetti et al. (1995) proposed a methodology for stratigraphic remote sensing using an example from arid northern Somalia. They integrated air photos, Thematic Mapper (TM) imagery, and field measurements to create a chronostratigraphic framework for mapping and correlating the Jurassic through Eocene section (Figure 6.11 and 6.12). Measured sections in widely separated areas were used as control points for photohorizons mapped on 1:50,000 airphotos and regional TM images. These horizons were defined on the basis of colors, absorption features, stratal patterns (massive versus thin beds, cross bedding), and surfaces bounded by contrasting erosional styles. The fact that the photostratigraphic surfaces were laterally extensive and could be seen to change facies (e.g., reduced number of sandstones within a unit suggesting a change from a near shore to an offshore environment in a transgressive sequence) suggests that bounding surfaces may be equivalent to sequence stratigraphic surfaces.

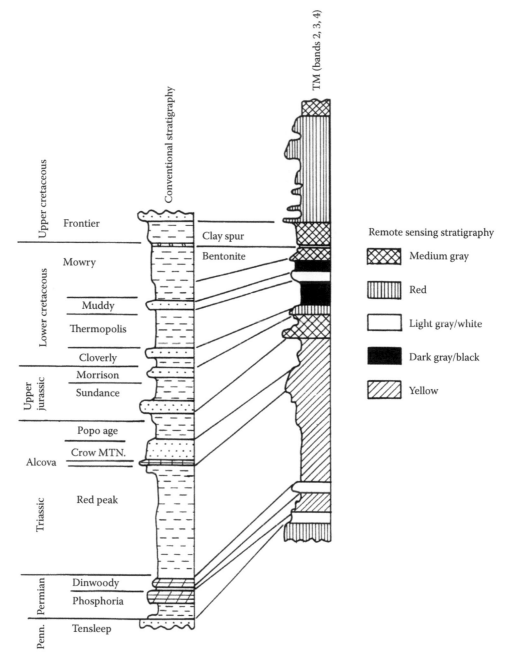

FIGURE 6.10

Correlation of Landsat TM spectral stratigraphy and conventional stratigraphy in the northern Casper arch, Wyoming. (From Lang, H.R. et al. 1987. *Bull. Am. Assn. Pet. Geol.* 71: 389–402.)

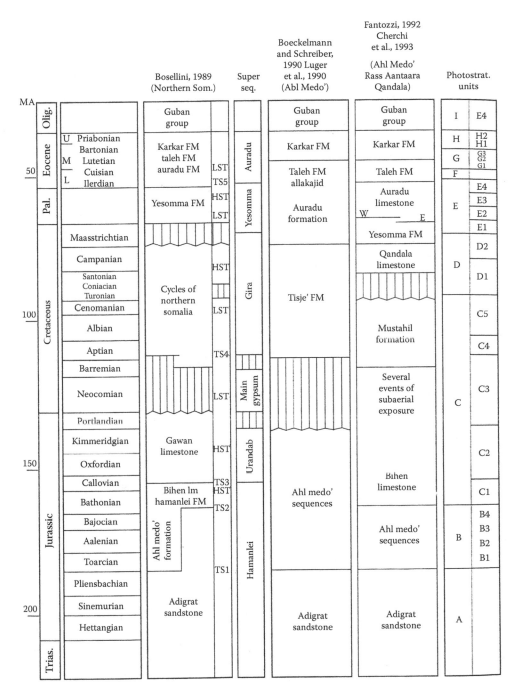

FIGURE 6.11

Photostratigraphy for northern Somalia. Stratigraphic framework. (From Sgavetti, M. et al., *Bull. Am. Assn. Pet. Geol.*, 79, 1571–1589, 1995.)

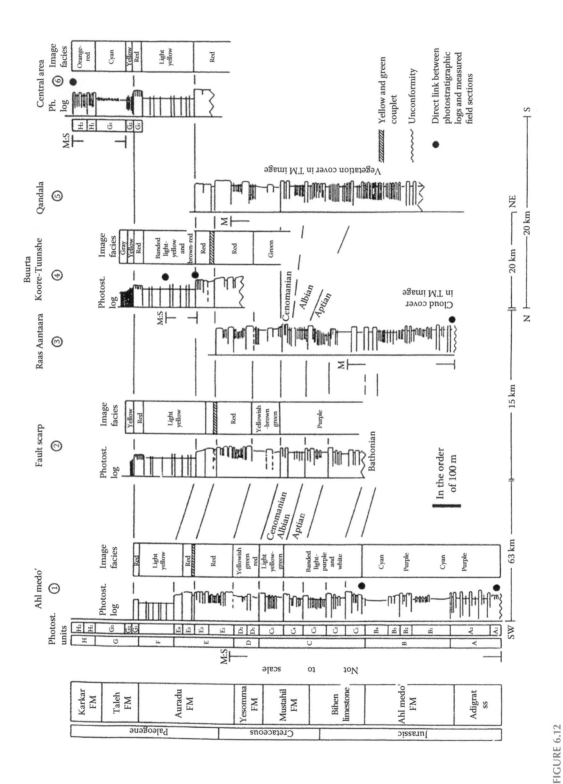

FIGURE 6.12

Photostratigraphy for northern Somalia. Correlation scheme. (From Sgavetti, M. et al., *Bull. Am. Assn. Pet. Geol.*, 79, 1571–1589, 1995.)

Using Radar to Map Stratigraphy

An example of stratigraphic mapping in the arctic using radar imagery is given by Hanks and Guritz (1997). ERS-1 synthetic aperture radar was used to map rock types in the Porcupine Lake area of the northeastern Brooks Range, Alaska. The region contains tundra vegetation in the valleys and is essentially barren on higher slopes. The northern part of the area contains pre-Mississippian metamorphics and Mississippian to Pennsylvanian carbonates, whereas the southern part of the test site comprises thrusted Mississippian to Jurassic carbonates and clastics. Carbonate slopes in this area consist of angular rubble that appears rough (bright) on the C band radar. Shales form valley bottoms or smooth slopes that are generally covered by vegetation or water. Sandstones form tabular rubble of intermediate roughness and brightness. These units can be distinguished using radar on the basis of their surface roughness and dielectric constant, which are a function of lithology and moisture content, respectively.

The ERS-1 C band radar (6.3 cm wavelength) was acquired during the summer and was radiometrically calibrated to convert radar brightness to a backscatter image. Terrain correction was performed using a digital elevation model to move pixels to their correct topographic position and minimize geometric distortions. The final product was a normalized incidence angle image. This assumes a surface of uniform roughness, and uses the digital elevation model to calculate the theoretical backscatter of each pixel based on the incidence angle of the radar beam on the surface at that point. The resulting artificial image is then subtracted from the terrain-corrected radar image (Figure 6.13). Backscatter in the normalized incidence angle image now represents changes in surface properties (roughness, dielectric constant). This image, together with some knowledge of the rock types in the area, was used to map lithologic units over large regions.

Hyperspectral Outcrop and Core Analysis

A little closer to the outcrop, Kurz et al. (2011a,b) used a 240-channel portable near-infrared hyperspectral imager to map mineralogy and texture of outcrops. Pixel resolution is 3 cm to 50 m, and the system is capable of correctly identifying carbonate and siliciclastic rocks, providing mineral content for each pixel. These classified images are merged with Lidar scans of the outcrop and conventional digital photos to generate 3D outcrop geometry and composition.

An interesting application of close-range hyperspectral techniques is in the description of drill core and cuttings. The oil industry has used ultraviolet light to illuminate drill core to map fluorescent minerals (mainly calcite) and hydrocarbon residue. Kruse (1996) presented a technique using a portable field spectrometer (PIMA) to measure infrared reflectance of split core in the range 1.3–2.5 μm. The spectrometer has 600 channels and generates real-time image cubes. This allowed extraction of individual mineral spectra, linear unmixing, and comparison to a spectral library in order to make mineral maps of the core. The technique decreased cost and turnaround time for core descriptions during both mineral and hydrocarbon exploration programs.

Ragona et al. (2007) reviewed a method for automatic mapping of sedimentary stratigraphy in the field or in drill core using short wave infrared (SWIR) hyperspectral imaging and neural networks to classify sediments with similar composition and grain size. Their portable scanner has 245 channels in the range 960 nm to 2.404 μm. A training set was created using several hundred reflectance spectra from core grouped into eight classes. The best model resulted in over 98% correct identification of stratigraphic units. SpecIm (Oulu, Finland) routinely uses hyperspectral scanning to characterize core minerals for the mining industry (Figure 6.14).

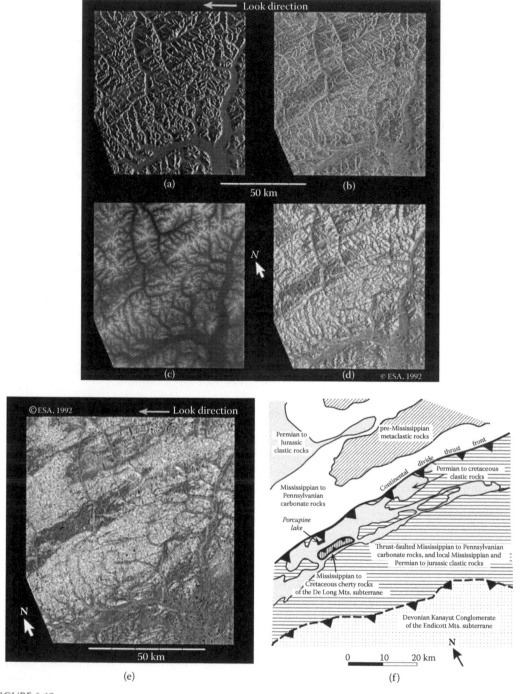

FIGURE 6.13

Porcupine Lake area, Alaska. (a) Simulated Synthetic Aperture Radar (SAR) image made from a DEM; (b) Radiometrically-calibrated ERS-1 image; (c) Digital elevation model; (d) Terrain-corrected SAR image; (e) Incidence angle image; (f) interpreted geologic map. (From Hanks, C.L., R.M. Guritz. 1997. *Bull. Am. Assn. Pet. Geol.* 81: 121–134.)

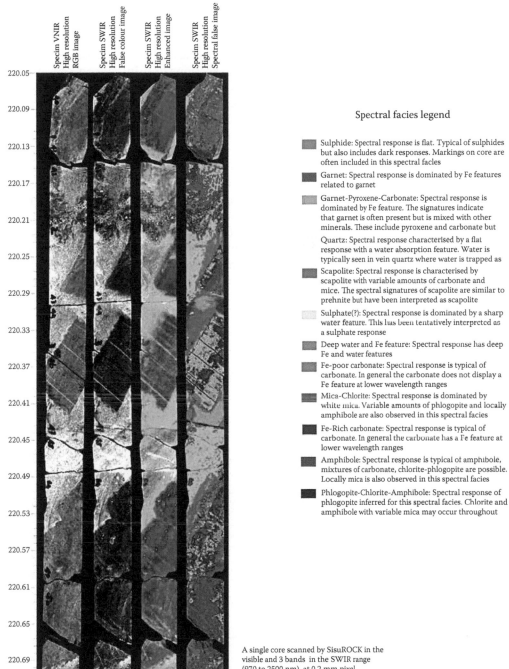

Specim VNIR High resolution RGB image

Specim SWIR High resolution False colour image

Specim SWIR High resolution Enhanced image

Specim SWIR High resolution Spectral false image

Spectral facies legend

Sulphide: Spectral response is flat. Typical of sulphides but also includes dark responses. Markings on core are often included in this spectral facies

Garnet: Spectral response is dominated by Fe features related to garnet

Garnet-Pyroxene-Carbonate: Spectral response is dominated by Fe feature. The signatures indicate that garnet is often present but is mixed with other minerals. These include pyroxene and carbonate but

Quartz: Spectral response characterised by a flat response with a water absorption feature. Water is typically seen in vein quartz where water is trapped as

Scapolite: Spectral response is characterised by scapolite with variable amounts of carbonate and mice. The spectral signatures of scapolite are similar to prehnite but have been interpreted as scapolite

Sulphate(?): Spectral response is dominated by a sharp water feature. This has been tentatively interpreted as a sulphate response

Deep water and Fe feature: Spectral response has deep Fe and water features

Fe-poor carbonate: Spectral response is typical of carbonate. In general the carbonate does not display a Fe feature at lower wavelength ranges

Mica-Chlorite: Spectral response is dominated by white mica. Variable amounts of phlogopite and locally amphibole are also observed in this spectral facies

Fe-Rich carbonate: Spectral response is typical of carbonate. In general the carbonate has a Fe feature at lower wavelength ranges

Amphibole: Spectral response is typical of amphibole, mixtures of carbonate, chlorite-phlogopite are possible. Locally mica is also observed in this spectral facies

Phlogopite-Chlorite-Amphibole: Spectral response of phlogopite inferred for this spectral facies. Chlorite and amphibole with variable mica may occur throughout

A single core scanned by SisuROCK in the visible and 3 bands in the SWIR range (970 to 2500 nm), at 0.2 mm pixel resolution. Scanning speed 20 mm/second.

FIGURE 6.14
Image of a core scanned by SisuROCKI (SpecIm, Finland) in various visible-near infrared and short-wave infrared (SWIR) combinations (range between 970 nm and 2.50 μm). 0.2 mm resolution. Core on the far left is imaged with visible light. Spectral facies legend goes with far right image. (Courtesy of SpecIm.)

QEMSCAN (http://www.fugro-robertson.com/products/qemscan) has applied scanning technology to determine porosity (type and percent) and quantitative mineralogy for both the mining and energy industries (Figure 6.15). This exotic remote sensing technique is an automated method that combines a scanning electron microscope with multiple energy-dispersive x-ray spectrometers. Thin sections of mineral concentrates, drill cuttings, core, ore samples, rocks, coal, and even volcanic ash and moon rocks have been evaluated using this technique. QEMSCAN uses a focused electron beam to image the sample and delineate "sample" from "background." It then steps the beam across the sample collecting x-rays at each point. This means that at each step, there is an x-ray spectrum and a backscattered electron (BSE) brightness. The x-ray spectrum at each point is modeled and a normalized elemental abundance derived and matched against a library of known mineral compositions including entries for overlapping compositions. All this is done on-the-fly so a mineralogical image of the sample is built up very rapidly. Output includes mineralogy, mineral attributes such as size, shape, and association together with porosity and matrix density. Data for particulate samples (e.g., mineral concentrates or cuttings) can be extracted as a bulk average for all particles in a sample or for specific particle types. With further image processing, grain size, pore shape, connectedness and pore type can be derived (Matthew Power, SGS QEMSCAN, email 6 December 2012).

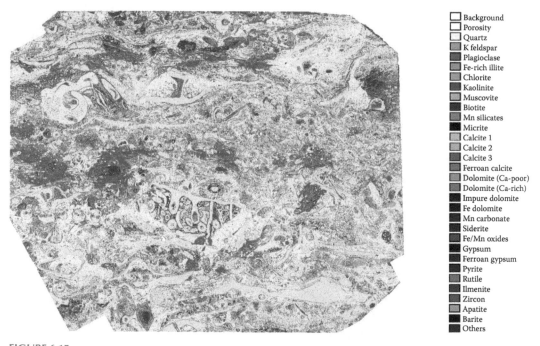

FIGURE 6.15

Hyperspectral mineral map of a limestone thin section. Sample size is about 20 mm across and was imaged at 10 μm (each pixel is a spot analysis that includes a backscattered electron (BSE) and an x-ray determination). Multiple calcite entries in the mineral list relate to different BSE gray scales (same x-ray composition) and allow microporous calcite to be imaged. (Courtesy of Matthew Power, SGS QEMSCAN.)

Source Rock Mapping

Udo and Etuk (1990) demonstrated that it is possible to evaluate the quality and thermal maturity of organic matter in rock extracts in the laboratory using infrared spectroscopy. Rowan et al. (1992, 1995) studied Landsat TM images over eastern Nevada and western Utah and related spectral reflectance of the organic-rich Mississippian Chainman Shale to its maturity as measured on fresh and weathered field samples using vitrinite reflectance and Rock-Eval pyrolysis. They found that the shape of the shale reflectance curve changes from concave-downward to nearly flat (i.e., decreasing reflectance) as thermal maturity increases (Figure 3.10). The absolute reflectance decreases in all wavelengths between 400 nm and 2.5 µm, with a greater decrease at longer wavelengths. They were able to map differences between mature and supermature areas using the ratio of TM band 4/5 (after first minimizing the effect of vegetation using a 4/3 ratio). A decrease in reflectance in band 5 and increasing 4/5 ratio values indicate increasing thermal maturity. They were able to screen out limonitic sandstones, which had similar ratio values, by using the diagnostic reflectance curves of these iron-bearing units. Weathering of immature and mature Chainman Shale causes increasing reflectance in TM 5, but has no effect on the reflectance of supermature samples (vitrinite reflectance greater than 2.0). An attempt to correlate reflectance in the TM bands, as well as various ratios, to the percent total organic carbon (TOC) was unsuccessful. It is felt that this is because of low TOC (≤1.57%). Richer source rocks may have more of a spectral response.

Smith (1978) and Aguilera (1980) mentioned that both gamma radiation and seeping light hydrocarbons increase with increasing organic carbon content in shale. This suggests that radiometers and sniffers should be able to locate high TOC source rock.

Seeps Mapping

Naturally occurring hydrocarbon seeps have been associated with offshore oil fields in the Gulf of Mexico, offshore California, in the Caspian Sea, and elsewhere (Johnson, 1971). It is relatively easy to map seeps offshore, since the background (water) is more-or-less homogeneous. Whereas water generally transmits or absorbs radiation in the visible and near-visible range, oils will fluoresce in the ultraviolet and reflect in the visible and near-infrared. Thus, if a seep is large enough to be detected with a given instrument, it should be obvious after the standard contrast enhancements. Offshore seeps are quite evident on radar imagery, where the oil suppresses capillary waves (cm scale ripples) and thus decreases water surface roughness. Instead of getting a speckled or medium gray radar return, one gets a dark patch (no return). For this to work, there must be sufficient wind to cause ripples to form. A smooth surface can also be caused by a total lack of wind, heavy rain, strong currents, or naturally occurring surfactants associated with marine organisms (Berry and Prost, 1999). These natural seeps have been sampled and consist of both light and heavy molecular weight oil fractions (Abrams and Logan, 2010). Offshore gas seeps can be detected using thermal images, since the gas bubbles bring up cooler bottom water as they rise. On thermal images, oil slicks appear either cooler or warmer than background, depending on their thickness and composition

(Goodman, 1989). Sonar can detect submarine mud volcanoes, gas vent craters, and chemosynthetic mounds related to hydrocarbon seepage (mostly gas) from the ocean floor (e.g., AAPG Explorer, 2002).

O'Brien et al. (2003) mapped naturally occurring marine oil slicks in the Bonaparte and Browse basins in the Timor Sea and North-West Shelf, Australia, using satellite radar. The RadarSat Wide 1 and ERS radar images each covered a 150 × 150 km area with a ground resolution of 20 m (minimum detectable slick size ~ 120 m). Radar imagery was integrated with some seismic data and petroleum system charge history. Hydrocarbons are thought to have migrated to the sea surface along faults and where seals are breached, especially in areas with active hydrocarbon generation and migration. Slicks are frequently located near or above discovered fields in these basins. ASTER data have been used to map offshore seeps in the Campos basin, Brazil, and Bay of Campeche, Mexico (Lammoglia and Souza Filho, 2012). Seeps are delineated on ASTER imagery by using an unsupervised, neural network fuzzy-clustering algorithm. Representative spectra are extracted from atmospherically corrected ASTER data (nine bands in the visible to SWIR) in the various classes. ASTER spectra are checked against a predicted °API by a partial least squares regression model. This model is based on spectra of oils with 13° to 47° API. Using this model, these workers estimated °API values of 19.6° ± 1.37° and 15.9° ± 2.9° for the seeps in the Campos Basin and the Bay of Campeche, respectively. Since oils produced in Campos and Campeche fields typically show °API values from 17–24° and 12–16.5°, respectively, the close match of these results indicates the potential of their method using ASTER data to map seeps and remotely infer the chemical properties of the seeping hydrocarbons.

Identification of seeps is not necessarily easy onshore. In most cases, one is more likely to find indirect indications of seepage such as mud volcanoes (Romania, Trinidad), altered sandstones (Cement field Oklahoma, Lisbon Valley field, Utah), or stressed vegetation (Patrick Draw, Wyoming). Few seeps are large enough to be visible from aircraft or orbital altitudes (Pitch Lake, Trinidad; La Brea, California). If they are large enough to be visible from aircraft or from orbit they could still be lost in background noise. It has been shown that tar sands have unique spectral curves (Andreoli et al., 2007; Cloutis, 1989), and it may be possible to locate these indicators of eroded reservoirs or ancient seeps if they are large enough (Figure 3.10). See Chapter 9, Mature Basin Exploration, for a case history that integrates onshore hyperspectral seeps mapping with other geophysical methods to locate undiscovered petroleum accumulations in the Ventura basin, California.

Other means of locating hydrocarbons onshore rely on image processing techniques to characterize areas that are known to be productive, and then compare them to areas thought to contain hydrocarbons in the subsurface. A patent by Phillips Petroleum (now ConocoPhillips), for example, describes discriminant probability functions that can be used to separate "dry" from "productive" pixels on Landsat TM images by comparing reflectance values of hydrocarbon producing areas to the reflectance of adjacent areas (Sundberg, 1990). Work by Lammoglia et al. (2008) was able to characterize areas of seepage using Landsat 7 ETM+ and ASTER/Terra imagery. The ETM+ data were processed using a pseudo-ratio technique that used principal components analysis. ASTER data were processed using the Spectral Angle Mapper and the Mixture Tuned Matched Filtering techniques adapted for this multispectral data set. In addition, ASTER data were classified using neural networks. All of the microseepage sites are characterized by an absence of green vegetation and ferric iron minerals and the presence of ferrous iron minerals and of clays and/or carbonates.

Stratigraphic Traps

It is obviously not possible to directly map stratigraphic traps in the subsurface, since most sensors do not penetrate the Earth beyond the surface veneer. Under certain circumstances one may, however, be able to deduce where such features are most likely to occur. For example, northwest-elongated sandstone reservoirs in the Cretaceous Mannville Group of Alberta were deposited as shoreline sands, as were the Pennsylvanian Goosegg-Minnelusa interval and Cretaceous Teapot-Parkman interval of the Powder River basin of Wyoming. In some areas, of course, the relationship of these sand bodies to surface features is hard to define or nonexistent. Yet in other areas, during specific time intervals, deposition was controlled by active structures, and these structures can be expressed at the present surface where recent tectonic activity or drape over the sandstone isopach thicks has revealed them (Slack, 1981; Marrs and Raines, 1984; Michael and Merin, 1986). The same can be said for some of the northeast trending fluvial sandstone reservoirs (such as the Cretaceous Muddy-Skull Creek interval) in the Powder River basin, where the rivers that deposited these sands apparently followed fracture zones that are expressed at the surface today (Figure 6.16).

Another type of stratigraphic play that may be evident at the surface involves reef reservoirs. Reefs are stratigraphic traps indirectly related to underlying structure: they favor growing on top of horst blocks, along the high edge of tilted fault blocks,

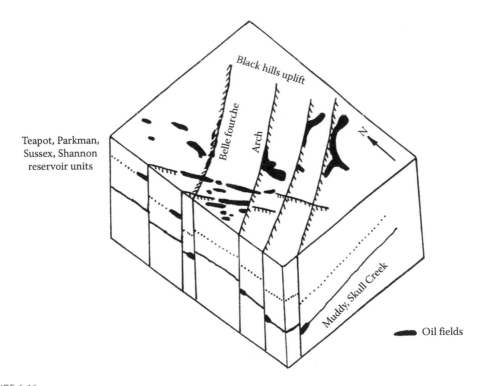

FIGURE 6.16
Schematic block diagram of the northern Powder River basin, Wyoming, showing paleotectonic control on the Cretaceous fields. Shoreface deposits and offshore bars (e.g., Teapot, Parkman, Sussex, and Shannon formations) trend northwest; river channel deposits (e.g., Muddy, Skull Creek formations) trend northeast. (Modified after Slack, P.B., *Bull. Am. Assn. Pet. Geol.*, 65: 730–743, 1981.)

or on the high corners of rotated blocks. For example, facies and thickness changes in the Devonian Duvernay and Cooking Lake Formations at Duhamel field, Alberta, occurred along a tectonic hinge line caused by possible fault movement (Figure 6.17). These facies changes in turn influenced the growth of the Leduc reef (Andrichuk, 1961). The distribution of Devonian reefs in the Western Canadian basin has been related to control by northwest and northeast trending faults (Greggs and Greggs, 1989): the reefs are elongated along these trends, and appear to be localized on subtle structural highs. In the Tangent area of Alberta dolomitized, fault-bounded carbonate reservoirs in the Devonian Wabamun were found to be similar to the Ordovician Albian-Scipio play in Michigan (Churcher and Majid, 1989). In both cases, dolomitization occurred above deep-seated fracture systems that allowed hydrothermal alteration of limestones to high-porosity dolomite reservoirs along linear trends (Hurley and Budros, 1990; Davies and Smith, 2006). In both these areas, the carbonates were dolomitized along fracture zones, and the dolomites provide both intercrystalline and fracture porosity for oil accumulations. The deep fracture zones have been mapped using High Resolution Aeromagnetics (HRAM).

Paleochannels, deltas, and turbidite deposits may exist in the subsurface basinward of present day drainage systems. This is because these types of deposits commonly occur as stacked vertical sequences in rapidly subsiding depositional environments such as the Gulf of Suez, the San Joaquin Valley of California, or the coast along the Gulf of Mexico (Fisher and McGowan, 1969). In addition, many of the river systems delivering sediments are now and have been for the past several million years incised into the topography or controlled by the position of fault blocks. In such cases, paleochannels should occur more-or-less vertically beneath or basinward of present-day channel systems.

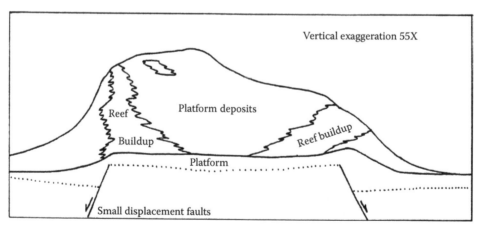

FIGURE 6.17

East-west cross section through the Devonian Swan Hills reef buildup, Alberta. Reefs developed above upfaulted platform margins. The small displacement faults that border the platform have 5 to 13 m vertical offset, and also appear to control the distribution of dolomite zones in the reef. (After Viau, C.A., A.E. Oldershaw. 1984. Structural controls on sedimentation and dolomite cement in the Swan Hills field, central Alberta. In *Carbonates in Subsurface and Outcrop. Canadian Soc. of Pet. Geol. Core Conf*: 103–131. CSPG ©1984, Reprinted by permission of the CSPG whose permission is required for further use.)

Stratiform Deposits

There are a number of economic minerals that occur as stratigraphic units or follow strata. Among these are coal, phosphates, uranium-vanadium, banded iron formations, chemical precipitates, stratiform deposits, replacement mantos, laterites, and placers. Each has features that may be recognized using remote sensing techniques.

Coal, Phosphate, Potash, and Uranium–Vanadium Deposits

Outcrops of coal are recognized and mapped on airphotos and imagery as dark bands, generally weathered out to a greater extent than adjacent sandstones and to the same degree as nearby shales. They may be difficult to distinguish from dark shales except where they have burned and form a distinctive clinker, or natural slag. These clinkers are red in color and form a relatively resistant rubble that, when eroded, resemble badlands.

Phosphate formations, such as the Permian Phosphoria in the northwestern United States, are not only recognized as excellent petroleum source rocks but also contain valuable amounts of phosphate and other metals (e.g., vanadium). Phosphate is an important ingredient in crop fertilizers. This phosphorus salt, dark when freshly broken, weathers to a characteristic light color that can often be followed for great distances along strike using photogeologic methods. Phosphate minerals such as apatite have characteristic absorptions at 9.0 to 9.5 µm that should be recognizable on hyperspectral imagery (Adler, 1964).

Potash, a potassium salt, is found in units such as the Devonian Prairie Evaporite of Saskatchewan in western Canada. Potash is used mostly in fertilizers. The main mineral, sylvite (KCl), usually occurs in bedded, light-colored evaporite deposits. Potash minerals such as sylvite have distinctive absorption spectra between 900 and 1000 nm and at 2 µm that can be mapped using hyperspectral imagery (Farifteh et al., 2008).

Uranium–vanadium deposits form at roll fronts in sandstones where oxidizing and reducing groundwaters are in contact. The metals, in solution in oxygen-rich groundwater, precipitate under reducing conditions (e.g., where there is an increase in organic matter, or in areas of seeping hydrocarbons). Roll fronts may be recognized, for example, where an outcropping red-bed sandstone is bleached at the surface. This represents the contact between oxidizing and reducing zones where the iron oxide minerals have been reduced to pyrite or mobilized in groundwater and moved out of the area (see Chapter 7, Altered Rock and Soil Related to Hydrocarbon Seepage, and Figure 7.2). Such is the case, for example, with uranium occurrences in the Triassic Chinle Formation red-bed sandstone at Lisbon Valley, Utah. Uranium at Lisbon Valley is concentrated above hydrocarbons of the Lisbon Valley field (Kerr and Jacobs, 1964; Conel and Niesen, 1981).

Base Metal Sedimentary Deposits and Chemical Precipitates

Banded iron deposits ranging from 15 to 35% iron are economically important in the Lake Superior district, in Brazil, Australia, and elsewhere. Iron minerals include magnetite, limonite, siderite, chlorite, specularite, hematite, and pyrite, among others. Typically laminated to thin bedded, these sedimentary deposits are almost always Precambrian or Paleozoic. Oolitic iron of the Silurian Clinton Formation in the eastern United States, or

Jurassic oolitic limonite between 30 and 35% iron in Luxembourg and Alsace-Lorraine, is found associated with limestone and silica. Iron minerals (limonite, siderite, hematite, and chlorite) are thought to have replaced the original calcareous ooze. Iron formations can be recognized on imagery as a result of (1) their overall resistance to erosion. In many areas, as in the Superior district, the iron ores are resistant. In some, like Clinton ores of the Birmingham district, Alabama, they are softer; (2) the characteristic iron absorption features of the iron minerals at 400 and 900 nm; and (3) the red colors (reflectance peak at 730 nm) of the oxidation products. Depending on the associated minerals, one can also look for absorption bands characteristic of carbonates (siderite) and silicates (chert, greenalite). Sulfide facies weather to easily recognized iron oxides at or near the surface. Airborne electromagnetic (EM) techniques are an appropriate compliment to imagery in exploring for or extending known iron deposits.

Manganese deposits are sometimes associated with banded iron formations and sometimes with volcaniclastics or deep marine precipitates. Ores range from pyrolusite in an oxidizing environment to rhodochrosite and hausmannite (intermediate Eh-Ph) to alabandite or manganosite in reducing environments. The ores may be recognized by their dark color (pyrolusite), but are more likely to be found as a result of more flamboyant adjacent units such as highly altered red and green andesitic tuffs in the Elqui River valley, Chile, the red-brown "bayate" jasperoid in Cuba or Haiti, or the more resistant iron formations of the Cuyuna range, Minnesota and Morro do Urucum, Brazil.

Manganese and iron oxide concretions are widely distributed on the seafloor. These nodules form by colloidal precipitation of Mn and Fe from seawater. The nodules appear to be forming at an annual rate of 6×10^6 metric tons in the Pacific Ocean (Mero, 1962). Acquisition of side-scan sonar images as part of a systematic seafloor exploration program enhances the ability to locate manganese nodule beds. Side scan sonar imagery has traditionally been interpreted visually. However, image processing techniques for enhancing and classifying sonar imagery are being developed. Automated textural analysis allows a more objective and cost-effective approach. The area between the Clarion and Clipperton fracture zones (NE equatorial Pacific) contains some of the highest concentrations of manganese nodules in the world (Kim et al., 2004). Seafloor sonar characteristics in this area have been classified into several textural facies (Lee and Kim, 2004). Manganese nodules are abundant, for example, in areas dominated by abyssal hill crests with thin sediment cover and in relatively flat areas draped by thin sediments. The close relationship between distribution of sonar facies and manganese nodule abundance implies that the difference in acoustic reflectivity of long range, 11–12 kHz sidescan sonar (with some ground truth), is useful for regional assessments of manganese nodule occurrence. The correlation between sonar imagery analysis of seafloor textures and manganese nodule abundance shows this to be a viable exploration tool (Lee and Kim, 2004).

The Kupferschiefer of northern Europe is a stratiform, metal-bearing bituminous calcareous shale of Mid-Permian age. Although it is less than a meter thick, it extends from northern England to Poland. It overlies a thin conglomerate and is overlain in turn by limestones. Ore minerals include bornite, chalcocite, chalcopyrite, galena, sphalerite, tetrahedrite, and pyrite, with copper being the most important metal mined. Likewise, Zambian copper ores exist in Precambrian carbonaceous shales and consist of chalcopyrite, bornite, and chalcocite. In Zambia, the entire sequence has been subjected to greenschist facies metamorphism. These two examples of stratiform base metal deposits show that if they are exposed at the surface they can be mapped like any other sedimentary unit, using airphotos or hyperspectral imagery. The main difference is that these deposits contain pyrite that weathers to iron oxides: these oxides can be easily mapped on digital imagery using band ratios, and the characteristic ore and gangue minerals can be identified using hyperspectral imagery.

Disseminated Sedimentary Ores and Replacement Mantos

Gold mineralization at Carlin, Nevada, is disseminated as microscopic grains in an altered and leached Devonian limestone (Figure 6.18). Ore minerals include gold, realgar, cinnabar, and stibnite associated with extensive silicification, argillic alteration, and pyritization. Gold, derived from epithermal solutions, precipitated in permeable and porous limestone associated with illite, organic matter, pyrite, and microcrystalline quartz. Multispectral and hyperspectral imagery can be used to map ore and alteration minerals occurring at the surface (see Chapters 7 and 10 for examples).

Mississippi Valley-type deposits occur where sulfide ore minerals, chiefly sphalerite and galena (with minor amounts of chalcopyrite, pyrite, and marcasite), replace carbonates. Lead and zinc deposits occur in individual horizons in the Mississippian Keokuk and Warsaw limestones in the Tri-State mining district of Missouri, Oklahoma, and Kansas. A dolomitic core is surrounded progressively outward by the main ore zone, jasperoid zone (silicified chert breccia), a shale zone, and unaltered crinoidal limestones. Prominent lineaments exist on the surface: these are thought to be basement faults that controlled mineralizing solutions. A shale seal exists above the mineralized unit, perhaps containing the fluids within the porous and reactive limestones and dolomites. In addition to multispectral/hyperspectral imagery to map ore minerals, jasperoid, and dolomitic alteration, the association of deposits with faults is an important clue in locating new deposits or extensions of existing mineral zones.

FIGURE 6.18
Manto and mine at Carlin, Nevada. Note the bleaching and iron oxide alteration adjacent to the mine. (True color image courtesy of Google Earth, © 2012 Google.)

In the Gilman district, Colorado, the Mississippian Leadville limestone was dolomitized prior to mineralization. The dolomite, being somewhat more porous and permeable, reacted readily with mineralizing fluids from nearby intrusives and became host to native gold, gold–silver tellurides, copper–silver sulfosalts, pyrite, sphalerite, galena, chalcopyrite, bornite, and pyrrhotite. Some of the mantos are up to 100 m wide by 50 m thick and several hundred meters long. The Magma (copper) mine at Superior, Arizona, has Devonian Martin limestone replaced by a deposit 7 m thick, 290 m wide, and 1500 m along bedding. Ore minerals include specularite, pyrite, chalcopyrite, bornite, chalcocite, sphalerite, galena, and gangue minerals include minor quartz, barite, and magnetite. While surrounding carbonates are relatively unaltered, the adjacent diabase and schist are sericitized and silicified. In both these examples, faults are thought to have played an essential role in moving fluids into the carbonates. Thus, faults, alteration zones, and diagnostic minerals are the key to mapping these types of deposits on imagery. Airborne potential fields surveys, including magnetic and EM methods, should highlight faults and sulfide minerals.

Fluvial and Marine Placer Deposits

Fluvial placer deposits can be important sources of gold, diamonds, tin, sand, and gravel (Figure 6.19). Minerals include native gold and platinum, zircon, cassiterite, chromite, rutile, ilmenite, and several gemstones. Fluvial placer deposits occur because the heavy minerals, including magnetite, fall out of suspension in areas where the velocity of water in a river or stream slows abruptly, such as below rapids or falls or where the stream gradient flattens. These are the same conditions that are conducive to braided streams, that is, bed load exceeds the stream's capacity to carry sediment. Logical places for placers to accumulate are in point

FIGURE 6.19
Elizabeth Bay placer diamond beaches, Namibia. After being deposited by rivers, the beach placers are reworked and concentrated by longshore currents. (True color image courtesy of GeoEye, TerraMetrics, and Google Earth, © 2012 Google.)

bars (Figure 6.20) on the inside curves of meanders, sand bars below rapids, and at deltas in lakes or along the coast adjacent to rivers (where longshore currents can then rework the sediments). Because most preserved placers occur at the surface and are of Cenozoic age, they are most amenable to remote sensing methods. If one knows where a lode deposit exists, it is logical to look for placers downstream. Digital elevation models derived from laser or radar altimeters can provide topography and make it easier to locate areas where stream channels widen after a steep stretch and where the gradient flattens. Airphotos, multispectral and hyperspectral imagery, and airborne magnetics can also help locate placer deposits.

In the case of gold paleo-placers, such as in the Sierra Nevada foothills of California or the Victoria field of Australia, old buried stream beds exist beneath or nearly beneath present-day streams. In both examples, the placers were also buried beneath lava flows. Since the lavas flowed down the stream valleys, they tend to be thicker there. After a period of erosion, the adjacent, thinner lavas were worn away and the thick, valley fill deposits remain as topographic highs capping the old placer deposits. These features are readily mapped on airphotos and imagery of an appropriate scale.

Alluvial diamond deposits are concentrated by fluvial or coastal marine processes following weathering and erosion of diamond-bearing kimberlite pipes, dikes, fissures, and lamproitic intrusions. These processes can produce large alluvial diamond accumulations. Alluvial diamond deposits are presently known to occur in South Africa, Namibia, Angola, Australia, Brazil, Democratic Republic of Congo, Central African Republic, Sierra Leone, Venezuela, and Russia.

The marine placers, such as the gold deposits at Nome, Alaska, the diamond deposits of Namibia, or the titanium deposits of Florida, are usually brought to the beach by rivers, then reworked and concentrated in narrow zones parallel to the coast by both wave and wind action. At Mineral City, Florida, for example, the main concentration of titaniferous sands occurs between sand dunes and the wave-swept beach where storm waves have

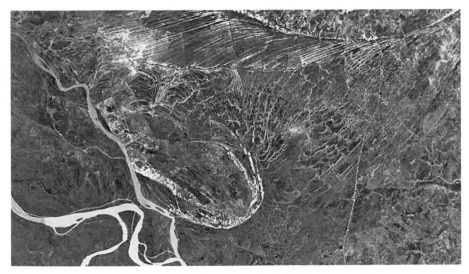

FIGURE 6.20
Sand and gravel deposits can be found in active and abandoned river channels such as these along the Rio Parana Near Baradero, Argentina. Note the characteristic scroll bars (point bars) on abandoned channels. (True color image courtesy of CNES/SPOT Image and Google Earth, © 2012 Google.)

reworked the sand dunes, removing the fines and lighter minerals. As sea level has risen since the last ice age ended, some of these deposits are now slightly offshore in submarine bars. Magnetite is a common heavy mineral in these deposits, so aeromagnetics is a logical tool to use along with airphotos and/or hyperspectral imagery.

Sand and gravel are important constituents in concrete and are also used for roadbeds, building site preparation, and other construction purposes. They are usually easy to recognize and map on airphotos and images because they occur in river channels, beach deposits, and sand dunes. When covered by vegetation, they will support plant communities such as grasses or pine trees that favor well-drained soils and substrate (Figure 6.20).

Laterites

Oxidation and leaching of surface rocks can lead to gossans when sulfide deposits are at the surface, or can lead to laterite, the insoluble residue of iron, aluminum, manganese, or nickel-rich parent rocks. Enrichment takes place where the oxidized product is stable and other constituents are selectively leached away. Iron, manganese, and aluminum form oxides and hydroxides that are relatively insoluble at the surface. Lead forms a stable sulfate; copper, lead, and zinc can form stable carbonates in some environments. Copper, zinc, nickel, and chromium can form stable silicates and oxides.

In tropical climates, iron and aluminum form metal oxide-rich soils known as "laterites." Iron laterites form over ferromagnesian rocks, especially in gentle topography where siliceous constituents can dissolve without eroding the soils. Many iron laterites form over serpentinite. Aluminum laterites, called bauxite, tend to form over syenites and nepheline syenites. Common minerals include boehmite, gibbsite, and diaspore, which are often associated with clay minerals. Bauxites also form over carbonate rocks associated with residual clays in a red soil called *terra rosa*. Manganese laterites, consisting of pellets of manganese oxides, are known as "granzon." Nickel laterites form over serpentines, peridotites, and dunnites, as in the deposits of New Caledonia, which contain 6–10% nickel and lesser amounts of chromium and cobalt.

The common factor in all these weathering-enriched deposits is that they occur in tropical or semi-tropical environments. The laterite generally forms a hard soil horizon that is not agreeable to many plants, so that the forest canopy will change over these deposits: plants become stunted, or the soil is barren. This change should be detectable using multispectral imagery or airphotos. The association with clay minerals and hydroxides means that, where exposed, these minerals can be detected by their absorption bands between 2.2 and 2.4 μm. *Terra rosa* soils are recognized by their red color, and iron laterites by the iron oxide mineral absorptions and red color. Hyperspectral imagery should be able to recognize individual diagnostic minerals.

References

AAPG Explorer. 2002. Marine Geohazards. *AAPG Explorer*. 23: 36.

Abrams, M.A., G. Logan. 2010. Geochemical evaluation of ocean surface slick methods to ground truth satellite seepage anomalies for seepage detection (abs.). *AAPG Search and Discovery* #40604: 1 p.

Abrams, M.J., R.P. Ashley, L.C. Rowan, A.F.H. Goetz, A.B. Kahle 1977. Use of imaging in the 0.46–2.36 μm spectral region for alteration mapping in the Cuprite mining district, Nevada. U.S. *Geol. Survey Open File Rept.* 77–585: 19 p.

Adler, H.H. July–August, 1964. Infrared spectra of phosphate minerals: symmetry and substitutional effects in the pyromorphite series. *Am. Mineral.* 49: 1002–1015.

Aguilera, R. 1980. *Naturally Fractured Reservoirs*. Tulsa: PennWell Books: 720 p.

Andreoli, G., B. Bulgarelli, B. Hosgood, D. Tarchi. 2007. *Hyperspectral Analysis of Oil and Oil-Impacted Soils for Remote Sensing Purposes*. Institute for the Protection and Security of the Citizen. European Commission Joint Research Centre. Luxembourg: European Communities: 36 p.

Andrichuk, J.M. 1961. Stratigraphic evidence for tectonic and current control of Upper Devonian reef sedimentation, Duhamel area, Alberta, Canada. *Bull. Am. Assn. Pet. Geol.* 45: 612–632.

Aspinall, R.J., W.A. Marcus, J.W. Boardman. 2002. Considerations in collecting, processing, and analyzing high spatial resolution hyperspectral data for environmental investigations. *J. Geograph. Syst.* 4: 15–29.

Berry, J.L, G.L. Prost. 1999. Hydrocarbon Exploration. In A.N. Rencz (ed.), *Remote Sensing for the Earth Sciences, Manual of Remote Sensing* v. 3. New York: John Wiley Sons: 490–501.

Bierwirth, P.N. 1990. Mineral mapping and vegetation removal via data-calibrated pixel unmixing, using multispectral images. *Int. J. Remote Sens.* 11: 1999–2017.

Boardman, J.W. 1991. *Sedimentary facies analysis using imaging spectrometry: A geophysical inverse problem*. PhD Dissertation, Boulder: University of Colorado: 212 p.

Boardman, J.W., F.A. Kruse. 1994. Automated spectral analysis: a geological example using AVIRIS data, northern Grapevine Mountains, Nevada. *Proceedings of the 10th Thematic Conference on Geologic Remote Sensing v. 1*. Ann Arbor: ERIM: I–407 to I–418.

Churcher, P.L., A.H. Majid. 1989. Similarities between the Tangent-Wabatum type play of the Alberta Basin and the Albion-Scipio type play of the Michigan Basin. *Bull. Canadian Pet. Geol.* 37: 241–245.

Clark, R.N., T.V.V. King, M. Klejwa, G.A. Swayze, N. Vergo. 1990. High spectral resolution spectroscopy of minerals. *J. Geophys. Res.* 95 B8: 12653–12680.

Cloutis, E.A. 1989. Spectral reflectance properties of hydrocarbons: remote sensing implications. *Science.* 245: 165–168.

Conel, J.E., H.R. Lang, E.D. Paylor, R.E. Alley. 1985. Preliminary spectral and geologic analysis of Landsat-4 Thematic Mapper data, Wind River basin area, Wyoming. *IEEE Trans. Geoscience Rem. Sens.* GE–23: 562–573.

Conel, J.E., P.L. Niesen. June 8–10, 1981. Remote sensing and uranium exploration at Lisbon Valley, Utah. In *International Geoscience and Remote Sensing Symposium*, Washington, DC, v. 1. (A83-10001 01-42). New York: Institute of Electrical and Electronics Engineers: 318–324.

Conel, J.E., R.O. Green, G. Vane, C.J. Bruegge, R.E. Alley. 1987. AIS-2 radiometry and a comparison of methods for the recovery of ground reflectance. In G. Vane (ed.), *Proceedings of the 3rd Airborne Imaging Spectrometer Data Analysis Workshop*. Pasadena: Jet Propulsion Laboratory Publication 87–30: 18–47.

Corbeanu, R.M., K. Soegaard, R.B. Szerbiak, J.B. Thurmond, G.A. McMechan, D. Wang, S. Snelgrove, C.B. Forster, A. Menitove. 2001. Detailed internal architecture of a fluvial channel sandstone determined from outcrop, cores, and 3-D ground-penetrating radar: examples from the middle Cretaceous Ferron Sandstone, east-central Utah. *Bull. Am. Assn. Pet. Geol.* 85: 1583–1608.

Corbeanu, R.M., M.C. Wizeich, J.P. Bhattacharya, X. Zeng, G.A. McMechan. 2004. Three-dimensional architecture of ancient lower delta-plain point bars using ground-penetrating radar, Cretaceous Ferron Sandstone, Utah. In T.C. Chidsey, Jr., R.D. Adams, T.H. Morris (eds.), *Analog for Fluvial-Deltaic Reservoir Modeling: Ferron Sandstone of Utah* v.50, *AAPG Studies in Geology*. Tulsa, OK: 427–449.

Crosta, A.P., C. Sabine, J.V. Taranik. 1998. Hydrothermal Alteration Mapping at Bodie, California, using AVIRIS Hyperspectral Data. *Remote Sensing Environ.* 65: 309–319.

Davies, G.R., L.B. Smith Jr. 2006. Structurally controlled hydrothermal dolomite reservoir facies: An overview. *AAPG Bulletin.* 90: 1641–1690.

Farifteh, J., F. van der Meer, M. van der Meijde, C. Atzberger. 2008. Spectral characteristics of salt-affected soils: a laboratory experiment. *Geoderma.* 145: 196–206.

Fisher, W.L., J.H. McGowan. 1969. Depositional systems in Wilcox Group (Eocene) of Texas and their relation to occurrence of oil and gas. *Bull. Am. Assn. Pet. Geol.* 53: 30–54.

Gaffey, S.J. 1985. Reflectance spectroscopy in the visible and near-infrared (0.35–2.55 μm): applications in carbonate petrology. *Geology.* 13: 270–273.

Gao, B.-C., C.O. Davis, A.F.H. Goetz. 2006. A review of atmospheric correction techniques for hyperspectral remote sensing of land surfaces and ocean color. Geoscience and Remote Sensing Symposium. *IGARSS IEEE International Conference* July 31 to Aug 4, 2006: Denver, CO: 1979–1981.

Gao, B.-C., K.H. Heidebrecht, A.F.H. Goetz. 1993. Derivation of scaled surface reflectances from AVIRIS data. *Remote Sens. Environ.* 44: 165–178.

Goetting, H.R., R.J.P. Lyon. 1987. A knowledge-based software environment for the analysis of spectroradiometer data. *Proceedings of the 5th Thematic Conference on Remote Sensing for Exploration Geology.* Ann Arbor: ERIM: 513–520.

Goetz, A.F.H., G. Vane, J.E. Solomon, B.N. Rock. 1985. Imaging spectrometry for Earth remote sensing. *Science.* 228: 1147–1153.

Goodman, R.H. 1989. Application of the technology in North America. In A.E. Lodge, ed. *The Remote Sensing of Oil Slicks.* Chichester, UK: John Wiley and Sons: 39–65.

Greggs, R.G., D.H. Greggs. 1989. Fault-block tectonism in the Devonian subsurface, western Canada. *J. Pet. Geol.* 12: 377–404.

Hanks, C.L., R.M. Guritz. 1997. Use of Synthetic Aperture Radar (SAR) for geologic reconnaissance in Arctic regions: an example from the Arctic National Wildlife Refuge, Alaska. *Bull. Am. Assn. Pet. Geol.* 81: 121–134.

Henderson, F.B. III, B.N. Rock. eds. 1983. *Frontiers for Geological Remote Sensing from Space.* Fourth Geosat Workshop. Falls Church, VA: American Society of Photogrammetry.

Hook, S.J., J.E. Dmochowski, K.A. Howard, L.C. Rowan, K.E. Karlstrom, J.M. Stock. 2005. Mapping variations in weight percent silica measured from multispectral thermal infrared imagery—examples from the Hiller Mountains, Nevada, USA and Tres Virgenes-La Reforma, Baja California Sur, Mexico. *Remote Sens. Environ.* 95: 273–289.

Hunt, G.R., J.W. Salisbury. 1970. Visible and near-infrared spectra of minerals and rocks: I. Silicate minerals. *Mod. Geol.* 1: 283–300.

Hunt, G.R., J.W. Salisbury. 1971. Visible and near-infrared spectra of minerals and rocks: II. Carbonates. *Mod. Geol.* 2: 23–30.

Hunt, G.R., J.W. Salisbury. 1976. Visible and near-infrared spectra of minerals and rocks: XI. Sedimentary Rocks. *Mod. Geol.* 5: 211–217.

Hurley, N.F., R. Budros. 1990. Albion-Scipio and Stoney Point Fields—U.S.A., Michigan Basin. In E.A. Beaumont, N.H. Foster, eds., *Stratigraphic Traps I: AAPG Treatise of Petroleum Geology.* Atlas of Oil and Gas Fields: Tulsa, OK: 1–37.

Jinhai, X. 1990. Formation of strata-bound ore deposits in China: studies on fluid inclusions. *Chinese J. Geochem.* 9: 169–177.

Johnson, T.C. 1971. *Natural oil seepage in or near the marine environment: a literature survey.* Coast Guard Office of Research and Development, Project No. 714141/002: Washington, D.C.: 30 p.

Kahle, A.B., L.C. Rowan. 1980. Evaluation of multispectral middle infrared aircraft images for lithologic mapping in the East Tintic mountains, Utah. *Geology* 8: 234–239.

Kerr, P.F., M.B. Jacobs. 1964. Argillic alteration and uranium emplacement on the Colorado Plateau. *Proceedings of Clays and Clay Minerals Symposium.* New York: Pergamon Press: 111–128.

Keshava, N. 2003. A survey of spectral unmixing algorithms. *Lincoln Lab. J.* 14: 55–78.

Kim, H., C. Park, J. Park, K. Kim. Fall Meeting 2004. Digital image processing techniques for enhancement and classification of MR1 side scan *Sonar* imagery and preliminary results of *Manganese* nodule occurrence between the Clarion and Clipperton fracture zones, NE Equatorial Pacific (abs.). *Am. Geophys. Union,*#OS33B–0586.

Krohn, M.D. 1986. Spectral properties (0.4 to 25 microns) of selected rocks associated with disseminated gold and silver deposits in Nevada and Idaho. *J. Geophys. Res.* 91 B1: 767–783.

Krohn, M.D., M.J. Abrams, L.C. Rowan. 1978. Use of imaging in the 0.46–2.36 μm spectral region for alteration mapping in the Cuprite mining district, Nevada. *U.S. Geol. Survey Open File Rept.* 78–585: 66 p.

Kruse, F.A. 1996. Identification and mapping of minerals in drill core using hyperspectral image analysis of infrared reflectance spectra. *Int. J. Remote Sens.* 17: 1623–1632.

Kruse, F.A., A.B. Lefkoff, J.B. Dietz. 1993a. Expert system-based mineral mapping in northern Death Valley, California/Nevada, using airborne visible/infrared imaging spectrometer (AVIRIS). *Remote Sens. Environ.* 44: 309–336.

Kruse, F.A., A.B. Lefkoff, J.W. Boardman, K.B. Heidebrecht, A.T. Shapiro, P.J. Barloon, A.F.H. Goetz. 1993b. The spectral image processing system (SIPS)—interactive visualization and analysis of imaging spectrometer data. *Remote Sens. Environ.* 44: 145–163.

Kruse, F.A., A.B. Lefkoff. 1993. Knowledge-based geologic mapping with imaging spectrometers. *Remote Sens. Rev.* 8: 3–28.

Kruse, F.A., K.S. Kierein-Young, J.W. Boardman. 1990. Mineral mapping at Cuprite, Nevada with a 63-channel imaging spectrometer. *Photogram. Eng. Rem. Sens.* 56: 83–92.

Kurz, T.H., S.J. Buckley, J.A. Howell. 2011. Close-range hyperspectral imaging for mapping outcrop composition: (abs.). *AAPG Search Discov* # 90135: 1 p.

Lammoglia, T., C.R. Souza Filho, R.A. Filho. 2008. Characterization of hydrocarbon microseepages in the Tucano Basin (Brazil) through hyperspectral classification and neural network analysis of advanced spaceborne thermal emission and reflection radiometer (ASTER) data. *Intl Arc. Photogram., Remote Sens. Spatial Inf. Sci.* Vol. XXXVII. Part B8: 1195–1200, Beijing.

Lammoglia, T., C.R. Souza Filho. 2012. Mapping and characterization of the API gravity of offshore hydrocarbon seepages using multispectral ASTER data. *Remote Sens. Environ.* 123: 381–389.

Lang, H.R., M.J. Bartholemew, C.I. Grove, E.D. Paylor. 1990. Spectral reflectance characterization (0.4 to 2.5 and 8.0 to 12.0 μm) of Phanerozoic strata, Wind River basin and southern Bighorn basin areas, Wyoming. *J. Sed. Pet.* 60: 504–524.

Lang, H.R., S.L. Adams, J.E. Conel, B.A. McGuffie, E.D. Paylor, R.E. Walker. 1987. Multispectral remote sensing as stratigraphic and structural tool, Wind River basin and Big Horn basin areas, Wyoming. *Bull. Am. Assn. Pet. Geol.* 71: 389–402.

Lee, S.H., K.H. Kim. 2004. Side-scan *sonar* characteristics and *manganese nodule* abundance in the Clarion—Clipperton fracture zones, NE equatorial Pacific. *Marine Georesources. Geotechnol.* 22(1–2): 103–114.

Marrs, R.W., G.L. Raines. 1984. Tectonic framework of Powder River Basin, Wyoming and Montana, interpreted from Landsat imagery. *Bull. Am. Assn. Pet. Geol.* 68: 1718–1731.

Mero, J.L. 1962. Ocean-floor manganese nodules. *Econ. Geol.* 57: 747–767.

Michael, R.C., I.S. Merin. 1986. Tectonic framework of Powder River basin, Wyoming and Montana, interpreted from Landsat imagery: DISCUSSION. *Bull. Am. Assn. Pet. Geol.* 70: 453–455.

Mustard, J.F., C.M. Pieters. 1987a. Quantitative abundance estimates from bidirectional reflectance measurements. *J. Geophys. Res.* 92 B4: E617–E626.

Mustard, J.F., C.M. Pieters. 1987b. Abundance and distribution of ultramafic microbreccia in Moses Rock dike: quantitative application of mapping spectroscopy. *J. Geophys. Res.* 92 B10: 10,376–10,390.

O'Brien, G.W., G. Lawrence, A.K. Williams. November, 2003. Assessing controls on hydrocarbon leakage and seepage. *World Oil.* 224: 49–56.

Parente, M., A. Plaza. 2010. A survey of geometric and statistical unmixing algorithms for hyperspectral images. *IEEE.* 48: 4 p.

Prost, G.L. 1980. Alteration mapping with airborne multispectral scanners. *Econ. Geol.* 75: 894–906.

Ragona, D.E., B. Minster, T. Rockwell. 2007. Automated classification of sedimentary units in drill core and outcrops using high-resolution hyperspectral imaging and neural networks (abs.). *AAPG Search Discover* #90063, AAPG Ann. Mtg. Long Beach: 1p.

Rowan, L.C., F.G. Poole, M.J. Pawlewicz. 1995. The use of visible and near-infrared reflectance spectra for estimating organic matter thermal maturity. *Bull. Am. Assn. Pet. Geol.* 79: 1464–1480.

Rowan, L.C., M.J. Pawlewicz, O.D. Jones. 1992. Mapping thermal maturity in the Chainman Shale near Eureka, Nevada, with Landsat thematic mapper images. *Bull. Am. Assn. Pet. Geol.* 76: 1008–1023.

Rowan, L.C., P.H. Wetlaufer, A.F.H. Goetz, F.C. Billingsley, J.C. Stewart. 1974. *Discrimination of Rock Types and Detection of Hydrothermally Altered Areas in South-Central Nevada by the Use of Computer-Enhanced ERTS Images.* U.S. Geol. Survey Professional Paper 883: Washington, D.C.: U.S. Geological Survey: 35 p.

Sgavetti, M. 1992. Criteria for Stratigraphic Correlation Using Aerial Photographs: Examples from the South-Central Pyrenees. *Bull. Am. Assn. Pet. Geol.* 76: 708–730.

Sgavetti, M., M.C. Ferrari, R. Chiari, P.L. Fantozzi, I. Longhi. 1995. Stratigraphic correlation by integrating photostratigraphy and remote sensing multispectral data: an example from Jurassic-Eocene strata, northern Somalia. *Bull. Am. Assn. Pet. Geol.* 79: 1571–1589.

Slack, P.B. 1981. Paleotectonics and hydrocarbon accumulation, Powder River basin, Wyoming. *Bull. Am. Assn. Pet. Geol.* 65: 730–743.

Smith, E.C. 1978. A practical approach to evaluating shale hydrocarbon potential. *Second Eastern Gas Shales Symposium.* Morgantown, WV: US Department of Energy: 73 p.

Sundberg, K.R. 1990. Spectral data processing method for detection of hydrocarbons. Phillips Petroleum Company. U.S. Patent 4,908,763: 59 p.

Szerbiak, R.B., G.A. McMechan, R.M. Corbeanu, C.B. Forster, S.H. Snelgrove. 2001. 3-D characterization of a clastic reservoir analog: from 3-D GPR to a 3-D fluid permeability model. *Geophysics* 66: 1026–1037.

Udo, O.T., E.E. Etuk. 1990. Application of infrared spectroscopy for rapid petroleum source rock evaluation. *J. Geochem. Explor.* 37: 285–300.

Viau, C.A., A.E. Oldershaw. 1984. Structural controls on sedimentation and dolomite cementation in the Swan Hills field, central Alberta. In Eliuk, L., ed., *Carbonates in Subsurface and Outcrop. Canadian Soc. of Pet. Geol. Core Conf:* Calgary, AB: 103–131.

Zaini, N. 2009. Calcite-dolomite mapping to assess dolomitization patterns using laboratory spectra and hyperspectral remote sensing: a case study of Bedaroieux mining area, SE France. MSc thesis. *Enschede: Intl. Inst. For Geo-Information Science and Earth Observation:* 81 p.

Additional Reading

Ernst, W.B., E.D. Paylor II. 1996. Study of the Reed Dolomite aided by remotely sensed imagery, central White-Inyo Range, easternmost California. *Bull. Am. Assn. Pet. Geol.* 80: 1008–1026.

Goetz, A.F.H. 1992. Imaging spectrometry for earth observations. *Episodes.* 15: 7–14.

Kahle, A.B., A.F.H. Goetz. 1983. Mineralogic information from a new airborne thermal infrared multispectral scanner. *Science* 222: 24–27.

Kahle, A.B., D.P. Madura, J.M. Soha. 1980. Middle infrared multispectral aircraft data: analysis for geological applications. *Appl. Optics.* 19: 2279–2290.

Kurz, T.H., S.J. Buckley, J.A. Howell, D. Schneider. 2011. Integration of panoramic hyperspectral imaging with terrestrial Lidar data. *Photogram. Rec.* 26(134): 212–228.

Kurz, T., S. Buckley, J. Howell. April, 2012. Close-range hyperspectral imaging and Lidar scanning for geological outcrop analysis: Workflow and methods. *Int. J. Remote Sens.* in press.

Lammoglia, T., C.R. Souza Filho. 2011. Spectroscopic characterization of oils yielded from Brazilian offshore basins: potential applications of remote sensing. *Remote Sens. Environ.* 115: 2525–2535.

Lang, H.R. 1999. Stratigraphy. *Manual of Remote Sensing* v. 3, Chap. 7. New York: John Wiley and Sons: 357–374.

Taranik, J.V., A.P. Crosta. 1996. Remote sensing for geology and mineral resources: an assessment of tools for geoscientists in the future. *Intl. Archiv Photogram. Rem. Sens.* XXXI B7: 689–698.

Uhlir, D.M. February, 1995. Hyperspectral Imagery: On the Brink of Commercial Acceptance. *Earth Observation Magazine.* 3 p.

Wrucke, C.T. 1974. Geologic map of the Gold Acres—Tenabo area, Shoshone Range, Lander County, Nevada. U.S. Geol. Survey Map MF–647.

7

Remote Geochemistry

Chapter Overview

"Diagenesis" is the sum of all physical and chemical changes in minerals or sediments after accumulation and burial (Jackson, 1997). It can include adding or removing minerals, recrystallization, and replacement. For remote sensing purposes, we are most concerned with processes such as "leaching" (e.g., removal of feldspars), "alteration" (conversion of feldspars and micas to clay minerals), "silicification" (introduction of silica cement through groundwater or hydrothermal fluids), and "dolomitization" (converting limestone to dolomite by adding hydrothermal magnesium). These processes are key indicators of mineralization of igneous or metamorphic rocks by hydrothermal fluids.

These processes affect both igneous and sedimentary rocks. For example, Lower Cretaceous sandstone of the Kamik Formation in the Richardson Mountains, Yukon Territory (Canada), were silicified as a result of burial to depths of 7 km or more as indicated by vitrinite reflectance. This burial diagenesis, indicated at the surface by silica cementation (quartz overgrowths), would be evident as resistance to erosion and as a strong silica response on hyperspectral imagery. The conclusion from identifying this diagenesis at the surface would be that the Kamik sandstone in this area had been deeply buried and is not likely to be a good hydrocarbon reservoir.

This chapter will discuss alteration associated with mineral deposits and hydrocarbon accumulations and how the unique mineralogical, chemical, magnetic, and radiometric changes involved with this alteration can be detected. We will also review remote detection of gases associated with mineral and hydrocarbon occurrences.

Altered Rock and Soil Related to Hydrocarbon Seepage

Tone and color changes related to the topographic effects of buried structure were mentioned in Chapter 5. Another type of tonal anomaly is due to changes in mineralogy related to diagenetic alteration over leaking hydrocarbons (Figure 7.1). These changes usually occur in the subsurface and are later exhumed and appear at the surface, although they can also form at the surface. Three geochemical effects have been noted:

1. Mobilization of iron minerals because of a reducing environment caused by seeping hydrocarbon gas, primarily methane. This causes "bleaching" of red beds and induced polarization anomalies.

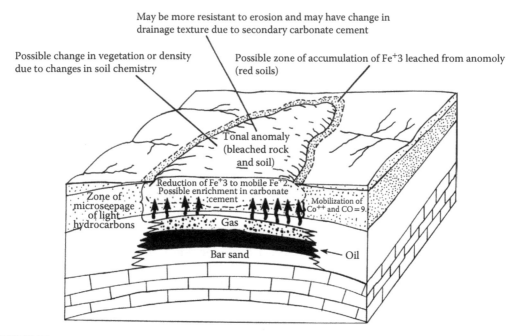

FIGURE 7.1
Diagenetic soil chemistry changes related to leaking hydrocarbons and the tonal and topographic anomalies that may be associated with them.

2. Precipitation of calcium carbonate in a reducing environment over oil and gas fields causing light soil tones, an erosion-resistant surface cement or "caliche," and high soil resistivity.

3. Mobilization of uranium and potassium minerals away from the reducing environment causing radiometric lows over petroleum accumulations.

Donovan (1974) noted at Cement field, Oklahoma, that iron-cemented red bed sandstones are reduced over areas of petroleum microseepage. Iron oxides are either reduced or mobilized in a reducing environment and tend to move away from the crest of the producing anticlinal structure. Depending on the pH and Eh conditions, the sequence of alteration under reducing conditions is as follows:

$$\text{Hydrated ferric oxides} \rightarrow \text{Hematite} \rightarrow \text{Magnetite}$$

Reduction of ferric iron generates pyrite (2%–4% in thin sections) in red bed sandstones with a minimum pH of 8.3 at the Velma, Eola, and Chickasha anticlinal fields, Oklahoma (Ferguson, 1975, 1979). In all cases, this caused a "bleached" appearance in the red beds overlying production.

Ferrimagnetic pyrrhotite exists in well samples above oil and gas reservoirs at Cement field, and does not exist off the field (Reynolds et al., 1990). A major source of sulfide in the iron sulfide minerals may have been microbial sulfate reduction, where the sulfate-reducing bacteria derive their metabolic energy from leaking hydrocarbons. Machel and Burton (1991) observed that magnetite and pyrrhotite are formed in diagenetic environments containing hydrocarbons and that hematite is generally dissolved or replaced.

Mineral alteration and bleaching have been documented in red beds of the Permian Cutler and Triassic Wingate Formations over the Lisbon and Little Valley fields, Utah

FIGURE 7.2

True color image showing bleached areas of red bed sandstone (arrow) in the Triassic Wingate Formation, Lisbon Valley, Utah. Bleaching is believed caused by alteration related to leaking hydrocarbon gases and is associated with uranium deposits in the near surface. (Image courtesy of GeoEye and Google Earth, Copyright [2012] Google.)

(Segal et al., 1984; Conel and Alley, 1985; Morrison and Parry, 1986; Segal et al., 1986). The bleaching coincides with occurrences of near-surface uranium, also believed to have been localized at least in part by seeping hydrocarbons. The light-toned anomalies (Figure 7.2) result from the presence of carbonate cement and feldspars in the sandstone, a relative abundance of clay minerals (primarily kaolinite and chlorite-smectite), and an absence of pyrite. This model for carbonate development requires bacterial metabolism to reduce sulfate to sulfide as well as oxidation of methane to bicarbonate (Berner, 1971):

$$\underset{\text{methane}}{CH_4} + \underset{\substack{\text{sulfate}\\\text{ion}}}{SO_4^-} \rightarrow \underset{\text{bicarbonate}}{HCO_3^-} + \underset{\text{bisulfide ion}}{HS^-} + \underset{\text{water}}{H_2O}$$

Bleaching of the iron oxide–rich Triassic Chugwater sandstone at Spence Dome oil field near Sheep Mountain, Wyoming, is associated with uranium mineralization in the near surface and oil production at depths less than 1750 m (Malhotra et al., 1989). These authors also noted absence of kaolinite and decrease in pH in soils developed on the Cretaceous Cloverly shale above production.

Isotopically distinct carbonate cements were reported over the Recluse field, Wyoming (Dalziel and Donovan, 1980). Seeping hydrocarbons had been oxidized by microbes in the near surface, and the resulting carbon dioxide went into solution and later precipitated over the field as unique isotopic carbonate cement depleted in carbon 13 and either enriched or depleted in oxygen 18. The authors reported the same from the Davenport field in Oklahoma.

The manganese content of surface carbonate cements increases dramatically over the Recluse field, Wyoming, and toward the center of a halo-type anomaly over production near Boulder, Colorado (from 120 to 5400 ppm) (Donovan et al., 1975; Dalziel and Donovan, 1980). Apparently, Mn^{++} has a wide range of redox stability compared to iron and is relatively enriched in soils over areas of petroleum microseepage. Also over the Recluse field, sage leaves (*Artimesia tridentata*) and pine needles (*Pinus ponderosa*) were found to contain anomalously high concentrations of iron and manganese. This is thought to be a result of reducing soil conditions above seeping hydrocarbons; reducing conditions allow divalent iron and manganese to be complexed or adsorbed on soil particles, where they are easily taken up by plants.

Lilburn and Al-Shaieb (1983) summarized several of these observations:

1. Red bed sandstones are bleached as a result of a reducing environment caused by leaking hydrogen sulfide, are relatively enriched in clay minerals, primarily kaolinite and corrensite, and often are associated with uranium deposits.

2. Oxidation of hydrocarbons caused caliche or carbonate-cemented sandstones (isotopically distinct and often manganese rich) in the near surface, and caused calcite replacement of gypsum.

3. Pyrite and pyrrhotite formed over production as a result of reduction of iron oxide by hydrogen sulfide gas.

Starting with the premise that hydrocarbon reservoirs leak, and that some fraction of this leakage will reach the surface, then under the proper conditions (i.e., no impermeable seals; presence of porous, iron-rich near-surface rocks), there should be detectable features related to reduction of mobile iron to pyrite (Sternberg, 1991). Pyrite is polarizable and thus detectable using induced polarization surveys. It has also been shown that calcite is associated with these zones: calcite cement buildup is detectable using surface or airborne resistivity methods. Obviously, some fields will have no measurable anomalies because they do not have the proper preconditions, whereas other fields could have false anomalies.

Several oil fields in the United States have been evaluated using the U.S. Department of Energy's National Uranium Resource Evaluation (NURE program) surveys to determine whether aerial gamma ray data can help locate hydrocarbon accumulations (e.g., Saunders et al., 1993; similar surveys have been acquired in Canada and Australia, among others). These surveys measure uranium, potassium, and thorium spectra. The thorium content is used to normalize the uranium and potassium readings and to correct for altitude, cosmic radiation, lithology, soil moisture, surface water, vegetation cover, and atmospheric radon. Potassium concentrations appear to be significantly lower over oil and gas fields, and uranium was found to be slightly concentrated in profiles over Alabama Ferry, Leona, O.S.R., and Lonesome Dove fields, Texas. The Aguarita and Dark Horse oil fields and Selden gas field, Texas, were actually discovered using gamma ray spectral data in conjunction with soil gas sampling, magnetic susceptibility measurements, and subsurface geology. The model used to explain this effect is that the clay minerals (mainly illite) are destroyed by carbonic acid and organic acids. The clays release potassium ions to the groundwater where they are removed by leaching. Meanwhile, uranium migrates from an oxidizing environment to a reducing one and tends to concentrate there over time. Thus, low concentrations of uranium in surface sediments may be clues to subsurface hydrocarbons (Figure 7.3).

Hyperspectral imagery has been used to map diagenetic alteration in the Jurassic Navajo Sandstone of southern Utah (Bowen et al., 2007). Results indicate that a combination of handheld and airborne measurements can identify reduction of iron minerals (bleaching), iron oxidation and precipitation (red–ocher–yellow staining), areas where potassium feldspars have been altered to kaolinite and illite, and areas where calcium carbonate has been dissolved or precipitated. The massive eolian Navajo Sandstone happens to be well

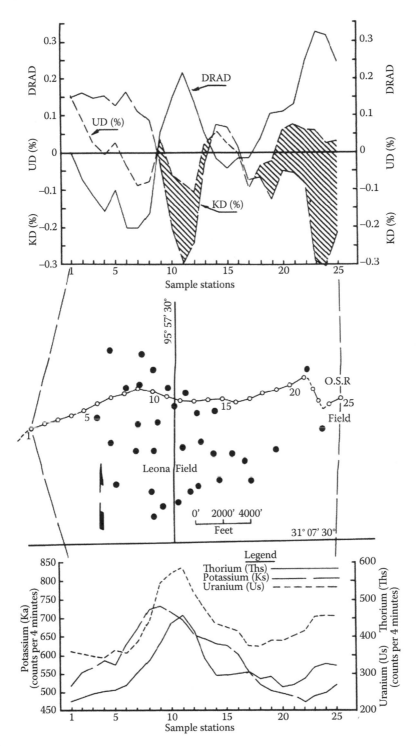

FIGURE 7.3
National Uranium Resource Evaluation radiometric survey showing uranium is slightly concentrated in profiles over Leona and O.S.R. fields, Leon County, Texas. KD% and UD% are deviations from the ideal potassium
and uranium values; DRAD is UD–KD. (From Saunders, D.F. et al., *SEG 45th Annual Midwest Meeting*, 1993.)

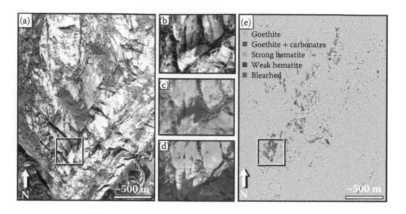

FIGURE 7.4

Diagenetic mineral variation in the East Kaibab area, northern Arizona. Images show altered rock, and end-member maps show alteration mineral distribution. Both show the influence of fluid flow pathways in altering the Jurassic Navajo Formation sandstone. (a) Approximate true color image (HyMap bands 15, 8, 4) of reaction front zone. Box indicates area of reaction front shown in (b) to (d). (b) Approximate true color image. (c) Minimum noise fraction transform of the visible-near infrared HyMap data (first 38 bands) shows changes in iron distribution. (d) Minimum noise fraction on the entire range of HyMap data (into the short wave infrared) shows changes in the distribution of iron, carbonates, and clays. Red and orange = hematite; green = goethite; blue = bleached; and yellow = transition zone in iron, clay, or carbonate content. (e) Spectral end-member distribution derived from the Matched Filter partial unmixing algorithm (ENVI). (From Bowen, B.B. et al., *Bull. Am. Assn. Pet. Geol.*, 91, 173–190, 2007.)

exposed in this semiarid region. Mineral variation controlled by diagenetic processes is seen to have a significant effect on kilometer-scale flow properties (porosity and permeability). Images showing various combinations of altered rock, as well as end-member maps showing alteration mineral distribution, clearly show the influence of fluid flow pathways in altering this unit that serves as a hydrocarbon reservoir in the subsurface (Figure 7.4).

Landsat-7 Enhanced Thematic Mapper imagery, supported by magnetic susceptibility measurements and geochemical soil sampling, has reportedly identified diagenetic carbonate soil anomalies associated with producing and presumed hydrocarbon accumulations at Fula'erji in the Songliao Basin, China (Zhang et al., 2009). The surface in this area comprises floodplain and meadows (mainly sagebrush and grass), along with corn fields. The producing fields contain high molecular weight (and low °API gravity) hydrocarbons at depths starting around 350 m and below. Anomalies that are not over production are assumed to be over undiscovered shallow oil accumulations.

Image enhancement techniques at Fula'erji that highlighted alteration include principal components analysis and band ratioing. The best four principal components were combined with ETM 3/1, 4/3, and 7/5 ratios. Anomalies related to subsurface hydrocarbons are expressed as faint tonal "blobs" in the Holocene sediments. Coincident with the tonal anomalies are high magnetic susceptibilities. It is thought that leakage of light hydrocarbons and the resulting reducing environment promoted formation of pyrite and precipitation of magnetic iron oxides and sulfides including magnetite, maghemite, pyrrhotite, and greigite. Near-surface diagenetic carbonates and carbonate cements are common manifestations of hydrocarbon seepage (Schumacher, 1996; Meer et al., 2002). Soil samples collected over one of the Fula'erji tonal anomalies suggest diagenetic carbonate is concentrated over the center of the soil tonal anomaly and decreases outward. It is felt that this combination of soil tone, magnetic susceptibility, and soil chemistry are all related to leakage of light hydrocarbons from this heavy oil accumulation (Figure 7.5).

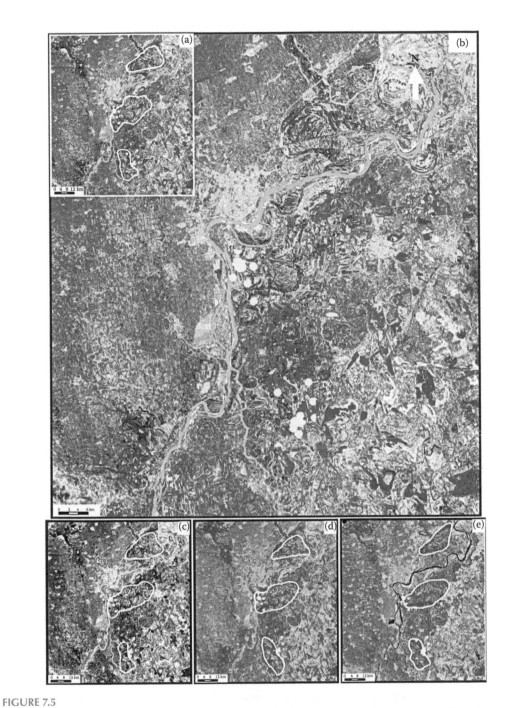

FIGURE 7.5

Four false-color composite images consisting of ratios and principle components (PCs). The white outlines indicate tonal anomalies. (a) The composite image of 3/1 (R), 1357-PC3 (G), and 123457-PC4 (B). (b) Enlarged composite of 3/1 (R), 1357-PC3 (G), and 123457-PC4 (B). (c) The composite of 7/5 (R), 1357-PC3 (G), and 123457-PC4 (B). (d) The composite of 4/3 (R), 7/5 (G), and 3/1 (B). (e) The composite of 3/1 (R), 3457-PC4 (G), and 1345-PC3 (B). Soil samples collected over tonal anomalies at Fula'erji, Songliao Basin, China, suggest diagenetic carbonate exists over the center of the soil tonal anomalies and decreases outward. Soil tone, magnetic susceptibility, and soil chemistry anomalies all appear to be related to leakage of light hydrocarbons above this heavy oil field. (From Zhang, G. et al., *Bull. Am. Assn. Pet. Geol.*, 93, 31–49, 2009.)

Following up on work by Richers et al. (1982), chemical, mineralogical, and perhaps vegetation changes were found above the Patrick Draw field, Wyoming (Khan and Jacobson, 2008). This oil field has a small gas cap and produces from a stratigraphic trap in the Cretaceous Almond Fm. at a depth of 1700 m. The surface consists of Pleistocene to Holocene soil and alluvium with stunted sagebrush cover. Faults have been mapped from the producing sandstones to the surface, and light hydrocarbons have been identified in soil gas sampled along these faults.

Hyperion hyperspectral imagery was used to map alteration at Patrick Draw field. A supervised classification defined end-members for spectral unmixing and resulted in a map of likely minerals and their abundance for each pixel. Areas of anomalous mineral occurrence were identified by the classification (Figure 7.6). X-ray diffraction analysis identified more clay minerals in the anomalous areas and more feldspar in unaffected areas. This is consistent with the concept of feldspars weathering more readily to clay in the presence of oxidizing hydrocarbons that are leaking along faults. Field samples were collected and analyzed for carbon isotopes. $\delta^{13}C$ isotopes in the range -2.88 to -45.32 indicate that this carbon is derived from crude oil and/or methane.

The mineral changes described here as a result of hydrocarbon microseepage can form over any type of hydrocarbon trap (structural or stratigraphic). Schumacher (2010) compared hydrocarbon microseepage surveys with subsequent drilling results for 2700 U.S. and international wells in frontier and mature basins, onshore and offshore, and at depths from 300 to 4900 m. His results indicate that 82% of wells drilled on positive microseepage anomalies were completed as commercial wells, whereas only 11% of wells drilled without a seepage anomaly were discoveries. These results emphasize the usefulness of remote sensing surveys to detect hydrocarbon seepage and guide geochemical sampling programs.

Altered Rock and Soil Related to Mineral Deposits

"Alteration" is any change in mineral composition of a rock by physical or chemical means, in particular by hydrothermal solutions or weathering, for example, supergene alteration (Jackson, 1997). Alteration associated with mineral deposits varies depending on the composition of the host rock and the chemistry of the mineralizing fluids. Alteration generally requires a combination of permeability (e.g., through fractures) and chemical reactivity of the host rock in contact with the hydrothermal solution. It should be noted that not all alteration is associated with mineralization, nor do all mineral deposits have prominent alteration zones. Common alteration types include silicification, argillization, sericitization, propylitization, carbonitization, and hydration. Generally, the more extensive the alteration zone, the more susceptible the country rock has been to ore deposition, since the hydrothermal fluids have obviously penetrated the rocks extensively.

The wall rocks around deep ore bodies are relatively impermeable and have little or no alteration unless there is a strong chemical contrast between the host and country rock (e.g., acid-rich fluids in a carbonate host rock). Contact zones around pegmatites may be either narrow and sharp or wide and gradational (Park and MacDiarmid, 1970). Alteration minerals associated with pegmatites include beryl, monazite, sphene, tantalite, phlogopite, topaz, zircon, fluorite, allanite, microcline, and biotite. The alteration zone is generally less than 1 m wide but can be up to 7 m wide. Some pegmatites are characterized by bleaching because of removal of iron and conversion of biotite to muscovite.

Alteration in metamorphic wall rocks contains a conspicuous suite of minerals including grossularite, andradite, and almandite garnets, wollastonite, epidote, pyroxenes, ilvaite, idocrase, serpentine, spinel, and scapolite, among others. Shales, otherwise impermeable, can become permeable after silicification and fracturing. Shales in an igneous or metamorphic contact zone develop the sugary texture of hornfels with clusters of epidote, chlorite,

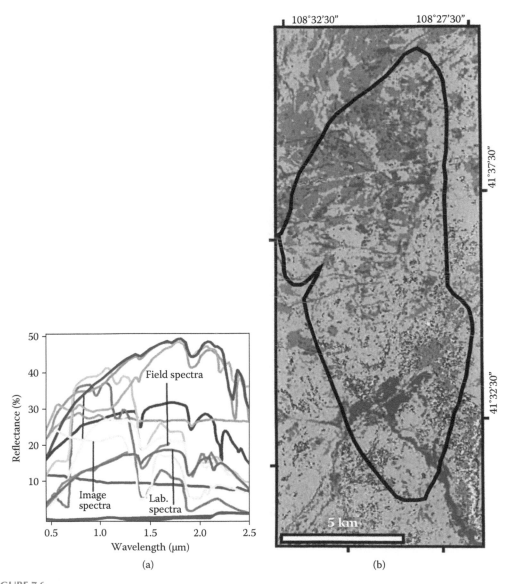

(a)   (b)

FIGURE 7.6
A supervised classification of Hyperion hyperspectral imagery over Patrick Draw field, Wyoming, generated a map of likely minerals and their abundance for each pixel. Anomalous mineral occurrences were identified by the classification. (a) Spectral reflectance of the end-members used in the classification. Color of the spectral curve corresponds to color of the class. Field and image spectra are added for comparison. (b) Classification based on laboratory spectroscopy. (From Khan, S.D. et al., *GSA Bull.*, 120, 96–105, 2008. Published with permission of the Geological Society of America.)

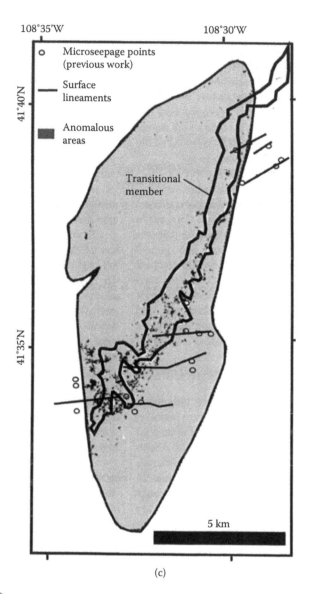

(c)

FIGURE 7.6 *(Continued)*
A supervised classification of Hyperion hyperspectral imagery over Patrick Draw field, Wyoming, generated a map of likely minerals and their abundance for each pixel. Anomalous mineral occurrences were identified by the classification. (c) Patrick Draw field showing anomalous areas, surface lineaments, and microseepage control points. (From Khan, S.D. et al., *GSA Bull.*, 120, 96–105, 2008. Published with permission of the Geological Society of America.)

andalusite, garnet, and cordierite. Silica-rich fluids injected into a marble host rock generate the minerals wollastonite and idocrase (Park and MacDiarmid, 1970). Intruded carbonates can be metasomatized to form skarns consisting of silicate and oxide minerals. Some limestones and dolomites are simply recrystallized as a result of heat and pressure. The process of recrystallization may drive out impurities and lead to a much lighter rock. Gray limestone at Hannover, New Mexico, is coarsened and lightened because of expulsion of carbonaceous material. Silicification is common at igneous–metamorphic contacts, with the silica derived either from the mineralizing fluids or from the country rock. The North Lily deposit at East Tintic, Utah, was found at least partly because of alteration associated with the limestone replacement ore body: hydrothermal dolomite and minor chlorite were followed by dickite, kaolinite, halloysite, beidellite, and rutile; then by allophane, quartz, and barite; and finally by mineralization accompanied by sericite (McKinstry, 1948).

Sericite is the most extensive alteration mineral in mesothermal deposits (Park and MacDiarmid, 1970). Calcite, dolomite, siderite, rhodochrosite, and ankerite are also common. Chlorite is often found paired with sericite around veins. Fine-grained silica in the form of jasperoid can be abundant. Pyrite may be conspicuous, as well as feldspars and clay minerals. Porphyry copper deposits are frequently characterized by overlapping or concentric zones of alteration from an inner "potassic" (quartz-sericite-potassium feldspar) to intermediate "phyllic" (quartz-sericite-tactite), "argillic" (quartz-sericite-kaolinite), and outer "propylitic" (chlorite, calcite, sericite, and montmorillonite) assemblages (Abrams and Brown, 1984; Lepley et al., 1984) (Figure 7.7). The deposits at Bingham, Utah, and Miami, Arizona, could have been located purely on the basis of outcrops of altered porphyry containing quartz, sericite, and adularia (McKinstry, 1948).

Chlorite is perhaps the most abundant epithermal alteration mineral. Sericite is common, and other shallow alteration minerals include alunite, zeolites, chalcedony, opal, calcite, dickite, kaolinite, illite, and montmorillonite. The country rock may be extensively altered for several hundred meters from a deposit. Intermediate and mafic volcanics commonly contain propylitic alteration characterized by chlorite, calcite, and epidote.

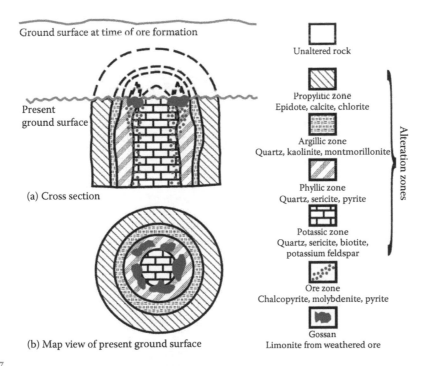

FIGURE 7.7
Porphyry copper deposits are frequently characterized by overlapping or concentric zones of alteration from inner potassic (quartz-sericite-potassium feldspar) to intermediate phyllic (quartz-sericite-tactite), argillic (quartz-sericite-kaolinite), and outer propylitic (chlorite, calcite, sericite, and montmorillonite) assemblages. (From Ott, N. et al., *Geosphere*, 2, 236–252, 2006., modified after Lowell, J.D. et al., *Econ. Geol.*, 65, 373–408, 1970.)

Mississippi Valley–type lead–zinc deposits are characterized by calcite, dolomite, marcasite, chalcedony, and opal. The calcite and dolomite should be evident on multispectral or hyperspectral imagery as carbonates, and the chalcedony and opal should be recognizable on thermal imagery as silicates (Bellian et al., 2006).

Supergene enrichment occurs in the oxidized zone above deposits where the water table is deep and minerals have a chance to leach. Sulfides decompose to sulfates and leave behind native metals, silicates, carbonates, and oxides. Limonite, clay minerals, and silica in the form of jasperoid are characteristic of the alteration assemblage (Figure 7.8).

Tones and colors associated with alteration include bleaching, darkening, and various colored halos, particularly pastel colors (Park and MacDiarmid, 1970). Clay minerals can be white or shades of green or gray. Argillization is generally recognized by its bleached appearance, but the addition of chlorite and epidote causes a green color. Pyritization of red bed sandstones causes a bleached appearance because of reduction of iron. Oxidation of pyrite, on the other hand, produces the conspicuous red to red brown zones known as gossans (Lovering, 1949; Blanchard, 1968).

Bleached or iron oxide–stained outcrops are easily mapped using color photography or multispectral imagery (Rowan et al., 1974; Prost, 1980; Abrams et al., 1983; Podwysocki et al., 1983; Segal, 1983). Hyperspectral imagery may be required to map suites of alteration minerals (see Mineral Exploration case histories, Chapter 10).

Alteration associated with mineralizing fluids has characteristics other than tone, color, and unique spectra. Resistant minerals such as silica in veins or iron-rich gossans regularly form topographic knobs or rises. The outcrop at Broken Hill, Australia, forms a conspicuous ridge visible for miles in the otherwise flat plains (McKinstry, 1948). Cerro Negro stands out above the deposit at Parral, Chihuahua (Mexico), as do the quartz veins at Oatman, Arizona. Veins lacking quartz, and especially containing sulfide deposits, may appear as depressions because of the ease of weathering of the soft, altered materials (carbonate gangue and

FIGURE 7.8
Supergene alteration of mainly Oligocene volcanic and intrusive rock at Red Mountain, south of Ouray, Colorado. (True color image courtesy of Google Earth, Copyright [2012] Google.)

clays). Depressions have also been attributed to the oxidation and attendant shrinkage of deposits. At Bisbee, Arizona, apparent shrinkage cracks extend 200 m above the deposit to the surface where they outline an irregular, cracked body of ground (McKinstry, 1948). Resistant iron ores in the Lake Superior range are associated with hills and ridges.

Remote Sensing of Gases, Radiation, and Aerosols

All petroleum basins leak hydrocarbons (Schumacher, 2012). Seepage can be manifested as anomalous hydrocarbon concentrations in soil or the atmosphere just above the soil. Microseepage anomalies help confirm an active hydrocarbon system in frontier basins, tell us whether a trap has been charged (which in many settings seismic cannot), help identify bypassed pay in abandoned fields, and can help identify sweet spots in unconventional resource plays (basin-centered gas or fractured shales).

Case History: Sniffing the Air

On March 22, 2010, *Time Magazine* ran an advertisement for Shell titled "Sniffing the air" (Time, 2010). In it, they discussed airborne methane (and soon perhaps ethane) detection at concentrations as low as 100 parts per trillion by an instrument flying at an altitude of 300 m. The system has been tested over the Algerian desert, and while no discoveries were attributed to the new technology, they felt that combining it with airborne gravity and magnetic measurements taken on the same aircraft will enable Shell to identify "hot spots" over forests, swamps, or agricultural lands.

Fast-forward to 2012. Shell has now deployed "LightTouch," a system that can measure methane from a light aircraft flying 150 m above terrain (Hirst et al., 2012; Hirst et al., in press). This direct hydrocarbon detection technique maps hydrocarbon gas fluxes over areas up to ~5000 km² per flight. The method can detect, locate, and quantify the emission rate of a calibrated emission source as well as correctly locate previously unknown sources. Two recent surveys show the method can detect live hydrocarbon systems, help interpret migration pathways, and provide information on seal integrity. LightTouch has been deployed simultaneously with gravity and magnetic surveys. The method uses proprietary ultrasensitive gas sensors to directly measure gas molecules that naturally seep from subsurface hydrocarbon systems into the air.

The sensor automatically and continuously logs concentration measurements along with global positioning system data, altitude, barometric pressure, air temperature, and wind velocity. Flight data are subsequently merged with meteorological data and auxiliary data such as the air sample transit time from the wing-mounted inlet to the sensor. The resulting data set comprises concentrations, positions, and meteorological parameters.

Measured concentrations are ambiguously linked to the position of the gas sources on the ground. There are numerous parameters that affect the measured concentration, and these parameters change significantly during each flight. The measurement location is influenced by the gas emission rate, the wind speed and direction, the rate of plume spreading as described by a gas dispersion model, and other atmospheric factors.

The first of two surveys (Area "A") was flown over an extremely dry and remote desert area with sparse vegetation cover. The exploration concession comprised ~10,000 km² and contained undeveloped dry gas fields in the middle of an otherwise remote, unpopulated region. More than 100 hours of airborne methane gas concentration data were collected during 13 flights. It was immediately obvious from the raw concentration data that there were major emission sources within the concession.

The second survey, of Shell's Metouia block, Tunisia, in late 2008 covered over 5000 km² of the concession in a single flight. Gravity and magnetic data were collected as well. The 34 flights provided multiple coverage of the concession. Downwind flybys of a persistent emission source, a conveniently located flare in a gas-processing plant near the airport, were included in each day's flight. The flare provided a persistent methane source that was readily detectable to the sensors from a range of several kilometers. Multiple flybys of the flare allowed calibration of dispersion models, confirmation of air sample transit time, and evaluation of different wind data sets by comparison with the gas plume's orientation.

Shell was able to identify several regions of significant methane emissions within both concessions that they consider to be natural (Figure 7.9). The strongest of these sources have methane flux rates in the range of 60–80 kg/(h·km²). The methane flux levels detected are large compared to the maximum emission rates of the even most prolific sources of biogenic methane (Mitsch and Gosselink, 2007); hence, useful methane surveys over heavily vegetated areas should be possible. The average regional methane flux levels reported for Tunisia are much greater than the current best available estimates of anthropogenic methane flux for the region.

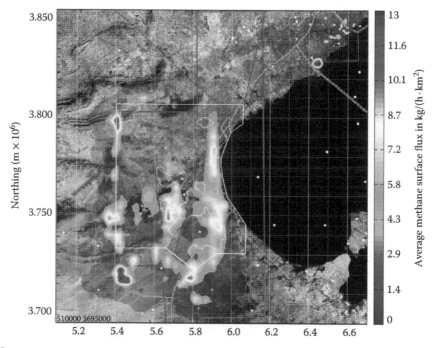

FIGURE 7.9

Shell LightTouch survey of the Metouia block, Tunisia, indicates several areas that appear to be naturally occurring hydrocarbon emissions. This average methane flux map was derived from all eight LightTouch flights. The display includes all sources of methane emissions, but average fluxes less than 4.5 kg/(h·km²) are suppressed. The white dotted lines outline the major estimated structural highs. The solid white line shows the limit of structural information. Anthropogenic sources have not been removed. (Provided by Bill Hirst; Shell Global Solutions International BV. From D. Schumacher and R. K. Warren, eds., *Non-Seismic Detection of Hydrocarbons: Assumptions, Methods, and Exploration Case Histories*: Soc. Exploration Geophysicists, Special Publication, in preparation.) (Copyright Shell Global Solutions International BV; author bill.hirst@shell.com.)

Detection of Gases, Radiation, and Aerosols Associated with Mineral Deposits

"There are no gases uniquely associated with mineral deposits ...." according to Hale (2000). This is because many of the same gases that emanate from mineral deposits can be the result of biological processes in the soil. Some gases associated with mineralization, like CO_2, are also biogenic and found in relatively large concentrations in normal atmosphere. Nevertheless, and taking these considerations into account, soil gas geochemistry is a viable exploration method. Gases associated with mineral deposits also have a minor role in biogenic activity and are normally found in trace amounts in the atmosphere. They include mercury vapor, H_2S, helium, and radon. These gases and vapors are most likely released to the atmosphere during volcanism or ongoing volcanogenic exhalative processes. Similar to gases associated with hydrocarbon accumulations, these gases have to be collected, concentrated, and detected. Although there are companies that do such surveys (e.g., Tekran Instruments for mercury; RKI Instruments for H_2S; Pfeiffer-Vacuum for helium), gas detectors are generally used during ground-based surveys and are not usually flown on aircraft or helicopters. Barringer did invent and patent an airborne mercury spectrometer that has been used to discover mercury deposits, contamination, and hot springs (Jepsen, 1973).

As mentioned in Chapter 3 (under the section Instrument Systems), airborne radiometric surveys can detect gamma rays emitted by near-surface uranium, thorium, and potassium associated with granitic bedrock, potash deposits, and uranium accumulations. Gamma rays are also used to locate alteration zones in acidic and intermediate intrusions and in soil mapping, since different soils frequently have different radiometric signatures (Moon et al., 2006).

"Radon gas," formed during radioactive decay of uranium and thorium, is commonly associated with uranium deposits. Radon gives off alpha particles that can be detected using alpha cups or alpha cards, usually placed on the surface or buried in the subsurface. Gamma rays have, on occasion, been detected remotely using an alpha particle x-ray spectrometer. This instrument was on the lunar Pathfinder and Martian rover missions (NASA a,b; APXS, online; Alpha Particle Spectrometer, online).

Under certain conditions, the vegetation respiration process may allow remote geochemical detection of select elements associated with mineral deposits (Curtin et al., 1974; Barringer, 1977; Beauford et al., 1977). Curtin et al. (1974) sampled volatile compounds exhaled from Lodgepole pine (*Pinus contorta*), Engelmann spruce (*Picea engelmannii*) and Douglas fir (*Pseudotsuga menziesii*) at 19 subalpine locations in Colorado and Idaho. The condensed exhalations were filtered and evaporated, and then the residue was ashed and analyzed using a spectrometer. The residues of these exudates contained lithium, beryllium, boron, sodium, magnesium, titanium, vanadium, chromium, manganese, iron, cobalt, nickel, copper, zinc, gallium, arsenic, strontium, yttrium, zirconium, molybdenum, silver, lead, bismuth, cadmium, tin, antimony, barium, and lanthanum. They concluded that anomalous concentrations of metals found by air sampling and analysis of aerosols derived from plant respiration may be a useful tool for remote geochemical mineral exploration.

Further work on the release of heavy metals to the atmosphere from vegetation was carried out by Beauford and Barber at Imperial College (Beauford et al., 1977). Their work was adapted to exploration by Barringer (1977) to cover large areas quickly. Atmospheric particulates were collected using an aircraft that flew close to the ground; these aerosols were analyzed for elements associated with mineral deposits. The Barringer system, called AIRTRACE, was mounted on either a fixed wing aircraft or a helicopter. It scooped up airborne particles and concentrated them on sticky tape. The location of the tape was recorded using a flight path camera, and the tape was later analyzed in a laboratory. The pattern of

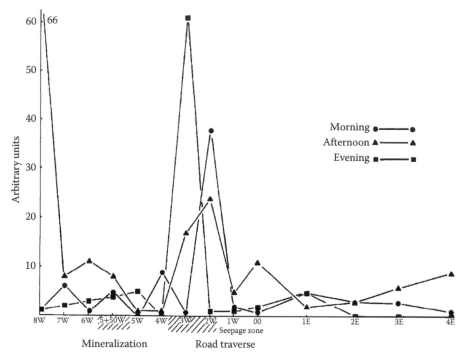

FIGURE 7.10
Barringer's AIRTRACE MK III double peak anomaly for copper, Limerick prospect, Bancroft Ontario. (From Barringer, A.R., *USGS Professional Paper*, 1015, 231–251, 1977.)

element responses over mineralization was typically a "double-peak," and anomalies were recorded only in the immediate vicinity of deposits (Figure 7.10) (Barringer, 1975, 1977).

In summary, we have shown that rock and soil over mineral and hydrocarbon deposits can be affected by the underlying accumulations. Alteration and diagenesis related to these deposits can be detected using remote sensing technologies, including photography, multispectral and hyperspectral sensors, potential fields and resistivity instruments, gas detectors, radiometers, and aerosol detectors. Although "remote" may not be distant in the classical sense, in that some of these detectors are flown within meters or a few tens of meters of the ground, they still allow the explorer to cover large areas quickly and economically.

References

Abrams, M.J., D. Brown. 1984. *Joint NASA/Geosat Test Case Project*. Tulsa: American Association of Petroleum Geologists: 4-1 to 4-73.

Abrams, M.J., D. Brown, L. Lepley, R. Sadowski. 1983. Remote sensing for copper exploration in southern Arizona. *Econ. Geol.* 78: 591–604.

Barringer, A.R. 1975. AirTrace: an airborne geochemical technique. Pre-print of paper presented at the 1st William T. Pecora Memorial Symposium, Sioux Falls, SD: U.S.Geological Survey: 32 p.

Barringer, A.R. 1977. AIRTRACE—an airborne geochemical exploration technique. *USGS Professional Paper* 1015: 231–251.

Beauford, W., J. Barber, A.R. Barringer. 1977. Release of particles containing metals from vegetation into the atmosphere. *Science.* 195: 571–573.

Bellian, J.A., C. Kerans, R.A. Beck, Y. Price, K. Nedunuri. 2006. Calcite-dolomite delineation using airborne hyperspectral data for Ordovician paleokarst mapping (abs.): *AAPG Discovery Digest* Article #90052: *Am. Assn. Pet. Geol.* Annual Convention, 15: 8.

Berner, R.A. 1971. *Principles of Chemical Sedimentology.* New York: McGraw-Hill Book Co.: 300 p.

Blanchard, R. 1968. Interpretation of leached outcrops. *Nevada Bur. Mines Bull.* 66: 196 p.

Bowen, B.B., B. A. Martini, M.A. Chan, W. T. Parry. 2007. Reflectance spectroscopic mapping of diagenetic heterogeneities and fluid-flow pathways in the Jurassic Navajo Sandstone. *Bull. Am. Assn. Pet. Geol.* 91: 173–190.

Conel, J.E., R.E. Alley. 1985. Lisbon Valley, Utah uranium test site report. In: Abrams M.J., J.E. Conel, H.R. Lang, H.N. Paley (eds.), *Joint NASA/GEOSAT Test Case Project* Sect. 8. Tulsa: American Association of Petroleum Geologists: 8-1 to 8-158.

Curtin, G.C., H.D. King, E.L. Mosier. 1974. Movement of elements into the atmosphere from coniferous trees in subalpine forests of Colorado and Idaho. *J. Geochem. Explor.* 3: 245–263.

Dalziel, M.C., T.J. Donovan. 1980. Biogeochemical evidence for subsurface hydrocarbon occurrence, Recluse oil field, Wyoming; preliminary results. *U.S. Geological Survey Circular* 837: 11 p.

Donovan, T.J. 1974. Petroleum microseepage at Cement, Oklahoma: evidence and mechanism. *Bull. Am. Assn. Pet. Geol.* 58: 429–446.

Donovan, T.J., R.L. Noble, I. Fiedman, J.D. Gleason. 1975. A possible petroleum-related geochemical anomaly in surface rocks, boulder and weld counties, Colorado. *U.S. Geological Survey Open File Report No.* 75–47: 11 p.

Ferguson, J.D. 1975. The subsurface alteration and mineralization of Permian red beds overlying several oil fields in southern Oklahoma. M.Sc. thesis. Stillwater: Oklahoma State University: 95 p.

Ferguson, J.D. 1979. The subsurface alteration and mineralization of Permian red beds overlying several oil fields in southern Oklahoma, part II. *Shale Shaker.* 29: 200–208.

Hale, M. 2000. *Geochemical Remote Sensing of the Subsurface.* Amsterdam: Elsevier: 549 p.

Hirst, B., P. Jonathan, M. Padgett, T. Krayenbuhl. In press. Airborne surface gas flux mapping for hydrocarbon exploration. In: Schumacher, D., R.K. Warren (eds.), *Non-Seismic Detection of Hydrocarbons: Assumptions, Methods, and Exploration Case Histories.* Soc. Exploration Geophysicists, Special Publication, in preparation.

Hirst, B., P. Jonathan, T. Krayenbuhl, F. Gonzalez Del Cueto. 2012. Airborne surface gas flux mapping for onshore frontier exploration (abs.): *AAPG Search and Discovery* Article #90142. Long Beach, CA: *Am. Assn. Petr. Geol.* Annual Convention: 1 p.

Jackson, J.A. 1997. *Glossary of Geology*, 4th ed. Alexandria: American Geological Institute: 769 p.

Jepsen, A. 1973. Trace Elements in the environment. *Adv. Chem.* 123: 81–95.

Khan, S.D., S. Jacobson. 2008. Remote sensing and geochemistry for detecting hydrocarbon microseepages. *GSA Bull.* 120: 96–105.

Lepley, L., M.J. Abrams, L. Readdy. 1984. *Joint NASA/Geosat Test Case Project.* Tulsa: American Association of Petroleum Geologists: 6-1 to 6-95.

Lilburn, R.A., Z. Al-Shaieb. 1983. Geochemistry and isotopic composition of hydrocarbon-induced diagenetic aureoles (HIDA), Cement Field Oklahoma: Part I. *Shale Shaker.* 34: 40–56.

Lovering, T.S. 1949. *Rock Alteration as a Guide to Ore—East Tintic District, Utah.* Urbana: Economic Geology Publishing Company: 64 p.

Lowell, J.D., J.M. Guilbert. 1970. Lateral and vertical alteration-mineralization zoning in porphyry ore deposits. *Econ. Geol.* 65: 373–408.

Machel, H.G., E.A. Burton. 1991. Causes and spatial distribution of anomalous magnetization in hydrocarbon seepage environments. *Bull. Am. Assn. Pet. Geol.* 75: 1864–1876.

Malhotra, R.V., R.W. Birnie, G.D. Johnson. 1989. Detection of surficial changes associated with hydrocarbon seepage, Sheep Mountain anticline, WY. *Proceedings of the 7th Thematic Conference on Remote Sensing for Exploration Geology, Calgary*, October 2–6. Ann Arbor: ERIM: 1097–1110.

McKinstry, H.E. 1948. *Mining Geology.* Englewood Cliffs: Prentice-Hall, Inc.: 233–276.

Meer, F., P. Dijk, H. van der Werff, H. Yang. 2002. Remote sensing and petroleum seepage: a review and case study. *Terra Nova.* 14: 1–17.

Mitsch, W.J., J.G. Gosselink. 2007. *Wetlands*, 4th ed. New York: John Wiley: 600 p.

Moon, C.J., M.K.G. Whateley, A.M. Evans. 2006. *Introduction to Mineral Exploration*, 2nd ed. Oxford: Blackwell Publishing: 481 p.

Morrison, S.J., W.T. Parry. 1986. Formation of carbonate-sulfate veins associated with copper ore deposits from saline basin brines, Lisbon Valley, Utah. *Geol. Soc. Am. Prog. with Abs., Rocky Mtn. Sect. Mtg.* 18: 397 p.

NASA a, APXS. http://msl-scicorner.jpl.nasa.gov/Instruments/APXS/ (accessed October 3, 2012).

NASA b, 2001. Alpha Particle Spectrometer. http://lunar.arc.nasa.gov/results/alpha.htm (accessed October 3, 2012).

Ott, N., T. Kollersberger, A. Tassara. 2006. GIS analyses and favorability mapping of optimized satellite data in northern Chile to improve exploration for copper mineral deposits. *Geosphere.* 2: 236–252.

Park, C.F. Jr., R.A. MacDiarmid. 1970. *Ore Deposits*. San Francisco, CA: W.H. Freeman and Co.: 145–164.

Pfeiffer-Vacuum. 2013. Sniffing for helium. http://www.pfeiffer-vacuum.com/products/leak-detectors/sniffing/container.action (accessed June 19, 2013).

Podwysocki, M.H., D.B. Segal, and M.J. Abrams. 1983. Use of multispectral scanner images for assessment of hydrothermal alteration in the Marysvale, Utah, mining area. *Econ. Geol.* 78: 675–687.

Prost, G.L. 1980. Alteration mapping with airborne multispectral scanners. *Econ. Geol.* 75: 894–906.

Reynolds, R.L., N.S. Fishman, R.B. Wanty, M.B. Goldhaber. 1990. Iron sulfide minerals at Cement oil field, Oklahoma: implications for magnetic detection of oil fields. *Bull. Geol. Soc. Am.* 102: 368–380.

Richers, D.M., R.J. Reed, K.C. Horstman, G.D. Michels, R.N. Baker, L. Lundell, R.W. Marrs. 1982. Landsat and Soil-Gas Geochemical Study of Patrick Draw Oil Field, Sweetwater County, Wyoming. *Bull. Am. Assn. Pet. Geol.* 66: 903–922.

RKI. 2013. Instruments for H_2S. http://www.rkiinstruments.com/?gclid=COCV8YCu-bACFWQ0Q-god8Cr-TA (accessed June 19, 2013).

Rowan, L.C., P. H. Wetlaufer, A.F.H. Goetz, F. Billingsley, J.H. Stewart. 1974. Discrimination of rock types and detection of hydrothermally altered areas in south-central Nevada by the use of computer-enhanced ERTS images. *U.S. Geological Survey Professional Paper* 883: 35 p.

Saunders, D.F., K.R. Burson, J.F. Branch, C.K. Thompson. 1993. Relation of thorium-normalized surface and aerial radiometric data to subsurface petroleum accumulations. *Geophysics:* 58: 1417–1427.

Schumacher, D. 1996. Hydrocarbon-induced alteration of soils and sediments. In: Schumacher, D., M.A. Abrams (eds.), *Hydrocarbon Migration and Its Near-Surface Expression* 66. AAPG Memoir: Tulsa: 71–89.

Schumacher, D. 2010. Integrating hydrocarbon microseepage data with seismic data doubles exploration success. *Proceedings Indonesian Petroleum Association, 34th Annual Conference. and Exhibition:* 11 p.

Schumacher, D. 2012. Hydrocarbon microseepage: a significant but underutilized geologic principle with broad applications for petroleum exploration and production (abs.). *AAPG Search and Discovery* Article #90142: *Am. Assn. Petr. Geol.* Annual Convention, Long Beach: 1 p.

Segal, D.B. 1983. Use of Landsat multispectral scanner data for the definition of limonitic exposures in heavily vegetated areas. *Econ. Geol.* 78: 711–722.

Segal, D.B., M.D. Ruth, I.S. Merin. 1986. Remote detection of anomalous mineralogy associated with hydrocarbon production, Lisbon Valley, Utah. *Denver: Rocky Mtn. Assn. Geols., The Mountain Geologist.* 23: 51–62.

Segal, D.B., M.D. Ruth, I.S. Merin, H. Watanabe, K. Soda, O. Takano, M. Sano. 1984. Correlation of remotely detected mineralogy with hydrocarbon production, Lisbon Valley, Utah. *Proceedings of the 3rd ERIM Thematic Conference on Remote Sensing for Exploration Geology.* Ann Arbor: ERIM: 273–292.

Sternberg, B.K. 1991. A review of some experience with the induced polarization/resistivity method for hydrocarbon surveys: success and limitations. *Geophys.* 56: 1522–1532.

Tekran. 2013. Instruments for Mercury. http://www.tekran.com/products/ambient-air/overview (accessed June 19, 2013).

Time. March 22, 2010. Sniffing the air. *Time Magazine*: p. 48.

Zhang, G., L. Zou, X. Shen, S. Lu, C. Li, H. Chen. 2009. Remote sensing detection of heavy oil through spectral enhancement techniques in the western slope zone of Songliao Basin, China. *Bull. Am. Assn. Pet. Geol.* 93: 31–49.

8

Modern Analogs

Chapter Overview

Remote sensing imagery provides some of the best visualization of ancient processes and environments that we as geoscientists are trying to reconstruct and understand. Whether we are working with a Cretaceous carbonate bank in the Tampico Basin of Mexico or fluvial sand reservoirs in the Western Canadian Sedimentary Basin, imagery of modern depositional environments can help us understand ancient sedimentary processes, and the distribution, geometries, and scale of ancient geobodies (those sand or carbonate bodies that comprise hydrocarbon reservoirs). They can assist us in developing our mental model of a depositional environment (where barrier bars are with respect to fluvial sand sources; where a fore-reef is with respect to the back reef lagoon), and they can provide direct input (object size, orientation, and facies relationships) to object-based geomodels of reservoir units.

In addition to depositional processes, remote sensing imagery can show the effects of modern structural and erosional processes. The Afar Triangle area in Eritrea, Djibouti, and Ethiopia allows us to visualize an ancient onshore triple junction in the western Gulf of Mexico or at the Benue Trough, Nigeria. Images of the aftermath of a tsunami or hurricane allow us to better understand near-shore seismogenic and storm deposits. Modern landslides help us recognize ancient gravity flows. Imagery of present-day flooding should help us understand the magnitude of flood-based erosion. Modern volcanism provides insights to ancient explosive activity.

Finally, Earth-based analogs help us understand what we are seeing on other planets, and vice versa.

Depositional Environments

The ability to show a distribution of facies that may be analogous to an ancient system lends itself to multiple applications. Perhaps the most common is to visualize the depositional architecture of a subsurface reservoir unit. The present-day depositional environment provides us with a mental model of where things should be, and we can use this model when exploring for or developing a field.

One of the finest compilations of modern analogs, and one that shows the early interest of the petroleum industry, is the lavishly illustrated "Satellite Images of Carbonate Depositional Settings" (Harris and Kowalik, 1994). Detailed descriptions of modern carbonate environments are provided along with scaled overlays of oil fields producing from carbonate reservoirs. Various workers have classified these carbonate environments into depositional facies and depth settings using visible and radar imagery and bathymetry (Rankey, 2002; Purkis et al., 2005). This characterization allows identification of body geometries, heterogeneities, and energy environments (Harris et al., 2011), which can then be used in building reservoir geomodels (Harris, 2010).

Taking this a step further, one should consider multiple working hypotheses regarding the original depositional environment of a particular reservoir. For example, several environments are examined in turn to determine the most likely depositional setting for isolated linear marine sands of the Cretaceous Shannon Sandstone exposed in the Salt Creek anticline and adjacent subsurface Hartzog Draw Field, Wyoming (Suter and Clifton, 1999). The authors compare these Shannon sands to those in the foreland basin of the Holocene Fly River of Papua, New Guinea (Figure 8.1); to incised Quaternary valleys of the Louisiana continental shelf, Gulf of Mexico; to wave-dominated highstand shorelines of the Quaternary Gulf coast of Texas (Figure 8.2); and to tidal sand ridges of the East China Sea, among others.

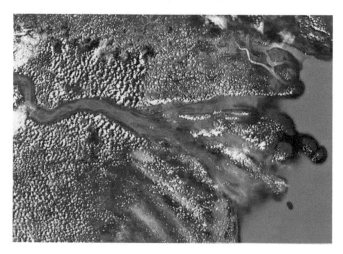

FIGURE 8.1
The Fly River delta, Papua-New Guinea, has been used as a depositional analog for the Cretaceous Shannon Sandstone reservoir of Wyoming. (True color image courtesy of TerraMetrics and Google Earth, © 2012 Google.)

FIGURE 8.2
Matagorda Island wave-dominated barrier bar, Texas Gulf coast. (True color image courtesy of Google Earth, © 2012 Google.)

Imagery is being used to show present-day depositional environments: Landsat (suspended sediments), EOS MODIS (sediment plumes), GOES-12 (storm systems), SeaWIFS (coastal sediment plumes, currents, chlorophyll concentrations; Figure 8.3), AVHRR (sea surface temperatures showing currents), and multibeam bathymetric images (bottom morphology) are used along with airphotos (movement of longshore currents and associated barrier islands and offshore sand bars; Figure 8.4) and digital bathymetry (bottom gradients, morphology) to illustrate a number of reports detailing present-day coastal environments (Suter, 2006; Boyd et al., 2007).

A second reason for using modern analogs is as a guide to reservoir architecture when building reservoir geomodels. Geomodels are three-dimensional (3D) representations of the subsurface composed of cells (volume elements, or voxels) that have some x–y–z dimension and can be assigned multiple properties such as porosity, permeability, saturation,

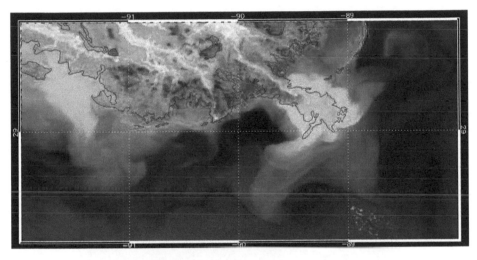

FIGURE 8.3
SeaWIFS highlights sediment plumes off the Mississippi River delta, Louisiana. (Courtesy of U.S. Geological Survey.)

FIGURE 8.4
Longshore currents move sand on barrier islands and sandbars. Image of Galveston Bay and barrier bars. (True color image courtesy of TerraMetrics and Google Earth, © 2012 Google.)

and/or facies. Reservoir models are used to generate multiple 3D representations of the subsurface, for example, showing hydrocarbon reservoir distribution in order to direct drilling or calculate volumetrics. Input data come from seismic volumes (depth surfaces, faults) and well logs. Seismic data generally lack the spatial resolution desired by a geologist, whereas wells are generally widely spaced and sample only a small volume. Something more is required to extrapolate geobodies in the space between wells and with finer detail than is available on seismic data. The modern analog is ideal for this role. Geologists use imagery of modern systems to derive the geometry, scale, and juxtaposition relationships of facies such as meander belts (Buehler et al., 2011; Rittersbacher et al., 2011), lacustrine deltas (Li et al., 2012), coastal plains (Leckie, 2003) (Figure 8.5), and carbonate banks (Grammer et al., 2004; Eisinger and Jensen, 2010) (Figure 8.6). The Brazos River, Texas, and the Yazoo River, Mississippi, have been used as analogs for meandering channel systems such as those found in the Cretaceous McMurray Formation oil sands reservoir of Alberta, Canada

FIGURE 8.5
Coastal plain, coastal marsh, and tidal flats north of Hilton Head, South Carolina. (True color image courtesy of TerraMetrics and Google Earth, © 2012 Google.)

FIGURE 8.6
Carbonate lagoon, Playa del Carmen, Mexico. (True color image courtesy of TerraMetrics and Google Earth, © 2012 Google.)

(Figure 8.7). Willapa Bay, Washington, is used as an analog to ancient estuaries (Figure 8.8). The coastal marshes of South Carolina are used as analogs for tidal flat environments. Dune fields in the Sahara are used as an analog to the Jurassic Navajo and Nugget eolian reservoirs of the western United States (Figure 8.9). Modern analogs are not only derived from aircraft or satellite imagery; new techniques such as Lidar-hyperspectral fusion are merging airborne imagery and ground- or helicopter-based Lidar scans of outcrops (Enge et al., 2007; Buckley et al., 2010; Rittersbacher et al., 2011). Geometric information derived from the Lidar includes slope, aspect, size, and location. These data are then used as realistic input to constrain an integrated geometric, geophysical, and geological interpretation of a reservoir system. The result is one or more computer-based realizations of the subsurface that explain the connectivity, spatial relationships, and fluid flow characteristics between facies (geobodies) and help the geologist understand the sedimentary and flow architecture of the units.

FIGURE 8.7
Channel scroll bars and meanders, Yazoo River near Yazoo, Mississippi. (True color image courtesy of USDA Farm Service Agency and Google Earth, © 2012 Google.)

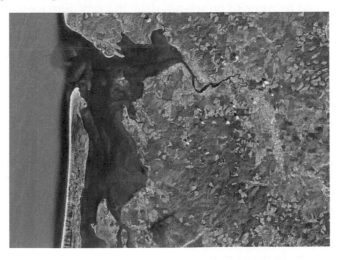

FIGURE 8.8
Willapa bay is used as an analog for estuaries, as in the Cretaceous McMurray Formation of Alberta. (True color image courtesy of TerraMetrics and Google Earth, © 2012 Google.)

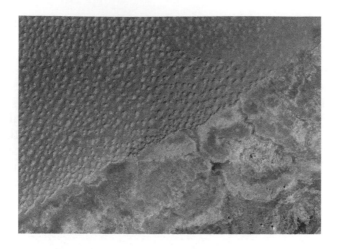

FIGURE 8.9
Dune fields such as those seen along the east edge of the Grand Erg Oriental of Algeria are used as analogs for the Jurassic eolian sandstone reservoirs of the Navajo and Nugget Formations, western United States. (True color image courtesy of CNES/SPOT Image and Google Earth, © 2012 Google.)

Another use for modern analogs as seen on imagery is to assist in the identification of features (geobodies) interpreted on 3D seismic. Rabelo et al. (2007) describe using geometries derived from meandering rivers in Surinam, particularly meander loops and point bars, to interpret high-resolution seismic data. The various interpretations thus generated are used to create stochastic (rather than deterministic) reservoir models.

The usefulness of modern analogs is only as good as the appropriateness of the modern setting to the ancient depositional environment. If the wrong analog is used, the conclusions drawn from the analog will likely be incorrect. As well, although we are taught that the present is the key to the past, we also know that present processes are not in all cases the same as past processes because of evolution of plants, animals, and the atmosphere; changing climate; and so forth. Finally, the process of incorporating the analog into seismic interpretations or geomodels also involves making assumptions about things that are not seen in outcrop, on imagery, or on seismic data. In particular, we usually carry a mental model that allows us to merge all aspects of the geology of an area. That mental model, itself subject to revision, may not be accurately transferred to a computer model. Once again numerous assumptions must generally be made about which algorithm provides the most realistic outcome. Limitations of the software (such as the number of cells or amount of upscaling required) can introduce artifacts to any model. All of these limitations should be borne in mind when using analogs for building reservoir geomodels.

Storm Deposits

Storms cause flood deposits, overwash fans, gravity flows, and beach erosion. Large amounts of coarse sediment are moved into mud-dominated environments. Unstable slopes fail. Fluvial channels shift, whereas meanders become cut off and abandoned. Barrier bars are rearranged, moved, or eliminated. Delta lobes may avulse and become abandoned and chenier ridges will shift or disappear (Figure 8.10). Shoreface deposits can be moved offshore because of the larger waves associated with storms (erosion down to "storm base" depths as opposed to normal "wave base"). Storm surges can cause large-scale vegetation damage and transport large amounts of coarse clastic and organic material into deep water environments. Modern analogs can be used to assist in reconstructing

FIGURE 8.10
Abandoned LaFourche delta lobe and chenier ridges, Louisiana coast. (True color image courtesy of TerraMetrics, Digital Globe, and Google Earth, © 2012 Google.)

ancient storm settings and deposits. Remote sensing imagery allows one to determine quantitatively as well as qualitatively the effects of storms on deposition, erosion, and landforms; and determine the degree of inundation and plant mortality (Liu, 2004).

Imagery presents a unique perspective on storm-related changes in that it provides spatial context, scale, and associations. Visible, near infrared, hyperspectral, and Lidar-based digital elevation model (DEM) images offer the means to map the distribution of geomorphic, topographic, hydrologic, and vegetation changes associated with storms. Change mapping based on multitemporal images reveal the extent of geomorphic, bathymetric, and floral changes. Vegetation indices and land cover classifications provide a visual (qualitative) and measurable (quantitative) determination of the distribution and diversity of plant communities before and after storms (Bianchette, 2007). The redistribution of sediments (e.g., coarse sand layer in a barrier lagoon, gravity flows) and changing landforms (e.g., channel scours, slumps) can be used to assist in the interpretation of the geologic record on the outcrop, in well logs and core, and in seismic lines or volumes.

Erosional Environments

As with depositional settings, modern examples of fluvial erosion, coastal erosion, and mass wasting processes can be used to interpret what is seen in the rock record. Areas subject to active erosion reveal the kinds of landforms associated with denudation: steep cutbanks, meander scars, incised lowstand valleys, breached levees, tidal inlet channel scours, and wave cut terraces are all examples evident on imagery, bathymetry, and DEMs (Figure 8.11). Their stratigraphic expression includes scours, unconformities, and coarse clastics (breccias, rip-up clasts, conglomerates, and turbidites). Change detection techniques can be applied to show surface differences that have occurred over various time periods. These are particularly apt for mapping the extent and scale of channel switching, floodplain erosion, and coastal erosion (e.g., Mars and Houseknecht, 2007).

Erosion susceptibility mapping is an important aspect of land condition monitoring (Thomas, 2001). In South Australia, imagery is used to monitor wind and water erosion,

FIGURE 8.11
Wave cut terraces, Portuguese Point, California. This tectonically emergent coastline has an uplifted wave cut bench and is currently undergoing active wave erosion along the sea cliffs. (Dr. Bruce Perry, Department of Geological Sciences, CSU Long Beach. http://geology.campus.ad.csulb.edu/people/bperry/AerialPhotosSoCal/PortugueseBendtoLongPoint.htm.)

assist in modeling erosion susceptibility, and determine areas prone to erosion. Soil type, moisture, acidity, salinity, slope, vegetation cover, and land use (primarily grazing patterns) are determined, and the amounts and types of erosion and conditions favorable for erosion are predicted. Benefits of using imagery for erosion susceptibility programs include instantaneous coverage of large areas, consistency, and repeatability.

Structural Analogs

Imagery showing examples of geologic structures can be used as analogs during exploration for subsurface structural traps, to guide building geomodels so that they incorporate realistic structural styles, and to help interpret geologic maps and seismic sections.

Rifting along the present-day East African Rift, Gulf of Suez onshore, and Basin-and-Range province as revealed on imagery has been used as analogs to assist mapping the distribution of structures, source rock, and reservoir rock in the North Sea, the North American mid-Continent rift, and the Gulf of Suez offshore, among other rift settings (Figure 8.12). For example, the Gulf of Suez rift has been intensively studied by academics and oil companies as an analog for rift basins in general. This is due not only to the good exposure of the onshore part of the rift, but also to the existence of oil fields and subsurface well and seismic data within the rift itself (Nelson et al., 1992; Landon, 1994).

Overlapping normal faults and associated relay zones and structural ramps in the Permian Cedar Mesa Sandstone of The Grabens, Canyonlands National Park, Utah, and of the Eocene Roda Sandstone in the Spanish Pyrenees have been used as analogs in building structural reservoir models (Figures 5.43 and 8.13). These models are subsequently used to simulate flow between injector and producer wells in similar subsurface settings (Enge et al., 2007).

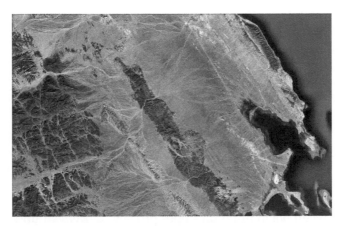

FIGURE 8.12
Rifting expressed as northwest-trending blocks near Hurghada and Gebel Zeyt, Egypt. Rifts such as the Gulf of Suez or Rio Grande rift are tectonic analogs to the North Sea basin. (True color image courtesy of CNES/SPOT Image and Google Earth, © 2012 Google.)

FIGURE 8.13
Relay ramp near Delicate Arch, Arches National Park, Utah. (a) View east up the relay ramp, center of the figure. The ramp is bounded on both sides by normal faults that die out along the ramp. (b) TerraMetrics/Google Earth image showing the same relay ramp (arrow) between two east–west faults. Hydrocarbons use such ramps in the subsurface to migrate from downthrown blocks to upthrown blocks. (True color image courtesy of Google Earth, © 2012 Google.)

Similarly, any present-day structural setting (strike-slip deformation along the San Andreas system; thrusting in the Himalayas) can be used as analogs to interpret and understand ancient structural settings.

Analogs for Hazards Planning

Mass wasting is particularly important in areas with heavy cultural development such as housing tracts or road networks. Landslides and downslope soil creep affect traffic, housing, dams, and other infrastructure. Imagery with examples of active mass wasting can help transportation departments and developers determine when not to cut into the toe of an active slide, or when not to build on slopes prone to movement under the influence of gravity or seismic shocks (Figure 8.14).

Imagery showing the aftermath of seismic events, including tsunamis, allows determination of intensity and extent of destruction. This kind of information is essential for planning urban development, building defensive structures, and safe placement of infrastructure, among others (see Chapter 16).

Imagery of modern active volcanoes provides analogs to volcanic activity near areas of human habitation. Mapping the distribution of pyroclastic flows, mudflows, and ash falls from analogous eruptions allows urban disaster planners and emergency responders to determine the potential extent of destruction and disruption and prepare an appropriate response (Figure 8.15).

FIGURE 8.14
Aerial view of the 1995 La Conchita slide and debris flow along highway 101 south of Santa Barbara, California. Note the scoop-shaped trace of the bright escarpment and the tongue-like debris flow. An older, overgrown escarpment appears near the top, left. (Photograph by R.L. Schuster, U.S. Geological Survey.)

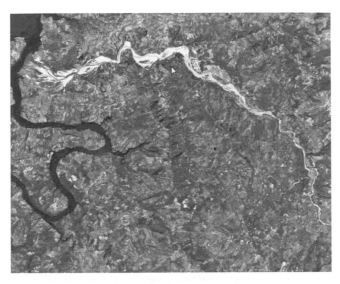

FIGURE 8.15
Volcanic debris flow (white, in river valley; see arrow) from the 1982 eruption of El Chichón volcano, Chiapas, Mexico, can still be seen 30 years later. (True color image courtesy of CNES/SPOT Image, GeoEye, and Google Earth, © 2012 Google.)

Planetary Analogs

The use of Earth analogs for interpretation of planetary features goes back to the arguments regarding the volcanic versus impact origin of craters on the moon, and the arguments still resound regarding features on Earth (see, e.g., Meteorite.org, online; Utah Geological Survey, online; Buchner and Kenkmann, 2007). Lava fields near Flagstaff, Arizona, and on Hawaii are considered good enough analogs to the moon that they were used by Apollo astronauts training to use the lunar rover (Lunar and Planetary Institute, online; NASA a, online).

Using similar lines of thought, cold and dry arctic environments on Svalbard (Norway) and Devon Island (Canadian high arctic) have been used to train astronauts, field test equipment, and study landforms expected to be encountered on Mars (Figure 8.16) (NASA b, online; NASA c, online; Europlanet online). Utah is a particularly alluring location for Martian analogs (Bowen et al., 2007; Chan et al., 2011). The arid climate combined with eolian, fluvial, glacial, and evaporite environments, along with volcanic landforms and impact craters, make it a fertile landscape for planetary analogs. Soil development, diagenetic concretions, patterned ground, crusts, and seepage features provide clues to similar features seen on Mars missions. Hematite concretions in the Navajo Sandstone have been suggested as analogs for potentially similar diagenetic systems identified on Mars using hyperspectral instruments (Chan et al., 2004, 2005). Other areas have also laid claim to being Martian analogs. The Flinders Ranges in South Australia provide rugged terrain in an arid climate with a diverse regolith showing evidence of glaciation, asteroid impacts, ancient and modern hydrothermal systems, and mineral alteration. For these reasons, it is considered a fitting analog for hyperspectral studies of Martian environments (Thomas et al., 2012). Not only are the landscapes appropriate, but the rocks record the evolution of Earth's atmosphere and biosphere during the Cryogenian, Ediacaran, and Cambrian periods. Dry lake deposits in the Badwater Basin of Death Valley, California, have been studied as terrestrial analogs to Martian evaporite deposits (Baldridge and Farmer, 2004).

FIGURE 8.16
Haughton Crater, Devon Island, Canada (circular dark area, center) is being used by NASA for Mars living experiments. The island is vegetation free, very cold, and contains this 39 million year old, 23 km diameter impact crater. (True color image courtesy of TerraMetrics and Google Earth, 2012.)

The high-resolution MODIS/ASTER airborne simulator (MASTER) imagery over Death Valley was degraded to the spatial resolution of the Mars Global Surveyor thermal emission spectrometer and Mars Odyssey thermal emission imaging system, both of which have been mapping Mars from orbit. The simulator has been used to detect sulfates and carbonates: MASTER was found to correctly map these soil constituents in Death Valley 90% of the time, lending confidence to the Martian spectrometer analyses.

Images of the Kenyan portion of the East African Rift, the Afar region, and Central European Rift have been used as analogs to help interpret the Tempe Fossae rift on Mars (Hauber et al., 2010). These Martian features have lengths of up to 1400 km, widths in tens of kilometers, and depths to 3000 m. Tempe Fossae is characterized by segmentation with changing style, development of asymmetric basins with changing polarity, and accommodation zones between basins. Most of these Martian structures are associated with volcanism. Based on these observations and a comparison to terrestrial continental rifts, these apparent grabens are interpreted as rift systems. The radial orientation of some rifts with respect to the Tharsis magmatic province is consistent with rifting related to regional extensional stress. Modern tectonic and erosion features on Earth can be and have been used as analogs for the terrestrial planets (Mercury, Venus, and Mars) as well as for ancient Earth.

References

Baldridge, A.M., J.D. Farmer. 2004. Mars remote-sensing analog studies in the Badwater Basin, Death Valley, California. *J. Geophys. Res.* 109(E12006): 18 p.

Bianchette, T. 2007. Using hurricane Ivan as a modern analog in paleotempestology: lake sediment studies and environmental analysis in Gulf Shores, Alabama. MSc. Thesis. Louisiana: State University: 124 p.

Bowen, B.B., B.A. Martini, M.A. Chan, W.T. Parry. 2007. Reflectance spectroscopic mapping of diagenetic heterogeneities and fluid-flow pathways in the Jurassic Navajo Sandstone. *Bull. Am. Assn. Pet. Geol.* 91: 173–190.

Boyd, R., K. Rumming, T. Boyd, P. Davies, B. Eyre, A. Short. 2007. The physical environment. In: Rule, M., A. Jordan, A. McIlgorm (eds.), *The Marine Environment of Northern New South Wales: A Review of Current Knowledge and Existing Databases*. Coffs Harbour, Australia: National Marine Science Centre: Northern Rivers Catchment Management Authority: 5–41.

Buchner, E., T. Kenkmann. 2007. Upheaval Dome, Utah, USA: impact origin confirmed. *Geology* 36: 227–230.

Buckley, S.J., E. Schwartz, V. Terlaky, J.A. Howell, R.W. Arnott. 2010. Combining aerial photogrammetry and terrestrial Lidar for reservoir analog modeling. *Photogram. Eng. & Rem. Sen.* 76: 953–963.

Buehler, H.A., G.S. Weissmann, L.A. Scuder, A.J. Hartley. 2011. Spatial and temporal evolution of an avulsion on the Taquari River distributive fluvial system from satellite image analysis. *J. Sed. Res.* 81: 630–640.

Chan, M.A., K. Nicoll, J. Ormö, C. Okubo, G. Komatsu. 2011. Utah's geologic and geomorphic analogs to Mars—an overview for planetary exploration. In: Garry, W.B., J.E. Bleacher (eds.), *Analogs for Planetary Exploration*, v. 483. Geological Society of America Special Papers: Boulder, CO: 349–377.

Chan, M.A., B. Beitler, W.T. Parry, J. Ormö, G. Komatsu. 2004. A possible terrestrial analogue for hematite concretions on Mars. *Nature* 429: 731–734.

Chan, M.A., B.B. Bowen, W.T. Parry, J. Ormö, G. Komatsu. 2005. Red rock and red planet diagenesis: comparisons of Earth and Mars concretions. *Geol. Soc. Am., GSA Today* 15: 4–10.

Enge, H.D., S.J. Buckley, A. Rotevatn, J.A. Howell. 2007. From outcrop to reservoir simulation model: workflow and procedures. *Geosphere* 3: 469–490.

Eisinger, C., J. Jensen. 2010. Object modeling for reservoir characterization in carbonates: CSPG GeoCanada 2010, extended abstract, Canadian Society of Petroleum Geologists 2010 Annual Meeting, Calgary, AB: 4 p.

Europlanet, Svalbard. A Terrestrial Cold-Climate Analog for Mars. http://europlanet.dlr.de/node/index.php?id=484 (accessed 21 Oct. 2012).

Grammer, G.M., P.M. Harris, G.P. Eberli. 2004. Integration of outcrop and modern analogs in reservoir modeling: overview with examples from the Bahamas. *AAPG Memoir* 80: 1–22.

Harris, P.M., W.S. Kowalik. 1994. Satellite Images of Carbonate Depositional Settings—Examples of Reservoir- and Exploration-Scale Geologic Facies Variation, v. 11. AAPG Methods in Exploration Series: Tulsa, OK: 147 p.

Harris, P.M. 2010. Delineating and quantifying depositional facies patterns in carbonate reservoirs: insight from modern analogs. *Bull. Am. Assn. Pet. Geol.* 94: 61–86.

Harris, P.M., S.J. Purkis, J. Ellis. 2011. Analyzing spatial patterns in modern carbonate sand bodies from Great Bahama Bank. *J. Sedimentary Res.* 81: 185–206.

Hauber, E., M. Grott, P. Kronberg. 2010. Martian rifts: structural geology and geophysics. *Earth Planet. Sci. Lett.* 294: 393–410.

Landon, S.M. 1994. Interior rift basins. *AAPG Memoir* 59: 276 p.

Leckie, D.A. 2003. Modern environments of the Canterbury Plains and adjacent offshore areas, New Zealand—an analog for ancient conglomeratic depositional systems in nonmarine and coastal zone settings. *Bull. Canadian Petroleum Geol.* 51: 389–425.

Li, S., X. Yu, S. Li. 2012. The architecture of braided deltas in modern Daihai Lake, northern China: implications for 3-D sedimentation models of rift lakes (abs.): *AAPG Search and Discovery* Article 50603, Long Beach, CA: Am. Assn. Petr. Geols. Annual Convention: 1 p.

Liu, K. 2004. Paleotempestology: geographic solutions to hurricane hazard assessment and risk prediction. In: Janelle, D., B. Warf, K. Hansen (eds.), *WorldMinds: Geographical Perspectives on 100 Problems*. Dordrecht: Kluwer Academic Publishers: 443–448.

Lunar and Planetary Institute. Black Point Lava Flow, Arizona. http://www.lpi.usra.edu/lunar/analogs/blackpoint/ (accessed 21 Oct. 2012).

Mars, J.C., D.W. Houseknecht. 2007. Quantitative remote sensing study indicates doubling of coastal erosion rate in past 50 yr along a segment of the Arctic coast of Alaska. *Geology* 35: 583–586.

Meteorite.org. Upheaval Dome. http://meteorite.org/upheaval-dome.shtml (accessed 21 Oct. 2012).

NASA a. Moon/Mars Analog Mission Campaigns. http://www.nasa.gov/centers/ames/research/technology-onepagers/moonmars.html (accessed 21 Oct. 2012).

NASA b. Arctic Mars Analog Svalbard Expedition. http://www.nasa.gov/mission_pages/mars/news/amase/index.html (accessed 21 Oct. 2012).

NASA c. Haughton-Mars Project (HMP) Devon Island. http://www.nasa.gov/centers/ames/research/expeditions/haughton.html (accessed 21 Oct. 2012).

Nelson, R.A., T.L. Patton, C.K. Morley. 1992. Rift-segment interaction and its relation to hydrocarbon exploration in continental rift systems. *Bull. Am. Assn. Pet. Geol.* 76: 1153–1169.

Purkis, S.J., B.M. Riegl, S. Andréfouët. 2005. Remote sensing of geomorphology and facies patterns on a modern carbonate ramp (Arabian Gulf, Dubai, U.A.E.). *J. Sediment. Res.* 75: 861–878.

Rabelo, I.R., S.M. Luthi, L.J. van Vliet. 2007. Parameterization of meander-belt elements in high-resolution three-dimensional seismic data using the GeoTime cube and modern analogs (abs.). In: Davies, R.J., H.W. Posamentier, L.J. Wood, J.A. Cartwright (eds.), *Seismic Geomorphology: Applications to Hydrocarbon Exploration and Production*. Geological Society of London: Geological Society Special Publication: 121–137.

Rankey, E.C. 2002. Spatial patterns of sediment accumulation on a Holocene carbonate tidal flat, northwest Andros Island, Bahamas. *J. Sediment. Res.* 72: 591–601.

Rittersbacher, A., J.A. Howell, S.J. Buckley. 2011. Using helicopter-based laser scanning to analyze controls on fluvial channel belt architecture in the Blackhawk Formation, eastern Utah, USA. *Geophys. Res. Abs.* 13: EGU2011–11432.

Suter, J.R. 2006. Facies models revisited: clastic shelves. *SEPM Special Pub.* 84: 339–397.

Suter, J.R., H.E. Clifton. 1999. The Shannon Sandstone and isolated linear sand bodies: interpretations and realizations. *SEPM Special Pub.* 64: 321–356.

Thomas, M. 2001. *Remote Sensing in South Australia's Land Condition Monitoring Project*. Adelaide, Australia: Government of South Australia: 102 p.

Thomas, M., J.D.A. Clarke, V.A. Gostin, G.E. Williams, M.R. Walter. 2012. The flinders ranges and surrounds, South Australia: a window on astrobiology and planetary geology. *Episodes* 35: 226–235.

Utah Geological Survey. Utah's Belly Button, Upheaval Dome. http://geology.utah.gov/surveynotes/geosights/upheaval_dome.htm (accessed 21 Oct. 2012).

9

Remote Sensing in Petroleum Exploration

Chapter Overview

The purpose of this chapter is to review some examples where remote sensing has contributed to the discovery of hydrocarbons. Case histories are given for basins in different stages of exploration maturity, for structural and stratigraphic traps, and using different sensor systems. Of necessity this is an incomplete list, and readers are referred to the literature for more extensive readings. It is important to repeat here that remote sensing in and of itself is not a stand-alone exploration tool. None of the discoveries reviewed here are exclusively the result of a remote sensing/photogeologic study. Indeed, in most exploration programs the contribution of remote sensing comes at the start of the program; that is, it is used to define an area wherein to stake claims, to acquire leases, or to design follow-up surveys. Therefore, although there are some cases of discoveries made using only remote sensing surveys, in most cases remote sensing is but one early step in a long process of integrating multiple data sets. The case histories herein are culled from the remote sensing and industry literature and are meant to serve as examples of different approaches to exploration programs.

Frontier Petroleum Exploration

Case History: Offshore Barbados Petroleum Exploration

In 1996, ConocoPhillips acquired a license to explore offshore Barbados (Dolan et al., 2004). The area lies within the 20 km thick Barbados accretionary prism associated with the Lesser Antilles arc at the leading edge of the Caribbean tectonic plate. Neogene turbidites had deposited reservoir sands in an area distal to the proto-Orinoco delta. Folding and thrusting provide trapping structures. Source rocks equivalent to the Upper Cretaceous La Luna facies of Venezuela would have generated oil and/or gas.

Several data sets were used in this evaluation, among them high-resolution multibeam sonar, backscatter sonar, synthetic aperture radar (SAR), two-dimensional (2D) and three-dimensional (3D) seismic, and drop cores (Figure 9.1). Sea bottom seeps were expected to have a different backscatter signal than the normal seabed because of buildup of carbonates around chemosynthetic communities. Multibeam bathymetry would pinpoint these anomalous areas. Drop cores provided positive identification of sea bottom seeps and geochemical analysis showed the oil and gas seeps to be mature, thermogenic hydrocarbons

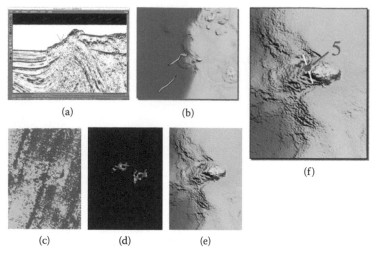

FIGURE 9.1
Several types of data were integrated to select the drop core locations, offshore Barbados. (a) Seismic data showing the possible conduit for seep. (b) Slicks from SAR. (c) Raw backscatter. (d) Processed backscatter. (e) Merged backscatter and multibeam bathymetry. (f) Final location for drop core survey. (From Dolan, P. et al., *AAPG International Conference and Exhibition*, October 24–27, Cancun, Mexico, 2004.)

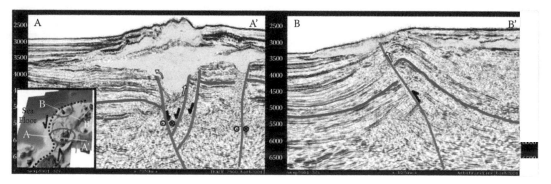

FIGURE 9.2
Seismic profiles illustrating mud volcanoes (A-A′) and seabed shale flows (B-B′). (From Dolan, P. et al., *AAPG International Conference and Exhibition*, October 24–27, Cancun, Mexico, 2004.)

derived from Type II (marine algal) kerogen. SAR mapped recurring sea surface slicks related to seepage. Dredging recovered fine-grained quartz sandstone along the reservoir subcrop. 2D seismic lines revealed the structure to be thrusted folds, and a prospect-specific 3D survey suggested that the reservoir consists of deepwater sheet sands.

Seismic revealed multiple episodes of folding and thrusting with northeast–southwest fold axes and northwest–southeast transpressional faults consistent with east–west shortening (Figure 9.2). Seal integrity is paramount in areas with multiple, complex deformation episodes. Conformance of seismic amplitudes to structures was considered the best indicator of seal integrity. Because oil was preferred, and due to relatively low temperatures in the section, presence of biogenic gas was considered the highest risk.

In 2002, the Sandy Lane-1 was drilled on the best prospect. Though not a commercial discovery, this well confirmed the presence of thick, well-developed sands in sheet-turbidite fans. No oil shows were seen, but methane was encountered: stable carbon isotopes

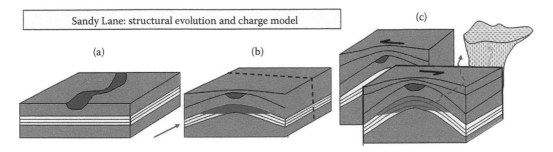

FIGURE 9.3
The post-well structural interpretation of the Sandy Lane prospect. (a) Early pliocene, (b) Late pliocene-early pleistocene, and (c) Late pleistocene. (From Dolan, P. et al., *AAPG International Conference and Exhibition*, October 24–27, Cancun, Mexico, 2004.)

confirmed low saturation thermogenic gas on the structure. Post-well appraisal determined the most likely failure mechanism was fault breaching of the fold trap that allowed hydrocarbons to leak off (Figure 9.3). And yet the authors feel that integration of swath bathymetry and backscatter sonar to identify sea bottom seeps, combined with SAR identification of surface slicks, drop core geochemistry and seismic interpretation of structure all provided insight to, and minimized the exploration risk of this structurally complex region.

Case History: Unraveling the Trap at Trap Springs, Nevada

The Trap Springs field in Railroad Valley, Nevada, was discovered by Northwest Exploration Company on the basis of a detailed study of a known field and possible analogs followed by photogeology, field work, and a seismic program (Figure 9.4). This is a classic example of photogeology playing a major role in the discovery of an oil field (Foster, 1979; Duey, 1979).

The geologists involved in this discovery began by analyzing the Eagle Springs field, discovered in Railroad Valley in 1954. Through analysis of wells and seismic data over the Eagle Springs field, they determined that the major factor controlling entrapment was the erosional and fault truncation of the reservoir units, that is, the Oligocene Garrett Ranch volcanics and the Eocene Sheep Pass Formation (Figure 9.5). This truncation occurred against a fault-bounded uplift at the valley margin.

Their work showed that the reservoir beds were present in a structural low, and that the structural highs along the basin margin were "bald" Paleozoic knobs that had been stripped of Tertiary reservoir units. This exploration model differed from previous (unsuccessful) exploration efforts in that all previous programs had drilled the structural highs in the mistaken belief that the Eagle Springs field was located on a high block rather than in the low adjacent to it. These other programs invariably drilled from basin fill immediately into Paleozoics.

Once they understood the trapping mechanism, Northwest Exploration set about mapping structures on airphotos and in the field. It turned out that the mountain front was not the location of the main basin-bounding fault, because the front had been eroded back and a pediment had formed. The faults could be seen on airphotos as linear tonal and vegetation anomalies. These anomalies are caused by moist soil and springs localized along the basin-bounding faults (Figure 9.6). Furthermore, the highs are bordered by faults that extended from the valley into the adjacent ranges (Figure 9.7). This system of faults, then, delineated a series of high and low blocks that were interpreted as analogs to the Eagle Springs field.

The photointerpretation was used to guide a field program that identified the mountain front and main graben-bounding faults. One problem the geologists encountered in

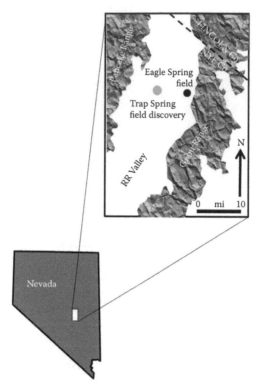

FIGURE 9.4

Index map of the study area at Railroad Valley, Nevada. (From Foster N.H., *Basin and Range Symposium and Great Basin Field Conference*, Denver, CO, Rocky Mountain Association of Geologists, 1979. Reprinted with permission of the Rocky Mountain Association of Geologists.)

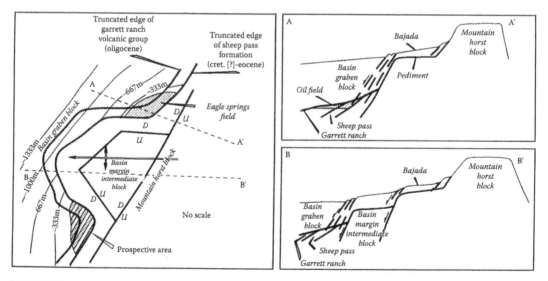

FIGURE 9.5

The Eagle Springs field is located on an intermediate block adjacent to a "bald" high. Cross section AA′ cuts across the Eagle Springs field; section BB′ cuts across the high block and shows how the reservoir has been removed there. (From Foster N.H., *Basin and Range Symposium and Great Basin Field Conference*, Denver, CO, Rocky Mountain Association of Geologists, 1979. Reprinted with permission of the Rocky Mountain Association of Geologists.)

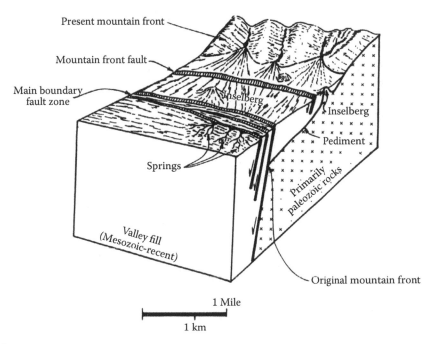

FIGURE 9.6
Block diagram showing that the mountain front in eastern Railroad Valley does not coincide with the main mountain front fault or main graben-bounding fault. (From Foster N.H., *Basin and Range Symposium and Great Basin Field Conference*, Denver, CO, Rocky Mountain Association of Geologists, 1979. Reprinted with permission of the Rocky Mountain Association of Geologists.)

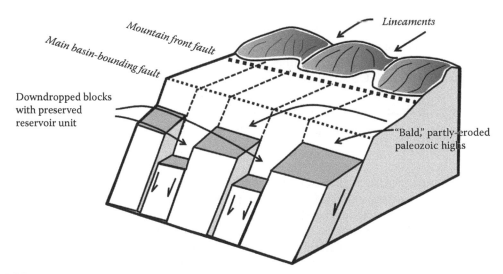

FIGURE 9.7
The downdropped fault blocks in Railroad Valley are broken along crosscutting faults that extend from the basin into the mountains. The reservoir units were eroded off the tops of the intermediate blocks but were preserved on downthrown blocks adjacent to them. (From Foster N.H., *Basin and Range Symposium and Great Basin Field Conference*, Denver, CO, Rocky Mountain Association of Geologists, 1979. Reprinted with permission of the Rocky Mountain Association of Geologists.)

the field was old lake shorelines that in some areas were confused with faults during the photointerpretation. They differentiated between these by noting that faults cut across topographic contours, whereas paleoshorelines followed the contours.

After field work and a preliminary seismic program, Northwest Exploration acquired the acreage. A further detailed seismic survey was obtained. The discovery well was drilled in November, 1976. About 14 million barrels of oil had been produced by the end of 2007 (Anna et al., 2007).

Mature Basin Petroleum Exploration

Case History: An Integrated Approach Brings New Life to an Old Basin

The following example is provided by Cara Hollis, NEOS GeoSolutions.

The onshore portion of the Ventura Basin, California, has been extensively explored, studied and produced since 1861. Approximately 100 oil and gas fields have been discovered, with cumulative production from the basin exceeding four billion barrels of oil equivalent. Although activity has slowed in recent decades, several E&P operators believe discoveries remain to be made and are working to identify by-passed targets masked by Ventura's complex geology, challenging topography, heavy cultural overprint, and environmental concerns.

Because of the challenging environment, ground-based seismic acquisition is not a practical alternative in most parts of the basin. In 2000, Ellis et al. described a program that used hyperspectral imagery in southern California for directly and indirectly mapping onshore hydrocarbon seeps. They were able to confirm seeps, soil mixed with oil, altered soil, and a unique vegetation assemblage associated with a heavily vegetated anticline. In 2011, Houston-based NEOS GeoSolutions acquired a 3330 km^2 (1300 mi^2) basin-scale airborne survey. Data from this survey were integrated with all available geological, geophysical, geochemical, and petrophysical data to provide an integrated, multimeasurement interpretation that is being used to identify new opportunities in this heavily explored basin.

To complement existing 2D seismic lines, well logs, and geological maps that are available in the public domain, NEOS acquired a series of new geophysical measurements, including

- Gravity: To delineate deep basin architecture and basin-scale structural features
- Magnetic: To map regional fault and fracture networks
- Hyperspectral: To detect oil seeps and microseepage impacts on surface vegetation

Additional data sets, such as radiometric and electromagnetic surveys, are also acquired on some basin-scale projects, though they were not in the Ventura Basin survey.

These new data sets delivered insights to the program's underwriters, even when interpreted individually. For instance, the gravity data (Figure 9.8) showcased the presence of a much deeper than expected sediment column in portions of the basin that had limited well penetrations. In addition, gravity measurements highlighted several unknown and untested minibasins along the edges of the main basin depocenter.

Analysis of the magnetic data provided further exploration insight. The vast majority of the fields in the Ventura Basin are aligned with identifiable magnetic anomalies that correspond to previously unmapped deep-seated fault systems (Figure 9.9), as might be expected in a tectonically active, structurally driven basin.

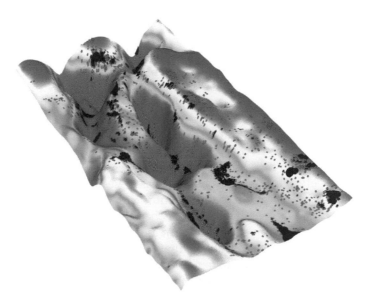

FIGURE 9.8

Gravity data indicate a much thicker than expected sediment column in portions of the Ventura Basin with sparse well penetrations. Gravity measurements highlight several minibasins along the margin of the main depocenter. Red colors are gravity highs; gravity lows are purple. Blue dots are dry holes; black dots are producing wells. (Courtesy of Cara Hollis and NEOS GeoSolutions.)

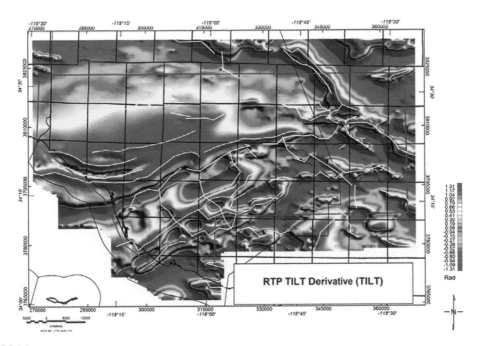

FIGURE 9.9

The majority of the fields in the Ventura Basin are associated with magnetic anomalies that correspond to probable deep-seated faults. Red areas are magnetic highs; purple areas are magnetic lows. White lines are previously unmapped faults interpreted from magnetic-derived data. (Courtesy of Cara Hollis and NEOS GeoSolutions.)

Analysis of hyperspectral imagery highlighted a large number of potential seeps, shown as red dots in Figure 9.10. A field program confirmed that the hyperspectral survey not only identified visible seeps, such as pools of oil at road cuts and in rivers, but also seeps that could not be seen with the naked eye. These "invisible" seeps are areas where hydrocarbons are mixed with the surface soils. The hyperspectral survey provided an accurate and robust seep map that was then correlated with faults and fractures that penetrate the near surface, confirming an active hydrocarbon system and providing insights into migration pathways and the potential charge (and discharge) of the reservoirs below. In addition to locating seeps, the hyperspectral data were used to perform a "geobotanical" analysis of vegetative health throughout the basin. The end member spectra illustrate the reflective difference between the healthy and stressed vegetation (Figure 9.11). These differences were assigned a color scale to document the distribution of healthy and stressed vegetation over the area. The green color shows healthy plants and the red color denotes stressed plants. This geobotanical survey provides valuable input into environmental baseline surveys. The analysis clearly shows that most vegetation stress in the basin is not caused by oil seeps but could be due to factors associated with soil or moisture changes or the impact of heavily populated areas surrounding the oil fields. Such insights may prove a valuable form of "insurance" in the future, should a landowner assert that vegetation is under stress because of oil field development activities versus other man-made or naturally occurring causes.

Armed with the location of new gravity-defined depocenters and minibasins, the ability to identify regional fault systems using magnetic data, and the presence of direct and indirect hydrocarbon indicators on the surface from hyperspectral imaging, explorationists are now able to qualitatively identify new leads and play types throughout the Ventura Basin.

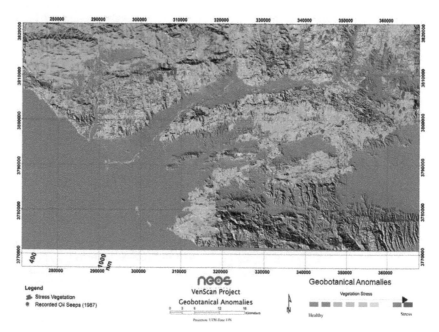

FIGURE 9.10

Hyperspectral imagery highlights vegetation stress. Seeps are shown as red dots. Data show that seeps are not the primary cause of vegetation stress. Urban and developed areas have been masked in gray. (Courtesy of Cara Hollis and NEOS GeoSolutions.)

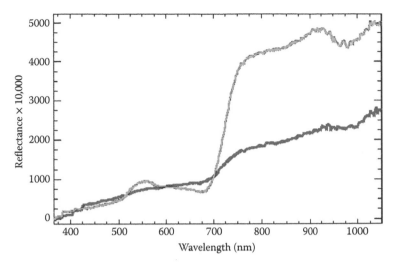

FIGURE 9.11

Reflectance spectra for healthy (green) and stressed (violet) vegetation, Ventura basin. Reflectance curves falling between these end members are assigned varying colors in Figure 9.10. (Courtesy of Cara Hollis and NEOS GeoSolutions.)

The multimeasurement methodology used to analyze the Ventura Basin provides analysts with an additional tool to aid in the search for hydrocarbons. This tool is based on a geostatistically driven, software-enabled search for the unique measurements and attributes that correlate with known fields (or high-production-rate wells) in a basin or designated area of investigation. Once these "correlative anomalies" (anomalies associated with known fields or wells) are identified, a pattern recognition algorithm can be used to identify similar anomalies in areas without well control. In essence, the algorithms search for unexploited parts of the basin that share the same set of gravity, magnetic, and hyperspectral features as observed at the area's known fields and highest producing wells. In the Ventura basin, nearly 50 raw and calculated attributes were considered, including items like Bouguer gravity and the first vertical derivative of reduced-to-pole magnetic data. This integrated, multimeasurement methodology applied to the Ventura data sets helped to determine the statistical relevance of each measurement and attribute, eliminate measurements and attributes that were not relevant, and mathematically determine weighting factors to apply to each statistically relevant data set and attribute. The result is an objective, mathematically driven map of the entire basin, highlighting areas that should be relatively more prospective or productive. As with any measurement or tool, the analyst does not blindly follow the output, but instead uses the insights provided to de-risk exploration concepts and to identify possible new leads.

NEOS applied its proprietary geostatistical methodology to four separate geological horizons that are producing today in the Ventura Basin, including the shallow Pico formation, the Sespe, the Miocene (effectively the Monterey Shale zone), and the deep Eocene. As one would expect, the shallow zones had more well penetrations and discoveries than the deeper intervals. Nonetheless, even the shallow Pico contains several high-potential exploration leads that have yet to be drilled (Figure 9.12). The results become even more interesting as one goes deeper, where the number of well penetrations into geostatistically identified anomalies is smaller and, therefore, the corresponding exploration potential is even higher.

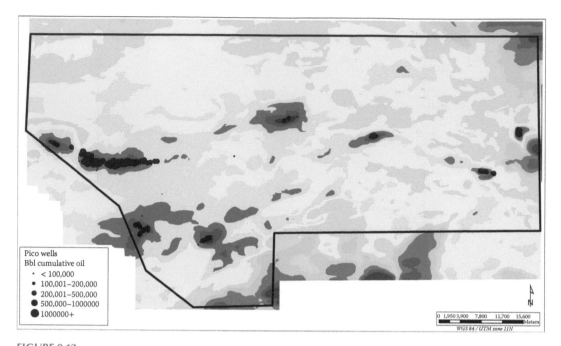

FIGURE 9.12

Geostatistical assessment of the probability of liquid hydrocarbons in the shallow Pico formation, Ventura County, California. Hot colors correspond to a high probability of oil. Black dots are wells producing from the Pico Fm. (Courtesy of Cara Hollis and NEOS GeoSolutions.)

The multimeasurement interpretation methodology used in the Ventura Basin is providing the region's operators with the additional data and insights needed to unleash a new wave of exploration. The underwriters of this multiclient shoot are using this information to plan higher resolution airborne surveys over new prospective areas, highlighted by the geostatistical analysis, in deeper less explored areas of the basin.

Case History: Stratigraphic Traps in the Paradox Basin, Utah

An attempt to understand the controls on petroleum accumulation and production in the Paradox Basin of southeastern Utah (Figure 9.13) led to the conclusion that most existing fields are stratigraphic traps, which appear to be structurally controlled in the sense that the main reservoir facies, Pennsylvanian algal mounds, are located along fault-controlled paleobathymetric hinge lines or along the margins of faulted blocks (Merin and Michael, 1985). This provided the incentive for an integrated Landsat/potential fields/field mapping study of major structures in the basin.

Landsat imagery was acquired and interpreted to help locate these reservoirs by mapping the regional structural framework and local structural details associated with producing fields and trends. The satellite structure map was compared to gravity and magnetics maps and existing isopach data to determine which surface features are basement-related. The major features were checked using field mapping to confirm or deny the interpretation. The goal was to test this technique and provide a strategy and process for future exploration efforts.

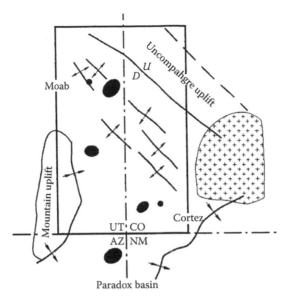

FIGURE 9.13
Index map of the Paradox Basin stratigraphic trap play area. Major fields are shown in black. (From Merin, I.S., Michael, R.C., *Proceedings of the Fourth Thematic Conference for Remote Sensing in Exploration Geology*, Ann Arbor, MI, Environmental Research Institute of Michigan, 1985.)

Pennsylvanian algal mounds in the Paradox Formation are lens-shaped, partly dolomitized carbonate muds with porosity enhanced both by subareal leaching during or shortly after deposition and later diagenetic dolomitization. These mounds grade laterally and vertically into tight micrites.

The Landsat images were interpreted at a scale of 1:250,000. All evidence of structure, including folds, faults, and fracture zones were mapped. The authors found that northeast and northwest lineaments (probable basement fault zones) are dominant throughout the area. The interpretation was checked by mapping structures in the field. Those lineaments that crossed outcrops were found in most cases to have either measurable offset or zones of lineament-parallel joints. Many of these zones were also found to coincide with gravity and magnetics lineaments, suggesting that they originated as basement offsets.

The results of this investigation show that individual carbonate mounds trend northwest, yet many are abruptly terminated or segmented by northeast-oriented structures (Figures 9.14 and 9.15). Isopach maps reveal that the algal buildups occur on northwest-oriented paleostructural highs, and that within the mounds individual pods may be oriented northeast. Several fields, including the Ismay, Bug, Papoose Canyon, Tricentral, Gothic Mesa, and Cache follow this pattern.

The methodology proposed as a result of this work is to first learn what controls the distribution of known production. In this case, stratigraphic traps were indirectly controlled by structure since the biohermal reservoirs are located on structurally controlled paleobathymetric highs. The next step was to determine the structural framework of the area by analyzing Landsat images and comparing the structure interpretation to isopach, gravity, and magnetic maps to learn where the bathymetric highs had been and to provide confidence in the satellite mapping. This step is used to eliminate those structures less likely to have exerted control (facies or thickness changes) on the desired stratigraphic intervals. Field mapping in critical areas provided confidence that the features being mapped were indeed faults and fracture zones. A map showing the distribution of existing fields was used to identify those producing trends that could be extended into new areas, and to suggest areas without production that appear analogous to producing fields.

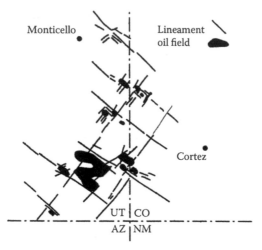

FIGURE 9.14
Regional lineaments and oil fields in the Paradox Basin area. (From Merin, I.S., Michael, R.C., *Proceedings of the Fourth Thematic Conference for Remote Sensing in Exploration Geology*, Ann Arbor, MI, Environmental Research Institute of Michigan. 1985.)

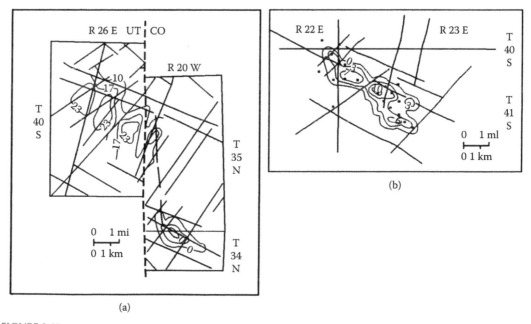

FIGURE 9.15
The correlation of isopachs and Landsat lineaments shows how the Pennsylvanian algal mounds appear structurally controlled. (a) Ismay isopach (in meters), Ismay field, and Cache field. (b) Lower Desert Creek isopach and Gothic Mesa field. (From Merin, I.S., Michael, R.C., *Proceedings of the Fourth Thematic Conference for Remote Sensing in Exploration Geology*, Ann Arbor, MI, Environmental Research Institute of Michigan, 1985.)

Resource Plays

Over the past decade tight sand, carbonate, and shale plays have become the new frontier in North American oil and gas and, increasingly, elsewhere. Between 2000 and 2011, shale gas production in the United States has grown to be approximately 30% of total dry gas production. Gas resources comprise about 827 TCF in the lower 48 states and another 611 TCF in Canada (Waldo, 2012). The major plays are the Eagle Ford, Barnett, Woodford, Fayetteville, Haynesville, Marcellus, Utica, Bakken, Niobrara, Lewis, Mesaverde, Montney, and Horn River. Some of these are tight sandstones, limestones, or siltstones rather than shales, but all have in common that they require fracture stimulation (Hemborg, 2000).

A review of the literature shows an interesting ambivalence toward the role of natural fractures in finding sweet spots in these plays. Historically, shale reservoirs were considered worthless unless naturally fractured zones could be found and exploited. These fractures were concentrated in zones around faults (the damage zone), along the crest of folds (zones of maximum curvature), and above karst collapse features (Steward, 2011). Seismic acquisition was focused on these areas of likely intense fracturing, and seismic data was processed to enhance attributes such as coherence and curvature (Elebiju et al., 2010). According to Dr. Mark Zoback of Stanford University, hydraulic fracturing can, in addition to inducing fractures, activate existing faults in a reservoir that, by slipping slowly, will contribute to production (Brown, 2013).

Another way of looking at these reservoirs has emerged recently. Some workers feel that the key attribute for a productive reservoir is maturity related to depth of burial, followed by composition (brittleness versus ductility), and especially the combination of interbedded or laminated brittle and ductile units (Slatt, 2011). Others say that the main drivers for well performance are stimulation technique, reservoir temperature gradient, and lack of intense faulting (Gilman and Robinson, 2011). In this view the presence of faults only acts to steal propant energy from the stimulation: the more faults, the more potential leak-off zones. Instead, they insist, one should look for unfaulted regions and drill long laterals with lots of hydraulic fracturing stages. Finally, one ought to consider the regional stress orientation and magnitude (active and residual), present day burial depth and its effect on vertical and confining pressures, and pore pressure.

At this time, the debate regarding the role of remotely identified fracture zones in resource plays remains unresolved. Regardless whether you believe that natural fractures help by providing the primary conduits for fluid movement, or hinder stimulation by draining hydraulic fracturing fluids, imagery and potential fields data can help you locate fractured zones.

Case History: Niobrara Fractured Carbonate–Shale Reservoir, Denver Basin, Colorado–Wyoming

The Cretaceous Niobrara Formation in the Denver Basin (northeast Colorado and southeast Wyoming) is a fractured reservoir and source rock that consists primarily of limestones, chalks, and interbedded calcareous shales (Figure 9.16). Total organic carbon averages about 5%, but can be as high as 8%. It is also an oil-prone, mature source rock that has been in the oil generation window since the Eocene. The Niobrara in the Denver Basin is 90–130 m thick, lies at depths between 1800 and 2100 m, and has matrix permeabilities generally less than 0.01 millidarcy (mD) and porosities less than 10%. This is a poor reservoir unless

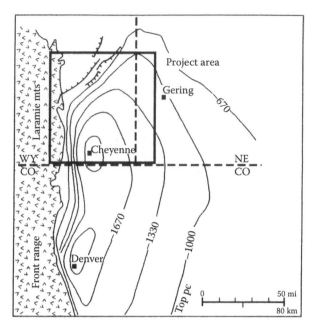

FIGURE 9.16
Index map of the fractured Niobrara play area, Colorado-Wyoming. Structure contour on the Precambrian. Contour interval = 330 m. (From Merin I.S., Moore, W.R., *Bull. Am. Assn. Pet. Geol.*, 70, 351–359, 1986.)

enhanced by fracturing. "Natural fracturing ... is the single most critical component of successful Niobrara oil production for all reservoir types ..." (Redden, 2012).

Three types of fracture-controlled oil production have been described for the Niobrara Formation: (1) production from fractures associated with flexures and folding along the west side of the basin (e.g., Loveland and Berthoud fields); (2) production from fractures in relatively undeformed parts of the basin (e.g., Silo field); and (3) production from randomly oriented and layer-confined "polygonal faults" that are restricted to the Niobrara and Pierre formations and are thought to comprise minor normal faults resulting from fluid expulsion and formation compaction (Sonnenberg, 2012). An understanding of the structural segmentation of the basin, rock mechanical properties, and present-day stress regime is critical to predicting prospective areas and well performance.

A program to define fractured production fairways within the basin used Landsat multispectral scanner images at a scale of 1:125,000 (Merin and Moore, 1986). The image interpretation was compared to structure contour, isopach, and resistivity maps. In addition, changing crustal compressive stresses were analyzed to better understand the fracture history of the basin.

Landsat imagery was examined and structure maps were made using standard photointerpretation techniques (Figure 9.17). Lineaments assumed to be surface indicators of fracture zones mostly consist of aligned stream segments, tonal patterns, and geomorphic features. The interpretation was later digitized and lineaments were plotted as length-weighted and frequency-weighted rose diagrams to help identify significant trends. These lineaments most likely represent extensions of previously mapped structures or previously unrecognized structures because the lineaments are parallel or subparallel to known faults and tectonic zones. The basin appears to be a mosaic of tectonic blocks bounded by northeast and northwest fault zones that can be mapped at the surface as lineaments.

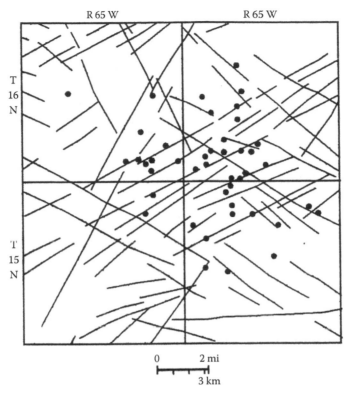

FIGURE 9.17
Distribution of Landsat lineaments and oil wells, Silo field area, Laramie County, Wyoming. (From Merin I.S., Moore, W.R., *Bull. Am. Assn. Pet. Geol.*, 70, 351–359, 1986.)

Well logs were examined to extract information for a basin-wide Niobrara resistivity map that shows where this reservoir unit is oil saturated (≥ 60 ohm contour) (Figure 9.18). The logs were used to make isopachs, and Niobrara thins were correlated with paleohighs and areas of increased organic carbon content (richer source rock) or higher geothermal gradient. Neither of these data sets correlated well with the surface fracture zones except along the northwest margin of the basin. In most cases you should not expect them to: what they show is the location of rich source rock.

The stress history of the basin reveals that early Laramide (Late Cretaceous through Paleocene) maximum horizontal compressive stresses were oriented east-northeast, whereas later Laramide (Eocene) compressive stresses were aligned northeast (Figure 9.19). Assuming that fractures in the basin are dominantly extensional, they likely formed parallel to the maximum compressive stress. As these stresses rotated through time successive new fractures formed as a result of natural hydraulic fracturing related to overpressures associated with generation of Niobrara oil. This model predicts that northeast fractures acted as conduits and formed storage capacity (fracture porosity) for an otherwise tight formation. The northeast fractures were preferentially open and filled with hydrocarbons during late Laramide east-northeast shortening. The best reservoirs, then, should be in highly fractured rock at the intersection of surface lineaments or along braided and intersecting northeast fracture zones.

A statistical study of producing wells versus distance to surface lineaments showed that 79% of oil-producing wells in the Niobrara are located within 0.16 km of a lineament, and

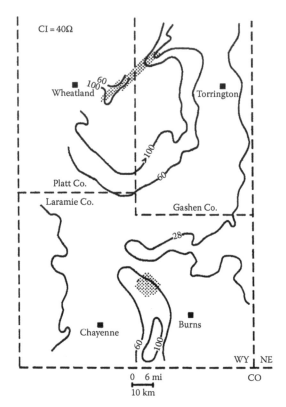

FIGURE 9.18
Resistivity map of the Niobrara. Contour interval = 40 ohm. Shading indicates areas where wells produce oil or have shows. (From Merin I.S., Moore, W.R., *Bull. Am. Assn. Pet. Geol.*, 70, 351–359, 1986.)

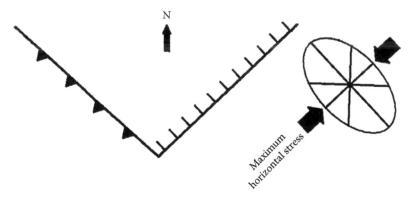

FIGURE 9.19
Orientation of late Laramide (Eocene) stresses in the Denver Basin, and probable fracture sets associated with these stresses. Northeast joints are most likely to be open and contain oil. (From Merin I.S., Moore, W.R., *Bull. Am. Assn. Pet. Geol.*, 70, 351–359, 1986.)

that about two-thirds are located along northeast lineaments. This is supported by core that shows oil is being produced from zones with open vertical fractures.

The results of this work suggest that productive fairways can be defined for naturally fractured reservoir plays using careful remote sensing structural interpretations in conjunction with subsurface data. To the extent that the surface lineaments reflect the boundaries of basement blocks, they may also reveal the structural segmentation of the basin and perhaps limits of differing styles of fracture-controlled production.

Case History: The Marcellus Fractured Shale Play

The "next big thing" in petroleum exploration is the resource play, that is, extracting oil and gas from fractured shales (source rock), tight siltstones, and tight sandstones. This has come about because of a revolution in drilling and hydraulic fracturing technology that permits extended reach horizontal wells to encounter large volumes of rock, followed by multistage massive hydraulic fracturing that exploits and props open natural fractures or, in their absence, generates open fractures that allow gas and oil to drain from low-porosity and low-permeability rock matrix.

An example of such a gas play is the Middle-Upper Devonian Marcellus Shale of the Appalachian basin in the northeastern United States. The Marcellus Shale consists of five facies: (1) argillaceous mudstone, (2) calcareous mudstone, (3) organic-rich mudstone, (4) calcareous organic-rich mudstone, and (5) fossiliferous limestone (Zhou et al., 2012). The organic-rich facies are thought to have been deposited in either deep (thousands of meters) or shallow (as little as 45 m) water under anoxic conditions (Carter et al., 2011). Organic-rich shales are the gas exploration targets: shales have up to 12% total organic carbon (Nyahay et al., 2008). Yet the low porosity (around 10%) and low permeability (>5 mD [Soeder, 1986]; 100–500 nD [Carter et al., 2011]) make stimulation of natural fractures and/or creation of artificial fractures necessary to drain the matrix.

There are three sets of joints in the Marcellus: the J1 (oldest) is parallel to the original east-northeast maximum horizontal compressive stress. This joint set has been rotated from vertical during subsequent folding. These joints are restricted mainly to organic-rich black shales and can have high densities. The J2 set is oriented northwest to north-northwest and is found preferentially in Marcellus gray shales and in coarser clastics as much as 2000 m above the Marcellus. The J3 set (youngest) is also oriented east-northeast but is vertical; that is, it has not been rotated by folding (Engelder et al., 2009; Carter et al., 2011; Engelder, 2011). The main productive J1 trend is thought to result from overpressures and natural hydraulic fracturing related to peak burial and generation of hydrocarbons in the organic-rich facies. The J2 trend is perpendicular to fold axes: whereas this trend was controlled by contemporary stress orientation, the location of J2 joints may have been a result of later, high-pressure gas plumes or leakage chimneys extending upward from the source rock (Engelder et al., 2009; Carter et al., 2011; Tan et al., 2012).

It appears that industry generally stays away from Marcellus fractures caused by folds and faults, unlike exploration in the Barnett Shale of the Fort Worth basin, Texas, which targets these features (Carter et al., 2011; Elebiju et al., 2010). The consensus in the Marcellus play is that primary flow is along joint sets (J1 and J3) parallel to present-day east-northeast maximum horizontal stress, and that the best completions take advantage of all joint sets. The most successful wells are drilled perpendicular to these fractures, that is, in a north-northwest or south-southeast direction. The result is that a horizontal well will encounter the maximum number of open J1 and J3 fractures by drilling in these directions. Once the J1 and J3 fractures are stimulated the J2 set joints that intersect them will start to open (Engelder et al., 2009; Carter et al., 2011). Microseismic testing supports this conclusion: in one case the first three of seven stages of hydraulic fracture stimulation showed northeasterly trends; stages four through six were perpendicular to the J1 and J3 trends (Williams-Stroud et al., 2012). In at least some areas, strong linear trends appear visible on imagery (Figure 9.20). Field mapping is needed to confirm these trends are related to jointing.

FIGURE 9.20
Strong north-northeast jointing between Lexington and Roxbury, New York, is expressed topographically as aligned ridges and drainages (arrow). (True color image courtesy of TerraMetrics and Google Earth, Copyright 2012 Google.)

FIGURE 9.21
Lineaments related to "cross-strike structural discontinuities" could be useful in gas exploration in the Appalachian overthrust belt (Wheeler, 1980). An east-southeast CSD (dotted line) is aligned perpendicular to structural trends near Chambersburg, Pennsylvania, as indicated by fold terminations, water gaps, and offset ridges. (True color image courtesy of TerraMetrics and Google Earth, Copyright 2012 Google.)

As suggested by Engelder et al. (2009), the productive trend may be known, but the location of dense fracture zones must be determined. Wheeler (1980) suggested that lineaments related to "cross-strike structural discontinuities" could be useful in gas exploration in the Appalachian overthrust belt (Figure 9.21). These discontinuities (fracture zones, faults) are more or less perpendicular to the strike of bedding and are thought to be related to basement faults that controlled tear faults and lateral ramps in the overthrust sheets. A study for the U.S. Department of Energy (1998) determined that an important aspect of detecting naturally fractured gas reservoirs was using remote sensing and nonseismic geophysical methods, particularly during regional exploration. Important information was available using imagery and aeromagnetics to locate regional linear structures (fracture zones), determine depth to basement, and map basement-controlled faults and fracture zones. Side-looking airborne radar (SLAR) has been shown to be an effective way to identify fracture zones that can localize groundwater and natural gas, and has been used to map the surface distribution of the Marcellus Shale based on its texture (Figure 9.22) (Altamura, 1999). The Rock Fracture Group, SUNY Buffalo, and Norse Energy Corp found

FIGURE 9.22
Side-looking radar image of part of the Valley and Ridge province of central Pennsylvania. Devonian units, including the Marcellus Shale, are deformed into thrusted anticlines and synclines. Cross-strike discontinuities (dotted line) and faults/joint zones (arrows) are apparent as structural breaks. (From Altamura, R.J., *Penn. Geology*, 30, 2–13, 1999.)

it useful to evaluate Landsat, SLAR, hyperspectral imagery, airphotos, DEMs, aeromagnetics and gravity to map lineaments (faults and fractures) in the Marcellus and Utica Shales of the northern Appalachian basin (Jacobi, 2011). These lineaments were correlated to joints and faults on seismic data, in outcrop, and in core. Elsewhere, integrated programs incorporating aeromagnetics, satellite and airphoto imagery, and SLAR have been used to explore for resource plays such as the MesaVerde tight sands of the Piceance basin, Colorado (Kuuskraa et al., 1996; URS Corporation, 2006).

References

Altamura, R.J. 1999. Geologic mapping using radar imagery in the Ridge and Valley province. *Penn. Geology* 30: 2–13.

Anna, L.O., L.N.R. Roberts, C.J. Potter. 2007. *Geologic Assessment of Undiscovered Oil and Gas in the Paleozoic–Tertiary Composite Total Petroleum System of the Eastern Great Basin, Nevada and Utah.* U.S. Geological Survey Digital Data Series DDS–69–L, 50 p.

Brown, D. 2013. *Looking deeper into fracturing's impacts.* AAPG Explorer, March 2013: p. 26.

Carter, K.M., J.A. Harper, K.W. Schmid, J. Kostelnik. 2011. Unconventional natural gas resources in Pennsylvania: the backstory of the modern Marcellus Shale play. *Environ. Geosci.* 18: 217–257.

Dolan, P., D. Burggraf, K. Soofi, R. Fitzsimmons, E. Aydemir, O. Senneseth, L. Strickland. 2004. Challenges to exploration in frontier basins—the Barbados accretionary prism. In: *AAPG International Conference and Exhibition*, October 24–27, Cancun, Mexico: 6 p.

Duey, H.D. 1979. Trap Spring Oilfield, Nye County, Nevada. In: G.W. Newman and H.D. Goode (eds.), *Basin and Range Symposium and Great Basin Field Conference.* Denver, CO: Rocky Mountain Association of Geologists: 469–476.

Elebiju, O.O., G.R. Keller, K.J. Marfurt. 2010. Investigation of links between Precambrian basement structure and Paleozoic strata in the Fort Worth basin, Texas, U.S.A., using high-resolution aeromagnetic (HRAM) data and seismic attributes. *Geophysics* 75: B157–B168.

Ellis, J.M., H.H. Davis, M.B. Quinn. 2000. Airborne hyperspectral imagery for the petroleum industry (abstract). *AAPG Search and Discovery* Article No. 90914: AAPG Annual Convention, New Orleans, LA: 1 p.

Engelder, T. 2011. The role of strain in controlling orientation of natural hydraulic fractures in gas shales (abstract). *AAPG Search and Discovery* Article No. 90124: AAPG Annual Convention, Houston, TX: 1 p.

Engelder, T., G.G. Lash, R.S. Uzcátegui. 2009. Joint sets that enhance production from Middle and Upper Devonian gas shales of the Appalachian basin. *Bull. Am. Assn. Pet. Geol.* 93: 857–889.

Foster, N.H. 1979. Geomorphic exploration used in the discovery of Trap Spring Oilfield, Nye County, Nevada. In G.W. Newman, H.D. Goode (eds.), *Basin and Range Symposium and Great Basin Field Conference*. Denver, CO: Rocky Mountain Association of Geologists: 477–486.

Gilman, J., C. Robinson. 2011. Success and failure in shale gas exploration and development: attributes that make the difference (abstract). *AAPG Search and Discovery* Article No. 80132, http://www.searchanddiscovery.com/documents/2011/80132gilman/ndx_gilman.pdf (accessed October 21, 2012).

Hemborg, H.T. 2000. *Gas Production Characteristics of the Rulison, Grand Valley, Mamm Creek, and Parachute Fields, Garfield County, Colorado.* Resource Series, Vol. 39. Denver, CO: Colorado Geological Survey Division of Minerals and Geology, Department of Natural Resources: 30 p.

Hollis, C. NEOS GeoSolutions. www.neosgeo.com (accessed October 21, 2012).

Jacobi, R. 2011. *Marcellus and Utica in the Field: Looking at Faults, Fractures and Folds That Affect the Sedimentary Units of the Northern Appalachian Basin.* AAPG online webinar: 147 p. http://www.pttc.org/aapg/marcellusutica.pdf (accessed 21 June 2013).

Kuuskraa, V., D. Decker, S. Squires, H.B. Lynn. 1996. Naturally fractured tight gas reservoir detection optimization: Piceance Basin. *The Leading Edge* 15: 947–948.

Merin, I.S., R.C. Michael. 1985. Application of structures mapped from Landsat imagery to exploration for stratigraphic traps in Paradox basin. In: *Proceedings of the Fourth Thematic Conference for Remote Sensing in Exploration Geology*. Ann Arbor, MI: Environmental Research Institute of Michigan: 183–192.

Merin, I.S., W.R. Moore. 1986. Application of Landsat imagery to oil exploration in Niobrara Formation, Denver Basin, Wyoming. *Bull. Am. Assn. Pet. Geol.* 70: 351–359.

Nyahay, R., J. Leone, L. Smith, J. Martin, D. Jarvie. 2008. *Update on the Regional Assessment of Gas Potential in the Devonian Marcellus and Ordovician Utica Shales in New York.* AAPG Eastern Section Meeting, Lexington, Kentucky, September 16–18, 2007: 68 p. http://www.searchanddiscovery.com/documents/2007/07101nyahay/images/nyahay.pdf (accessed October 21, 2012).

Redden, J. 2012. Niobrara shale—hoping complex play becomes next Bakken. *World Oil*: 233: 82–93.

Slatt, R.M. 2011. Important geological properties of unconventional resource shales. *Shale Shaker*. 62: 224–242.

Soeder, D.J. 1986. Geologic controls on gas production from Appalachian basin Devonian shales (abstract). *AAPG Search and Discovery* Article No. 91043: AAPG Annual Convention, Atlanta, GA: 1 p.

Sonnenberg, S.A. 2012. Polygonal fault systems—a new structural style for the Niobrara Formation, Denver Basin, CO (abstract). *AAPG Search and Discovery* Article No. 90142: AAPG Annual Convention, Long Beach, CA: 1 p.

Steward, D.B. 2011. The Barnett Shale oil model of north Texas (abstract). *AAPG Search and Discovery* Article No. 110151, http://www.searchanddiscovery.com/documents/2011/110151steward/ndx_steward.pdf (accessed October 21, 2012).

Tan, Y., T. Johnston, T. Engelder. 2012. Testing the gas plume hypothesis using fracture distribution above Marcellus (abstract). *AAPG Search and Discovery* Article No. 90142: AAPG Annual Convention, Long Beach, CA: 1 p.

URS Corporation. 2006. *Phase I Hydrogeologic Characterization of the Mamm Creek Field Area in Garfield County.* Board of County Commissioners, Garfield County, Colorado: 2-1 to 2-17.

U.S. Department of Energy. 1998. *Detection and Analysis of Naturally Fractured Gas Reservoirs: Summary and Synthesis*: Prepared by Blackhawk Geometrics Inc., Golden, CO: 45 p. http://www.boulder .swri.edu/~grimm/synth_summ_biblio.pdf (accessed 21 June 2013).

Waldo, D. 2012. A review of three North American shale plays: learnings from shale gas exploration in the Americas (abstract). *AAPG Search and Discovery* Article No. 80214, http://www .searchanddiscovery.com/documents/2012/80214waldo/ndx_waldo.pdf (accessed October 21, 2012).

Wheeler, R.L. 1980. Cross-strike structural discontinuities—possible exploration tool for natural gas in Appalachian overthrust belt. *Bull. Am. Assn. Pet. Geol.* 64: 2166–2178.

Williams-Stroud, S., R. Zhou, B. Hulsey. 2012. Fracture mechanics interpreted from stress inversion analysis on microseismic event source mechanisms in the Marcellus Shale (abstract). *AAPG Search and Discovery* Article No. 90142: AAPG Annual Convention, Long Beach, CA: 1 p.

Zhou, J., P. Rush, A. Sridhar, R. Miller. 2012. The anatomy of Middle Devonian Marcellus shale, Appalachian basin: a scheme of mudstone classification and its implications for shale gas exploration (abstract). *AAPG Search and Discovery* Article No. 90142: AAPG Annual Convention, Long Beach, CA: 1 p.

Additional Reading

Berger, Z., T.H.L. Williams, D.W. Anderson. 1992. Geologic stereo mapping of geologic structures with SPOT satellite data. Bull. *Am. Assn. Pet. Geol.* 76: 101–120.

Berry, J.L., G.L. Prost. 1999. Hydrocarbon exploration. In: A.N. Rencz (ed.), *Remote Sensing for the Earth Sciences, Manual of Remote Sensing*, Vol. 3. New York: John Wiley & Sons: 449–508.

Doeringsfeld, W.W. Jr., J.B. Ivey. 1964. *Use of Photogeology and Geomorphic Criteria to Locate Subsurface Structure.* Mountain Geologists, Vol. 1. Denver, CO: Rocky Mountain Association of Geologists: 183–195.

Ellis, J.M. 1986. *Geologic Interpretation of Northwestern PPL-18, Papua New Guinea.* Papua New Guinea Department of Mining and Petroleum Open-File Report F1/R86-126.

Lamerson, P.R. 1988. *Photogeologic Study of PPL-101.* Papua New Guinea Department of Mining and Petroleum Open-File Report F1/R/88-52.

Nishidai, T., J.L. Berry. 1981. Geologic interpretation and hydrocarbon potential of the Turpan basin (NW China) from satellite imagery. In: *Proceedings of Eighth Thematic Conference on Remote Sensing*, Vol. 1. Ann Arbor, MI: Environmental Research Institute of Michigan: 373–389.

Penny, F.A. 1975. *Surface Expression of Deep Discoveries, Central Rockies.* Denver, CO: Rocky Mountain Association of Geologists: 55–61.

Reid, W.M. 1988. Application of Thematic Mapper imagery to oil exploration in Austin Chalk, central Gulf Coast basin, Texas. *Bull. Am. Assn. Pet. Geol.* 72: 239.

Rowan, L.C., M.J. Pawlewicz, O.D. Jones. 1992. Mapping thermal maturity in the Chainman Shale near Eureka, Nevada, with Landsat thematic mapper images. *Bull. Am. Assn. Pet. Geol.* 76: 1008–1023.

Valenti, G.L., J.C. Phelps, L.I. Eisenberg. 1996. Geologic remote sensing for hydrocarbon exploration in Papua New Guinea. In: *Proceedings of 11th Thematic Conference on Geological Remote Sensing*, Vol. 1. Ann Arbor, MI: Environmental Research Institute of Michigan: 97–108.

10

Remote Sensing in Mineral Exploration

Chapter Overview

Case histories are provided for hydrothermal gold, porphyry copper, and kimberlites/ diamond exploration. In all instances, these are integrated programs in the sense that air-photos or satellite imagery are combined with field spectra acquisition, field mapping, and analysis of other sensor system products, notably aeromagnetics and digital elevation models (DEMs). In mineral exploration, the emphasis in most cases is on mapping altera-tion rather than stratigraphy. Structure, however, remains important: faults and fracture zones provide the pathways for intrusion and hydrothermal fluids that carry the minerals of interest and cause the alteration. The following case histories show mineral exploration applications of remote sensing.

Exploration Applications of Alteration Mapping

Case History: Mapping the Grizzly Peak Caldera, Colorado

A PhD dissertation by Dave Coulter at Colorado School of Mines set out to determine the effectiveness of hyperspectral imagery to map alteration and natural acid drainage associ-ated with sulfide mineralization at the Grizzly Peak Caldera in central Colorado (Coulter, 2006). The 15 km by 20 km Grizzly Peak Caldera is an Oligocene explosive volcanic collapse structure located between 3500 and 4000 m elevation in a high-relief alpine environment on the Continental Divide 25 km southwest of Leadville. Much of the caldera is above timberline and well exposed, although lower areas are partly grass covered and some slopes are forested. Volcanic deposits consist of early rhyolitic dikes and flows, a rhyolitic to dacitic volcanic dome, and postexplosion rhyolitic to quartz–latite tuffs. Mineralization at this unexploited molybdenum–copper porphyry deposit consists of sulfides of molyb-denum (MoS_2), copper ($CuFeS_2$), and iron (FeS_2) that occur primarily within quartz stock-works and veins. Zones with up to 15% sulfide content have been identified.

Four types of hydrothermal alteration are recognized: (1) quartz–sericite alteration consist-ing of light tan to brown rocks with minor pyrite and intensely altered quartz–sericite zones, (2) quartz–sericite–pyrite, consisting of 2%–15% pyrite with quartz and sericite, (3) argillic alteration comprising kaolinite and sericite with minor montmorillonite within or peripheral to other alteration zones, and (4) propylitic alteration in the caldera characterized by chlorite, epidote, and pyrite (Cruson, 1973). Mineral mapping and identification concentrated on illite in the quartz–sericite alteration zones. Dominant minerals in the argillic alteration zone are kaolinite, dickite, and pyrophyllite, as well as jarosite, goethite, and hematite. Image acquisition and processing focused on these alteration minerals. Iron minerals are characterized by their visible to near-infrared spectral absorption at 900 nm. Illite is identified by an absorption at 2.20 µm; dickite has a double absorption at 2.18 and 2.21 µm; kaolinite has an absorption minimum at 2.20 µm; and pyrophyllite has a minimum at 2.15 µm (Figures 10.1 and 10.2).

Imagery acquired over the caldera included airborne National Oceanographic and Atmospheric Administration/Jet Propulsion Lab Airborne Visible/Infrared Imaging Spectrometer (AVIRIS) and HyperSpecTIR (HST) as well as color digital aerial photography flown by the National Agricultural Imagery Program (NAIP) for the U.S. Department of Agriculture. The AVIRIS instrument was flown at a nominal elevation of 4 km, providing

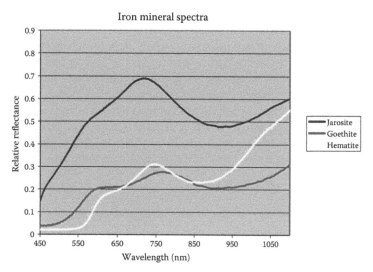

FIGURE 10.1
Visible and NIR spectra of iron minerals. (From Clark, R.N. et al., *U.S. Geological Survey Open File Report No. 93-592*, 1340, 1993.)

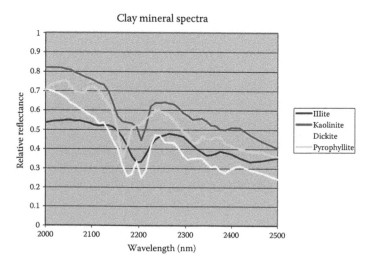

FIGURE 10.2
SWIR spectra of clay minerals. (From Clark, R.N. et al., *U.S. Geol Survey Open File Report No. 93-592*, 1340, 1993.)

a 4 m pixel. The HST imaging spectrometer, built and operated by SpecTIR Corporation of Reno, Nevada, collects 227 channels of data between 450 nm and 2.45 μm with a bandwidth of 10 nm. HST was flown at an elevation of 2.5 km above terrain and had a nominal pixel size of 2.5 m. Data from both instruments were radiometrically and geometrically corrected both during and after flight.

Satellite imagery was provided by the Advanced Spaceborne Thermal Emission and Reflection Radiometer (ASTER) instrument on the Terra satellite. Pixel size for this moderate resolution system varied from 15 m in the visible to 30 m in the short wave thermal infrared (SWIR) to 90 m in the TIR. For this reason, it was useful only for mapping broad mineral families. A DEM with 10 m resolution was provided by the National Elevation Dataset (USGS, 1999). The DEM, orthorectified NAIP airphotos, and ground control points were used for geometric corrections.

Data analysis involved inversion (unmixing) of the hyperspectral data into mineral classes and proportions. Since hyperspectral instruments average the spectra of all minerals within a pixel, it was necessary to determine which minerals are expected in the area of interest and use them as end-members for the mineral unmixing algorithm. End-member minerals were identified by gathering field spectra and by matching image spectra to the U.S. Geological Survey spectral library (Clark et al., 1993). End-members used in this work were the clay minerals pyrophyllite, dickite, kaolinite, low Al illite, medium Al illite, and high Al illite, as well as the iron minerals jarosite, goethite, and hematite. A "soil" end-member (with a muted clay signature) allowed unaltered soil to be discriminated from alteration. A visible near-infrared (VNIR) image was used for unmixing the iron minerals, whereas a SWIR image was used for unmixing the clay minerals.

The HST hematite–goethite–jarosite mixture image of the leach cap known as West Red shows the distribution of these minerals (Figure 10.3). Natural acid seeps are located at A, B, and C. A jarosite/goethite ratio image highlights the location of most intense

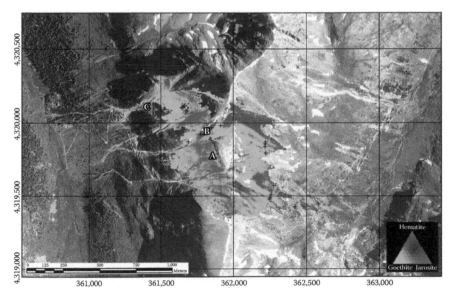

FIGURE 10.3
HST partial unmixing analysis of iron minerals for the leach cap known as West Red, Grizzly Peak Caldera, Colorado. Natural acid seeps are located at A, B, and C. Mineral key in lower right. (From Coulter, D., PhD dissertation, Golden, Colorado: unpub. Colorado School of Mines: 146 p, 2006.)

mineralization (Figure 10.4). The western slopes are dominated by jarosite and goethite. The eastern slopes contain mainly goethite and hematite. Maps derived from AVIRIS imagery for the most part are in good agreement with HST data.

Both argillic and sericitic (illite) alteration occur at West Red. Unmixing images are based on kaolinite, dickite, and pyrophyllite end-members. Pyrophyllite zones are identified in the same area (A) on both HST and AVIRIS mineral maps (Figures 10.5 and 10.6). Area B is

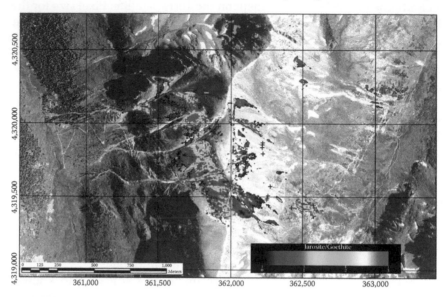

FIGURE 10.4
Jarosite/goethite ratio image of West Red superimposed on a true color airphoto. High jarosite shown as red; high goethite shown as violet. (From Coulter, D., PhD dissertation, Golden, Colorado: unpub. Colorado School of Mines: 146 p, 2006.)

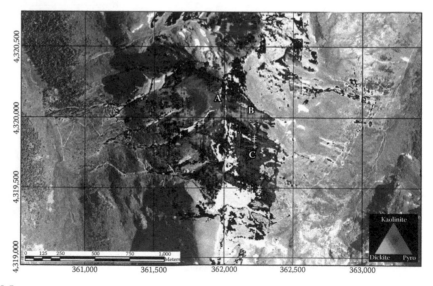

FIGURE 10.5
HST partial unmixing analysis of clay minerals showing argillic end-members, West Red. Minerals attributed to areas A, B, and C are compared in Figures 10.5 and 10.6. (From Coulter, D., PhD dissertation, Golden, Colorado: unpub. Colorado School of Mines: 146 p, 2006.)

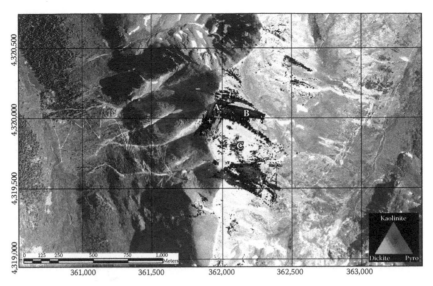

FIGURE 10.6
AVIRIS partial unmixing analysis of clay minerals showing argillic end-members, West Red. Minerals attributed to areas A, B, and C are compared in figures 10.5 and 10.6. (From Coulter, D., PhD dissertation, Golden, Colorado: unpub. Colorado School of Mines: 146 p, 2006.)

mapped as dickite on the HST image, but as a mixture of pyrophyllite and kaolinite on the AVIRIS map. Area C shows weak kaolinite on the HST map and no kaolinite on the AVIRIS map. These discrepancies are thought to be a result of either nonunique mixture modeling in the SWIR (area B) or a function of bands centered on versus slightly off the absorption minima (area C). Illite results for both systems are similar.

The hyperspectral imagery, processing, and resulting maps described in this work clearly show the distribution of end-member alteration minerals and areas of intense and minor alteration associated with a Mo–Cu porphyry system. Mineral and alteration maps, combined with alteration zoning patterns and structure mapping, highlight the caldera as an unexploited mineral deposit.

Case History: The Missing Link—Alteration Associated with Gold Mineralization at Goldfield, Nevada

The high price of precious metals drives the search for gold. Yet until now there has been no efficient or effective way to differentiate barren from productive areas. At Goldfield, Nevada, Spectral International Inc. determined to do just that (Hauff et al., 1999).

The Goldfield mining district, located in southern Nevada (Figure 10.7), has an arid climate with limited vegetation cover, mainly sagebrush, rabbit brush, cactus, Joshua trees, and grasses. Goldfield is a quartz–alunite-type epithermal gold deposit in silicified early Miocene intermediate volcanics. The structurally controlled deposit may be in a collapsed caldera. Mineralized veins, which also contain significant amounts of silver and copper, are known as ledges, and are associated with an arcuate fracture zone and intense hydrothermal alteration. Primary alteration minerals include quartz, alunite, kaolinite, dickite, diaspore, pyrophyllite, and illite. Some ledges are barren, whereas others are ore-bearing. The mine has produced over 4,000,000 ounces of gold since coming on production in 1910.

FIGURE 10.7
Index map of the Goldfield mining district, Nevada. (From Hauff, P. et al. *Proceedings of the 13th International Conference on Applied Geologic Remote Sensing*, March 1–3, Vancouver, Canada. Ann Arbor, MI: ERIM, 1999.)

Several open pits exist within the mining district and expose the alteration zoning. Minerals in the central ore zone include dickite with silica ± diaspore; alunite occurs along the edges and kaolinite in the argillic envelope. This work revealed that gold is likely to be present whenever dickite occurs in a quartz–alunite ledge. Apparently, alunite–kaolinite–quartz alone is not gold-bearing. Kaolinite is considered indicative of low-grade or unmineralized rock. An attempt was made to confirm these observations through the use of remote sensing data.

Four types of data were used in this project. These included Landsat Thematic Mapper (TM) data, portable field spectrometer data, AVIRIS, and SWIR Full Spectrum Imager (SFSI) datasets. First, a Landsat TM image was created to provide a regional overview (Figure 10.8).

This image was processed to a color ratio composite consisting of the band ratios 3/1 (red), 5/4 (green), and 5/7 (blue). Each band ratio was independently stretched to obtain the best contrast and color balance (Figure 10.9). The ratio composite was used to predict areas of alteration and the general composition of alteration in the mining district. Vegetation and water were masked to enhance the mineralization. Iron-rich minerals are red and orange; silica-rich rocks are green; clays are shades of blue; and the predicted alteration is shown as light shades of pink, yellow, and blue to white. Note the arcuate alteration zone east and north of Goldfield.

Ground spectra (Figure 10.10) were collected using the Portable Infrared Mineral Analyzer II (PIMA-II) portable field spectrometer. This instrument provides a spectral reflectance curve for small areas that have more or less homogeneous mineralogy and/or surface cover. It is used to calibrate the hyperspectral imagery by providing spectral control points where mineralogy, alteration, soil, and ground cover are known. This calibration database is used to generate hyperspectral image maps.

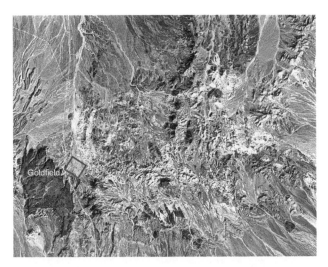

FIGURE 10.8
Landsat TM 7-4-1 (R-G-B) of Goldfield mining district. (Courtesy of Perry Remote Sensing.)

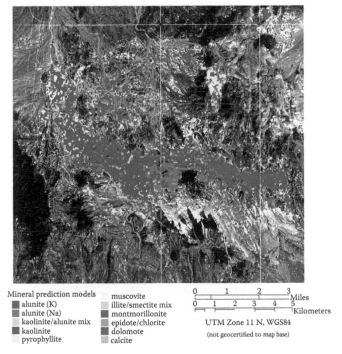

Mineral prediction models		
■ alunite (K)	muscovite	
■ alunite (Na)	illite/smectite mix	
kaolinite/alunite mix	■ montmorillonite	
■ kaolinite	■ epidote/chlorite	
pyrophyllite	■ dolomote	
	calcite	

0 1 2 3
Miles
0 1 2 3 4 5
Kilometers

UTM Zone 11 N, WGS84

(not geocertified to map base)

FIGURE 10.9
Landsat TM color ratio composite using bands 3/1 (red), 5/4 (green), and 5/7 (blue). Proposed mineral distributions are shown. (Courtesy of Perry Remote Sensing.)

Classification maps were made from AVIRIS and SWIR Full Spectrum Imager (SFSI) images. Classification was based on the ground control spectra. The AVIRIS 20 m resolution pixels and 12–15 nm band widths can discriminate mineral species, but recognizing the mineral dickite was difficult using AVIRIS, since the spectral curve is similar to that of kaolinite at the resolution of the AVIRIS instrument. The SFSI instrument was better able to resolve these

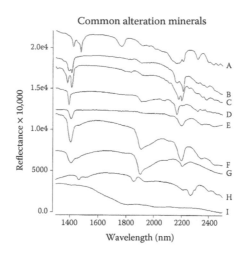

FIGURE 10.10

Ground spectra from the PIMA-II portable field spectrometer collected in the Goldfield district by Spectral International Inc. PIMA infrared spectra for the common alteration minerals are shown here: (A) alunite, (B) kaolinite, (C) dickite; (D) pyrophyllite; (E) phyllic illite; (F) hydrothermal illite; (G) montmorillonite, (H) jarosite, and (I) diaspore. (From Hauff, P. et al. *Proceedings of the 13th International Conference on Applied Geologic Remote Sensing*, March 1–3, Vancouver, Canada. Ann Arbor, MI: ERIM, 1999.)

FIGURE 10.11

Mineral map of Goldfield mining district. Black and green show dickite alteration. Dickite is considered the primary indicator of gold mineralization. (From Hauff, P. et al. *Proceedings of the 13th International Conference on Applied Geologic Remote Sensing*, March 1–3, Vancouver, Canada. Ann Arbor, MI: ERIM, 1999.) (Courtesy of Spectral International Inc.)

differences and predict the location of dickite alteration. The resulting images show black and green where gold is most likely to occur (Figure 10.11). These images show that dickite is a dominant alteration mineral in the gold-producing zones. The authors strongly feel, based on their field sampling and hyperspectral image classification, that dickite, rather than kaolinite or alunite, is the primary indicator of gold mineralization in the Goldfield district.

Exploration Applications of Structure Mapping

Case History: Structure Mapping as a Guide to Porphyry Copper Deposits, Northeast China

Too often the focus in remote sensing projects is on image processing alone as the answer to exploration needs. In fact one can locate many mineral exploration targets using combinations of alteration and structural indicators. An exploration program in China was carried out in an area where alteration mapping is largely ineffective because of heavy forest cover (Jiang et al., 1994).

The purpose of this study was to develop criteria for mapping porphyry copper deposits using Landsat TM imagery in a heavily forested area in northeastern China. Thick soil and vegetation cover make conventional mapping difficult. The project area is along the margin of the Daxinganling basin and contains Early Ordovician to Permian interbedded volcanics and terrigenous strata. Two large porphyry-associated copper deposits were discovered here during the 1970s.

Intrusives are Variscan (Late Devonian–Permian) granodiorite and granodiorite porphyry, although only the granodiorite is known to contain ore deposits. Field mapping determined that intrusives and country rocks are extensively hydrothermally altered, including potassium–feldspar/silicification (central), sericitic (intermediate), and propylitic (outermost) zones. The intrusives appear to be located along or at the intersection of regional faults.

Landsat TM images were processed to false color images using bands 4, 5, and 7 printed as red, green, and blue, respectively. Color ratio composites, principal component (PC) images, and vegetation index images were generated. Known copper deposits show up clearly on images because the intrusives are low-relief features with obvious textural contrast to adjacent volcanics and sediments. Thick vegetation cover makes it difficult to distinguish stratigraphy and alteration.

Structural features, in particular regional faults and fault intersections, appear to be the key to localizing both intrusives and mineralization. Faults provided the paths for migration of hydrothermal fluids and served as the locus for precipitation of mineralizing solutions. A series of nested circular structures (ring faults, ring dikes, alteration zones) associated with these intrusives are described for the first time (Figure 10.12). The circular structures are expressed as valleys, and have a maximum radius of about 12 km. The ring-like features are intersected by radial faults. It is proposed that, similar to resurgent calderas in Colorado (Bove et al., 2001), multiple intrusions over time formed the circular and radial fault patterns, whereas regional fault intersections localized the intrusives. Radial faults are expressed as linear drainages.

The circular structure at Daxinganling is developed in Ordovician submarine volcanics and Silurian marine sediments. The granodiorite stocks are thought to have intruded at a weak point developed at the intersection of regional faults. Intrusions penetrated the existing sediments and formed extrusive volcanic flows and tuff deposits on the paleosurface. Two large copper deposits exist at the southern and southwestern margins of the circular structure. Some mineralization also occurs along the southwestern margin at the intersection of northwest- and northeast-trending faults. Field observations suggest that northeast faults channeled hydrothermal fluids that controlled mineralization, whereas northwest faults constrained the shape of ore bodies. Both alteration and mineralization are influenced by the location and intensity of faulting.

This work shows that intrusions and associated porphyry-type mineralization can be found by mapping major regional fault intersections that coincide with topographic and

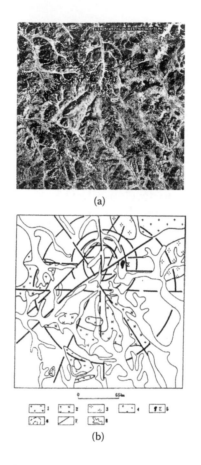

FIGURE 10.12

(a) Landsat TM image; (b) Structural interpretation showing ring fractures/dikes and radial fractures that appear to control porphyry copper mineralization in Devonian-Permian granodiorites of the Daxinganling basin, northeast China. (From Jiang, D., *Proceedings of the 10th International Conference on Geologic Remote Sensing*, v. 2, December 2002. Ann Arbor, MI: ERIM: II-611–II-618, 1994.)

textural indicators of intrusives. Radial and concentric fault patterns can reveal intrusives and mineralized zones, even in areas with heavy vegetation and soil cover.

Exploration Applications of Hyperspectral Mapping

Case History: Exploration for Diamond-bearing Kimberlites and Diatremes, Utah, Colorado, and Wyoming

A project to determine whether AVIRIS could characterize the mineral assemblages associated with kimberlites was undertaken by Fred Kruse and Joe Boardman (Kruse and Boardman, 1997, 1998).

Two sites were examined: one along the Comb Ridge monocline in southeast Utah and the other in the State-Line district of Colorado/Wyoming. Comb Ridge is an east-dipping, roughly north–south trending monocline consisting of Pennsylvanian to Jurassic age sandstones, shales, and limestones. The monocline is developed over a basement fault.

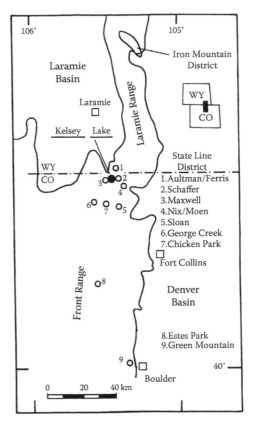

FIGURE 10.13

Location map, Kelsey Lake kimberlites, Colorado-Wyoming. (From Kruse, F.A., Boardman, J.W., *Proceedings of the 12th International Conference Applied Geologic Remote Sensing*. 1. Ann Arbor: ERIM: I-21–I-28, 1997; Kruse, F.A. and Boardman, J.W., *Proceedings of the 7th JPL Airborne Earth Science Workshop* 97-21, 1. Pasadena: Jet Propulsion Laboratory: 259 p, 1998.)

This part of Utah is arid, and vegetation away from rivers is sparse and restricted to some grass, sagebrush, and cactus.

The State-Line district along the border between Colorado and Wyoming occurs in Precambrian igneous and metamorphic rocks (Figure 10.13). The area is temperate, with conifer forests and grass in clearings. Diamonds (mostly industrial grade) have been found in kimberlites of the State-Line district.

Kimberlites are diamond-bearing mantle rocks (mica peridotites) that are explosively injected into the crust, usually along deep-going faults, and contain predominantly ultramafic minerals including olivine, phlogopite, perovskite, spinel, chromite, diopside, monticellite, apatite, calcite, and iron-rich serpentine. The gas-charged magma is ejected from the vent and a crater is formed that is usually filled with the explosion breccia. Alteration to serpentine and calcite is common. Xenoliths from the mantle, crust, and overlying sediments are found in the diatremes. A diatreme is the cone-shaped near-vertical pipe that contains the kimberlite. Their expression at the surface is typically circular to elliptical, with sharp contacts against the surrounding country rock, and the kimberlites often weather to topographic depressions that in many cases are filled by a pond or lake.

The AVIRIS instrument contains 224 channels with a 10 nm spectral resolution (see Chapter 3, Airborne Multispectral/Hyperspectral Scanners). Processing consisted of an atmospheric correction, selection and identification of spectral end-member minerals, and mapping of end-member occurrence and abundance. The results are presented as black/white images where brighter pixels represent higher abundances of a mineral, or as black background images showing only the distribution of targeted minerals.

The Utah site contains three diatremes: the Mule Ear diatreme to the north, the Moses Rock diatreme 6 km south, and the Cane Valley diatreme 3 km southwest of Moses Rock. These diatremes contain varying amounts of country rock and "serpentine tuff." Ten end-member minerals were defined using the AVIRIS bands from 400 nm to 1.3 µm. These include hematite, goethite, five soil classes, green vegetation, water, and shadow. Thirteen end-members were defined for the AVIRIS channels between 2.0 and 2.4 µm, including dolomite, calcite, kaolinite, illite, silica in sandstone, probable weathered serpentine, and possible iron-rich illite. These end-members were used to map the diatremes. AVIRIS data show the Mule Ear diatreme is characterized by two dolomites, minor calcite, illite, kaolinite, and goethite. The Cane Valley diatreme is similar. Dolomite and a mineral that may be weathered serpentine (absorption feature at 2.35 µm) are the primary minerals at the Moses Rock dike.

Diatremes in the State-Line district are localized along prominent joint and fault zones associated with the Virginia Dale ring-dike complex (Figure 10.14). The Virginia Dale complex is a roughly circular feature about 14 km in diameter in the northern Front Range of Colorado and Wyoming. The complex is a part of a large batholith (Precambrian Sherman Granite) intruding metamorphic rocks. An outer granite ring dike surrounds a zone of metamorphic rocks and mafic igneous rocks; quartz monzonite forms the core. Diatremes appear to be localized where the ring-dike complex is cut by radial faults (Eggler, 1968).

The kimberlite assemblage in the State-Line area consists of breccia containing kimberlite, serpentine, calcite, phlogopite, magnetite, perovskite, chlorite, talc, and hematite. Paleozoic

FIGURE 10.14
AVIRIS saturation-enhanced true color image of the State-Line kimberlite district, Colorado-Wyoming. (Processed by Analytic Imaging and Geophysics LLC.)

xenoliths of dolomite, conglomerate, and sandstone occur in the diatremes. The kimberlites are deeply weathered and altered to serpentine and calcite. They are poorly exposed, so that most spectral signatures are dominated by green and dry vegetation. In some cases, dolomite, calcite, serpentine, phlogopite, and kaolinite can be mapped associated with the diatremes.

Five end-member minerals were mapped using AVIRIS (Figure 10.15). The end-member distribution was then extracted from the image of the Kelsey Lake mine area (Figure 10.16). These end-member minerals were compared to the spectral library and found to conform closely to the spectra for serpentine and phlogopite (Figures 10.17 and 10.18).

This work shows that minerals associated with kimberlites, particularly serpentine, calcite, dolomite, clays, and iron oxides can be mapped using hyperspectral sensors. In areas with soil and vegetation cover, it may be necessary to supplement the mineral maps with structure mapping to locate diatremes at fault intersections and along ring dikes.

Case History: Mineral Exploration in Northern Chile

The Gravity group at the Frei Universität Berlin (Germany) processed Landsat TM and Landsat Enhanced TM + imagery over training sites that included massive porphyry copper deposits in northern Chile (Ott et al., 2006). Images and various data types (DEM, aeromagnetics) were classified into favorable versus nonfavorable areas, and the different "layers" of data were merged into a single "favorability map" using a geographic information system.

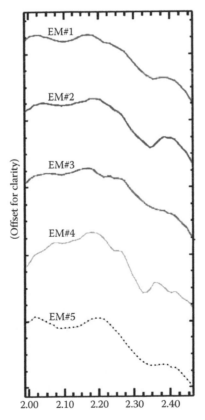

FIGURE 10.15

Five end-member mineral spectra mapped in the Kelsey Lake mine area, State-Line district. (Courtesy of Analytic Imaging and Geophysics LLC.)

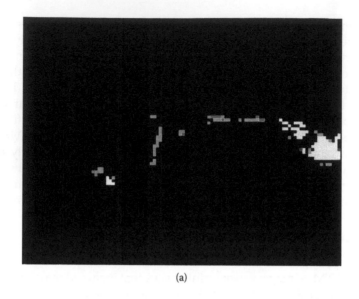

(a)

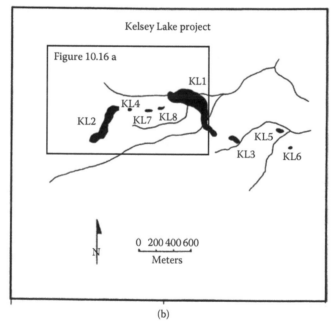

(b)

FIGURE 10.16

(a) AVIRIS prospects, Kelsey Lake mine area. Colors match end-members in Figure 10.15. (b) Kelsey Lake mine known kimberlite locations. (Processed by Analytic Imaging and Geophysics LLC.)

The Domeyko Cordillera of northern Chile contains a number of Mesozoic–Cenozoic igneous arcs. Conditions during the Eocene–Oligocene were optimal for formation of large porphyry copper deposits. Northeast convergence favored development of trench-parallel dextral transpression and trench-parallel faults. Emplacement of diorite to grano-diorite plutons and development of hydrothermal systems were enhanced by relaxation of transpression and eventual strike-slip reversal during early Oligocene. This was the main hypogene sulfide mineralization episode. By mid-Miocene the arid climate drove down

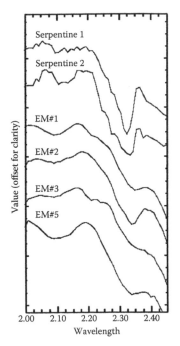

FIGURE 10.17
End-member spectra and reference library spectra for serpentine. (Courtesy of Analytic Imaging and Geophysics LLC.)

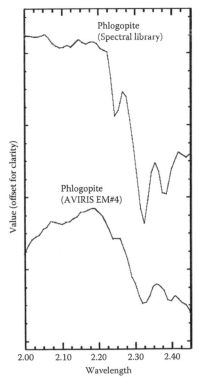

FIGURE 10.18
End-member spectra and reference library spectra for phlogopite. (Courtesy Analytic Imaging and Geophysics LLC.)

the water table and caused supergene oxidation, in situ copper oxide zones, and enriched sulfide blankets.

Sulfide minerals include chalcopyrite and bornite ± chalcocite, covellite, and enargite. Hydrothermal solutions caused early pervasive potassic alteration (K-feldspar and biotite), later quartz–feldspar and quartz–sericite alteration in veins and breccias near faults, and late, advance argillic alteration comprising pyrophyllite–alunite–quartz–pyrite. Subsequent supergene enrichment and leaching led to chalcocite blankets below the water table and highly leached deposits containing hydrated copper minerals along with alunite, iron oxides (jarosite, hematite, goethite), and kaolinite at the surface.

Exploration in this arid region is driven by the search for the leached cap of these deposits and their associated alteration minerals. La Escondida and Quebrada Blanca mining districts were used as training sites. La Escondida is particularly useful in that it has both pre- and post-mining imagery. Imagery was radiometrically corrected for atmospheric effects, and adjacent images were histogram matched so that the distribution of reflectance values in each image would cover the same range and have nearly identical means and standard deviations for each image. This is especially important to ensure similar results in all images subject to principal components analysis (PCA) and decorrelation stretching (see Chapter 2, section on Image Processing).

Landsat images were processed using PCA, decorrelation stretching, and band ratioing. The first PC shows most of the reflectance information in the image. The second PC is mostly topographic information (shading) and some lithology. The PCs 3 and 4 show good lithologic contrast. The fifth and seventh PC show mainly clay minerals. The PC image 5-4-3 represented as the colors red–green–blue (RGB) was useful for showing the distribution of clay and altered diorites (purple to red colors; Figure 10.19). The decorrelation stretch of PCs 5-3-1 as RGB shows altered intrusives as yellow and red colors (Figure 10.20). The ratio image used bands 5/7 (red) to highlight clays, 3/1 (green) for iron oxides, and 4/3 (blue) to emphasize healthy vegetation. Altered rocks in the color ratio composite image are shades of red to yellow (Figure 10.21). Finally, a black/white clay

Principal component image derived from Landsat TM data

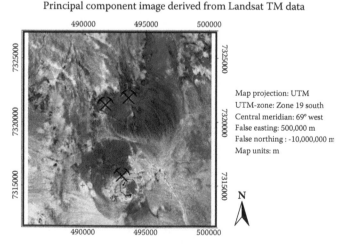

Map projection: UTM
UTM-zone: Zone 19 south
Central meridian: 69° west
False easting: 500,000 m
False northing : -10,000,000 m
Map units: m

FIGURE 10.19
PC image 5-4-3 as RGB shows the distribution of clay and altered diorites (purple to red colors), La Escondida mining district, northern Chile. (From Ott, N., *Geosphere*, 2, 236–252, 2006.)

Inverse principal component image derived from Landsat TM data

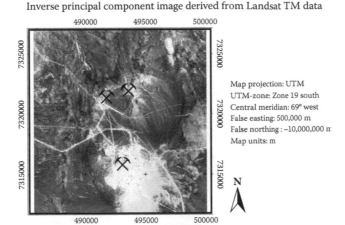

FIGURE 10.20
Decorrelation stretch (inverse PC image) of PCs 5-3-1 as RGB, La Escondida mining district. This figure shows altered intrusives as yellow and red colors. (From Ott, N., *Geosphere*, 2, 236–252, 2006.)

Color ratio image derived from Landsat TM data

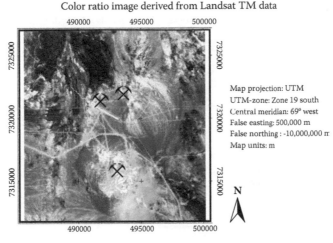

FIGURE 10.21
Ratio image uses bands 5/7 (red) to highlight clays, 3/1 (green) for iron oxides, and 4/3 (blue) to emphasize healthy vegetation. Altered rocks in this color ratio composite of the La Escondida mining district are shades of red to yellow. (From Ott, N., *Geosphere*, 2, 236–252, 2006.)

ratio image was generated using only the 5/7 band ratio. High-clay areas are white on this image (Figure 10.22).

Supervised classification of these images was run using before mining spectral properties of the mineralized areas at Escondida Norte as the training sites. Only two classes were used: altered (1) and not altered (0). Pixels on each image type (PCA, decorrelation stretch, color ratio, black/white ratio) were classified as 0 or 1, and values were summed. Favorability values in the final image ranged from 0 (not altered) to 4 (highly favorable alteration). The highly favorable areas coincided well with known mineralization in the La Escondida (two out of three current pits; areas 1 and 2 on Figure 10.23) and Quebrada Blanca (current open pit) training areas. Additional highly favorable altered areas are pointed out by this work (area 4, Figure 10.23). This fairly simple methodology illustrates one successful mineral exploration strategy.

Clay-band ratio image derived from Landsat TM data

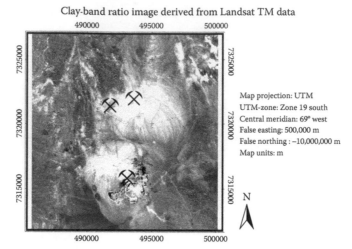

Map projection: UTM
UTM-zone: Zone 19 south
Central meridian: 69° west
False easting: 500,000 m
False northing : −10,000,000 m
Map units: m

FIGURE 10.22
Clay ratio image of La Escondida generated using the 5/7 band ratio. High-clay areas are white on this image. (From Ott, N., *Geosphere*, 2, 236–252, 2006.)

Favorability map of altered rocks at La Escondida mining district

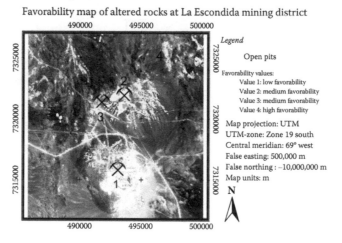

Legend

Open pits

Favorability values:
    Value 1: low favorability
    Value 2: medium favorability
    Value 3: medium favorability
    Value 4: high favorability

Map projection: UTM
UTM-zone: Zone 19 south
Central meridian: 69° west
False easting: 500,000 m
False northing : −10,000,000 m
Map units: m

FIGURE 10.23
Favorability map based on the PCA, decorrelation stretch, color ratio, and black/white ratio images. Red color indicates most favorable (mineralized) areas. (From Ott, N., *Geosphere*, 2, 236–252, 2006.)

References

Bove, D.J., K. Hon, K.E. Budding, J.F. Slack, L.W. Snee, R.A. Yeoman. 2001. Geochronology and Geology of Late Oligocene through Miocene Volcanism and Mineralization in the western Jan Juan Mountains, Colorado: U.S. Geol. Survey Prof. Paper 1642: Denver, CO: 30 p.

Clark, R.N., G.A. Swayze, A.J. Gallagher, T.V.V. King, W.M. Calvin. 1993. The US Geological Survey digital spectral library, version 1, 0.2 to 3.0 microns. *U.S. Geological Survey Open File Report No. 93-592*: 1340 p.

Coulter, D. 2006. Remote sensing analysis of alteration mineralogy associated with natural acid drainage in the Grizzly Peak Caldera, Sawatch Range. PhD dissertation, Colorado: unpub. Colorado School of Mines: 146 p.

Cruson, M.G. 1973. Geology and ore deposits of the Grizzly Peak caldron complex, Sawatch Range. PhD dissertation, Colorado: unpub. Colorado School of Mines: 180 p.

Eggler, D.H. 1968. Virginia Dale Precambrian Ring-Dike Complex, Colorado-Wyoming. *G.S.A. Bull.* 79: 1545–1564.

Hauff, P., R. Bennett, P. Chapman, G. Edmondo, G. Borstad, R. Neville, W. Peppin, S. Perry. 1999. Goldfield, Nevada: An old problem revisited with hyperspectral technology. In: *Proceedings of the 13th International Conference on Applied Geologic Remote Sensing*, March 1–3, Vancouver, Canada. Ann Arbor, MI: ERIM.

Jiang, D., P. Wang, F. Meng. 1994. Application of Landsat TM data into exploration for porphyry copper deposits in forested area. In: *Proceedings of the 10th International Conference on Geologic Remote Sensing, v. 2,*. Ann Arbor, MI: ERIM: II-611–II-618.

Kruse, F.A., J.W. Boardman. 1997. Characterizing and mapping of kimberlites and related diatremes in Utah, Colorado, and Wyoming, USA, using the Airborne Visible/Infrared Imaging Spectrometer (AVIRIS). In: *Proceedings of the 12th International Conference on Applied Geologic Remote Sensing*, v. 1. Ann Arbor, MI: ERIM: I-21–I-28.

Kruse, F.A., J.W. Boardman. 1998. Characterization and mapping of kimberlites and related diatremes using AVIRIS. In: *Proceedings of the 7th JPL Airborne Earth Science Workshop. 97–21, v. 1.* Pasadena: Jet Propulsion Laboratory: 259 p.

Ott, N., T. Kollersberger, A. Tassara. 2006. GIS analyses and favorability mapping of optimized satellite data in northern Chile to improve exploration for copper mineral deposits. *Geosphere.* 2: 236–252.

USGS. 1999, National Elevation Dataset: US Dept. Interior Geological Survey, Available at http:// ned.usgs.gov/ (accessed October 21, 2012).

Additional Reading

Agar, B., D. Coulter. 2007. Remote sensing for mineral exploration—a decade perspective 1997-2007. In: *Proceedings of Exploration 07: Fifth Decennial International Conference on Mineral Exploration*, B. Milkereit (ed.). The Leading Edge, Paper 7: 109–136.

Barniak, V.J., R.K. Vincent, J.J. Mancuso, and T.J. Ashbaugh. 1996. Comparison of a gold prospect in Churchill County, Nevada, with a known gold deposit in Mineral County, Nevada, from laboratory measurements and Landsat TM images. In: *Proceedings of the 11th Thematic Conference on Geologic Remote Sensing. v. 2.* February 27–29, Las Vegas, NV. Ann Arbor, MI: ERIM v. 2: 188–197.

Bennett, S.A. 1993. Use of Thematic Mapper imagery to identify mineralization in the Santa Teresa district, Sonora, Mexico. *Int. Geol. Rev.* 35: 1009–1029.

Bowers, T.L., L.C. Rowan. 1996. Remote mineralogic and lithologic mapping of the Ice River alkaline complex, British Colombia, Canada, using AVIRIS data. *Photogramm. Engr. Rem. Sens.* 62: 1379–1385.

Coulter, D.W., M.A. Sares, P.L. Hauff, D.A. Bird, D.C. Peters, F.B. Henderson III. 2009. Hyperspectral image analysis of natural acid drainage in the Grizzly Peak Caldera—implications for exploration and mining baseline studies. *Rev. Econ. Geol., Rem. Sens. Spectral Geol.* 16: 123–124.

Crósta, A.P., I.D.M. Prado, M. Obara. 1996. The use of Geoscan AMSS data for gold exploration in the Rio Itapicurú greenstone belt (BA), Brazil. In: *Proceedings of the 11th Thematic Conference on Geologic Remote Sensing v. 2*, February 27–29, Las Vegas, NV. Ann Arbor, MI: ERIM: 205–214.

Crósta, A.P., C. Sabine, J.V. Taranik. 1998. Hydrothermal alteration mapping at Bodie, California, using AVIRIS hyperspectral data. *Rem. Sens. Environ.* 65: 309–319.

Davidson, D., B. Bruce, D. Jones. 1993. Operational remote sensing mineral exploration in a semi-arid environment: the Troodos Massif, Cyprus. In: *Proceedings of the 9th Thematic Conference on Geologic Remote Sensing. v. 2,* 31 January-2 February, New Orleans, LA. Ann Arbor, MI: ERIM: 845–859.

Dick, L.A., G. Ossandon, R.G. Fitch, C.M. Swift, A. Watts. 1993. Discovery of blind copper mineralization at Collahuasi, Chile. In: *Program and Abstract Integrated Methods in Exploration and Discovery.* Littleton, CO: Society of Economic Geologists: AB21–AB23.

Harris, J.R., A.N. Rencz, B. Ballantyne, C. Sheridan. 1998. Mapping of altered rocks using Landsat TM and lithogeochemical data: Sulphurets-Bruce Jack Lake District, British Colombia, Canada. *Photogram. Engr. Rem. Sens.* 64: 309–322.

Ma, J., V.R. Slaney, J.R. Harris, D.F. Graham, S. B. Ballantyne, D. C. Harris. 1991. Use of Landsat TM data for mapping of limonitic and altered rocks in the Sulphurets area, British Colombia. In: *Proceedings of the 14th Canadian Symposiumon Remote Sensing,* May 6–9 Calgary, AB,: 419–422.

Miranda, F.P., A.E. McCafferty, J.V. Taranik. 1994. Reconnaissance geologic mapping of a portion of the rain-forest-covered Guiana Shield, northwestern Brazil, using SIR-B and digital aeromagnetic data. *Geophysics.* 59: 733–743.

Rencz, A.N., C. Bowie, B. Ward. 1996. Application of thermal imagery from Landsat data to identify kimberlites, Lac de Gras area, District of Mackenzie, N.W.T. In A.N. LeChaimant, D.G. Richardson, R.N.W. DiLabio, and K.A. Richardson (eds.), *Searching for Diamonds in Canada.* Geological Survey Open File Report. 3228: 255–257.

Rencz, A.N., J.R. Harris, S.B. Ballantyne. 1994. Landsat TM imagery for alteration identification. *Current Research Bull. Geol. Survey Canada*: 277–282.

Rowan, L.C., K. Watson, J.K. Crowley, C. Anton-Pacheco, P. Gumiel, M.J. Kingston, S.H. Miller, T.L. Bowers. 1993. Mapping lithologies in the Iron Hill, Colorado, carbonatite-alkalic igneous rock complex using thermal infrared multispectral scanner and airborne visible-infrared imaging spectrometer data. In: *Proceedings of the 9th Thematic Conference on Geologic Remote Sensing. v. 1.* Ann Arbor, MI: ERIM: 195–197.

Rowan, L.C., R.N. Clark, R.O. Green. 1996. Mapping minerals in the Mountain Pass, California, area using the airborne visible/infrared imaging spectrometer (AVIRIS). In: *Proceedings of the 11th Thematic Conference on Geologic Remote Sensing. v. 1.* Ann Arbor, MI: ERIM: 175–176.

Sabine, C. 1999. Remote Sensing Strategies for Mineral Exploration. *Manual of Remote Sensing* v. 3, 3rd ed. New York: John Wiley and Sons: 375–448.

Singhroy, V.H., F.A. Kruse. 1991. Detection of metal stress in boreal forest species using the 0.67 μm chlorophyll absorption band. In: *Proceedings of 8th Thematic Conference on Geologic Remote Sensing. v. 1,* April 29–May 2, Denver, CO, Ann Arbor, MI: ERIM: 361–372.

Spatz, D.M., R.T. Wilson. 1994. Exploration remote sensing for porphyry copper deposits, western America Cordillera. In: *Proceedings of the 10th Thematic Conference on Geologic Remote. Sensing. v. 1.* Ann Arbor, MI: ERIM: 227–240.

Spatz, D.M., R.T. Wilson. 1997. Remote sensing characteristics of the volcanic-associated massive sulfide systems. In: *Proceedings of the 12th Thematic Conference on Geologic Remote Sensing v. 1.* Ann Arbor, MI: ERIM: 1–12.

Watson, K., F.A. Kruse, S. Hummer-Miller. 1990. Thermal infrared exploration in the Carlin trend, northern Nevada. *Geophysics.* 55: 70–79.

Windeler, D.S., R.J.P. Lyon. 1991. Discrimination dolomitization of marble in the Ludwig skarn near Yerington, Nevada, using high-resolution airborne infrared imagery. *Photogramm. Engr. Rem. Sens.* 57: 1171–1178.

Section III
Exploitation, Hydrologic, and Engineering Remote Sensing

Introduction

At first glance, it might not appear obvious that remote sensing would have a role to play in oil field exploitation, mine development, or engineering projects. After all, can one obtain the detail needed to pick well locations or map veins using satellite imagery? Yet, there are many potential applications for both detailed airphoto and high-resolution satellite analyses. The simplest use for imagery is to serve as a map base when topographic maps are not available, when more detail is needed than is available on existing topographic or planimetric maps, or when maps are out of date and thus useless for engineering purposes. Most commonly roads and streams have changed since the original maps were made, or development has occurred, or landslides and floods have changed the landscape. These image maps can be uncorrected (i.e., no standard projection) or they can be rectified to eliminate most of the distortions inherent in images and photos and converted to a useful map projection.

Exploitation Applications

In addition to base maps, imagery is useful for change detection, whether it is subsidence over a producing field or build out of a tailings pond. Other exploitation applications include the following:

1. Helping choose infill and stepout well locations based on the extension of known surface trends
2. Assisting with enhanced oil recovery (e.g., waterflood) programs
3. Mapping the location and extent of coalbed methane and fractured reservoir sweet spots

4. Designing horizontal well trajectories for a development drilling program
5. Locating mineralized zones outboard from known deposits
6. Solving mine safety problems such as the locations of likely roof falls (underground) or unstable slopes (open pit)

On occasion, remote sensing data have been used successfully to select a well site. In these cases, airphotos or high resolution satellite imagery have usually provided the detail that is required to map a structural trap. This particular application of remote sensing was more common in the past; today it tends to be used mainly in poorly explored or remote areas. In remote areas, it may actually be less expensive to drill wells than to run a seismic program, making remote sensing an effective, low-cost solution to site selection. In certain plays, specifically, fractured reservoirs; or when exploration is in areas with high-velocity surface units such as carbonates, volcanics, or permafrost; or areas with salt solution or karst collapse features, mapping structure on imagery is as reliable and effective as interpretation of conventional seismic surveys. In areas with deep weathering, near-surface volcanics, subsurface salt, or near vertical units, seismic data can be difficult to obtain and process properly, and remote sensing becomes a valuable adjunct to the interpretation of geophysical data.

Much mineral exploration is the extension of known deposits or trends into the surrounding region. Many of these areas are remote or contain rugged topography and are not easily traversed. Field mapping is both time consuming and expensive. It would be useful to have a tool that directs mapping and sampling programs to the most favorable areas. Multispectral/hyperspectral imagery and potential fields data do this.

Mining is inherently dangerous. In addition to roof falls in underground mines, slope stability is a critical issue in open-pit mining. After mining has been completed, there is the issue of subsidence of old, abandoned mines, particularly in rapidly growing urban areas where the location of the old mines has been lost.

Hydrologic Applications

Surface and groundwater are vital for agriculture, industry, and urban development, and are also integral parts of wilderness ecosystems. Typical remote sensing projects involving surface and groundwater involve the following:

1. Locating sources of surface and/or groundwater
2. Monitoring surface water (quality, abundance, temperature, and flow rates)
3. Locating geothermal resources
4. Monitoring erosion, including fluvial and coastal erosion
5. Flood control

Remote sensing by itself usually cannot locate sources of groundwater, determine water quality, or identify geothermal hot spots. Yet, when used together with ground measurements, sampling, surface mapping, geophysical surveys, and drilling, imagery can play a major role in accomplishing all of these.

Engineering and Logistical Applications

Engineering/logistical applications include the following:

1. Planning access into and out of an area (road building, barging)
2. Choosing facilities sites
3. Planning pipeline, slurry, conveyor belt, and powerline routes
4. Locating sources of aggregate for drill pads or artificial islands and construction projects
5. Locating old well sites, old seismic lines, and old mine shafts and adits
6. Mapping the distribution of permafrost, soft ground, or outcrops
7. Mapping ice movements in the Arctic
8. Siting offshore drilling platforms
9. Accurately locating shorelines (at both high and low tide)
10. Mapping water depths for port facilities or to avoid dangerous shoals
11. Locating sites of potential landslides that could affect housing developments, highways, powerlines, dams, and other infrastructure

Chapters 11 through 13 deal with these issues by showing how remote sensing has been used to make these operations cost effective, time efficient, and safe.

11

Remote Sensing in Resource Exploitation Projects

Chapter Overview

The problems that one faces during exploitation of a field or deposit include making production more efficient and cost effective, adding reserves, and finding ways to extend the mine or field life by finding bypassed resources. Airphotos, airborne Lidar, and high-resolution satellite imagery can generally be used to plan and carry out these objectives. Several examples are provided to show remote sensing-based approaches to locating and exploiting new reserves. Cases from the petroleum industry illustrate using imagery to help locate infill and stepout wells (Bravo Dome, New Mexico), producing from fractured reservoirs (Austin Chalk, south Texas), coalbed methane (Piceance basin, Colorado), and waterfloods (Cottonwood Creek field, Wyoming). An example from the Huancavelica mining district, Peru, shows the same process applied to the hard rock mining industry. Finally, remote sensing contributions to mine safety and slope stability are discussed.

Choosing Infill and Stepout Well Locations

Infill well locations are usually determined by a detailed examination of existing well data and extrapolating trends such as thickness changes, facies changes, and structural position. If detailed seismic data exist, that data can be used to assist in the evaluation of the field, for instance, suggesting where the high point on a structure is located, where lateral facies changes occur, the depth of a gas–water contact, or where there could be fault closure. Detailed seismic data often do not exist within a field, however. In such cases, it may be possible to use surface information provided by imagery to (1) map structural closure, (2) map faults and joint zones, and (3) provide some idea as to the dominant trend of fracturing within the field and how it changes across the area. To the extent that unit thickness and facies were controlled by paleostructure, it may be possible that these structures are expressed at the surface through reactivation or settling and they could suggest depositional patterns (e.g., Chapter 9, Case History: Stratigraphic Traps in the Paradox Basin, Utah).

In stratigraphically trapped fields, the distribution of the reservoir is a function of paleo-topography. Reservoir sands, for example, may be deposited as an apron or sheet around basement highs and missing over the uplifts. Under such circumstances, it would be a waste of effort to drill on these highs, even though they appear interesting on seismic data. Because the overlying section settles and compacts through time, the basement highs may now appear as surface highs (Figure 11.1). These subtle rises can be mapped on imagery using techniques described in Chapter 5.

Present day surface

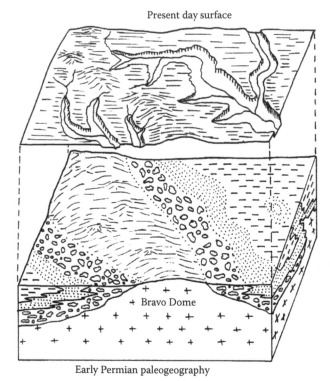

Early Permian paleogeography

FIGURE 11.1
Diagram showing granite wash and sand sheets distributed around basement highs at Bravo Dome, New Mexico. Drape of younger units as a result of compaction causes present day topographic and drainage indications of structure.

Case History: Developing the Bravo Dome CO_2 Field, New Mexico

An example of paleotopography expressed on the present day surface can be seen at Bravo Dome, a CO_2 field in New Mexico (Broadhead, 1990). The principal reservoir is the Permian Tubb sandstone, which lies at depths from 580 to 900 m. The dome itself is a southeast-plunging projection of the Pennsylvanian Sierra Grande uplift (Figure 11.2). By Early Permian time arkosic sands derived from the granitic uplift were being deposited as clastic wedges around the flanks of the uplift. The CO_2 is produced from the finer outlying sands rather than the granite wash closest to the Permian inselbergs. The Cretaceous cover was gently folded during Laramide deformation in this area, and some of the small folds that are now at the surface are probably a result of Laramide movements. Although Bravo Dome itself is a subsurface feature that is not revealed by outcrop patterns at the surface, it is expressed as radial drainage. Some of the small buried Precambrian knobs (1–5 km in diameter) that lack Tubb sandstone have been detected at the surface using airphotos to map drainage and tonal patterns.

Oil and gas companies have a vested interest in keeping their operating costs down and would rather not drill a well on or near one of the granite knobs where the reservoir is missing. By looking at structure contour maps derived from existing wells and a 5 km (3 mi.) seismic grid, and comparing this map to gravity and a remote sensing surface structure map, project geologists at Amoco determined that ~50% of the circular anomalies identified by remote sensing geomorphic analysis at Bravo Dome are related to actual

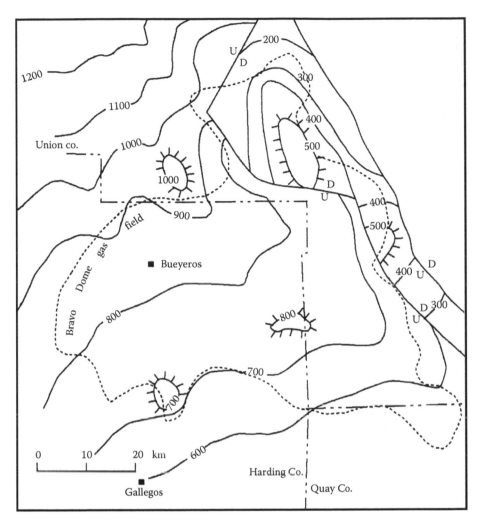

FIGURE 11.2

Structure map on the Precambrian, Bravo Dome, New Mexico. Contours in meters. Note that the dome is a southeast-plunging nose with scattered knobs. (From Broadhead, R.F., Bravo Dome carbon dioxide gas field, In: Beaumont E.A., N.H. Foster (eds.), *Structural Traps 1, Treatise of Petroleum Geology Atlas of Oil and Gas Fields*, Tulsa, American Association of Petroleum Geologists, 213–232, 1990.)

basement structures. The surface work adds confidence to the other data sets in that it helps fill in detail on the size and location of knobs where other data sets are weak.

In other areas, a reservoir unit may have been deposited against a fault or may have been offset vertically or laterally by a fault. These faults can not only bound reservoir units, but often also provide reservoir closure on one or both sides of the fault (Figure 11.3). Mineral deposits are also frequently cut by faults and parts of the ore body have been moved some distance vertically and/or laterally. In such cases, it would be desirable to be able to map the location of the fault and the sense of displacement, and predict where the mineralization should be located.

Whereas evidence for such faults is not always visible at the surface, one can frequently see some indication of where the section was faulted, and perhaps even get a sense of the displacement direction from the present surface (Figure 11.4). This information assists

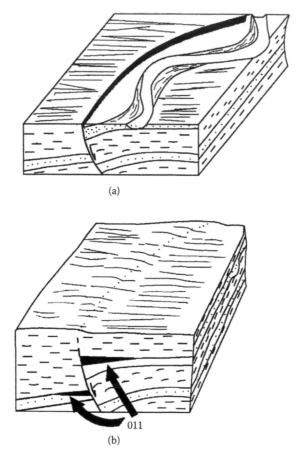

(a)

(b)

011

FIGURE 11.3
Diagram showing how surface indications of a fault can help locate a reservoir unit and can provide the updip trapping mechanism as well. (a) Paleogeography showing structural control of channel sands. (b) Present day surface. The fault trace provides the clue to reservoir trend.

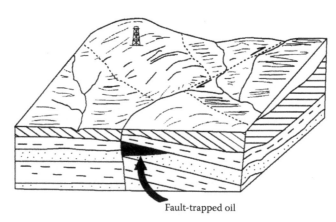

Fault-trapped oil

FIGURE 11.4
Surface indicators of offset can help determine where to drill an offset to existing production.

with planning a core drilling program to find extensions of veins, reefs, or mineralized bodies. The same holds for drilling infill and stepout wells, particularly when the hydrocarbon reservoir consists of a channel sand, barrier bar, or other hard to follow target. If a channel, for example, was controlled by a fault location and trend, the present indications of the fault location and extent could well indicate the distribution of the channel fill. If the channel has been offset in a normal sense and all the units dip uniformly, it should be possible to predict where the offset sand might be encountered and where stepout wells should have the best chance of hitting the objective (Figure 11.5). Likewise, if there are indications of lateral offset along a zone based on offset surface features, fracture patterns, or en echelon structures, for example, one should be able to better predict the location of the offset mineral deposit or reservoir sand at depth.

Case History: Developing a Fractured (Conventional) Reservoir, Austin Chalk, Texas

Exploitation of fractured reservoirs requires knowledge of the location, density, and trend of subsurface fractures. In many cases, these units would not be reservoirs at all were it not for the capacity of the fractures to hold and deliver hydrocarbons. The Cretaceous Austin Chalk in Texas has porosities that range from 3% to 9%, but unless the rock is fractured, the matrix permeability averages less than 0.1 mD (Snyder and Craft, 1977). Yet, ultimate

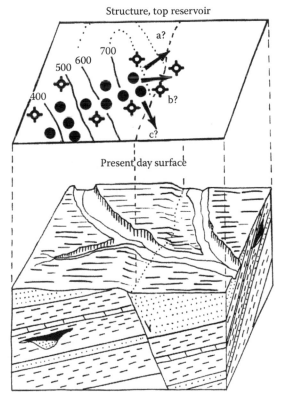

FIGURE 11.5
Surface mapping can suggest where to look for offset reservoirs. In the example shown here, the structure map based on well data could suggest a fold, but surface indicators suggest a normal fault. The channel sand reservoir was displaced in the "a" direction.

recovery has been estimated at four to eight billion barrels of oil, making this a target worth pursuing (Ewing, 1983). The fractured and overpressured Mississippian Mission Canyon limestone in the Williston basin is estimated to contain 100 to 500 million barrels of recoverable oil in the Mondak field alone (Parker and Hess, 1980). This accumulation is not associated with any structural closure. Normal matrix porosity is between 2% and 4%, and permeability is generally less than 0.01 mD. There would be little or no production without a fractured reservoir. The Cretaceous Niobrara in the Denver basin has matrix porosities generally less than 10% and permeabilities of 0.01 mD or less at depths greater than 1800 m (see Chapter 9, Resource Plays). Yet, initial production rates as high as 865 barrels of oil per day have been reported out of fractured zones (Rountree, 1984). Fractures provide primary porosity and deliverability in many world-class fields, such as those around Lake Maricaibo, Venezuela, and are essential for production of coalbed methane since coals have effectively no primary porosity.

The procedure for evaluating fracturing within an area consists of mapping all indications of faults and joint zones using techniques discussed in Chapter 5. Since most existing maps do not indicate joints, and may not show faults unless significant offset is measured, image interpretation can usually provide much new information. Interpreted fractures cannot, in many cases, be identified as faults or joints, and in most cases the dip of the fracture surface cannot be determined from imagery alone. Fractures dipping less than 45° tend to have irregular traces and thus are seldom mapped. Outcrops have more obvious fractures than unconsolidated materials, and brittle units have more and better expressed joints than ductile units. Thus, it is obvious that airphoto, satellite, Lidar, and magnetics interpretations can locate many, but not all, fractures. Analysis of these sensors can provide estimates of fracture system lateral lengths and azimuths, but at least some ground measurements (surface mapping, azimuthal seismic) are required as follow-up to determine the inclination, vertical lengths, and continuity of joints and joint trends from layer to layer.

The upper Cretaceous Austin Chalk of south Texas produces oil from fractured zones along a 15–20 km wide zone extending northeast–southwest across the coastal plain from Crystal City to Bryan-College Station. Production is at depths of 2000–3000 m. Dip is uniformly southeast and no structural trap (other than fractures) is involved. Matrix porosity is 3%–7% and permeabilities are up to 0.1 mD. Oil was generated in place or migrated up from the underlying Eagle Ford Fm. Fractures are thought to be related to normal faults that form horsts and grabens parallel to the Gulf Coast. The objective of this project was to determine the orientation and intensity of natural fractures. Several Landsat Thematic Mapper images with 30 m resolution were processed to false color images.

Fracture maps are the best way to display the location of faults and major joint zones (Figure 11.6). Fractures can be digitized to simplify the manipulation of fracture data. Rose diagrams are a convenient way to display the orientation of fracture sets. The usual technique is to superimpose a grid of predetermined size on a fracture map and count the number of interpreted fractures in a given azimuth range, for example, 15° increments from west to north to east (Figure 11.7). This gives a frequency-based rose diagram for each grid cell. It is also possible to count the cumulative lengths of fractures in every 15° increment per grid cell. This provides a length-weighted rose diagram. Frequency-weighted rose diagrams tend to overemphasize the importance of short fractures, since all fractures have the same importance. Length-weighted rose diagrams give more importance to long fractures. In both cases, the roses are a statistical device (circular histogram) that allows one to lump many fractures together and see what the principal trends are and how they

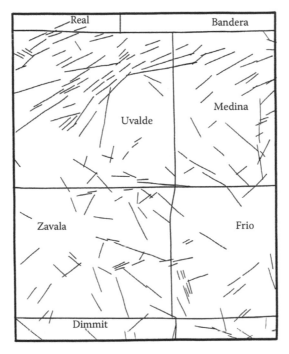

FIGURE 11.6
Fracture map interpreted from Landsat thematic mapper imagery over part of south Texas. Fractures have been broken into straight line segments for input into a fracture digitizing program.

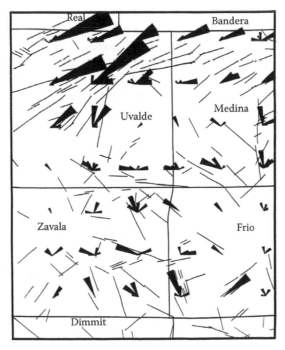

FIGURE 11.7
Length-weighted rose diagrams for each 10 × 10 km grid cell in the area of Figure 11.6. Rose petals are for 15° increments of azimuth.

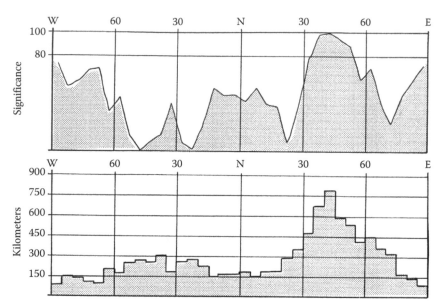

FIGURE 11.8

Significance plot (above) shows statistically significant trends at the 80% confidence level. (35–55E). Note that an absence of fracturing in a given direction can also be significant. Histogram (below) of total line-kilometers of fractures in each 5° azimuth range for the area of south Texas shown in Figure 11.6.

change across an area. It is also possible to run measures of statistical significance on large numbers of fractures to determine which fracture trends (or trends that lack fractures) are significant at a given confidence level (Figure 11.8).

One may be interested in finding the areas of most intensely fractured rock, zones that facilitate migration of fluids (e.g., hydrothermal solutions, oil, or groundwater). In this case, it is useful to generate a map of fracture intensity for all fractures, either frequency- or length weighted (Figure 11.9) (Sawatzky and Raines, 1981). One can also generate a map that contours the intensity of fracture intersections based on the assumption that areas with the most intersections will be most intensely fractured (Figure 11.10). We may suspect, based on well-bore breakouts or regional stress measurements, that one fracture set is more likely to be open and have good permeability, or have economic metals or minerals, whereas other sets are more likely to be filled with barren veins or closed. In this case, we can make a fracture intensity contour map of just the trend of interest (Figure 11.11).

Examination of cumulative production maps in oil fields that produce from fractured reservoirs indicate that adjacent wells can have highly variable production. The same holds true for gas fields and groundwater. This suggests that drilling in a fracture zone can improve the chances of good production, but it is essential to hit an open fracture and tap into the reservoir. Since most fractures below a few hundred meters depth are near vertical, horizontal drilling is a very effective way to improve the odds of tapping into the fracture system.

There are indications that fracturing is fractal (see, e.g., Barton and La Pointe, 1995; Ortega et al., 2006; Zazoun, 2008). Patterns of fracturing appear self-similar at various scales. The relative spacing between sets (the spacing ratio) and orientations of fracture sets should be the same at all scales from satellite interpretations to outcrop maps to core. This assumption, although imprecise, allows us to use a nomograph developed by Nolen-Hoeksema and Howard (1987) to estimate the optimum direction for drilling horizontal

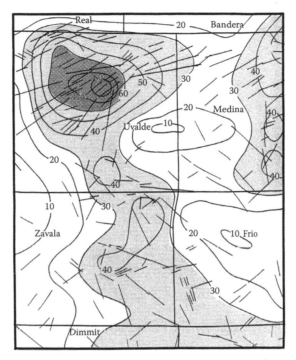

FIGURE 11.9
Map of fracture intensity for the area shown in Figure 11.6. Contours are kilometers of fractures per 10 × 10 km grid cell.

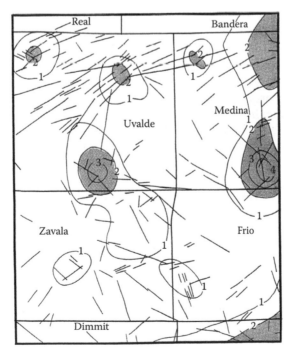

FIGURE 11.10
Fracture intersection density map for the area of Figure 11.6. Contours are the number of intersections per 10 × 10 km grid cell.

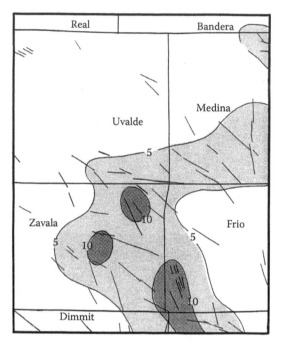

FIGURE 11.11
Fracture intensity map of all northeast-trending fractures in the area of Figure 11.6. Contours are the line-kilometers of fractures per 10 × 10 km grid cell.

wells (Figure 11.12). It is necessary to determine the primary and secondary fracture sets from the imagery, determine the closest spacing between fractures of the primary set (S^*_1), the closest spacing between fractures of the secondary set (S^*_2), the azimuth of the primary set (θ_1), and the azimuth of the secondary set (θ_2). The optimum drilling direction derived from the nomograph is the orientation of the horizontal well that will intersect the maximum number of fractures perpendicular or near-perpendicular to the well bore.

The most reliable information obtained from fracture maps is orientation data, since it is a statistical composite. In most cases, one can map surface fracture patterns and expect that these fracture orientations will exist at depth, although any one of the surface sets may be dominant at a particular depth or in a particular horizon.

The location of a surface fracture may not indicate its exact position at depth, since the fracture may represent a diffuse or inclined fracture zone, and in any case most individual fractures do not extend from the surface to great depths (if they did, the hydrocarbons would all escape!). Thus, locating a vertical well on a surface fracture will not guarantee hitting the fracture at the reservoir level. However, drilling a horizontal well perpendicular to the surface indications of fracturing, particularly that fracture set parallel to the present day principle horizontal stress, should vastly improve the chance of hitting open, conductive fractures.

Fracture intensity maps are fully valid only under conditions of uniform surface cover (the same lithology) and horizontal or gently inclined strata. Changing lithologies will cause changing fracture intensities and trends, since these features are strongly dependent on the mechanical properties of the rock. Moderate to steeply dipping beds will cause a succession of narrow outcrop bands of different lithologies across an area, resulting in bands of varying fracture intensity. Under such conditions, the fracture intensities should be compared within a unit along strike, but not from one unit to another.

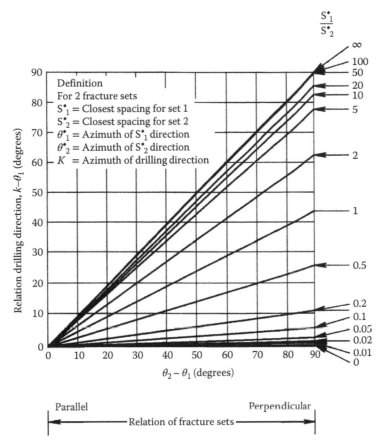

FIGURE 11.12
Nomograph suggesting the optimum direction to drill horizontal wells based on the orientation of primary and secondary fracture sets. (From Nolen-Hoeksema, R.C., J.H. Howard, 1987, *Bull. Am. Assn. Pet. Geol.*, 71, 958–966.)

Coalbed Methane

Coalbed methane is a significant source of natural gas. The Colorado Geological Survey, for example, estimated coalbed methane resources of 878 billion cubic meters (31 tcf) in the south half of the Piceance basin (Tremain et al., 1981); 700 billion cubic meters (25 tcf) of gas in-place has been estimated for coals in the San Juan basin (Bell et al., 1985). Important coalbed gas resources in North America exist in the Black Warrior basin of Alabama and in the Powder River basin of Wyoming and Montana, among others. Detecting fracture systems in coal horizons is crucial to producing this gas efficiently.

Cleat in coal is essential to producibility of gas. Coal fractures consist of nearly parallel, thoroughgoing joints, known as the face cleat, and a shorter, less continuous joint set known as the butt cleat, generally perpendicular to and terminating against the primary set. Fractures provide permeability and surfaces that allow gas to desorb from the coal particles. Although coal can form fractures early during lithification in a prevailing stress field (Ting, 1977), many workers describe coal cleat as generally parallel to, but better developed than, joints in adjacent rock (tectonic fractures; e.g., Boreck and Strever, 1980). The following case history describes a project involving the evaluation of a coalbed methane prospect using remote sensing techniques.

Case History: Developing Coal Bed Methane, Piceance Basin, Colorado

Amoco Production Company performed a detailed photointerpretation over three exploratory coal bed methane well sites in the Piceance basin, Colorado during the mid-1980s. The surface is rolling hills covered by forests and meadows, and the units are gently dipping near the basin synclinal axis. The area was divided into a grid and length-weighted rose diagrams of fracture trends were generated for every 10 km^2 area. This work was done prior to drilling and evaluating the area with vertical test wells.

In the first area, Muddy Creek Canyon, airphoto lineaments interpreted from 1:58,000 color-infrared stereo airphotos showed a dominant fracture trend of N60W. Surface fractures were measured at the future site of the Powers Federal #1. The Ohio Creek sandstone is at the surface, 710 m stratigraphically above the top of the Cretaceous Williams Fork Formation coal. Fracture measurements at two surface outcrop locations provided two dominant trends: north to N15E, and N60E to N75E (Figure 11.13). Oriented cores taken in the Williams Fork Formation at depths between 725 and 727 m identified face cleat at five places on the core. This cleat varied from N36W to S43W, perhaps as a result of the inaccuracies inherent to core orientation technology (W.B. Hanson, oral communication, 1988). Cleat frequency (inverse of spacing), also measured (map, below) at each of the five places on the core, was used to "weight" the orientations. This yielded a dominant face cleat direction of N45-85W. In this area, the joint sets interpreted from the airphotos better represented the cleat trends in core than did measurements on outcrops about 1.6 km from the well site.

Airphoto fracture sets in the second area, around the Electric Mountain #1 well, trend N60-75E, N20-30W, and N60-80W (Figure 11.14). Surface fracture measurements were made near the well site in sandstones and mudstones of the Eocene Wasatch Formation. Outcrop measurements indicate that a primary fracture set trends N60-75E and a secondary set trending N30-45W. Oriented core was acquired from 1477 to 1485 m depth in the Williams Fork Formation. On the basis of one coal sample, the face cleat was determined to be N74E and butt cleat approximately N16W. Both outcrop and airphoto measurements predicted the subsurface cleat trends accurately.

Analysis of airphoto fracture trends near the Ruth Mountain #1 well indicate that the dominant set trends N30-40E and a secondary set trends N20-30W (Figure 11.15). No outcrops were available near this well site for surface measurements. Oriented core in the Williams Fork coal at a depth of 2199 m provided a face cleat azimuth of N25W. Core taken at 2334 m had a face cleat of N23W. In this case, the dominant fracture trend from imagery corresponded to the secondary fracture trend at the targeted coal horizon, while the secondary surface orientation turned out to be the primary, or face cleat, in coals.

One may conclude from these examples that coal cleat orientation can in many cases be predicted on the basis of surface (airphoto, Lidar, or outcrop) measurements. These predictions can be used to help plan well locations along fracture zones, and to lay out directional well trajectories to best drain a coal gas reservoir. Although one was able to predict fracture trends in the subsurface, it was not possible to predict which of the surface joint sets would be dominant (face cleat) in the coal horizon at depth.

Secondary Recovery and Waterfloods

Remote sensing can be used to help evaluate waterflood or other secondary recovery operations where there is the likelihood of fractures controlling the flow of injected fluids or leading to premature breakthrough along the waterflood front. In some cases, pulse tests have shown no fluid communication between injection wells and producing wells on opposite sides of lineaments. In other cases, water or CO_2 from the secondary recovery operation enters an open fracture system and there is premature breakthrough of the flood front. For example, at the greater Aneth field in southeast Utah, horizontal wells are

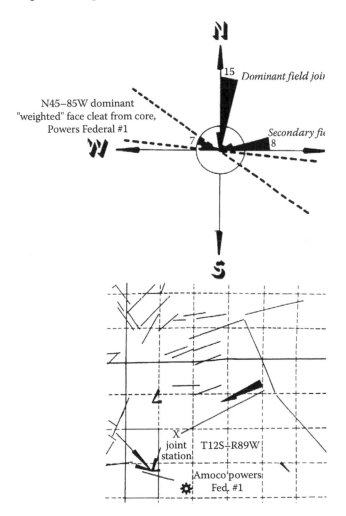

FIGURE 11.13
Airphoto interpretation map with rose diagrams for a 10 km² grid, and field and oriented core measurements of fractures at the Amoco Powers Federal #1, Gunnison County, Colorado. The rose diagram (above) shows joints measured on outcrops in the field and face cleat from oriented core. The number of field-measured joints in the dominant sets are noted on the rose. In this area, photogeologic joint sets near the well give a better indication of the face cleat in coals (725–727 m depth) than surface measurements taken at an outcrop 1.6 km north of the well.

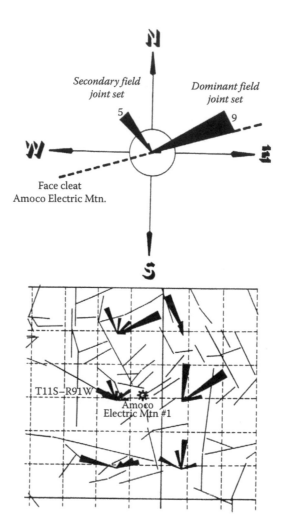

FIGURE 11.14

Airphoto interpretation map with rose diagrams for a 10 km² grid, and field and oriented core measurements of fractures at the Amoco Electric Mountain #1, Delta County, Colorado. The rose diagram (above) shows joints measured on outcrops in the field and face cleat from oriented core. The number of field-measured joints in the dominant sets are noted on the rose. Face cleat in coals (1477–1485 m depth) corresponds to the dominant outcrop fracture set, and to one of several major photogeologic fracture trends around the well.

oriented parallel to fault/fracture zones to prevent rapid CO_2 breakthrough (Chidsey et al., 2008). At Aneth, it was found that the fracture zone acts either as an impermeable barrier to fluid flow or as a conduit that carries fluids along the zone and prevents fluid flow across the zone.

Case History: Understanding a Waterflood at Cottonwood Creek Field, Wyoming

Airphotos were used to evaluate a secondary recovery program in a fractured reservoir at the Cottonwood Creek field in the Bighorn basin, Wyoming (Figure 11.16). Cottonwood Creek is a stratigraphic trap in the west-dipping Permian Phosphoria Formation at depths from 1538 to 3077 m. The trap is the result of a porous dolomite grading updip into an impermeable shale-anhydrite facies (Willingham and McCaleb, 1967). The Phosphoria is

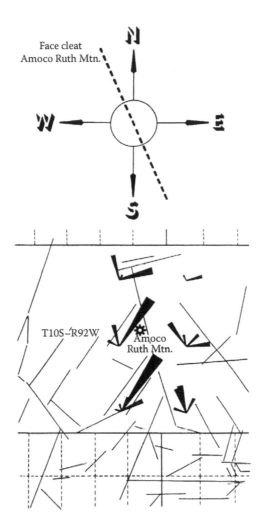

FIGURE 11.15
Airphoto interpretation map with rose diagrams for a 10 km² grid, and field and oriented core measurements of fractures at the Amoco Ruth Mountain #1, Mesa County, Colorado. The rose diagram (above) shows face cleat from oriented core. Face cleat in the coal (2199 and 2334 m depth) matches the secondary photogeologic fracture set. No outcrops were available for fracture measurements within several kilometers of the well.

~100 m thick in the field, and porosity averages 11%, with permeability averaging 13 mD. Engineers first suspected a fractured reservoir when a gas injection program initiated in 1958 resulted in rapid movement of gas to producing wells and oil production declined. A water injection program a year later had similar results. Jointing proved to be a disadvantage in this case because there was premature breakthrough of the injected gas and fluid along probable fracture zones.

Remote sensing was used to map fracture zones and determine if surface lineaments had any relation to subsurface reservoir properties (Prost, 1996). This information would be used to plan the drilling of new wells. Lineaments (inferred fractures) were mapped on color-infrared stereo airphotos at a scale of 1:58,000 (National High Altitude Program photos). The lineaments were then digitized and the density of inferred fractures was contoured in kilometers of fractures per 10 km² area (Figure 11.17).

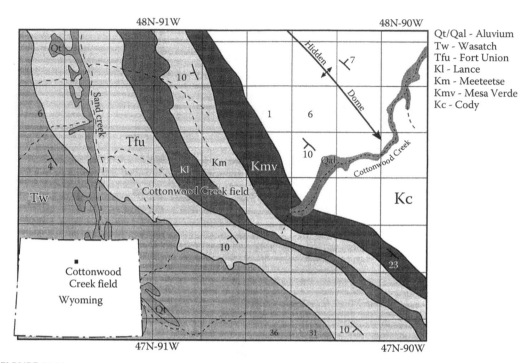

FIGURE 11.16
Surface geologic map of the Cottonwood Creek field area, eastern Bighorn basin, Wyoming. The field is a stratigraphic trap down dip from Hidden Dome.

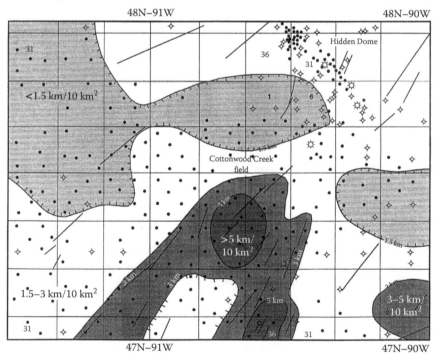

FIGURE 11.17
Map showing the distribution of wells in the Cottonwood Creek field superimposed on the northeast fracture density contours (kilometers of fractures per 10 km² grid) and northeast interpreted fractures.

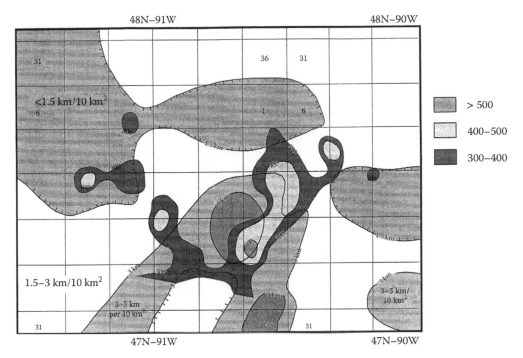

FIGURE 11.18

Contour map of porosity-feet, derived from well data, superimposed on the northeast fracture density contours. Porosity-feet is a measure of the reservoir volume above a specified porosity cutoff, in this case between 7% and 8.5% (e.g., 100 ft. of 8% porosity yields 8 porosity feet). Note the northeast alignment of the high-porosity zone.

The map of total fracture density did not correlate well with production data. Bedding dips west–southwest in this area, and many interpreted north–northwest lineaments may have been related to the strike of bedding. The next step was to determine if a correlation existed between production and fractures with a specific trend. Indeed, the map of northeast fracture density showed areas of more intense fracturing that corresponded to zones of increased total porosity-feet and high cumulative production (Figures 11.18 and 11.19). The regional maximum horizontal compressive stress is also oriented east-northeast in this area. This would tend to keep northeast fractures open. Open fractures provide fracture porosity and lead to both higher cumulative production and premature breakthrough of waterfloods along the northeast fracture zones. The results of this exercise allowed production engineers to take into account the role of fracturing, specifically northeast fractures, when planning future secondary recovery programs in this field.

Extending Known Mineral Deposits

As with oil fields, one may extend a known mineral trend and/or increase reserves by stepping out from existing deposits. The procedure is to map the characteristics associated with a known occurrence, usually a specific rock type, alteration assemblage, or fault/vein trend, then look for the same characteristics in the surrounding countryside. The following example is taken from the work by Agar and Villanueva (1997).

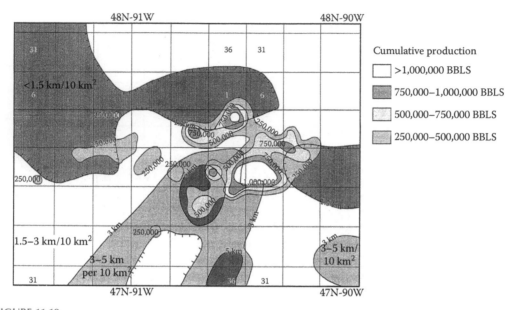

FIGURE 11.19

Contour map of cumulative production to 1986 (in barrels) superimposed on the northeast fracture density contours. Note the northeast alignment of high cumulative production.

Case History: Extending Known Deposits in the Huancavelica Mining District, Peru

The Andes of south-central Peru are remote and contain high-relief topography that makes it difficult for field parties to map on the surface. The high, arid, vegetation-free terrain is, however, well-suited for airborne remote sensing. Cia. De Minas Buenaventura S.A. operates several mines in the Huancavelica area ~250 km southeast of Lima. The northeastern part of the district contains Paleozoic basement exposed in the cores of domes and the Triassic–Jurassic Pucará group (including carbonates, bituminous shales, and phosphates) and Cretaceous–Tertiary sediments in flanking synclines. The southwestern part of the district contains Tertiary volcanic sequences overlying the older rocks. The volcanics are intruded by penecontemporaneous andesites and dacites. Most of the mineral deposits are associated with extensively altered volcanics, although lead–zinc–silver mineralization occurs within the Pucará group. Field mapping in this area is difficult, expensive, and suboptimal due to restricted access.

Sixty-three channel hyperspectral data were acquired using the digital airborne imaging spectrometer operated by Geophysical and Environmental Research (GER). Imagery was acquired over 5000 km² south of Huancavelica during October of 1996. Ground resolution varied between 10 and 14 m. The instrument has 27 channels in the visible and near infrared between 410 nm and 1.048 μm; 30 channels in the short-wave infrared from 1.40 to 2.42 μm; and six channels in the thermal infrared between 8.95 and 10.98 μm.

Ground spectra were gathered over hydrothermally altered areas prior to the overflights. These control points contain spectra of soil, fresh rock, and weathered rock. Weathered surfaces were found to contain spectra similar to, but less intense than, the fresh samples. Likewise, spectral curves acquired during the survey correlated well with spectra from the USGS spectral library (Figure 11.20) and ground control points (Figure 11.21).

Images were evaluated for areas of hydrothermal alteration, and favorable "target" areas were chosen and ranked based on the presence of argillic, propylitic, iron oxide,

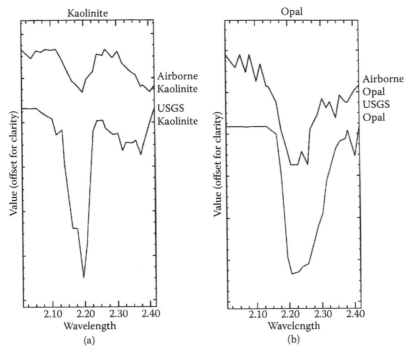

FIGURE 11.20
Geophysical and Environmental Research digital airborne imaging spectrometer airborne spectra versus USGS spectral library. (a) Kaolinite; (b) Opal. (From Agar, R.A., R. Villanueva. 1997, *Proc. of the 12th Intl. Conf. Applied Geologic Rem. Sens.*, Ann Arbor, ERIM, I-13–I-17.)

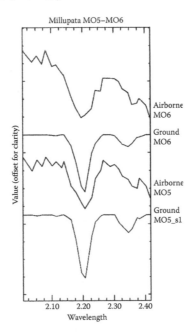

FIGURE 11.21
Geophysical and Environmental Research digital airborne imaging spectrometer airborne spectra versus ground spectra of the same area, samples M05 and M06, Millupata area. (From Agar, R.A., R. Villanueva. 1997, *Proc. of the 12th Intl. Conf. Applied Geologic Rem. Sens*, Ann Arbor, ERIM, I-13–I-17.)

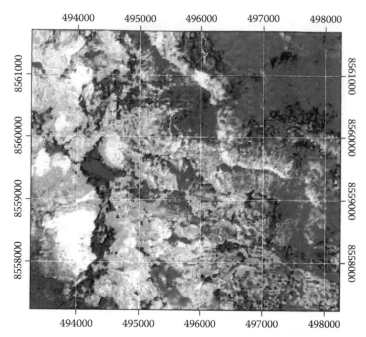

FIGURE 11.22
Hyperspectral alteration image of the Huancavelica area. Colors are designed to enhance propylitic alteration (chlorite, epidote, calcite), shown as white areas. (Processed by A.G.A.R.S.S. Pty. Ltd. in conjunction with Cia. Buenaventura.)

and siliceous alteration as well as faulting (Figure 11.22). All of the known mines and deposits contain these alteration minerals and were found to occur within these targets. Some targets, such as Arcopunco, were on unclaimed land and were quickly claimed after the initial overflight and before work continued. Evaluation of other areas was put on hold because of a lack of evident alteration at the surface. Narrow, veinlike targets in the Pucará limestone may represent Mississippi Valley-type mineralization. Areas considered most prospective for epithermal gold were subjected to further remote sensing image processing.

Mineral distribution maps were created in high-priority areas such as Arcopunco. A group of key alteration minerals was selected based on prior work. Images were generated for these reference minerals showing their distribution and where they are most intense (Figure 11.22). These reference minerals then became the end-members for spectral unmixing (Boardman, 1989; Boardman and Kruse, 1994). Spectral unmixing was used to determine the proportions of these end-member minerals. The images produced by unmixing provided an estimate of the abundance and distribution of each mineral. The resulting image maps were used to locate altered targets. Ground spectra confirmed the accuracy of the images.

This process was successful. Alteration at producing mines was characterized by acquiring ground spectra. These spectral curves were used to calibrate an airborne survey that mapped similar zones in the adjacent region. The resulting mineral alteration maps were used to direct the filing of claims and to prioritize properties already held. Epithermal gold targets were mapped, and a new class of target, the Mississippi Valley lead–zinc–silver association, was proposed.

Slope Stability and Mine Safety

There are a number of mining-related safety issues that can be addressed and possibly mitigated using remote sensing. These safety issues include subsidence over active and abandoned underground mines, slope stability in open pit mines, and roof falls in underground mines.

Mine subsidence is recognized as topographically low areas that generally are not located along streams and do not have drainage developed into them because of their young age (see Subsidence Over Abandoned Coal Mines Case History, Chapter 16). They may be filled with water or moist (darker) soil. They can be round, elongate, or branching, and range in size from less than one hectare to several square kilometers. A series of depressions adjacent to tailings is a good clue that a mine has collapsed. Factors affecting the size and shape of surface collapse over underground mines include tectonic and residual stresses, faults and foliation, width and height of the excavation, presence of groundwater, and strength and thickness of the overburden (Lee et al., 1976).

Just as surface subsidence can result from geologic discontinuities (weaknesses such as faults), so, too, mine roof falls result from these features. Lineament (interpreted fault) concentrations were determined to be related to coal mine roof falls in Virginia (Milici et al., 1982) and elsewhere. A vertical decompression of 7000 psi and subsequent subsidence was caused by underground mine excavation across a fault in metasediments at ~107 m depth near Idaho Springs, Colorado (Lee et al., 1976).

Fracture length (or the length of a zone of fracture segments) is more significant than density, as pervasive and apparently random short fractures suggest a surficial effect, whereas longer fractures are likely to originate deeper in the subsurface (Davison, 1994; Clark and Cox, 1996; Morley, 1999). As previously stated, a series of short faults along a trend, or a set of en echelon faults, may in fact be surface manifestations of a single large fracture at depth.

Slope stability in open pit mines (or highway road cuts, housing developments, dam abutments) can be adversely affected by the presence of joints and faults (Call, 1972; Savely, 1972; Montazer, 1978; Ness, 1982). The least damaging fractures are perpendicular to the strike of the slope, whereas the most unstable conditions occur when fractures are parallel to the slope and may be undercut by excavation or erosion, or lubricated by rain or melting snow. Interpretation of fault and joint orientation and location from high-resolution satellite images, Lidar, or medium- to low-altitude airphotos can help direct remediation by rock bolting, terracing, and draining the outcrop, thereby reducing the risk of catastrophic slope failure.

References

Agar, R.A., R. Villanueva. 1997. Applications of multispectral GER DIAS 63 band data in exploration of deposits at Huancavelica, Peru. In: *Proceedings of the 12th International Conference on Applied Geologic Remote Sensing*. Ann Arbor, MI: Denver, CO: ERIM: I-13–I-17.

Barton, C.C., P.R. La Pointe (eds.). 1995. *Fractals in the Earth Sciences*. Berlin: Springer: 288 p.

Bell, G.J., J.C. Seccombe, K.C. Rakop, A.H. Jones. 1985. Laboratory characterization of deeply buried coal seams in the western U.S. *Soc. Petrol. Eng. SPE* 14445: 1–11.

Boardman, J.W. 1989. Inversion of imaging spectrometry data using singular value decomposition. In: *Proceedings of IGARSS'89, 12th Thematic Canadian Symposium on Remote Sensing*, v. 4, Ann Arbor, MI: ERIM: Denver, CO: 2062–2069.

Boardman, J.W., F. A. Kruse. 1994. Automated spectral analysis: a geological example using AVIRIS data, northern Grapevine Mountains, Nevada. In: *Proceedings of 10th Thematic Conference on Geologic Remote Sensing*, v. I, Ann Arbor, MI: ERIM: I-407–I-418.

Boreck, D.L., M.T. Strever. 1980. *Conservation of Methane from Mined/Minable Coal Beds, Colorado.* Colorado Geol. Survey Open-file Rept. 80–5: Denver, CO: 95 p.

Broadhead, R.F. 1990. Bravo Dome carbon dioxide gas field. In: Beaumont E.A., N.H. Foster (eds.), *Structural Traps 1, Treatise of Petroleum Geology Atlas of Oil and Gas Fields*. Tulsa: American Association of Petroleum Geologists: 213–232.

Call, R.D. 1972. Analysis of geologic structure for open pit slope design. PhD dissertation. Tucson: University of Arizona: 201 p.

Chidsey, T.C. Jr., C.D. Morgan, D.E. Eby. 2008. *Major oil plays in Utah and vicinity*. Quarterly Technical Progress Reports, DOE Contract No. DE-FC26-02NT15133: 45 p.

Clark, R.M., Cox, S.J.D. 1996. A modern regression approach to determining fault displacement–length relationships. *J. Struct. Geol*. 18: 147–152.

Davison, I. 1994. Linked fault systems; extensional, strike-slip and contractional. In: Hancock P.L. (ed.), *Continental Deformation*. New York: Pergamon: 121–142.

Ewing, T.E. 1983. Austin/Buda fractured chalk. In: Galloway, W.E., T.E. Ewing, C.M. Garrett, N. Tyler, D.G. Bebout (eds.), *Atlas of Major Texas Oil Reservoirs*. Austin, TX: Texas Bur. Econ. Geol.: 41–42.

Lee, F.T., J.F. Abel, Jr., T.C. Nichols, Jr. 1976. *The Relation of Geology to Stress Changes Caused by Underground Excavation in Crystalline Rocks at Idaho Springs, Colorado. U.S. Geol. Survey Prof. Paper 965: 47 p.

Milici, R.C., T.M. Gathright, II, B.W., Miller, R. Gwin. 1982. Geologic factors related to coal mine roof falls in Wise County, Virginia. Virginia Div. Mines Res. Rept. ARC-CO-7232-80-1-302-0206: 101 p.

Montazer, P.M. 1978. Engineering geology of upper Bear Creek area, Clear Creek County, Colorado. MSc. thesis. Golden: Colorado School of Mines: 202 p.

Morley, C.K. 1999. Patterns of displacement along large normal faults: implications for basin evolution and fault propagation, based on examples from East Africa. *Bull. Am. Assn. Pet. Geol*. 83: 613–634.

Ness, M.S. 1982. A slope stability analysis of the Pinto Valley mine, Miami, Arizona. MSc. thesis. Golden: Colorado School of Mines: 92 p.

Nolen-Hoeksema, R.C., J.H. Howard. 1987. Estimating drilling direction for optimum production in a fractured reservoir. *Bull. Am. Assn. Pet. Geol*. 71: 958–966.

Ortega, O.J., R.A. Marrett, S.E. Laubach. 2006. A scale-independent approach to fracture intensity and average spacing measurement. *Bull. Am. Assn. Pet. Geol*. 90: 193–208.

Parker, J.M., P.D. Hess. 1980. The Mondak Mississippian oil field, Williston basin. *U.S.A. Oil. Gas. J*.: 78: 210–216.

Prost, G. L. 1996. Airphoto fracture analysis of the Cottonwood Creek field, Bighorn basin, Wyoming. SPE Preprint 35289. International Petroleum Conference and Exhibition, Villahermosa, Mexico: 25–32.

Rountree, R. 1984. Fractured Niobrara, Denver basin. *Western Oil Reporter*. 41: 31–40.

Savely, J.P. 1972. Orientation and engineering properties of jointing in the Sierrita pit, Arizona. MSc. thesis. Tucson: University of Arizona: 134 p.

Sawatzky, D.L., G.L. Raines. 1981. Geologic uses of linear-feature maps derived from small-scale images. In: O'Leary D.W., J.L. Earle (eds.), *Proceedings of the 3rd International Conference on Basement Tectonics*, Denver, CO: Basement Tectonics Committee, : 91–100.

Snyder, R.H., M. Craft. 1977. Evaluation of Austin and Buda formations from core and fracture analysis. *Gulf. Coast. Assn. Geol. Soc. Trans*. 27: 376–443.

Ting, F.T.C. 1977. Origin and spacing of cleats in coal beds. *J. Pressure. Vessel. Tech. Trans. of Am. Soc. Mech. Engrs*. 5: 624–626.

Tremain, C.M., D.L. Boreck, B.S. Kelso. 1981. Methane in Cretaceous and Paleocene coals of western Colorado, in: Western slope Colorado— western Colorado and eastern Utah. In: Epis R.C., J.F. Callender (eds.), *New Mexico Geological Society Guidebook, 32nd Field Conference*: Grand Junction, CO: 241–248.

Willingham, R.W., J.A. McCaleb. 1967. The influence of geologic heterogeneities on secondary recovery from the Permian Phosphoria reservoir, Cottonwood Creek, Wyoming. *Soc. Petrol. Engrs*. Paper SPE 1770: 12.

Zazoun, R.S. 2008. The Fadnoun area, Tassili-n-Azdjer, Algeria: Fracture network geometry analysis. *J. African Earth Sciences*. 50: 273–285.

12

Hydrology

Chapter Overview

The price of commodities goes up and down, as anyone in the oil and mining industries can attest. Yet water is one commodity that is increasingly in demand just about everywhere. An oil company drilling a well may be able to use brackish or saline water to mix their drilling mud, but fresh potable water is required for communities and agriculture and is not always available. Already we see competition between communities and farmers for scarce water resources. Remote sensing can help locate sources of groundwater, measure the amount and distribution of surface water, and in some cases can infer whether it is fresh or brackish. It is important to be able to monitor surface water for changes in salinity (agriculture), temperature (weather prediction), or turbidity (human consumption), both onshore and in marine settings. Water quality is addressed in Chapter 14, and flooding and erosion in Chapter 16. In this chapter, we look at how remote sensing can be used to locate groundwater, surface water, and measure soil moisture.

Geothermal energy is one component of any future energy mix. Essentially all obvious geothermal resources in North America have been identified, yet many small thermal springs and near surface heat sources may not have been catalogued. In most of the world, economic thermal resources have never been mapped. We look at how remote sensing provides a unique opportunity to identify and map geothermal systems.

Locating Sources of Groundwater

There are a number of ways one can use remote sensing to help look for sources of water. The simplest is to use airphotos or satellite imagery, either color or black/white, to locate dense and verdant vegetation (Taylor et al., 1980). This could be on a hillside, where a spring is discharging groundwater, on a pediment or alluvial fan, or in the bottom of gullies or dry creek beds. In temperate areas, trees and grasses are greener where groundwater is near the surface. In arid or semiarid areas, shrubs and trees are greener and denser where groundwater is near the surface. Vigorous plant growth is especially obvious on color infrared images, where the plants appear bright red.

"Phreatophytes" are plants that have deep roots that are able to reach the water table. Phreatophytes live in areas with standing or running water, in arid areas and along riverbeds and places that are apparently dry, but where the water table is near the surface. Some phreatophytes have a low tolerance for salt, indicating a fresh water aquifer. These plants can be a guide to the location of drinking and agricultural water in arid and semiarid areas. Examples of phreatophytes include honey mesquite (*Prosopis glandulosa*) and trees like the Ash, Alder, Willow, Cottonwood, and the Poplar. These trees grow in freshwater aquifers, where the water table is not more than 10 m below the surface.

Certainly if springs occur locally, these will show up on airphotos or satellite imagery as dark (moist) soil tones associated with dense, healthy green vegetation. Healthy plants that appear green on true color photography will be bright red on color-infrared imagery or photography because their red color is in stark contrast to the darker background plants, rocks, and soil.

Groundwater may be dammed against faults, forming alignments of springs that can be seen on imagery, usually expressed as lush vegetation where water is near the surface (Figure 5.59). Groundwater also collects in fracture zones that provide increased porosity and enhanced permeability (Lattman and Parizek, 1964). Fracture zones are also more easily eroded and tend to form topographic low spots that act as natural collection areas for surface and near-surface water. Near-surface groundwater often appears on NIR imagery as areas containing more vigorous vegetation in the summer and fall, as areas of early leafout in the spring, and delayed leaf senescence in the fall. In general, one is more likely to encounter groundwater close to the surface in topographically low areas, especially if they occur along a fault. Detecting groundwater by mapping lineaments (fault zones) has been shown to be an effective means of ensuring a fresh water supply, particularly in crystalline terrains (Dalati, 1997, 2004). Some faults have a greater capacity for groundwater storage than others: this may be the result of open fractures and faults parallel to the maximum regional compressive stress.

Thermal imagery can also help locate groundwater. The effects of solar illumination will overpower any signal derived from evaporative cooling in areas with moderate or high relief, but in areas with low relief moist soils appear cooler than surrounding dry areas. On predawn imagery, the effects of differential solar heating are subdued. Areas of moist soil appear warm on nighttime imagery due to the high thermal inertia of water.

Abandoned or ancient, buried river channels may be excellent sources of groundwater. These channels may be covered by thin sand sheets or alluvium and may have a slight topographic depression or a slightly different texture at the surface. Thermal imagery can detect the telltale dendritic pattern of cooler soils over the old channels. Radar has been credited with detecting the dark (moist) patterns of paleochannels in "hyperarid" regions by penetrating the sand cover to depths of 1 or 2 m (Figure 12.1) (McCauley et al., 1982; Robinson et al., 2000; Ghoneim et al., 2007). Ground-penetrating radar (GPR) surveys can detect the water table as well as contacts between fresh water and saline water or water and other liquids such as gasoline (Battelle Labs, 1983; Graf, 1990). This is caused by an electrical conductivity contrast between subsurface fluids and appears similar to an acoustic impedance contrast on seismic lines. A GPR survey was used to locate groundwater in unconsolidated sand and gravel at depths up to about 60 m in the arid Sinai near St. Catherine's monastery (Elfouly, 2000).

An integrated groundwater exploration program over the Serowe area of eastern Botswana using airborne magnetics and a Very Low Frequency (VLF) electromagnetic survey is described by Bromley et al. (1994). The Ntane Sandstone aquifer is confined between a mudstone below and basalt above, and the entire area is blanketed by 20–60 m of unconsolidated sand and/or the Kalahari Sandstone. The survey was flown at 20 m elevation above ground using 400 m line spacing. The instruments included (1) a three-component fluxgate magnetometer for measuring the total magnetic field; (2) a VLF system tuned to a transmitter in Australia and used to enhance conductive zone anomalies, and (3) a dual frequency (910 and 7040 Hz) coaxial EM system with a horizontal axis and 17 m coil separation. The EM surveys were confounded by low resistivity layering in the Kalahari beds

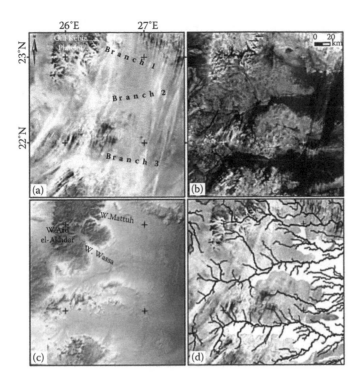

FIGURE 12.1
Radar and topography reveal old river channels buried beneath sand in the eastern Sahara. (a) The three wadi branches of the Gilf Kebir area are barely depicted in the Landsat ETM+ image. These drainage networks are clearly visible in both the RadarSat-1 (b) and Shuttle Radar Topography Mission (SRTM) data (c). (d) The RadarSat dark areas (whited out) have SRTM-derived channels (dark lines) overlain. The derived channels lie on the north side of the radar-dark feature, suggesting that the radar-dark signals may not be directly related to fluvial deposits, but rather are due to wind-blown sand and fine materials, that is, a texture difference. (From Ghoneim, E. et al. 2007. *Int. J. Remote Sens.* 28: 1759–1772. Reprinted with permission from Taylor & Francis Ltd, www.tandfonline.com.)

that made it impossible to separate the overlying units from the aquifer. The VLF survey revealed structural features (northwest-oriented fracture zones) that are crucial to locating groundwater (Figure 12.2). The results guided a subsequent drilling program that proved up 35,000 m³/day water sustainable for 25 years. The highest yields are associated with fracture zones identified by the VLF anomalies.

In some areas, water is more valuable than oil. As part of an application for an oil exploration permit in Bahrain, Amoco was asked if fresh water could be located as part of the exploration effort (Prost, 1988). Day and night Landsat Thematic Mapper thermal images were processed to determine if sources of fresh water could be found. Spatial resolution is 120 × 120 m, and thermal resolution is estimated at about 2°C. Nighttime imagery was acquired around 9:30 PM. Thermal and visible imagery was interpreted for evidence of faulting. Onshore and offshore areas were examined and it was determined that nighttime imagery showed cool spots off the east coast of the island near probable faults (Figure 12.3). It was felt that these cool areas might indicate subsea freshwater springs. Submarine springs would be cooler than the warm waters of the shallow Persian Gulf and would rise to the surface due to the lower density of fresh water. Pearl divers and others have encountered fresh water emanations offshore Bahrain (Taniguchi et al., 2002), but it is not known whether the cool spots mapped in this study were ever tested for fresh water.

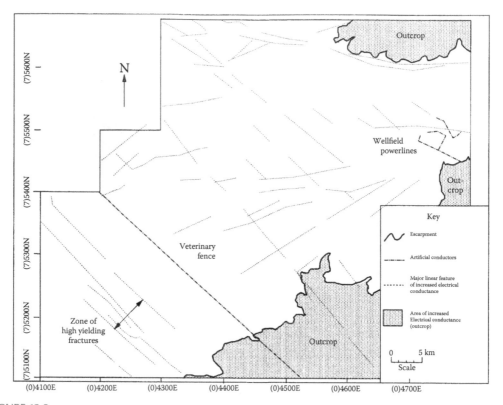

FIGURE 12.2
VLF interpretation map showing location of major fractures discerned by the VLF survey. (From Bromley et al 1994. *Ground Water* 32: 79–90. Reproduced with permission of Dr. John Bromley, Oxford University Centre for the Environment (OUCE), Centre for Water Research, Oxford.)

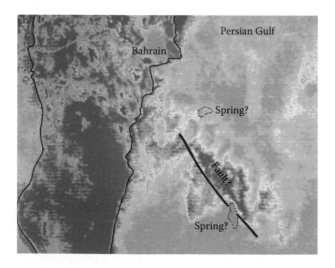

FIGURE 12.3
Cool, possibly fresh water (blue to cyan) in warmer Persian Gulf waters (green to red) off the east coast of Bahrain (cool blue area inside outline). An abrupt break in temperatures suggests a northwest-southeast-trending offshore fault that controls the water depth. Offshore fresh water springs are suggested by local cool spots. Landsat TM band 6 nighttime image. (Image processed by Amoco Production Company.)

In a similar study, Nelson et al. (1991) used thermal imagery and sampling to pinpoint small springs in Red Eagle Lake, Montana. Imagery was acquired at 6:00 AM from a helicopter using a thermal infrared video sensor angled ~45° from horizontal and having ground resolution of about 4 m. One spring, estimated to be flowing 450 L/min, showed anomalous cool temperatures. Spring water was up to 3°C cooler than the lake temperature as far as 10.5 m from the spring.

Soil Moisture Determination

Soil moisture is the total amount of water in the unsaturated zone, that is, above the water table. Knowledge of soil moisture conditions is used for crop planning, making predictions about infiltration (aquifer recharge) and runoff (flooding, erosion, reservoir capacity, irrigation). Soil moisture also affects heat and water transfer between the land and atmosphere, and for that reason it is used as input to weather forecasting models (Kerr, 2006).

Soil moisture content is highly variable in both time and space (Famiglietti et al., 1999). Moisture variations appear as changes in soil tone on black/white, color, or color-infrared imagery: moist soil is generally darker, and dry soil is generally lighter in appearance (Lyon, 1987). Aerochrome infrared film is especially sensitive to soil moisture (Wallen et al., 1977). The expression of moist soil and surface water on thermal imagery varies depending on the time of day and amount of moisture (Figure 12.4). Moist soil generally appears cooler than dry background soils due to evaporative cooling on daytime and nighttime imagery. Thermal imagery has been used to map near-surface groundwater in fractures, and could be used to map leaks in pipelines or dams (Sabins, 1987; Weil and Graf, 1991).

Microwave measurements of soil moisture and evapotranspiration are valuable inputs to modeling groundwater recharge (Jackson, 2002) and rainfall runoff (Scipal et al., 2005; Brocca et al., 2010). Microwave imagery averages moisture values over large areas and, with satellite data, is repeated at specific intervals. Airborne and satellite radar is sensitive to changes in moisture content if an area is relatively flat, has low to moderate plant cover, and a homogeneous surface material and texture. This is because radar waves react to changes in the electrical properties of materials, and moisture has a large effect on the electrical

FIGURE 12.4
Thermal image of surface water (warm = white) and moist ground (cool = dark, as in cutoff meanders), near the junction of Tincup Creek and the Salt River, Bonneville County, Idaho. (Predawn image flown at 2100 m altitude by Mars Inc., Phoenix.)

properties of soil. Areas of moist soil generally have a brighter appearance (higher backscatter coefficient) on radar images (Sabins, 1987). Spaceborne (SIR-C) and airborne radar (AIRSAR) have been used to measure soil moisture with an RMS error less than 4.2% (Dubois et al., 1995). L band radar is generally considered best suited for measuring soil moisture (Wagner et al., 2003). A number of satellites now carry radar instruments capable of measuring soil moisture, atmospheric water vapor, and cloud liquid water. These include the AMSR-E instrument on the EOS AQUA satellite, the Microwave Imaging Radiometer with Aperture Synthesis (MIRAS) instrument on the SMOS satellite, the Microwave Radiometer (MWR) on ENVISAT, the Advanced Technology Microwave Sounder (ATMS) and Cross Track Infrared Sounder (CrIS) on the Suomi NPP mission, the GOES Sounder on NOAA weather satellites, the Advanced Microwave Sounding Unit (AMSU)-A on the NOAA-N' platform, and the AMSU on NOAA and ESA weather satellites. As of March 2012, the NASA airborne CARVE mission also carries a cross-polarized L band radar for soil moisture and snow studies.

Passive microwave sensing also benefits from being largely unaffected by solar illumination and cloud cover. Another advantage of passive microwave sensors is that, in the absence of significant vegetation cover, soil moisture is the dominant influence on the signal. Resolution is coarse (km) since the signal is weak. The most useful frequency range for soil moisture sensing is 1-5 GHz (Njoku and Entekhabi, 1994).

Much effort has gone into creating various soil moisture indices, usually related to vegetation and temperature measurements. Su et al. (2003) used NOAA/AVHRR (MWR) data to develop a drought severity index that relates relative soil moisture at root depths to relative evaporation. They claim that soil measurements confirm the effectiveness of this index. A normalized multiband drought index for monitoring soil moisture was developed by Wang and Qu (2007). This index uses hyperspectral channels centered at 860 nm (reference band), 1640 nm (water absorption) and 2130 nm (absorption):

$$\text{NMDI} = \frac{\left(R_{860\,nm} - \left(R_{1640\,nm} - R_{2130\,nm}\right)\right)}{\left(R_{860\,nm} + \left(R_{1640\,nm} - R_{2130\,nm}\right)\right)}$$

where R is the reflectance at the reference and absorption bands. The authors claim this index is well suited to estimate both soil and vegetation moisture.

Chinese researchers have been prolific in the creation of indexes to map and predict drought. Li and Li (2008) described in general terms a temperature vegetation dryness index derived from NOAA/AVHRR satellite data and land surface temperatures. They claimed that a good correlation exists between measured soil moisture at 10 and 20 cm depths and their TVDI. Other workers used the same index, derived from 8-day composite MODIS temperature data and 16-day composite MODIS Normalized Difference Vegetation Index (NDVI) data (Wei et al., 2008). Comparing satellite-derived data to simultaneous field measured data resulted in a mean error of 11.8% and a minimum error of 4.3%.

The Cropland Soil Moisture Index (CSMI) was developed to take into account both shallow and deep soil moisture to reduce the instability of indices that results from vegetation variability:

$$\text{CSMI} = \frac{\left(\left(R_{858\,nm} * R_{2130\,nm}\right) - \left(R_{645\,nm} * R_{1645\,nm}\right)\right)}{\left(\left(R_{858\,nm} * R_{1645\,nm}\right) - \left(R_{645\,nm} * R_{2130\,nm}\right)\right)}$$

where R is the reflectance of the MODIS bands (Chen et al., 2009). The authors state that they can predict soil moisture in the 0–50 cm layer using CSMI to a high degree of accuracy.

A Canadian study shows that soil moisture is of concern even in non-arid climates. A Temperature-Vegetation Wetness Index (TVWI) is derived from MODIS bands, surface temperature, atmospheric pressure, and elevation above mean sea level (among others). The TVWI is used to indirectly determine soil moisture (Akther and Hassan, 2011). Their results were within ±20% of ground-measured soil moisture 82% of the time.

Surface Waters and Bathymetry

Short wavelengths penetrate water farther than long wavelengths, and blue light penetrates farthest. In clear water, light has been shown to penetrate between 20 and 30 m (see Chapter 1, section on "Water-covered Areas"). The response at the sensor is a function of the water depth, suspended material (turbidity), and the color of the bottom material (e.g., white sand will reflect better than sea grass), and to some extent the incidence angle and surface roughness. Lyon and Hutchinson (1995) developed a radiometric model for evaluating shallow water bodies such as rivers using airborne scanners. A computer program simulated the flow of electromagnetic energy from the atmosphere through water, reflecting off the bottom and back to the surface. The program addressed the influence of the water, water surface effects, and the atmosphere on solar radiation transfer through time and space. Inputs included the solar constant, extinction coefficients, and bottom type reflectance. Water depths could be predicted to depths of 3.7 m. Interestingly, blue wavelengths gave poor results due to scattering of these wavelengths in water. Blue-green, green, and red wavelengths showed the best correlation with water depths.

Shallow water appears light gray on panchromatic film, black on black/white infrared images, light green to light blue on true color imagery, and light blue on color infrared images. Deep water appears dark gray on panchromatic images, black on black/white infrared images, dark blue to dark green on color images, and black on color infrared imagery. Because light is reflected from near-surface suspended particles, turbid water has the same appearance as shallow water. Radar wavelengths (microwaves) are either reflected by rough water or are reflected away from the antenna (black, no returns) on smooth water (Figure 12.5). Thermal

FIGURE 12.5
ENVISAT radar image over the Galapagos Islands showing details of the sea surface. (Data from Waugh, R., 2012. Out of this world: an awe-inspiring portrait of Earth planet to celebrate the tenth anniversary of pictures from observation satellite. http://www.dailymail.co.uk/sciencetech/article-2110455/Our-coloured-planet-Largest-earth-observing-satellite-shows-best-pictures-10-years-orbiting-planet.html; courtesy of ESA.)

imagery generally senses surface water as warmer than background at night and cooler than background during the day. The temperature of surface waters is readily monitored using thermal sensors (see Geothermal Resources section). On nighttime thermal imagery, surface water appears warmer than background as a result of the higher thermal inertia of water, that is, the capacity of water to retain heat longer than rocks and soils.

Turbidity is usually associated with flowing water, particularly flood waters or spring runoff. Rivers entering lakes or the ocean generally have higher turbidity than the body of water they are flowing into (e.g., Lakes Erie and St. Clair near Detroit, Figure 3.26). Lakes generally show a turbidity gradient from inlet to outlet that appears as light blue to black on color infrared images (e.g., Lake Cabora Bassa, Mozambique; Figure 12.6). Increased salinity can have the same effect as turbidity, increasing the reflectance of a water body (south San Francisco Bay, Figure 12.7; Great Salt Lake, Figure 14.21).

FIGURE 12.6
Light blue indicates turbidity on this color infrared TM image of Lake Cabora Bassa, Mozambique. (Image processed by EOSAT [now GeoEye/Digital Globe, Longmont, CO].)

FIGURE 12.7
Light blue, green, yellow-brown, and off white shades of water indicate increased salinity, turbidity, and shallow water in south San Francisco Bay near San Jose, California. (True color image courtesy of TerraMetrics and Google Earth, © 2012 Google.)

Bathymetric Mapping and Ocean Floor Topography

Laser bathymetry is used to measure water depths in lakes, rivers, and along the coast. Typically, systems are flown at low altitude (e.g., 400 m) on a fixed wing craft. A typical system would have a twin pulsed laser at green (530 nm) and IR (1060 nm) wavelengths capable of penetrating 40–50 m in clear water, or less than 20 m in turbid water (Irish, 2000). Vertical accuracy is in the order of 20 cm, and positional accuracy is of 1.5 m (Figure 12.8).

The first satellite bathymetry was derived from SeaSat radar altimeter data (e.g., Haxby and LaBrecque, 1982; Haxby, 1987). These maps should properly be called seafloor topographic maps rather than bathymetric maps. If the ocean surface was at rest, and no tides or winds were acting on it, the water surface would lie along the geoid, the surface of equal gravitational potential at zero elevation. Submarine features such as mid-ocean ridges and seamounts create a gravitational pull, and features such as ocean trenches have lower gravitational attraction. Variations in gravity influence the ocean surface and cause bulges and depressions in the geoid. The geoid is observed by measuring the position of the water surface above the Earth center, and the gravity field can be calculated through inversion. Centimeter-scale radar altimeter data are averaged over large areas, and repeat observations are stacked, or time-averaged, to reduce noise and minimize the effects of tides, currents, and storm waves. Satellite altimeter-based bathymetric maps have the advantage over sonar and sounding-based bathymetry in that they cover the entire ocean at the same resolution and with the same processing. These maps do not depend on merged surveys of different vintage, resolution, and mapping processes separated by large gaps in coverage.

Workers were able to use this sea surface data to make maps that accurately reflect the tectonic fabric of the seafloor. The SeaSat data offered the first accurate maps of seafloor

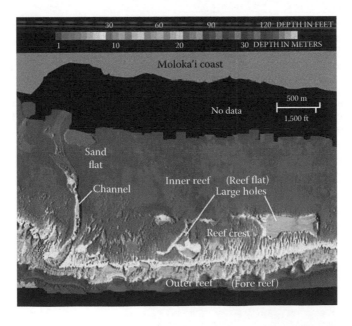

FIGURE 12.8
Laser bathymetry of U.S. Army Corps of Engineers SHOALS (Scanning Hydrographic Operational Airborne Lidar Survey) at Molokai, Hawaii, created by the USGS. Shoreline is at the top, shallow depths are lighter, and deeper depths are darker. (From Irish, J.L. 2000. An introduction to coastal zone mapping with airborne Lidar: the SHOALS system: 9. www.dtic.mil/cgi-bin/GetTRDoc?AD=ADA492313; courtesy of US Army Engineer Research and Development Center, Coastal and Hydraulics Laboratory, Joint Airborne Lidar Bathymetry Technical Center of Expertise, Mobile, AL; http://shoals.sam.usace.army.mil.)

topography at 60–70 km resolution (McAdoo, 2006). The mid-ocean ridges could be traced for thousands of miles, along with the offsetting transform faults, triple junctions, and ocean trenches (Figure 12.9). Scores of previously unknown marine volcanoes and sea-mounts were mapped. Tectonic topography could be seen even when buried beneath marine sediments (Smith, 1998).

The resolution of marine bathymetry has improved greatly with the advent of ERS-1 sat-ellite radar in 1991 followed by ERS-2, Envisat, Geosat, and others (Figure 12.10). Marine gravity-derived bathymetric maps now reveal structures on the order of 10 km. In fact, the fine resolution allowed Smith (1998) and Sandwell and Smith (2009) to determine that the

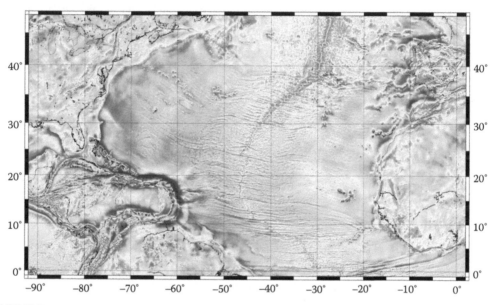

FIGURE 12.9
Marine seafloor topography provides relative bathymetric information. This map shows marine gravity anoma-lies derived from satellite altimetry by Sandwell and Smith (2009). (Courtesy of Dr. David Sandwell and Scripps Institution of Oceanography.)

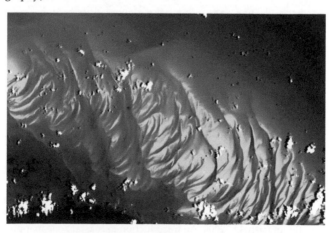

FIGURE 12.10
Carbonate shoals are revealed in exceptionally clear water of the Tarpum Bay southwest of Eleuthera Island, Bahamas. (True color ASTER image courtesy of NASA.)

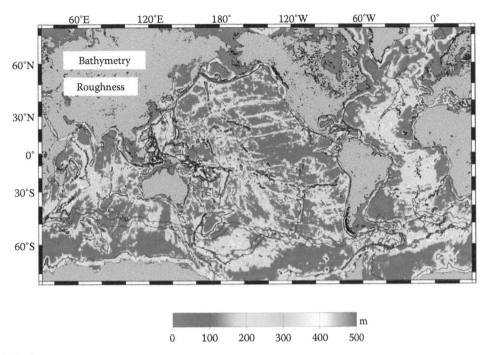

FIGURE 12.11
Marine gravity-derived topography reveals structures and allows determination of mid-ocean ridge relative spreading rates as a function of surface roughness (faster spreading provides a smoother surface). Color scale indicates seafloor roughness. (From Sandwell, D.T. et al. 2001. Bathymetry from space: white paper in support of a high-resolution, ocean altimeter mission. Scripps Institution of Oceanography, U.C. San Diego: 54 p. http://topex.ucsd.edu/marine_grav/white_paper.pdf; courtesy of Dr. David Sandwell and Scripps Institution of Oceanography.)

detailed roughness of the ocean floor is a function of spreading rate at the mid-ocean ridges. High spreading rates (e.g., East Pacific Rise) show low topographic and gravity amplitudes; low spreading rates (Mid-Atlantic Ridge) show high topographic and gravity amplitudes (Sandwell et al., 2001). Surface roughness as estimated from satellite altimetry can resolve features on the order of 30–50 m (Figure 12.11). In addition to aiding the mapping of tectonic features, satellite-derived bathymetric maps are useful for petroleum exploration (basin outlines, structural style, large scale traps, salt features), oceanic circulation models, understanding the distribution of volcanism, tsunami modeling, and transoceanic cable route planning.

Sub-Glacial Water

Lakes and rivers in Antarctica are not at the surface. They are buried under several kilometers of ice. Presumably, the same holds true for Greenland's ice sheets, ice sheets in Canada's high arctic, Siberia, and elsewhere. Several remote sensing techniques have been used to discover and map these buried lakes. The British began mapping Antarctica using airborne ice-penetrating radar (IPR) in the 1960s, and Lake Vostok was discovered by the Scott Polar Research Institute using these radar images in 1973. These investigations continue using the Active Microwave Instrument on ERS-1 (Stockman, 2012). NASA's MODIS and ICESat laser altimeter has already detected 124 lakes under Antarctic ice sheets (Figure 12.12; Smith et al., 2009). Lakes are mapped where the surface of the ice actively

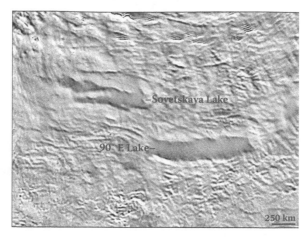

FIGURE 12.12
MODIS satellite image of two subglacial lakes, Antarctica. (Courtesy of NASA.)

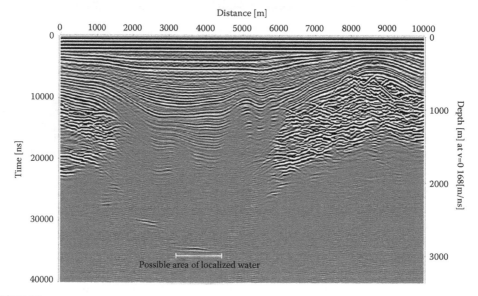

FIGURE 12.13
DELORES (Deep Look Radio Echo Sounder) radio-echo-sounding data over the catchment for subglacial Lake Ellsworth. Image is 10 km wide by 3.3 km high. Internal layer buckling can clearly be seen just right of center at depths of 0.75–1.5 km. Line is orientated southwest-northeast, providing a view northwest (down ice flow), toward Lake Ellsworth. (From Siegert, M. *Proposed Exploration of Subglacial Lake Ellsworth*. Antarctica: Final Comprehensive Environmental Evaluation. Lake Ellsworth Consortium: 87 p; 2012. http://www.antarctica .ac.uk/about_antarctica/environment/bas/subglacial_lake_ellsworth_final_cee.pdf, http://www.antarctica .ac.uk/afi/docs/2008-09_field_reports/AFI7-02_Siegert_field_report_0809.pdf, http://www.antarctica.ac.uk/about_ antarctica/environment/bas/subglacial_lake_ellsworth_final_cee.pdf; Image courtesy of Dr. Martin Siegert [University of Edinburgh; currently University of Bristol].)

moves up or down on a seasonal basis, indicating that the lakes are being filled or drained. Subglacial Lake Engelhardt, a 10 × 30 km feature, was revealed by an ice surface draw-down of 9 m over 3 years. The speed of movement of ice sheets can also be an indicator of sub-ice water lubrication. Airborne and ground-based "radio echo sounding" (RES) is an active remote sensing technique used to map water beneath the ice (Figure 12.13). RES, also known as IPR or radar sounding, transmits and receives electromagnetic energy in the megahertz range where ice is effectively transparent. Any surfaces with a dielectric contrast will create a reflection. The most common reflectors in glaciers and ice sheets are the ice surface, the ice–rock interface, and internal layer contrasts. Features such as subglacial lakes and thermal boundaries can also be detected. Strong RES reflectors indicate subglacial lakes (Bingham and Siegert, 2007; Siegert, 2012). RES can determine whether the ice–rock interface is wet or dry and can suggest areas where flowing water may exist (Zirizzotti et al., 2012).

Geothermal Resources

Lists of thermal springs exist for most developed areas (Berry et al., 1980). For those areas where such lists do not exist, there are several indicators of geothermal/hydrothermal systems. Among these are evidence of recent volcanism, alteration of bedrock and soils associated with springs, and indications of hot or warm water. Geothermal sources can be used for power generation, as at Geysers, California. They have been used for health, as at Bad Gastein, Austria, and for tourism, as at Bath, England. They can indicate that an area is volcanically active or has high near-surface heat flow, and thus may be suitable for generation of hydrocarbons in recent sediments (but may not be appropriate for hydrocarbon preservation).

Perhaps the most direct means of locating hot water is using thermal sensors. These sensors measure thermal flux rather than surface temperature (see Chapter 3, section describing "Airborne Thermal Scanners"), so it may be necessary to obtain ground control measurements to calibrate the imagery if temperatures are required. Thermal infrared scanners flown under predawn conditions reveal shallow water bodies that are warmer than the background soil and rocks. Point sources of warm water are a clue to warm or hot springs, since flowing water generally appears cool (Figure 12.14). Features that appear similar to hot springs on thermal imagery, and may be a source of confusion when analyzing the imagery, include black shales, basalts, or other dark rocks that retain heat at night.

Clues to thermal springs include areas of water in otherwise ice-covered rivers or lakes on winter imagery; ground fog over rivers and streams on cool, still mornings; and mineral deposits such as tufa, travertine, or siliceous sinter. Sulfur-secreting bacteria are common in some thermal springs and can cause brilliant colors (Figure 12.15). Green algae (bright red on color infrared images) thrives in water between 49 and 60°C; orange and red algae can be found in water between 60 and 71°C; white algae has been found in water as hot as 85°C. Although none of these criteria are unequivocal, all are indications of hot springs. Finally, recent cinder cones or lava flows, epithermal alteration (iron staining, argillization) of outcrops and soils, and vegetation stress, especially if near springs, provide evidence for near surface heat (Camacho et al., 1982). A proposed method to map plant stress associated with geothermal systems is to create an NDVI for spring or early summer imagery and later summer or fall imagery using Landsat TM data:

$$\mathrm{NDVI} = \frac{(\mathrm{TM4} - \mathrm{TM3})}{(\mathrm{TM4} + \mathrm{TM3})}$$

These seasonal indices are then subtracted from one another to determine where vegetation changed the most, thus indicating stress (Nash and Hernandez, 2001).

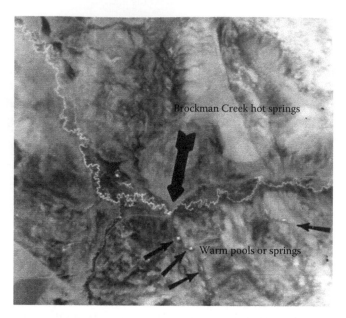

FIGURE 12.14
Predawn airborne thermal image of Brockman Creek hot springs, Idaho (large arrow) and numerous warm spots in nearby streams (small arrows). Brockman Creek flows from right to left. (Image processed by Motorola Airborne Remote Sensing [Mars] Inc., Phoenix.)

FIGURE 12.15
Grand Prismatic Spring, Yellowstone National Park, Wyoming. Vivid colors are a result of algae growing in this thermal spring. (Courtesy of Bing Images. http://www.bing.com/images/search?q=Grand+Prismatic+Spring%2c+Yellowstone+National+Park&qpvt=Grand+Prismatic+Spring%2c+Yellowstone+National+Park&FORM=IGRE#view=detail&id=78CE6B96B71887F5F2F5BD8CE9295B3F7D03AC1E&selectedIndex=22.)

High near-surface heat flow can be associated with near-vertical (linear) fault systems that tap deep crustal heat sources, particularly if there are young volcanics in the area. Work by the Lawrence Livermore National Laboratory/U.S. Department of Energy in conjunction with University of California Santa Cruz used hyperspectral imagery to map faults and paleo-fumaroles in the Long Valley caldera of eastern California (Pickles et al., 2001).

FIGURE 12.16
Terraced travertine pools at Mammoth Hot Springs, Yellowstone National Park, Wyoming, are characteristic of thermal springs. (Courtesy of The World Geography. http://www.theworldgeography.com/2012/11/terraced-pools.html and the WyomingTourism.org and http://traveldestinations.thundafunda.info/tourism-nature-pictures233.php.)

This area lies at the faulted boundary between the Sierra Nevada block and the Basin-and-Range province. Stressed and dead trees related to recent (1989) dike intrusion and high CO_2 flux in the soil were mapped using hyperspectral imagery. The hyperspectral sensors simultaneously mapped alteration assemblages (alunite, montmorillonite, kaolinite, hematite) related to the active geothermal system.

Travertine, especially as terraces, can be mapped using photography and hyperspectral imagery. It occurs in and around the hot springs at Mammoth in Yellowstone National Park, and other thermal areas (Figure 12.16).

In summary, remote sensing systems can locate areas likely to have near-surface ground water and may be able to determine if that water is fresh or saline; can detect the moisture levels in soils; can locate surface water and infer its depth and temperature; can help find bodies of water beneath glaciers, and can locate geothermal springs and areas of high near-surface heat flow.

References

Akther, M.S., Q.K. Hassan. 2011. Remote sensing based estimates of surface wetness conditions and growing degree days over northern Alberta, Canada. *Boreal Environ. Res.* 16: 407–416.

Battelle Labs. 1983. *Geophysical Survey Capabilities*. Richland, WA: Pacific Northwest Labs: 17 p.

Berry, G.W., P.J. Grim, J.A. Ikelman. 1980. Thermal Springs List for the United States. *National Oceanic and Atmospheric Administration Key to Geophysical Records Document*. USDA NOAA, Boulder, CO: 12: 59.

Bingham, R.G., M.J. Siegert. 2007. Radio-echo sounding over polar ice masses. *J. Environ. Eng. Geophys.* 12: 47–62.

Brocca, L., F. Melone, T. Moramarco, R. Morbidelli. 2010. Spatial-temporal variability of soil moisture and its estimation across scales. *Water Resour. Res.* 46: 14.

Bromley, J., B. Mannström, D. Nisca, A. Jamtlid. 1994. Airborne geophysics: application to a ground-water study in Botswana. *Ground Water* 32: 79–90.

Camacho, S., L. del Rio, L. Sanchez, J. Gonzalez. 1982. Impact of Hydrothermally Altered Soil on Vegetation as a Tool in Geothermal Exploration. *Proceedings of the 16th International Symposium on Remote Sensing of Environment*. Buenos Aires, Argentina: Ann Arbor, MI: ERIM: 145–153.

Chen, H., H. Zhang, R. Liu, W. Yu. 2009. Agricultural drought monitoring, forecasting and loss assessment in China. In C.M.U. Neale, A. Maltese (eds.), *Remote Sensing for Agriculture, Ecosystems, and Hydrology XI*. Proceedings of SPIE, Bellingham, WA: v. 7472: 19 p.

Dalati, M. 1997. Use of space imagery as a tool for groundwater resources management, case study: lineaments as groundwater exploration guides in Hama-Salamiyeh, Syria. Ann Arbor, MI: *11th ERIM International Conference on Applied Geologic Remote Sensing*, November 17–19, 1997. Denver, CO. Volume I: 376–383.

Dalati, M. 2004. An assessment of the space image capabilities to water resources management – case study: Al-Hammad semi arid plateau, Syria. *1st International Conference on Water Resources and Arid Environment*, December 5–8, 2004. Riyadh, Saudi Arabia: 9 p.

Dubois, P.C., J. van Zyl, T. Engman. 1995. Measuring soil moisture with imaging radar. *IEEE Trans. Geosci. Remote Sens.* 33: 915–926.

Elfouly, A. 2000. Faults and fractures intersections delineation as a tool for groundwater detection using remote sensing and ground penetrating radar techniques at Saint Catherine area, southern Sinai, Egypt; ICEHM. 2000. Cairo University, Egypt, September 2000: 293–310.

EOSAT. 1991. *EOSAT 1991* (calendar). Lanham, MD: Earth Observation Satellite Co. (now GeoEye/Digital Globe, Longmont, CO).

Famiglietti, J.S., J.A. Devereaux, C.A. Laymon, T. Tsegaye, P.R. Houser, T.J. Jackson, S.T. Graham, M. Rodell, P.J. van Oevelen. 1999. Ground-based investigation of soil moisture variability within remote sensing footprints during the southern Great Plains 1997 (SGP97) hydrology experiment. *Water Resour. Res.* 35: 1839–1851.

Graf, F.L. 1990. Using ground-penetrating radar to pinpoint pipeline leaks. *Matter Perform.* 29: 27–29.

Ghoneim, E., C. Robinson, F. El-Baz. 2007. Radar topography data reveal drainage relics in the eastern Sahara. *Int. J. Remote Sens.* 28: 1759–1772.

Haxby, W.F. 1987. Gravity field of the world's oceans, a portrayal of gridded geophysical data derived from SeaSat radar altimeter measurements of the shape of the ocean surface (map). *Lamont-Doherty Geol. Obs. Columbia Univ., Rept. MGG-3 and Data Announcement 87-MGG-04.* US National Geophysical Data Center, Boulder, CO.

Haxby, W.F., J. LaBrecque. 1982. Geotectonic imagery: the application of SeaSat altimetry to the tectonic evolution of the Indo-Atlantic basin (abs.). *EOS Trans. Am. Geophys. Union* 63: 908.

Irish, J.L. 2000. An introduction to coastal zone mapping with airborne Lidar: the SHOALS system: 9. www.dtic.mil/cgi-bin/GetTRDoc?AD=ADA492313 (accessed October, 23 2012).

Jackson, T.J. 2002. Remote sensing of soil moisture: implications for groundwater recharge. *Hydrogeol J.* 10: 40–51.

Kerr, Y.H. 2006. Soil moisture from space: where are we? Springer-Verlag online, http://www.springerlink.com/content/636414470gp33ukm/ (accessed October 23, 2012).

Lattman, L.H., R.R. Parizek. 1964. Relationship between fracture traces and the occurrence of ground water in carbonate rocks. *J. Hydrol.* 2: 73–91.

Li, C.Q., H.J. Li. 2008. The application of TVDI in drought monitoring for winter wheat in spring. http://en.cnki.com.cn/Article_en/CJFDTOTAL-YGJS200802006.htm (accessed October 23, 2012).

Lyon, J.G. 1987. The use of maps, aerial photographs and other remote sensing data for practical evaluation of hazardous waste sites. *Photogram. Eng. Rem. Sens.* 53: 515–519.

Lyon, J.G., W.S. Hutchinson. 1995. Application of a radiometric model for evaluation of water depths and verification of results with airborne scanner data. *Photogram. Eng. Rem. Sens.* 61: 161–166.

McAdoo, D. 2006. *Marine geoid, gravity and bathymetry: an increasingly clear view with satellite altimetry.* ESA International workshop, Venice: 7 p.

McCauley, J.F., G.G. Schaber, C.S. Breed, M.J. Grolier, C.V. Haines, B. Issawi, C. Elachi, R. Blom. 1982. Subsurface valleys and geoarchaeology of the Eastern Sahara revealed by Shuttle radar. *Science* 218: 1004–1020.

Nash, G.D., M.W. Hernandez. 2001. Cost-effective vegetation anomaly mapping for geothermal exploration. *Proceedings of the 26th workshop on geothermal reservoir engineering*, Stanford University, January 29–30, 2001: 8 p.

Nelson, J.E., S.E. Silliman, R.J. Wasowski. 1991. Analysis of thermal scanning imagery to investigate point inflows into surface water bodies. In D.B. Stafford (ed.), *Civil Engineering Applications of Remote Sensing and Geographic Information Systems*. Reston, VA: American Society of Civil Engineering: 179–188.

Njoku, E.G., D. Entekhabi. 1994. *Microwave remote sensing of soil moisture*. Jet Propulsion Lab, Pasadena, CA: 33 p.

Pickles, W.L., P.W. Kasameyer, B.A. Martini, D.C. Potts, E.A. Silver. August 26–29, 2001. *Geobotanical Remote Sensing for Geothermal Exploration*. San Diego, CA: Geothermal Resource Council 2001 Annual Meeting: 10 p.

Prost, G.L. 1988. Results of satellite bathymetric mapping, Qatar and UAE. Proc. of the French-Qatari regional symposium on remote sensing, February 23–25, 1988. Doha, Qatar: Scientific & Applied Research Centre, University of Qatar: 199–211.

Robinson, C., F. El-Baz, M. Ozdogan, M. Ledwith, D. Blanco, S. Oakley, J. Inzana. 2000. Use of radar data to delineate paleodrainage flow directions in the Selima Sand Sheet, eastern Sahara. *Photogram. Eng. Rem. Sens.* 66: 745–753.

Sabins, F.F., Jr. 1987. *Remote Sensing Principles and Interpretation*. New York: W.H. Freeman and Co.: 198–199.

Sandwell, D.T., W.H.F. Smith. 2009. Global marine gravity from retracked Geosat and ERS-1 altimetry: ridge segmentation versus spreading rate. *J. Geophys. Res.* 114: B01411.

Sandwell, D.T., W.H.F. Smith, S. Gille, S. Jayne, K. Soofi, B. Coakley. 2001. Bathymetry from space: white paper in support of a high-resolution, ocean altimeter mission. Scripps Institution of Oceanography, U.C. San Diego: 54 p. http://topex.ucsd.edu/marine_grav/white_paper.pdf (accessed October 23, 2012).

Scipal, K., C. Scheffler, W. Wagner. 2005. Soil moisture-runoff relation at the catchment scale as observed with coarse resolution microwave remote sensing. *Hydrol. Earth Syst. Sci. Discuss.* 2: 417–448.

Siegert, M. 2012. *Proposed Exploration of Subglacial Lake Ellsworth*. Antarctica: Final Comprehensive Environmental Evaluation. Lake Ellsworth Consortium: 87 p. http://www.antarctica.ac.uk/about_antarctica/environment/bas/subglacial_lake_ellsworth_final_cee.pdf. (accessed 22 June 2013).

Smith, B.E., H.A. Fricker, I.R. Joughin, S. Tulaczyk. 2009. An inventory of active subglacial lakes in Antarctica detected by ICESat (2003–2008). *J. Glaciol.* 55: 573–595.

Smith, W.H. 1998. Seafloor tectonic fabric from satellite altimetry. *Ann. Rev. Earth Planet. Sci.* 26: 697–738.

Stockman, J.D. 2012. The discovery of Lake Vostok. Yahoo! Voices. http://voices.yahoo.com/the-discovery-lake-vostok-10942110.html (accessed October 23, 2012).

Su, Z., A. Yacob, J. Wen, G. Roerink, Y. He, B. Gao, H. Boogaard, C. van Diepen. 2003. Assessing relative soil moisture with remote sensing data: theory, experimental validation, and application to drought monitoring over the North China plain. *Phys. Chem. Earth Pt. A/B/C* 28: 89–101. http://www.sciencedirect.com/science/article/pii/S147470650300010X (accessed October 23, 2012).

Taniguchi, M., W.C. Burnett, J.E. Cable, J.V. Turner. 2002. Investigation of submarine groundwater discharge. *Hydrol. Process.* 16: 2115–2129.

Taylor, R.S., R.P. Shubinski, T.S. George. 1980. Landsat imagery for hydrologic modeling. In R.W. Kiefer (ed.), Civil Engineering Applications of Remote Sensing. *Proceedings of the Specialty Conference of the Aerospace Division of the American Society of Civil Engineers*. Madison: University of Wisconsin, August 13–14, 1980.

Wagner, W., K. Scipal, C. Pathe, D. Gerten, W. Lucht, B. Rudolf. 2003. Evaluation of the agreement between the first global remotely sensed soil moisture data with model and precipitation data. *J. Geophys. Res.* 108: 15.

Wallen, V.R., H.R. Jackson, P.K. Basu, H. Baenziger, R.G. Dixon. 1977. An electronically scanned aerial photographic technique to measure winter injury to alfalfa. *Can. J. Plant Sci.* 57: 647–651.

Wang, L., J.J. Qu. 2007. NMDI: a normalized multi-band drought index for monitoring soil and vegetation moisture with satellite remote sensing. *Geophys. Res. Lett.* 34: 5.

Waugh, R. 2012. Out of this world: an awe-inspiring portrait of Earth planet to celebrate the tenth anniversary of pictures from observation satellite. http://www.dailymail.co.uk/sciencetech/article-2110455/Our-coloured-planet-Largest-earth-observing-satellite-shows-best-pictures-10-years-orbiting-planet.html (accessed March 5, 2012).

Wei, G.S., R.P. Shen, G.X. Ding. 2008. *Primary Studies on Estimating Soil Moisture in Deep Layer of Farmland by Remote Sensing in Yizheng.* CNKI. http://en.cnki.com.cn/Article_en/CJFDTOTAL-YGJS200801006.htm (accessed October 23, 2012).

Weil, G.J., R.J. Graf. 1991. Infrared thermography-based pipeline leak detection. In D.B. Stafford (ed.), *Civil Engineering Applications of Remote Sensing and Geographic Information Systems.* Reston, VA: American Society of Civil Engineering: 189–198.

Zirizzotti, A., L. Cafarella, S. Urbini. 2012. Ice and bedrock characteristics underneath Dome C (Antarctica) from radio echo sounding data. *IEEE Trans. Geosci. Remote Sens.* 50: 37–43.

Additional Reading

Govender, M., K. Chetty, H. Bulcock. 2007. A review of hyperspectral remote sensing and its application in vegetation and water resource studies. *Water SA* 33: 145–151.

Hajji, H., S. Sole, A. Ramamonjiarisoa. 1999. Analysis and prediction of internal waves using SAR image and non-linear model. Proc. of the CEOS SAR Workshop, Toulouse, October 26–29, 1999, ESA SP-450: 513–518.

Waring, G.A., R.R. Blankenship, R. Bentall. 1965. *Thermal Springs of the United States and Other Countries of the World – a Summary.* Washington, D.C.: US Government Printing Office: U.S. Geological Survey Prof. Paper 492: 383 p.

13

Logistics and Engineering

Chapter Overview

Logistics and engineering applications are among the most important functions of geologic remote sensing. Engineers, increasingly aware of the capabilities of remote sensing technology, are applying imagery and other remote surveys to mapping and measuring environmental factors. This chapter reviews some of the applications that have been used successfully by various engineering groups.

Engineering activities covered by remote sensing can be lumped broadly into the following three categories:

1. Infrastructure and reconnaissance mapping: This includes locating old tailings, well sites, seismic lines, facilities, roads, and pipelines, as well as finding sources of raw materials such as aggregate for construction.
2. Siting: Determining the safest, most cost-effective sites for plants, storage facilities, pipelines, roads, and port facilities.
3. Transportation: Finding the shortest and least expensive access to a drill site, determining the cheapest way to get products into or out of an area, determining ice movements and weather patterns that might interfere with truck or tanker traffic, and locating dangerous shoals.

The quantitative and repetitive nature of remote sensing lends itself to monitoring, measuring, and analysis. Although it may not seem obvious, these activities can be a matter of life and death.

Site Selection

Two and a half minutes before midnight on March 12, 1928, the Saint Francis dam collapsed. The dam, a curved concrete gravity structure built to contain a reservoir for Los Angeles, was 56.4 m high and held 45 million m³ of water. The flood wave was approximately 43 m high when it encountered the first victims, the dam keeper and his family, 400 m downstream in San Fracisquito Canyon. Five minutes later, having traveled 2.4 km at close to 30 km/h, the 33.5 m flood wave demolished a power plant and killed 65 workmen and their families. By the time, the waters reached the Pacific at 5:30 AM, 87 km from the dam, the flood was 3 km wide and traveling 8 km/h. It had destroyed six towns and killed at least 450 and perhaps as many as 600 people ... the exact number will never be known (Ventura County Star, 2008). Subsequent investigations and inquiries determined that the dam's foundation was inadequate. The Oligocene-Miocene Vaqueros Formation sandstone has a tendency to slake when saturated. The linear valley is underlain by the inactive San Francisquito fault that juxtaposes Precambrian Pelona Schist against the Oligocene Vasquez sandstone/conglomerate half way up the right dam abutment. The left abutment was anchored in an ancient bedrock landslide in the schist. Evidence suggests that the left abutment detached as the landslide reactivated due to a reduction in friction and the weight of water. Engineers who designed the dam were either unaware of the fault, landslide, and inclined foliation in the schist, or disregarded them (Wiley et al., 1928; Rogers, 2006). Modern high-resolution remote sensing should be able to detect these features.

The first step for any engineering application must be careful site selection and evaluation. Many factors go into selecting the site for plants, storage facilities, pipeline routes, and roads. Among these factors are construction suitability of the surface material, slope stability, storm or flood drainage, environmental sensitivity, and proximity to transportation networks. The goal of a site selection program is to evaluate surface materials to determine whether they are permeable, cohesive, and stable; to evaluate soils, their thickness, and suitability for agriculture or other competing uses; to evaluate the local drainage, bedrock, and land use. Remote sensing, whether satellite-derived, airphoto-based, or some other combination of sensors, provides an economic and rapid means of obtaining this information (Rinker and Frost, 1996). Maps can be generated showing surface materials, surface geology and soils, potential hazards, hydrographic drainage networks, plant cover, and land use. An analyst can use imagery and derivative maps to predict the impact on the local environment and prepare a surface sampling program to confirm the observations and test the strength of surface materials.

Image analysis provides a means for evaluating any region quickly and at low cost. Certain types of information are needed for site selection whether the plan is to build on the surface, tunnel through the Earth, drill into it, or excavate and mine it. Among these are data on the surface geology including the soils, rocks, and structure. Important data include the thickness, permeability, stability, and cohesiveness of soils; the type of subsurface and surface rocks, their attitude, and whether they are fractured. Local sources of sand and gravel are important for construction. Local hazards, such as near-surface caverns, landslides, or active seismic zones, are of paramount importance (see Chapter 16).

Geologic Structure

As failure of the St. Francis Dam illustrates, geologic structure is an essential part of determining the suitability of a construction site. This dam was anchored in a landslide on one side and a fault on the other. The dip of foliation in the schist, sliding when lubricated, may have played a role in the dam failure. The Diablo Canyon atomic power plant on the coast of California is adjacent to numerous faults thought to be inactive. The determination, usually by trenching, that a fault has been inactive for 10,000 years is no guarantee that it will remain inactive. Geologic structures (faults, joint sets, dip of bedding or foliation, and landslides) are features that can, in most cases, be identified and mapped using high-resolution imagery. Lidar can help determine the intensity and frequency of recent earthquakes (see Chapter 5).

Geologic structures are also important when siting offshore energy developments. Shallow water seismic surveys require knowledge of shoals, while platforms, subsea completions, and submarine pipelines require bottom stability information. Submarine slumps, faults, mud volcanoes, escarpments and steep slopes, outcropping hydrates, gas escape pockmarks, current scours, and shipwrecks can be detected and imaged using autonomous underwater vehicles (UAVs) equipped with remote sensing tools such as side-scan sonar, multibeam bathymetry systems, and frequency-modulated sub-bottom profilers (George, 2005). These systems have been proven effective in water depths up to 4500 m in the Gulf of Mexico.

Surface Materials

Facilities, roads, mine conveyor belts, powerlines, and pipelines should not be built in areas with potential hazards such as unstable slopes, river flooding, subsidence, and fluvial or coastal erosion. It is possible to determine the risk and extent of such hazards by obtaining multitemporal imagery and checking, for example, the scale of flooding over a period of a decade or more. Overgrown landslides and areas prone to brushfires can usually be detected by reviewing historical photos.

Generally speaking, one prefers to build on a solid, stationary surface. Bedrock is good in the sense that it is usually stable. One must determine, however, that bedding or foliation is not parallel to or undercut by the slope of the surface, as it may then initiate a detachment or slide. Soils are generally easy to work with as far as excavating or shaping the surface, but are subject to uneven settling and compaction, and thus may require special building techniques and future monitoring.

Different soils have diverse engineering properties that must be taken into account when planning site development. Sandy soils have good water infiltration and thus minimize the need to account for storm water runoff. On the other hand, spills can quickly find their way to the water table in sandy material. Sandy soils are less prone than clay-rich soils to mass downslope movements. Clay-rich soils can fail on slopes as shallow as one or two degrees when saturated. The ability to slide downslope as a result of gravity is enhanced by the types of particles and amount of water in the soil. Large, irregular-shaped particles (sands, conglomerates) generally have more resistance to sliding than fine-grained, flat particles (clays). All materials have less resistance to sliding when saturated with water, which acts as a lubricant (Krynine and Judd, 1957). Thus, using imagery to locate sites with good natural drainage can be helpful.

It should be possible to interpret sandy versus clay-rich soils on the basis of infiltration and relative drainage spacing, the steepness of drainage channels, and the type of vegetation growing in an area (see Chapter 4). Sandy soils have better infiltration and thus a more widely spaced drainage network than clay-rich soils. They also have V-shaped gullies, whereas silt and clay-rich soils tend to have U-shaped gullies. Clay-rich soils have grasses and shallow-rooted trees and shrubs, whereas sandy soils have deep-rooted plant communities (see Chapters 4 and 15). Areas with near-surface groundwater will appear as darker tones (moist) on visible or near infrared imagery and have more verdant vegetation in all soil types.

Subsidence

Some projects are in areas of active subsidence. This can be due to sinkholes forming in a karst environment, a result of groundwater or hydrocarbon withdrawal from the subsurface, or from collapse of abandoned underground mines (see Chapter 11, Mine Safety Issues, and Chapter 16, Subsidence Case History). In such regions, it is necessary to avoid areas prone to subsidence when planning infrastructure. Subsidence related to withdrawal of liquids from the subsurface can lead to undrained depressions or deep, open soil crevices or fissures. Identification of mine-related collapse depressions is aided by the presence of tailings in the vicinity. It is fairly simple to recognize areas that have subsided or collapsed: these are small, isolated depressions with ponded ground or surface water and no drainage development. It is difficult, if not impossible, to predict areas that will collapse unless one has a map of old workings or maps of cave systems in limestones. Caves are often developed along fracture systems.

Surface Slope

For the most part, one tries to avoid steep slopes when building facilities or roads, since site preparation costs go up with the amount of Earth that needs to be moved. Steep slopes are inherently less stable than gentle slopes, requiring more work to make the road embankment, pipeline right-of-way, open pit walls, or gas plant site safe. Relative slope is easy to observe on imagery. With stereocoverage or a digital elevation model, it is possible to determine the precise slope. DEMs can be generated from stereoimages using photogrammetric methods to determine elevation at various points, or they can be derived using satellite radar altimeter or airborne Lidar altimetric methods (Figure 13.1).

Permafrost

Knowing the distribution of permafrost is extremely important to anyone wishing to build in the arctic, shoot a seismic program, build a road, or drill a well. Its presence or absence dictates the construction methods to be used, how a seismic program is laid out, the time of year surface work can be done, and the kinds of insulation required for facilities foundations. One can determine the presence of permafrost by observing geomorphic features such as thermokarst lakes, patterned ground, ice wedge polygons, frost heave, hummocks, solifluction sheets, and pingos (Mollard and Janes, 1984; Figure 13.2). There are also subtle indicators such as tilted trees or thermokarsting along tracks. Trees tend lean in various directions due to thawing of the surface layer during the summer. Tracks across the tundra kill the reindeer moss that serves as insulation and result in melting of the near-surface permafrost. This self-perpetuating process leads to deep gullies known as thermokarst (Figures 13.3 and 13.4).

FIGURE 13.1
Digital elevation model of the Bingham mine southwest of Salt Lake City, Utah. (NEXTMap image courtesy of INTERMAP.)

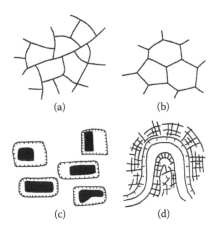

(a)   (b)

(c)   (d)

FIGURE 13.2

Permafrost soil patterns. (a) Irregular patterned ground and ice wedge polygons. (b) Polygonal patterned ground and ice polygons. (c) Thermokarst (rectangular) lakes. (d) Radial patterned ground around a river channel. (Modified after Mollard, J.D. and J.R. Janes. *Airphoto Interpretation and the Canadian Landscape,* Energy, Mines and Resources, Ottawa, Canada, 1984, Figure 7-5; Copyright Government of Canada. Reproduced with the permission of the Minister of Public Works and Government Services Canada, 2013.)

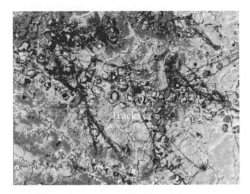

FIGURE 13.3

Tracks across the tundra kill the moss that insulates near-surface permafrost. This self-perpetuating process leads to deep gullies known as thermokarst. False color Landsat TM image over the Novy Port field, Yamal Peninsula, Siberia, shows tracks (red to black linear features) eroded into the permafrost. (Courtesy of Amoco Production Company.)

FIGURE 13.4

Thermokarst on a tributary of the Toolik River, Alaska. (Photo courtesy of Dr. Michael Gooseff, Department of Civil and Environmental Engineering, Pennsylvania State University.)

Expanding Clays

As anyone who has built a house on montmorillonite-rich soils can attest, unstable soils are not the desired foundation for a home. Cracks appear suddenly, and walls shift positions. Likewise, roads built on expanding clays without the proper bed preparation will quickly become uneven and possibly dangerous. The alternating swelling and contracting associated with wetting and drying causes clay-rich soils to form cracked and hummocky surfaces. Clay-rich rock and soils commonly form badlands, which are recognized by their intense gullying. It is possible to map changes in surface elevation over time with a high degree of accuracy (on the order of centimeters), and this may help determine the existence of unsuitable soils. Radar interferometry has been used for this type of mapping. For example, Gabriel et al. (1989) mapped the expansion of water-absorbing clays in the Imperial Valley of California using SeaSat radar altimeter data to an accuracy of 1 cm averaged over 10 m pixels.

The final site for a facility, route of a road or pipeline, or well pad location is the result of a series of tradeoffs. For example, when construction costs to build a pipeline through the arctic run several million dollars per kilometer, one would necessarily want to take the shortest, most direct route from point A to point B. Yet, one is also obligated to avoid lakes and river crossings that require special construction techniques that can run up costs even further. There are other environmental factors to consider as well: terrain sensitivity, fish or mammal migration routes, breeding grounds, and endangered animal habitats. These considerations are illustrated by the process involved in evaluation of a nuclear power plant site in India (see Kakrapar Power Plant Case History, this chapter).

Generating Appropriate Base Maps

Base maps are most often topographic maps showing infrastructure, that is, access roads, mines, or well locations, mine dumps, tailings, pipeline gathering systems, storage tanks, power lines, and facilities. A map that is more than a couple of years old is out of date in many areas of the world: new wells are drilled, new roads are built, new pipe is laid, and new mines are opened. An airphoto enlarged to 1:25,000, or satellite image at the same scale, can be acquired annually at a nominal cost and integrated into a geographic information system (GIS) to provide up-to-date maps of an area. Layers of information, such as roads and trails, powerlines, railroads, and towns, can be extracted from imagery. These layers can be added to a relational database that is called upon when, for example, an access map is required to move heavy equipment into an area. An archive of historical photos can be useful if there is ever a question of environmental liability regarding who caused a spill and when, when acid mine drainage began, where the drilling mud pits were buried 20 years ago, and when did pipeline or tailings pond leakage begin.

It is possible to generate orthophotos, eliminating tilt and radial distortion, and use them as maps for surveying or construction. Digital imagery can be rectified and plotted using any map projection and spheroid with a superimposed Universal Transverse Mercator or latitude–longitude grid (Figure 14.2). Topographic maps can be generated from airphoto or satellite stereo pairs. SPOT Pan and some radar imagery (e.g., RADARSAT, ERS-1) can be used to generate topographic maps with 15 m contours or better (Thompson and Mercer, 1996). Generation of topographic maps usually requires that ground control points are surveyed and that elevation control points are provided. Advances in global positioning system technology have made the acquisition of control points not only quick and inexpensive, but also accurate to plus or minus a few meters (non-differential) and to within 1 meter (differential). Keep in mind that control points must be readily identifiable on the imagery. Road intersections, bridges, large buildings, and the point at which a tributary enters a river or a river enters a lake make good control points. Survey crews use a large, high-contrast X where natural landmarks are missing. These points cannot assist in removing image distortion unless they can be accurately located on imagery. The effective use of GISs depends on accurate and up-to-date base maps.

Reconnaissance

Before going into an area, particularly in a foreign country, it is important to know what the infrastructure is, including where production occurs, where pipelines and storage facilities exist, and where the roads and powerlines, refineries and smelters are. This type of information is frequently not available, either because it does not exist or is not provided by the host country. One can learn, for example, something about the amount of oil produced in an area by counting oil storage tanks (Figure.13.5). It is even possible to estimate how full these tanks are by using thermal imagery that indicates the internal fluid levels: partially empty tanks are cooler than full tanks due to the cooler gases at the top (Lukosavich, 2012). Whether or not a pipeline exists is a crucial factor in determining the economics of investing in a specific region. If there are signs of a well having been drilled in an area, yet there is no obvious way to get the oil to market, this suggests that the well may have been an economic failure and that the area deserves higher exploration risk. Old tailings can reveal where mining activity occurred in the past: the lack of modern roads might tell you that nothing has happened there since the price of gold went over $32 per ounce.

When a new discovery is announced, it may not be accurately located on existing maps. Detailed satellite images can show not only exactly where the discovery is, but can also provide information such as whether a well was drilled on surface structure or if a deposit is localized by faulting. If there is no obvious surface structure, this suggests a certain

FIGURE 13.5
Oil storage tanks, piers, and tankers, southern Kharg Island, Iran. (True color image courtesy of NOAA, U.S. Navy, DigitalGlobe, TerraMetrics, and Google Earth, © 2012 Google.)

sophistication of the seismic (or other subsurface prospecting) program employed to locate the discovery. If there is surface structure, one can learn something about the kind of structures that produce in an area. This information can be used to explore for similar structures in the surrounding basin. If an ore body is located along a shear zone, it may be possible to explore along that zone, or to look for features analogous to the zone in the surrounding region. A lack of surface alteration suggests that potential fields methods or geochemical sampling was used to locate the deposit.

The number of wells in a field, and the degree to which they are interconnected by road and pipeline networks, provides important information to companies evaluating partnership deals and joint ventures (Figure 13.6). Likewise, the extent to which an area has prospect pits can reveal the extent of mineralized areas.

Part of many seismic programs involves permitting and cutting paths for the survey crew and equipment. In many parts of the world, it is fairly evident where previous seismic lines were cut because the vegetation has not completely recovered, even from surveys many years earlier. This is the case in Somalia, where seismic lines, cut through thorn bushes 35 years before imagery was acquired, are still visible as straight lines across the landscape (Figure 13.7). The same goes for parts of the arctic and other slow-changing landscapes.

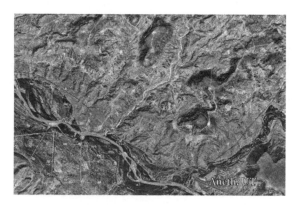

FIGURE 13.6
Part of the Aneth oil field, southern Utah, showing infrastructure including wells, pipelines, and roads. (True color image courtesy of GeoEye and Google Earth, © 2012 Google.)

FIGURE 13.7
Seismic lines in Somalia northwest of Mogadishu are still visible 35 years after they were cut. (True color image courtesy of CNES/SPOT Image, DigitalGlobe, and Google Earth, © 2012 Google.)

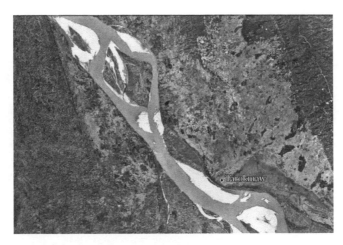

FIGURE 13.8

Sandbars in the Irrawaddy River near Tarokmaw, Myanmar. (True color image courtesy of GeoEye, CNES/ SPOT Image, and Google Earth, © 2012 Google.)

Gravel is an important resource for building drill pads and artificial islands in shallow water offshore areas. It is a key component in construction (as aggregate) and roadbeds. The cost of transporting gravel is a great incentive for finding local sources. Airphoto or satellite surveys can quickly locate gravel bars in or adjacent to rivers, glacial outwash, or coastal gravel banks that are excellent sources of granular materials (Figure 13.8) (Jolliffe and McLellan, 1980).

Water is necessary for drilling mud, for mining and smelting operations, to sustain field camps, and to mitigate dust in construction sites. Imagery can assist in locating the best local surface water or shallow groundwater (Chapter 12).

Transportation

One of the first things a drilling engineer asks in a new area is "how do we get to the drill site?" Easy and safe access to a wildcat well site or mineral prospect is important not only to keep costs down, but also to be able to drill at the exact spot pinpointed by surface mapping, seismic, and other surveys. Good airphotos or high-resolution satellite imagery can assist the road building effort by pointing out shortcuts, steep and shallow grades, landslide-prone areas, and so forth.

In some areas, it may be possible to gain access by barging supplies on rivers. In that case, it is helpful to have an up-to-date map of the river channel, sandbars, and water depths (Figures 13.8 and 13.9). Imagery is more accurate than most topographic maps, and it is certainly easier to obtain up-to-the-minute imagery than to have a new set of base maps produced. Likewise, it may be desirable to transport oil and ore by barge on water-ways instead of bringing it out by truck or building a pipeline. Lidar and multibeam echo sounders can provide river and offshore bathymetry.

One type of sonar, the Multibeam Echo Sounder, transmits sound energy and analyzes the return signal (echo) returned from the river bottom or seafloor to determine water depth and the nature of the bottom. Multibeam sonars transmit a broad acoustic fan-shaped pulse to produce coverage of the water bottom. These systems measure and record

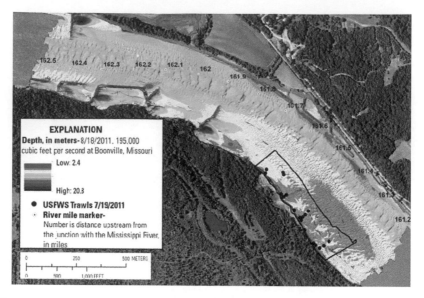

FIGURE 13.9
Bathymetry of the Missouri River just upstream of Hartsburg, Missouri. Mapped by the U.S. Geological Survey using a multibeam echo sounder on August, 18, 2011, when the discharge was 195,000 ft³/s at the USGS gage at Boonville, Missouri. (Courtesy of U.S. Geological Survey. http://www.usgs.gov/blogs/csrp/2011/10/19/mapping-baby-sturgeon-habitat-at-hartsburg/.)

the time for the acoustic signal to travel from the transmitter (transducer) to the river bottom or seafloor and back to the receiver. The coverage area is dependent on the depth of the water, and is typically two to four times the water depth (NOAA, online).

Increased drilling activity and the near certainty of future ice-breaking oil tankers make it essential to know the pattern of ice movement in the arctic. Offshore Alaska, the North Sea, offshore Siberia, and the Canadian and Scandinavian arctic are all areas of oil and gas exploration interest. Loading facilities and offshore well platforms must be able to withstand pressure and scouring from sea ice. All these require a thorough knowledge of the timing, direction, and rate of movement of ice floes. An excellent source of information is the historical archive of satellite images collected over the past 25–30 years. By looking at multitemporal imagery from a single season, one can determine the rate and direction of movement of individual floes. By looking at images from multiple years, one can begin to map the average and maximum extent of the pack ice, and the usual location and concentration of icebergs at various times of year. Daily airborne radar is being used extensively by the Canadians and Russians to map ice movements around offshore drill ships. Radar satellites make sea ice monitoring routine and much less expensive (Figure 13.10). Radar interferometry has been used to monitor sea ice movements (Meyer et al., 2011; Figure 13.11). Multibeam side-scan sonar is used to map sea bottom ice scour (Figure 13.12). It is essential to know where scour occurs, so that proper plans can be made for protecting sea bottom pipelines, subsea well completions, and for platform siting.

A final transportation issue is water-depth mapping. It is necessary to know the location of shoals in and around harbors and transportation routes. Many of these are mapped on Admiralty Charts or National Ocean Service Hydrographic bases. Still, in some areas, the shoals shift through time, and in other cases, they were not mapped accurately. Airborne water-penetration film (Chapter 1), the Coastal Zone Color Scanner, and Landsat TM all have excellent water-penetration capabilities. It is fairly simple to

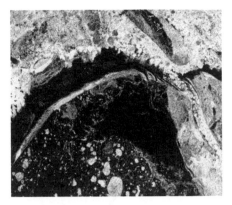

FIGURE 13.10
ERS-1 radar image of sea ice near the Kara Strait, offshore Siberia. The land area, bottom and center, has numerous frozen lakes. The ice offshore has different textures depending on its age. (Image processed by Amoco Production Company.)

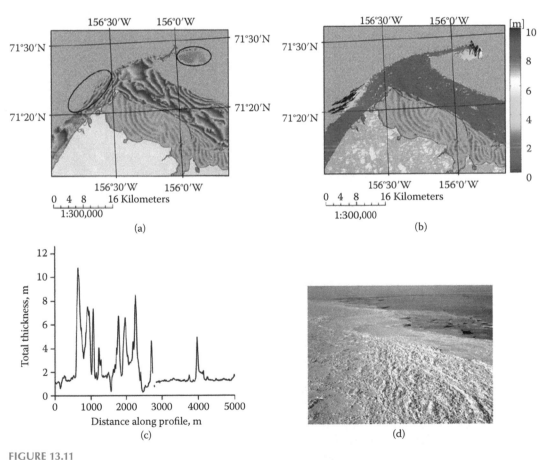

FIGURE 13.11
Examples of an analysis of fringe pattern (a) and two-dimensional ice surface displacements (b) observed in a radar interferogram over Barrow, Alaska. These measurements allow for the study of ice stability and deformation. A profile of total ice and snow thickness (c) was obtained from airborne measurements along the red line in (a) on 15 April 2008. Part (d) shows shear ridges corresponding to thick, deformed ice seen from the air. (From Meyer F.J. et al. *Rem. Sens. Environ.* 115, 3029–3043, 2011.)

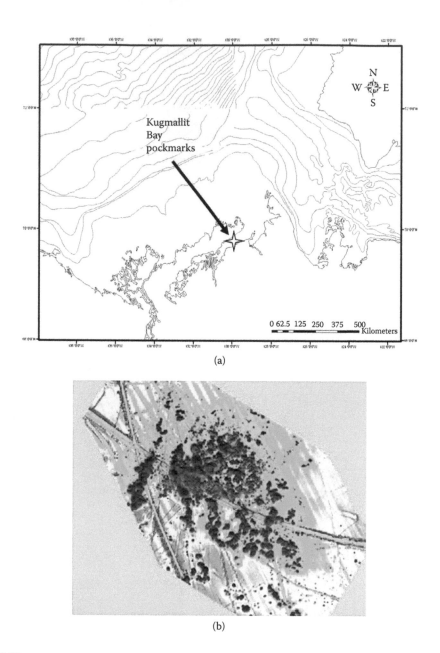

(a)

(b)

FIGURE 13.12
Multibeam side-scan sonar shows sea bottom ice scour (curving streaks) as well as gas vents and mud volcanoes (dark blue spots). (a) Location map, Kugmallit Bay, Northwest Territories, Canada. (b) Sonar image showing tracks of ice keels on the sea bottom. (Courtesy of Dr. Steve Blasco and Geological Survey of Canada.)

generate relative depth maps by density slicing or using various film densities to indicate areas that are shallow (Figure 13.13). It is slightly more difficult to generate good bathymetric maps, since many widely scattered control points are required. Multibeam sonar and airborne Lidar are being used for detailed shallow water bathymetric mapping. As mentioned in Chapter 12, apparent water depth is related to bottom reflectance and the amount of suspended sediment. If the bottom has both light and dark areas, caused

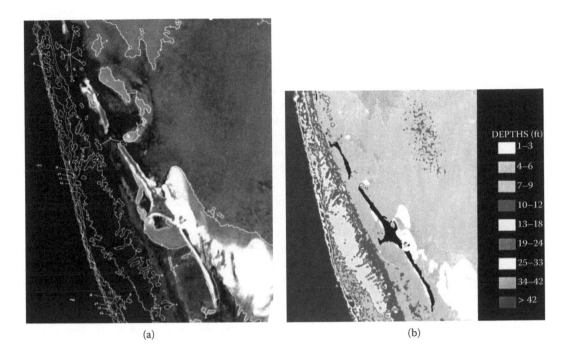

(a)   (b)

FIGURE 13.13
Contoured bathymetric image (a) and density-sliced bathymetric image (b) of North Cat Cay, Bahamas. (Landsat TM bands 1, 2, and 3 processed by ERIM [now Michigan Tech Research Institute of Michigan Technological University].)

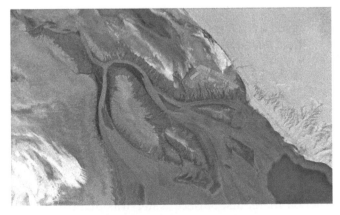

FIGURE 13.14
True color image of the Colorado River delta, Mexico, showing sediment laden waters entering the Sea of Cortez. (Courtesy of GeoEye, CNES/SPOT Image, DigitalGlobe, and Google Earth, © 2012 Google.)

by sea grasses or changing rock types, this will affect the apparent depth. Suspended particulates also reflect light as if they were the bottom. However, one can usually distinguish the pattern of sediment plumes from bathymetric patterns. Turbid water will totally obscure a river channel and form a featureless plume where the river flows into a lake or bay, whereas bathymetry will change across a channel and across a lake, bay, or delta (Figure 13.14).

Construction Projects

Large engineering projects require extensive planning. This planning should provide for minimal disturbance to the natural environment, the existing road, rail, and power infrastructure, and sources of water for cooling or construction. Planning must minimize and balance conflict with competing resource development in the area (e.g., agriculture, forestry, fisheries) and the interests of population centers. Seismicity, slope stability, and other factors must be accounted for. Power plants, in particular, require access to railroads for fuel and water for cooling. Nuclear power plants, in particular, are concerned about seismic activity that could disrupt operations (Lee et al., 1991). Pipeline planners, like road builders, prefer gentle, stable slopes, the least possible amount of earth to move and bedrock to blast, and as few river crossings as possible (see section "Case History: Pipeline Routing Using Remote Sensing and a Geographic Information System," Chapter 21). Tunnelers are concerned with seismicity, lithology, and fracturing in wall rocks. Dam construction requires local sources of sand and gravel for aggregate. Safe construction of dams must account for strike and dip of rock in which the dam is anchored, fracturing of wall rocks, and seismicity. Builders of dams are also concerned with porosity and permeability of the country rock, annual variations in runoff from snowpack and rainfall, erosion and sedimentation of the drainage basin upstream from the reservoir, and drowning of agricultural land and other resources. Image analysis can be a primary source of information to address these concerns. An example is given for the Kakrapar atomic power station in Gujarat State, India.

Case History: Site Selection for the Kakrapar Atomic Power Project, India

Satellite images and multiscale aerial photographs were used to prepare thematic maps that addressed surface geology, geomorphology, seismicity, water resources, land use, forest and mineral resources, availability of raw materials for construction, rail access, and environmental, health, and safety factors associated with the siting of an atomic power plant in Gujarat (Katti et al., 1993). The eastern half of the area is covered by Deccan basalts, whereas the western part of the area is mostly recent sediments. Dolerite and gabbro dikes of various ages intruded the area along east-northeast trends.

Regional scale satellite images were used to select a region of interest. This was followed by candidate site selection using finer resolution satellite images and airphotos to scales of 1:50,000 (Figure 13.15). Detailed analysis was performed using airphotos at scales of 1:25,000 with support from ground mapping and sampling.

Geomorphology was used to classify stable slopes, areas of rugged topography, and locations of construction materials. Surface water was mapped in lakes and drainages, and probable sources of near-surface groundwater were located. Land use was classified as cultivated, urban, forest, and "waste" land. Existing roads were mapped, and locations for new roads and power lines were evaluated. Forest cover was mapped to prevent degradation of an already depleting resource. Seismic activity was investigated to determine the location, frequency, and magnitude of earthquakes in the area. Sites closer than 5 km to active faults were not acceptable. Active faults are defined as having movement in the last 35,000 years, or multiple movements in the past million years. Image analysis, including low sun-angle photography, radar, and thermal imagery, was used to locate lineaments, recent fault scarps, shutter ridges, sag ponds, and aligned hot springs. A network of microseismic monitoring stations was established, and historical seismic activity was

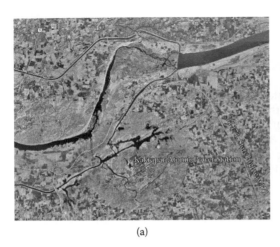
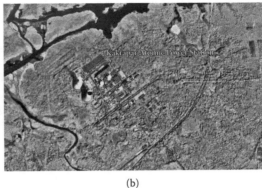

(a)   (b)

FIGURE 13.15
Kakrapar atomic power plant site. (a) Regional view. (b) Detail of the plant. (True color image courtesy of Digital Globe, GeoEye, and Google Earth, © 2012 Google.)

investigated. The age of the last active faulting was determined by dating disturbed and nondisturbed soil horizons. In summary, several remote sensing systems were used to determine the suitability of the site for a power plant, and to determine the probable impact on the environment around the plant.

Engineering applications of remote sensing are many and varied. As new problems arise, and new technologies become available, the number of applications will increase accordingly.

References

ERIM. 1992. *Discovering America (calendar)*. ERIM Image Processing Facility: Ann Arbor.

Gabriel, A.K., R.M. Goldstein, H.A. Zebker. 1989. Mapping small elevation changes over large areas: differential radar interferometry. *J. Geophys. Res.* 94 B7: 9183–9191.

George, R.A. April, 2005. AUV survey technology unveils seabed for deepwater developments. *World Oil*: 226: 99–102.

Jolliffe, I.P., A.G. McLellan. 1980. Prospectors for sand and gravel. *Geographic. Mag.* v. 52: 3 p.

Katti, V.J., Y.R. Satya Saradhi, S.N. Kak, D.C. Banerjee, R. Kaul. 1993. Application of remote sensing techniques for Nuclear Power Project Site Selection in India. *Int. J. Rem. Sens.* 14: 3291–3298.

Krynine, D.P., W.R. Judd. 1957. *Principles of Engineering Geology and Geotechnics*. New York: McGraw-Hill Book Co.: 730 p.

Lee, H., R. Wightman, R. Whittle. 1991. *Civil Engineering Applications of Remote Sensing and Geographic Information Systems*. New York: American Society of Civil Engineering: 199–208.

Lukosavich, N. July, 2012. Innovative thinkers. *World Oil*: 29 p.

Meyer, F.J., A.R. Mahoney, H. Eicken, C.L. Denny, H. C. Druckenmiller, S. Hendricks. 2011. Mapping arctic landfast ice extent using L-band synthetic aperture radar interferometry. *Rem. Sens. Environ.* 115: 3029–3043.

Mollard, J.D., J.R. Janes. 1984. *Airphoto Interpretation and the Canadian Landscape*. Ottawa, Canada: Energy, Mines and Resources: 117–136.

NOAA Office of Coast Survey. 2013. Multibeam Echo Sounders. http://www.nauticalcharts.noaa.gov/hsd/multibeam.html (accessed 21 June 2013).

Rinker, J.N., R.E. Frost. 1996. Remote sensing for engineering site selection. *J Comput Civil Engg.*: 10: 359–371.

Rogers, J.D. 2006. Lessons learned from the St. Francis Dam failure. *Geo-Strata* 6: 14–17.

Thompson, M.D., J.B. Mercer. March, 1996. Digital terrain models from RADARSAT. *Earth Observ. Mag.*: 22–25.

Ventura County Star. 2008. St. Francis Dam disaster: a tale of failure, tragedy and heroism. http://www.vcstar.com/news/2008/mar/12/the-st-francis-dam-disaster-80th-anniversary-a/#ixzz2ADaNfv4S (accessed 24 October 2012).

Wiley, A.J., G.D. Louderback, F.L. Ransome, F.E. Bonner, H.T. Cory, F.H. Fowler. 1928. *Report of the Commission to Investigate the Causes Leading to the Failure of the St. Francis Dam near Saugus, California*. Sacramento, CA: California State Printing Office: 79 p.

Additional Reading

Beaven, P.J. 1984. *Remote Sensing for Highway Engineering*. Proceedings of a Post-Graduate School, University of Dundee: 1129–137.

Foresman, T.W., R.C. Brown, Jr. 1982. *Desert construction siting utilizing remote sensing technology*. United States. Army. Corps of Engineers. Buffalo District: First Thematic Conference on Remote Sensing of Arid and Semi-Arid Lands. Cairo: 497–501.

Kiefer, R.W. (ed.). 1980. *Civil Engineering Applications of Remote Sensing*. New York: American Society of Civil Engineering: 195 p.

Legg, C.A. 1992. Remote Sensing and Geographic Information Systems: Geologic Mapping, Mineral Exploration and Mining. Chichester: Ellis Horwood: 256 p.

Lyon, J.G. 1995. Remote sensing and geographic information systems in hydrology. In Ward, A.D. and W.J. Elliott (eds.), *Environmental Hydrology*. Boca Raton: Lewis Publishers/CRC Press: 337–368.

Russell, J.D. 1991. Civil engineering applications of remote sensing London Docklands—a case study in applied GIS. In D.B. Stafford (ed.), *Civil Engineering Applications of Remote Sensing and Geographic Information Systems*. New York: American Society of Civil Engineering: 209–219.

Ungar, S. G., Merry, C. J., Irish, R., McKim, H. L., Miller, M. S. 1988. Extraction of topography from side-looking satellite systems—A case study with SPOT simulation data. *Rem. Sens. Environ.* 26: 51–73.

Section IV

Environmental Remote Sensing

The purpose of this section is to show how remote sensing technology is being used to resolve issues encountered by corporations, governments, and environmental organizations. Environmental applications of remote sensing include mapping the pattern of storm water runoff from plant sites, planning for spill containment, detecting onshore or offshore spills or pipeline leaks, and mapping environmentally sensitive areas such as wetlands and migratory bird nesting areas. Other uses for remote sensing include investigating historical archives to locate photos of old drill sites to determine whether there has been leakage from mud pits or if the pads have had an impact on local surface drainage. Imagery assists with monitoring changes in oil fields, mines, smelters, and refineries over a period of years. Remote sensing is monitoring the extent of hot water outflow from power plants and refineries and the impact of warm or nutrient-laden water on algal blooms. Environmental hazards, including landslides, earthquakes, and volcanic eruptions, can be identified and monitored using remote sensing technologies.

Environmental remote sensing is closely associated with urban planning and the development of Geographic Information System (GIS) databases. Remote sensing has served as both photo-documentation and as up-to-the-minute base maps for planning urban or industrial development and for mapping and outlining environmentally sensitive areas. The primary environmental uses for remote sensing in the petroleum and mining industries are in preventing or minimizing damage to sensitive development sites and plant locations, in spill contingency planning and monitoring, in monitoring pipelines, tailings, or platforms for leaks, in tracking spills and leaks, and in documenting past use (or misuse) of a parcel of land.

Environmental hazards include those things that are dangers to life, health, and habitation. They can be destructive to agriculture, forestry, fisheries, transportation, and navigation. Among the natural events that can be monitored and prepared for are landslides and avalanches, earthquakes and tsunamis, flooding and erosion, subsidence and sinkholes, shoals and sea ice, volcanoes, and the effects of global warming.

Whether or not it is eventually determined that humans are contributing to climate change, it is indisputable that global climate has been warming, on average, since the last ice age ended ~10,000 years ago. At the very least we know that sea levels have risen ~100 m (300 ft.) in that time. Remote sensing has been monitoring and will continue to monitor the effects of warming, on a global and local basis, to help humans adapt to a changing world. From the subsidence or submergence of coastal wetlands to a decrease in polar ice sheets to the northward migration of the pine bark beetle, imagery is available to help map these changes.

14

Environmental Baselines and Monitoring

Chapter Overview

A large part of environmental remote sensing is concerned with the establishment of baseline maps through a process of mapping various components of the surface such as vegetation, soil, water and drainage networks, and manmade features. It is possible to map animal habitats through a combination of slope, water, and vegetation mapping and field observations. Mine and tailings reclamation projects are assisted by regular, periodic monitoring. Surface and groundwater are monitored for evidence of pollution, areas of recharge and discharge, and for manmade input (e.g., power plant cooling water, fertilizers, acid mine water). Air pollution, although more difficult to monitor, can under certain circumstances be seen. Air pollution can be a result of fire, smelter plumes, dust storms, volcanism, and urban smog. All these maps and observations are derived in total or in part from an integrated analysis of various types of imagery.

Environmental Audits

The most suitable application for remote sensing technology, monitoring of the environment, is also the most routine and best demonstrated. An environmental audit examines a property prior to leasing or divestiture to determine whether surface damage exists (or existed), where and when it occurred, and whether it requires remediation. Environmental audits have become an indispensable part of assuring compliance with environmental regulations and ensuring the exercise of "due diligence" during real estate transfers. Lending institutions require this type of investigation to minimize their liabilities in future pollution cases. In the United States, the 1986 Superfund Amendment and Reauthorization Act established the "innocent purchaser" defense to environmental liability. This defense is applicable if the property owner exercises "all appropriate inquiry into the previous ownership and uses of the property" and conducts "appropriate inspections" at the time of acquisition (Lyon, 1987). Resource extraction companies would be wise to carefully evaluate leases to be sure they are aware of existing environmental damage that they may be required to remediate when relinquishing their lease.

The movement of hazardous waste across international borders is regulated by the Basel Convention of 1989 and the follow-on Rotterdam Convention of 1998. The Rotterdam Convention established responsibilities between importing and exporting countries for notification and monitoring of waste known to be harmful to human health and the environment. Remote sensing is one way of monitoring hazardous waste.

Hazardous Waste

The Love Canal neighborhood of Buffalo, New York, contained 21,000 tons of toxic waste that was buried 7 to 8 m deep and not properly contained (Blum, 2008). In the early 1950s, the site was covered and sold for residential construction. In the late 1970s, residents began having health problems that were eventually traced back to the waste site. When sampling began, none of the six boreholes drilled by investigators to locate the buried pollutants encountered anything of concern. A follow-up geophysical survey located the toxic plume between borehole locations (Schoer, 1999).

Hazardous waste is an inevitable result of manufacturing and other industrial processes. Substances may be manmade (e.g., batteries) or natural but concentrated by industry (lead, mercury). Hazardous is defined as either corrosive, ignitable, reactive, or toxic (Slonecker et al., 2010). A key step in remediation is knowing what you have, how much or how concentrated it is, and where it is. This is known as site characterization, site assessment, or an environmental audit.

The first stage of an environmental audit typically consists of an assessment of site conditions and the documentation of existing or potential environmental problems. This involves determining previous surface operations, noting present land use, observing surrounding land use, doing a site inspection, and interviewing those familiar with the site. Investigators review historical photos or images for any evidence of abnormal vegetation growth, discolored soil, trenching or landfills, surface structures, barrels and drums, and any signs of contamination due to spills, leaks, or dumps (Erb et al., 1980; White et al., 2008). Phase two of the environmental audit involves confirming the presence of contamination by sampling soil, soil gas, water, and plants. Sampling may still miss the pollution, as illustrated in the Love Canal neighborhood. Among the geophysical techniques recommended for locating such buried waste sites are magnetometry, electromagnetic (EM) surveys, resistivity surveys, and ground-penetrating radar (GPR) (Benson et al., 1984). Multispectral and hyperspectral scanner images have also been processed to enhance anomalous soil tones and vegetation patterns that may reveal the extent of any pollution.

Hazardous waste sites that can be monitored using airphotos, multispectral imagery, or hyperspectral imagery include the following: mine dumps and tailings; acid mine drainage; abandoned mine pits; leaks from cyanide leaching (gold) operations; concentrated runoff of agricultural pesticides; high natural, industrial, or agricultural levels of arsenic, selenium, cadmium, lead, and mercury; and organic chemicals such as benzene and toluene and some plastics. Thermal scanners are useful for detecting some heavy metals, organophosphates, and other heavy molecular weight compounds (Kahle et al., 1983; Hook et al., 1998). Fourier Transform Infrared (FTIR) is an active system that send out a pulse of energy and records the resulting absorption or emission at specific wavelengths. FTIR systems operate in the near visible and infrared parts of the spectrum and can be diagnostic for many chemicals considered hazardous wastes depending on the context (Slonecker et al., 2010).

Terrestrial Hydrocarbon Leaks and Spills

Leaks and spills related to industrial facilities and pipelines may occur on the surface or originate in buried pipelines or tanks. Leaks and spills can change soil tone and color. They can also affect vegetation type, density, and vigor (see Chapter 15, section Botanical Responses to Hydrocarbons). Hydrocarbon leaks and spills are detectable using hyperspectral imagery due to hydrocarbon absorption bands at 1.73 and 2.31 μm (Hörig et al., 2001). The 1.73 μm absorption is fairly unique to hydrocarbons, but other materials have absorptions near 2.3 μm. To avoid confusion with other materials, a hydrocarbon index (HI) has been defined as

$$HI = \frac{2(R_{1741} - R_{1705})}{3} + R_{1705} - R_{1729}$$

where R is radiance at 1705, 1729, and 1741 nm, respectively (Kühn et al., 2004). The HI was tested using a HyMap instrument flown at 1137 m (2.84 m resolution). The test successfully identified a 4×4 m patch of oil saturated sand (100 ml lubricating oil/kg sand) and another of slightly contaminated sand (25 ml oil/kg sand). It also identified plastic sheeting and artificial grass as hydrocarbons, but not plastic roofs.

A study at ITC (International Institute for Geo-Information Science and Earth Observation, Enschede, The Netherlands) investigated whether benzene-saturated bare soil could be identified using an Analytical Spectral Devices hyperspectral field spectrometer, the portable FTIR spectrometer, and a handheld gamma ray spectrometer (Cundill, 2008). Measurements were taken at the surface and on samples from 0.5 and 0.9 m depths. None of the data sets were able to differentiate the polluted from the clean samples.

On the other hand, some recent work indicates that reflectance spectroscopy, at least in the laboratory, can detect and quantify the amounts of hydrocarbons in soil. "Reflectance spectroscopy" consists of measuring the reflected electromagnetic energy from the soil samples in the visible-NIR-SWIR region (350–2500 nm), and modeling these spectra against samples with known concentration levels. Laboratory and *in situ* measurements of oil-contaminated soils are part of an effort to build a spectral library for use in hyperspectral remote sensing (Andreoli et al., 2007). Spectra were scanned over the 400–2500 nm interval with a spectral resolution of 1–2 nm in the visible-near infrared (VNIR) (400–1000 nm) and with 4 to 5 nm spectral resolution in the mid-wave thermal infrared (1.00–2.50 μm). Samples of oil-contaminated soil were approximated using pure sand and a loam composed of 78% sand, 20% silt, and 2% clay. Four types of hydrocarbons were used: diesel oil, used oil, and two types of crude oil, "Es Sider" light crude oil and "Iranian Heavy" heavy crude oil. Hydrocarbon absorptions at 1.730 and 2.310 μm were clearly visible in all of the oil-affected samples. Successive measurements made months and years later show unchanged results.

In another test, three types of soil consisting of various amounts of sand, silt, and clay were contaminated with varying concentrations (from 50 to 200 ppm) of several types of hydrocarbons including diesel, kerosene, Octane 95, and 418.1 EPA reference (Schwartz et al., 2012). Reflectance spectroscopy results obtained by the authors were compared with measurements from three commercial laboratories. The results are consistent and good predictors of contamination. If Octane 95 is removed (because of its volatility), the correlation coefficient is nearly perfect. Work is ongoing to see if these results can be duplicated using airborne systems.

Habitat Mapping

Other environmental uses for remote sensing include mapping environmentally sensitive areas as part of environmental impact statements prior to building roads, pipelines, or facilities in order to preserve the unique character of an area as well as to meet legal requirements. Most jurisdictions now require that mine or well sites be returned to their original condition after mining or drilling, and that the impact of access roads be minimized. This requires documentation of the original condition of the area, which is readily accomplished with archival and current airphotos or imagery. Sensitive areas such as wetlands or salmon spawning streams require that drainage is not interrupted and that discharge from a site be free of pollutants and/or particulates and at ambient temperature. Airphotos and high-resolution satellite images provide accurate and detailed base maps for survey teams and construction crews, and can improve the efficiency of cleanup and restoration efforts. Radar images can be used to map areas such as rainforest that are under permanent cloud cover. Thermal imagery can determine the temperature contrast between discharges and native waters. Remote sensing can help characterize a site before and after development (Kuehn et al., 2000).

Companies must take into account the impact of their facilities on neighboring urban areas. Many facilities were originally built in isolated areas. These areas subsequently attract development, and people living near these plants complain about the impacts of the facility. These effects can include noise, smell, effluent discharges, even blocking the scenery. These issues should be investigated and mitigated at the earliest possible date, both to guide local development and to lessen the impact of an industrial site on neighboring communities by taking into account wind patterns, surface runoff, and development patterns.

The process employed to evaluate an area or establish an *environmental baseline* (conditions at a point in time) involves acquiring imagery and digital elevation models (DEMs) and analyzing them for land surface relief, surface water, existing land use and/or disturbance, and mapping the distribution, density, vigor, and types of vegetation cover. Several dates and seasonal cycles should be analyzed. The combination of topography, water, and cover type determines the habitat for wildlife (Figure 14.1). If certain combinations of parameters are known, for example, that elk range requires a specific combination of vegetation, slope, and slope azimuth (aspect), it should be possible to characterize the habitat at a known location (the control point or "training site"). The training site is used to automatically map (classify) all areas with similar spectral characteristics (Huber and Casler, 1990; Koltun, 1993). Spectral classes are a function of discreet wavelengths reflecting off single or mixed cover types and the percent cover or vegetation density. Supervised classification is commonly used to map habitats and ecosystems once a set of control points has been established by field surveys. Prior to field surveys, it is possible to generate unsupervised classification maps by providing a finite number of categories and instructing the computer to perform multivariate analyses that lump all surface features into these categories (classes) based on similar reflectance characteristics. These categories then require field verification to determine their unique composition (Isaacson et al., 1982).

Images, particularly orthophotos or rectified imagery, can be used as base maps for field work, construction programs, or for regional orientation purposes. A latitude–longitude grid and other survey information is annotated on the images, thus providing highly accurate and up-to-date maps (Figure 14.2). One can go further still and generate topographic maps of areas that have stereo overlap of airphotos or satellite images (Chapter 13, section Generating Appropriate Base Maps). There are service companies that generate topographic maps from stereo imagery, stereo radar imagery, and radar or Lidar-derived DEMs. GoogleMaps provides a similar service, showing imagery or planimetric maps for any area.

Classification techniques are used in conjunction with DEMs and a GIS to categorize habitats on the basis of more than just imagery. The DEM provides information on elevation, slope, aspect, and surface roughness. GIS databases can be co-registered to an image and used to provide information concerning road networks, land ownership, well locations, drainage patterns, sample sites, rights of way, and tailings or dumps. Conversely, the image can be used to update a GIS database. Using imagery with the appropriate GIS, it should be a simple task to identify, for example, all south-facing slopes less than 15° without forest cover on private land and having good road access.

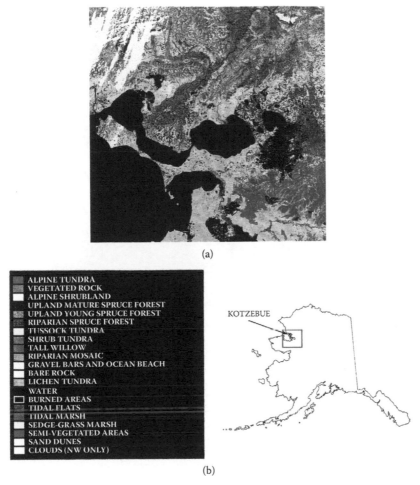

(a)

(b)

FIGURE 14.1
Caribou and grizzly bear habitat classification combines digital Landsat data and ground measurements near Kotzebue, Alaska. (a) Habitat map. (b) Legend and location map. (Processed by the General Electric Digital Image Analysis Lab, General Electric Company.)

Surface Disturbance and Change Detection

Mapping surface disturbance can be achieved using standard photointerpretation techniques and just about any kind of imagery. Unlike habitat mapping, the objects being mapped are generally manmade (roads, building sites, storage tanks or mud pits, etc.) and therefore, unless buried, are usually distinct from background. Sometimes, we are interested in mapping changes due to flooding, landslides, coastal erosion, or some other natural cause. There are cases where dumps, mud pits, pipelines, and other features have been buried and forgotten, leaving only subtle clues to their existence. These clues include features such as linear bulldozer scars over landfills, abrupt changes in the natural contour of slopes, and geometric vegetation patterns over a site (Williams, Jr., 1983). These are usually easier to see from the air than from the ground, because the contrast with background slope and vegetation is more apparent. Multitemporal images are useful for observing change, particularly

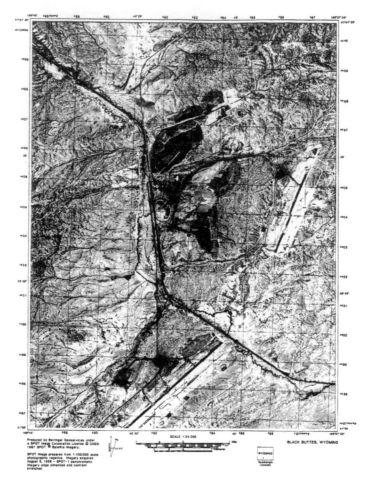

FIGURE 14.2
SPOT image map of Black Buttes, Wyoming, showing coal mining activities. (Generated by Barringer
Geoservices, Denver.)

manmade change in an area. Photos of the Love Canal toxic waste site taken over a period
of years readily show where old developments were and where more recent developments
have occurred (Lyon, 1987). Multispectral airborne video has proven to be particularly effective in monitoring hazardous waste sites over a period of time and should have widespread
application to many environmental monitoring activities (Marsh et al., 1991).

GPR is useful for locating buried waste or pipelines. GPR has been used as a reconnaissance tool by running regional survey lines, and has provided detail at a site once airphotos locate a disturbed area. The output appears much like a seismic line (Figure 14.3). GPR
can be processed to show either vertical or horizontal slices of the near surface.

Classification techniques, described in Chapter 2 under Classification, and earlier in this
chapter under Habitat Mapping, can be used to classify cultural features and development. One must be careful not to classify, for example, a rock outcrop or concrete building
the same as a rock quarry since they may have similar spectral characteristics. In such
cases, the geometry of the features frequently reveals whether or not they are manmade.

We may be interested in learning whether surface disturbance is increasing due to
development, illegal dumping, increased erosion, and so forth, or whether this disruption

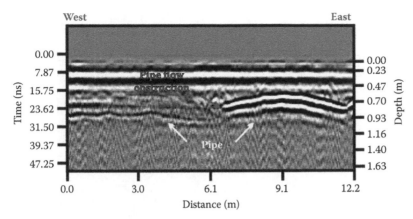

FIGURE 14.3

This GPR profile from data collected along trend directly over the top of a drainage pipe in saturated soil shows the abrupt transition from a water-filled pipe (west side) to an air-filled pipe (east side). This abrupt transition in the GPR response is due to a pipe obstruction that completely blocks the flow of water. (Courtesy of Barry Allred, USDA/ARS.)

is decreasing as a result of successful remediation. Change can be detected using image subtraction. Photos or images acquired at different times must be co-registered and, if necessary, resampled so that the resolution cell or picture elements (pixels) of each are the same size and represent the same surface features. Once this is accomplished, any changes that exceed a given threshold can be detected by subtracting the reflectance at one time from another. Those areas that show change are assigned an arbitrary color or pattern which can be superimposed on the more recent image (Figure 14.4).

A fast and relatively inexpensive way to monitor leaks, spills, illegal trespass, dumping, timber cutting, or mining on one's lease is to fly an aerial survey at given intervals (annual, semiannual, monthly) over all oil fields, facilities, pipelines, or undeveloped land. These images are evaluated in real time or in the office, and areas with indications of tampering or other changes can be quickly visited in the field. Many oil companies routinely fly periodic thermal infrared (TIR) surveys along oil pipelines to detect small leaks before they become a threat to the environment. Such surveys provide excellent documentation of current and past surface conditions. Satellites and unmanned aerial vehicles (drones) can be used in the same way.

Hyperspectral surveys have been proposed for monitoring industrial facilities such as nuclear power plants and waste repositories. Although primarily an environmental survey, it would also serve a security role in ensuring that no unanticipated changes have occurred. Just such a program was flown over the Olkiluoto power plant and repository in southwestern Finland during 2008 (Tuominen and Lipping, 2011). The main purpose of the hyperspectral survey was to gather environmental information as part of long-term repository monitoring. Hyperspectral imaging would locate radioactive contamination, if there were any, by its effect on vegetation. The effects of the radioactive contamination on vegetation include "radiotoxicity." where radionuclides are incorporated in plant tissue, and the effects of radiation from radionuclide concentrations in the soil. Radioactive contamination changes the rate of photosynthesis and thus the concentration of pigments in plant leaves. These changes can be detected using hyperspectral remote sensing techniques. Davids and Tyler (2003) presented results of *in situ* measurements of silver birch (*Bentula pendula*) and Scots pine (*Pinus sylvestris*) for several contamination levels in the Chernobyl exclusion zone. Results show that the chlorophyll red edge shifts to shorter wavelengths

FIGURE 14.4
Change detection, Louisville, Kentucky. Areas that show changes (airport runways, quarries) between 1988 and 1991 are shown in red (bright changes) and cyan (dark changes) on this SPOT Pan image (Image processed by Earth Satellite Corporation, Rockville, MD [now MDA Federal].)

(blue shift), the absolute reflectance decreases, and the Three-Channel Vegetation Index (TCHVI) correlates well with high concentrations of Sr-90 and Cs-137 in leaves.

$$TCHVI = \frac{((R-G)+(NIR-R))}{(|(R-G)|+|(NIR-R)|)}$$

The Olkiluoto site is dominated by forest and coastal water areas. Monitoring of benthic vegetation using hyperspectral remote sensing is part of ongoing surveys. Radioactive contamination affects submerged vegetation in the same manner as trees due to radionuclide incorporation into plant tissue.

The baseline survey at Olkiluoto was flown at an altitude of 1.9 km, providing a ground resolution of 2.5 m^2. The Piper aircraft carried an AISA imaging spectrometer that collected 481 spectral bands between 399 and 2452 nm. A simultaneous field campaign by the Geological Survey of Finland collected calibration data. As a follow-on to this program it was proposed that airborne campaigns be flown every third or fourth year to update the hyperspectral data. The optimal acquisition time is in July when the vegetation is verdant and the probability of good weather is highest.

Case History: Using Interferometric Synthetic Aperture Radar (InSAR) to Monitor Surface Deformation over a Steam-Assisted Gravity Drainage Project, Christina Lake, Alberta

At 5:15 AM on May 18, 2006, a steam geyser erupted through the floor of the boreal forest near the Joslyn Creek *in situ* oil sands development 60 km north of Fort McMurray, Alberta, Canada. Chunks of rock were thrown up to 300 m horizontally, and a 1 km long dust plume extended out from the release point. A 75 × 125 m crater was formed at the blowout. Fortunately, there was no loss of life or injury, and no harmful gases were released. Joslyn Creek is one of several thermal recovery developments, also known as steam-assisted gravity drainage (SAGD) projects, in the Athabasca basin near Fort McMurray (ERCB, 2010).

Christina Lake is an oil sands property in Alberta that is being developed by Cenovus Energy (operator) and ConocoPhillips using the SAGD process (Figures 14.5 and 14.6). Steam is injected into the subsurface at high pressure to melt bitumen in Lower Cretaceous McMurray Formation sands at an average depth of 400 m.

Disruption of the Christina Lake plant and pipelines due to surface deformation is undesirable and must be monitored. As well, caprock integrity is important to prevent an uncontrolled steam blowout such as that which occurred at the Joslyn Creek property in 2006 (Figure 14.7) (ERCB, 2010). As steam is injected, the surface above the injector wells is slowly displaced upwards up to several centimeters. One sign that pressures may be excessive would be an abrupt increase in surface heave.

Another reason for monitoring surface heave is to monitor steam chamber growth. Along with four-dimensional seismic and instrumented wells, surface heave can provide information as to where the steam chamber (that part of the reservoir that is accepting steam) is actively growing.

Among the methods being used to monitor surface displacement are conventional radar altimetry, coherent target monitoring (CTM) and radar interferometry (InSAR). Radar altimetry measures the time it takes microwaves from satellite radar (in this case RADARSAT-2) to reflect off the ground. CTM measures the time it takes for microwaves to reflect off buildings and corner reflectors installed across the field. Time is converted to distance and this is used to determine the relative change in surface elevation relative to a reference monument, and from one satellite pass to the next (in this case monthly). The change in elevation is then plotted on a displacement versus time chart (Figure 14.8).

Radar interferometry measures the distance between the surface and the radar satellite on multiple satellite overpasses and creates an interferogram that shows the change in surface elevation (Figure 14.9). The interferogram is used to generate a map showing areas of heave as well as subsidence (Figure 14.10).

These surface displacement maps show good agreement with maps derived from conventional radar altimetry and coherent reflector analysis (Figures 14.11 and 14.12), and indicate where steam chamber growth is occurring. One would expect maximum surface displacement over the actively growing steam chamber, and perhaps some subsidence where steam is no longer being injected. In fact, InSAR measurements have been used to validate geomechanical and thermal models by history matching steam injection and rock mechanics properties (linear thermal expansion coefficient, Young's Modulus, Poisson's Ratio) to surface heave (Chin et al., 2012).

Seasonal variations in surface elevations up to 20 mm have been attributed to frost heave in this mixed muskeg-boreal forest environment. At the time of writing (January 2013), there have been no indications of sudden or unexpected changes in surface displacement at the Christina Lake operations.

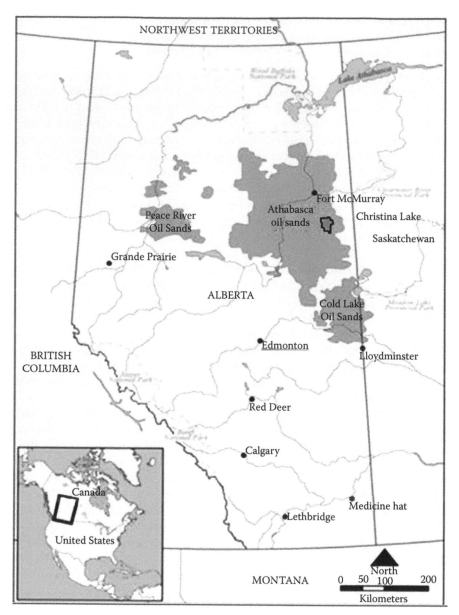

FIGURE 14.5
Location map showing the Christina Lake lease (green), northeastern Alberta.

Coal Mine Fires

In 1962, a coal bed was ignited by fire in a garbage dump on the outskirts of Centralia, Pennsylvania. As the fire worked its way along the underground seams, smoke began to rise out of new cracks in the ground, and in some areas the ground became hot to the touch. Local fire departments and the U.S. Bureau of Mines tried unsuccessfully to pump water into the ground, dig trenches, and use explosives to put out the fire. Sensors showed

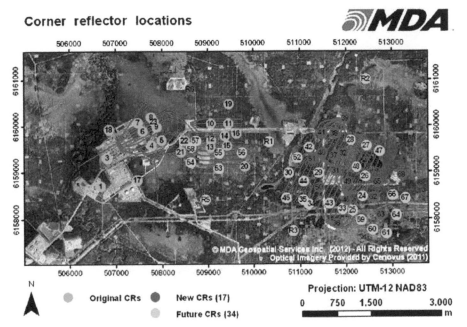

FIGURE 14.6
Airphoto with the location of the Christina Lake producing pads, pilot pad, and central processing facility. Horizontal wells shown in orange. Corner reflectors (CRs) are shown as colored dots. (Courtesy of Cenovus and MDA.)

FIGURE 14.7
Surface disturbance at the Joslyn Creek SAGD uncontrolled steam blowout, 2006. (Courtesy of Alberta ERCB; ERCB Staff Review and Analysis: Total E&P Canada Ltd., Surface Steam Release of May 18, 2006, Joslyn Creek SAGD Thermal Operation, February 11, 2010.)

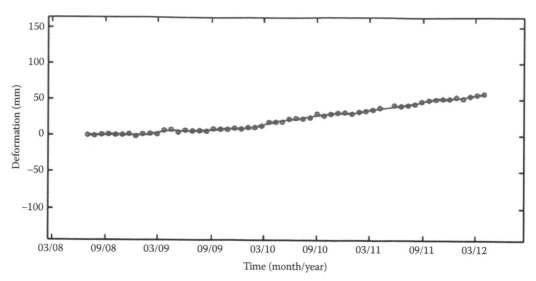

FIGURE 14.8
Displacement versus time chart, Christina Lake corner reflector #11, May 17, 2008, to March 31,2012. Maximum displacement is ~60 mm. A smooth curve, like this one, is the desired outcome. (Courtesy of Cenovus and MDA.)

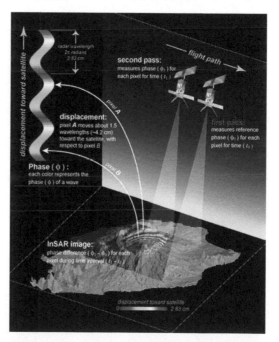

FIGURE 14.9
How a radar interferogram is generated. (Courtesy of Cenovus and MDA.)

temperatures close to 525°C less than 10 m below the surface. Soon homes had unhealthy levels of carbon dioxide, and sinkholes began opening in people's yards. In the 1980s, it cost the government $42 million to evacuate the town (Bryson, 1998).

Coal mine fires are rare, but widespread, and once started they can be devastating (Gangopadhyay, 2000; Voigt et al., 2004). They can be started by lightening or grass or forest fires that contact an outcrop, or can be accidentally started by man. Some fires have been

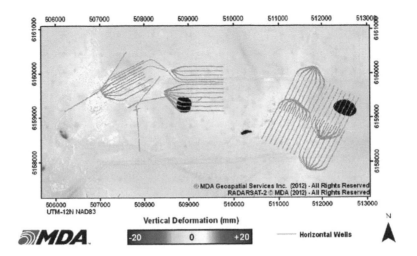

FIGURE 14.10
Map derived from the interferogram shows vertical deformation in mm, February 12 to March 7, 2012. Black areas are surface ponds. (Courtesy of Cenovus and MDA.)

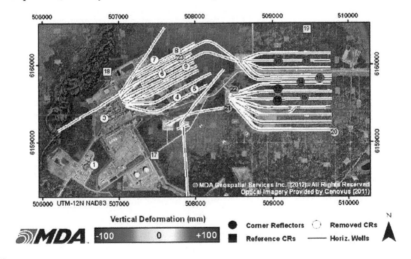

FIGURE 14.11
Cumulative corner reflector vertical deformation, July 2, 2008, to March 31, 2012. Numbers refer to corner reflectors; colors show displacement in mm. (Photos taken by Geographic Air Survey Ltd. and orthorectified by Aecom Edmonton; courtesy of Cenovus and MDA.)

burning for years, and they are difficult if not impossible to extinguish. These subsurface fires not only destroy a potentially valuable resource, but they can destroy a landscape and make it uninhabitable. The following example shows how coal fires can be monitored using remote sensing techniques (Bhattacharya et al., 1991; Mukherjee et al., 1991).

Case History: Monitoring the Jharia Coal Fire, India

Fire, possibly started by spontaneous combustion, was reported at the Jharia coal field, India, in 1916. Since that time an estimated 37,000,000 tonnes of high-grade coking coal have burned, and another 1,864,000,000 tonnes are at risk unless the fire can be extinguished.

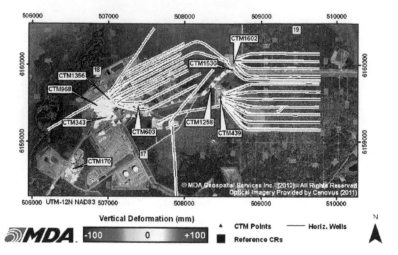

FIGURE 14.12
Cumulative coherent target vertical deformation, July 2, 2008, to March 31, 2012. (Courtesy of Cenovus and MDA.)

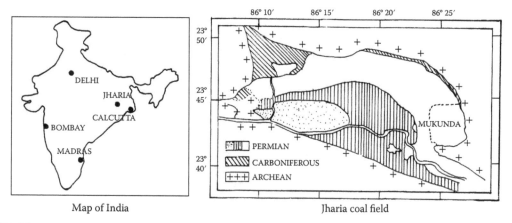

Map of India   Jharia coal field

FIGURE 14.13
Location map, Jeenagora colliery, Jharia coal field, India (From Bandyopadhyay, T.K. et al. 1994. Magnetic response and remote sensing for mapping coal seam fire-A study over Jharia coalfield, India. In *Proceedings of ERIM 10th International Conference on Geologic Remote Sensing. II.* Ann Arbor, MI: ERIM: II576–II584.)

Surface installations have been damaged, and the local environment has been degraded. The intensity and extent of the fires must be determined in order to plan how to extinguish the fires.

A study was undertaken in the Jeenagora colliery of the eastern Jharia coal field (Figure 14.13). The surface is covered by sand and overburden consisting of the Permian Baraklar Formation. Eighteen coal seams dip about 6° west and lie at depths of a few meters to 80 m.

A combination of airborne TIR imagery and a magnetic survey were acquired to help determine the extent and intensity of the fires. Predawn imagery was acquired in the 8.5–13.0 μm range with a ground resolution of 1.5 m at nadir and a thermal sensitivity of about 1°C. Two preset black bodies at 10°C and 65°C were used to calibrate the system. Isotherm maps of the surface were generated from relative radiance of the thermal imagery. Simultaneously, a magnetic survey consisting of 17 profiles between 50 and 130 m apart were acquired both over the known fire zone and over areas known to be unaffected by fire. The accuracy of the instrument was on the order of ±10 gammas. Laboratory work has shown that magnetic susceptibility of sediments is dramatically affected by heating. For example, the magnetic susceptibility of feldspathic sandstone and soils from the fire-affected zone was measured at less than 100×10^{-5} (SI), whereas the same sandstone in areas unaffected by fire had susceptibilities from 100×10^{-5} to 200×10^{-5} (SI).

Density slicing and color coding of the thermal imagery in conjunction with some surface temperature measurements showed where the surface has been heated by conduction or convection related to the underground fire. Temperatures varied mainly from 5°C to 15°C, with some readings as high as 45°C (Figure 14.14). However, the technique is limited by the ground conditions (surface cover, relief, disturbances).

Magnetic data were converted to vertical intensity and contoured, then smoothed (Figure 14.15). The areas affected by fires contain clinkers that have magnetic susceptibilities ranging from −1000 gammas to +2000 gammas. These highs and lows are thought to be influenced by the location of rooms and pillars in the coal as well as the location of fractures that allow groundwater and air movement to the coals. Unburned areas show almost no magnetic character, for example, south of Underground Barrier-1.

The combination of TIR imagery and ground magnetic surveys appears to be an effective way to monitor the location and intensity of underground coal fires.

Mine Reclamation

Many areas contain abandoned mines. Some are small and unmarked and there is danger of animals or people falling into them. Others involve large areas of disturbed ground, unstable slopes, groundwater seepage, and tailings or dumps that pose a danger of debris slides or heavy metal contamination in surface and groundwater. Satellite images and air-photos, combined with a GIS, can help develop efficient, cost-effective remediation and revegetation of mine tailings and dumps (Kremer et al., 1998; Fischer and Hermsmeyer, 1999). Airphotos, Landsat Thematic Mapper (TM), and Airborne Visible/Infrared Imaging Spectrometer (AVIRIS) data were used to find and map the distribution of tailings and other mine waste in the Cripple Creek district, Colorado, and Goldfield district, Nevada (Peters and Hauff, 2000). The following discussion describes mapping mine waste by charting the distribution of characteristic minerals (Hauff et al, 1999).

Case History: Mine Reclamation in the Couer d'Alene Mining District, idaho

A project to map the distribution of abandoned mines in the Couer d'Alene district of northern Idaho used AVIRIS hyperspectral imagery to locate tailings and dumps (Figures 14.16 and 14.17). The purpose was to identify areas prone to high concentrations of heavy metal contamination.

The Couer d'Alene district contains structurally complex sediments of the Precambrian Belt Supergroup, including argillites, quartzites, siltstones, and dolomites. Lead–zinc–silver mineralization is hosted in the clastic units as vein-type deposits containing sphalerite, galena, and tetrahedrite. The main toxic metals of concern include lead, zinc, and cadmium.

The approach used in this project was to learn as much as possible about the geology and history of the district, then examine the area during a field reconnaissance and document the results. Sites were located using global positioning system (GPS). Each site

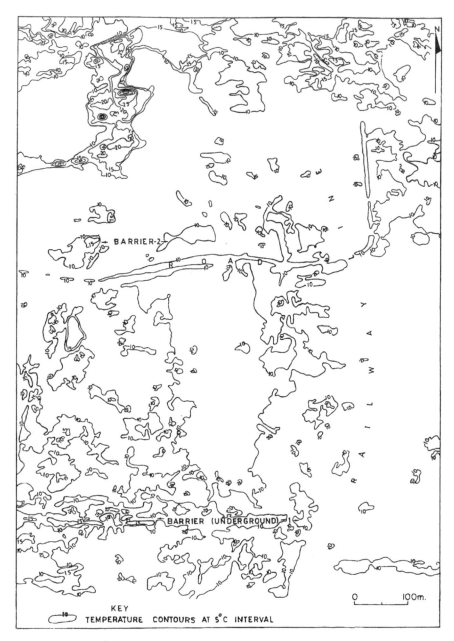

FIGURE 14.14
Surface temperatures, derived from thermal imagery, over the underground coal fire, Jharia coal field. (From Bandyopadhyay, T.K. et al. 1994. Magnetic response and remote sensing for mapping coal seam fire-A study over Jharia coalfield, India. In *Proceedings of ERIM 10th International Conference on Geologic Remote Sensing. II.* Ann Arbor, MI: ERIM: II576–II584.)

was photographed and rock samples were collected and analyzed using a portable x-ray fluorescence instrument. Rock, soil, vegetation, and solar spectra were obtained using a portable field spectrometer (Figures 14.18 and 14.19). The investigators acquired airborne imagery, ran laboratory analyses of field samples, and generated a spectral sample library. Image processing led to generation of mineral maps, which were confirmed in the field.

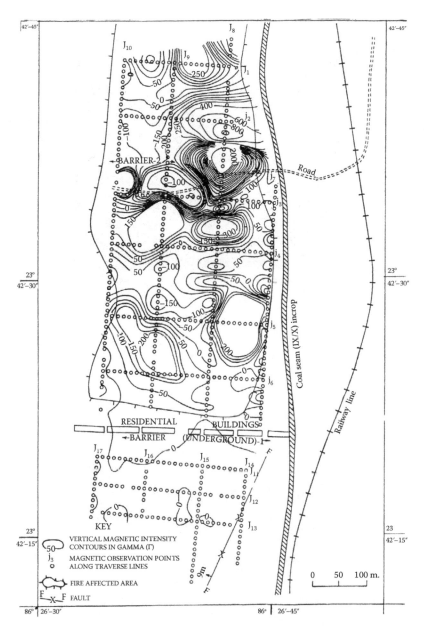

FIGURE 14.15

Magnetic field over underground coal fire, Jharia coal field. Highs and lows are influenced by the location of rooms and pillars as well as fractures (From Bandyopadhyay, T.K. et al. 1994. Magnetic response and remote sensing for mapping coal seam fire-A study over Jharia coalfield, India. In *Proceedings of ERIM 10th International Conference on Geologic Remote Sensing. II.* Ann Arbor, MI: ERIM: II576–II584.)

The high-resolution field spectrometer allowed the image processor to choose end-member spectra for use in spectral unmixing algorithms. The "pure pixel end-member method" (Boardman, 1993) was used. Raw data were calibrated to relative reflectance and the imagery was warped to GPS control points. A combination of channels was chosen to create false color images that highlighted the distribution of iron minerals, micas, and clays (muscovite and illite) (Figures 14.18 and 14.19). Mine dumps in the area are characterized by clay minerals (altered

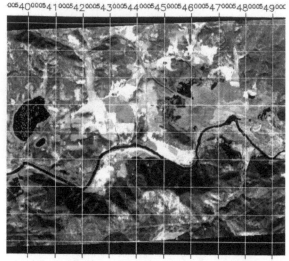

FIGURE 14.16
AVIRIS hyperspectral image over the lateral lakes of the Couer d'Alene River, Idaho. Rectified false color image by Spectral International and Borstad Associates. (From Hauff, P.L. et al. 1999. Hyperspectral evaluation of mine waste and abandoned mine lands: NASA- and EPA-sponsored projects in Idaho. In R.O. Green (ed.), *Summaries of the Eighth JPL Airborne Earth Science Workshop*, February 9–11, 1999: JPL Publication 99-17: 229–238.)

FIGURE 14.17
AVIRIS hyperspectral image over Canyon Creek and Nine Mile Creek drainages, Couer d'Alene mining district, Idaho. Rectified false color image by Spectral International and Borstad Associates. (From Hauff, P.L. et al. 1999. Hyperspectral evaluation of mine waste and abandoned mine lands: NASA- and EPA-sponsored projects in Idaho. In R.O. Green (ed.), *Summaries of the Eighth JPL Airborne Earth Science Workshop*, February 9–11, 1999: JPL Publication 99-17: 229–238.)

FIGURE 14.18

Classification image of the lateral lakes of the Couer d'Alene River area, Idaho. Tailings are indicated by the 2.2 μm absorption band color coded as yellow (best match to siderite and mica) to blue to magenta. Processed by Spectral International and Borstad Associates. (From Hauff, P.L. et al. 1999. Hyperspectral evaluation of mine waste and abandoned mine lands: NASA- and EPA-sponsored projects in Idaho. In R.O. Green (ed.), *Summaries of the Eighth JPL Airborne Earth Science Workshop*, February 9–11, 1999: JPL Publication 99-17: 229–238.)

FIGURE 14.19

Classification image of the Canyon Creek and Nine Mile Creek area, Idaho. Processing emphasizes iron minerals (red is best match; yellow moderate match). Processed by Spectral International and Borstad Associates. (From Hauff, P.L. et al. 1999. Hyperspectral evaluation of mine waste and abandoned mine lands: NASA- and EPA-sponsored projects in Idaho. In R.O. Green (ed.), *Summaries of the Eighth JPL Airborne Earth Science Workshop*, February 9–11, 1999: JPL Publication 99-17: 229–238.)

rock), whereas stream bottoms contain iron-coated cobbles concentrated where mine drainage enters the stream. X-ray diffraction work prior to image acquisition showed that tailings in the Mission Flats area are composed primarily of siderite, degraded muscovite, and silica. Pixels containing these minerals were easily identified using spectra extracted from the AVIRIS data. Field mapping following image processing suggests that there is a 90% match between minerals (and thus mine waste) mapped on imagery and actual minerals on the ground.

In this way, hyperspectral imagery helped highlight areas of the Couer d'Alene district that needed remediation.

Water Quality

Rivers, Lakes, and Ponds

Surface water is relatively easy to monitor because of the uniform background that clean water presents on imagery. Features that can be monitored include surface slicks and emulsions, suspended particulates, changes in water depth, water clarity, algal blooms, effluent, and thermal (hot or cold) discharges and mixing zones.

Equipment used to monitor surface water includes black-and-white (B/W), color, or color-infrared cameras and scanners, ultraviolet (UV) scanners, airborne laser fluorosensors, radar, thermal scanners, and Lidar. B/W, color, and color-infrared imagery can show sediment plumes due to riverbank or coastal erosion, algal blooms resulting from excess runoff of fertilizers or sewage discharge (leaks or intentional), and water clarity changes due to variations in salinity or turbidity (Figures 14.20 and 14.21). Ultraviolet scanners and laser fluorosensors reveal the presence of naturally occurring or manmade surface slicks and many types of organic effluent. Radar reveals surface roughness that might indicate where discharge is occurring. Thermal

FIGURE 14.20
Algal blooms (red) in the Adriatic Sea near Ravenna, Italy. Landsat TM infrared image from July 9, 1989. (Processed by Geospace, Inc., Bad Ischl, Austria for EOSAT [now GeoEye/DigitalGlobe, Longmont, CO].)

Water clarity changes as salinity increases north of the railroad trestle crossing Great Salt Lake, Utah. (Landsat-2 MSS 4-2-1 image processed by ERIM [now Michigan Tech Research Institute of Michigan Technological University].)

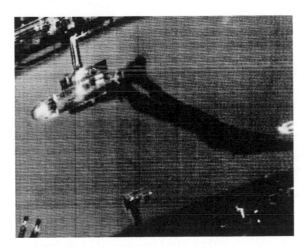

Thermal image of an accidental discharge (oil slick) from an offloading tanker. (Image acquired by Clyde Surveys, Maidenhead, United Kingdom.)

scanners can detect oil slicks (Figure 14.22) as well as hot water discharges from industrial sites such as refineries or power plants (Figure 14.23). Hyperspectral scanners can detect mineral precipitates on a riverbed resulting from mine discharges, as shown in the following example.

Case History: Acid Rock Drainage, Upper Arkansas River, Colorado

The Lake Creek watershed, in the upper Arkansas River basin, Colorado, has naturally occurring acid rock drainage (ARD) derived from a sulfide-rich porphyry system at its headwaters. High levels of metals and acid affect streams for up to 30 km from the source

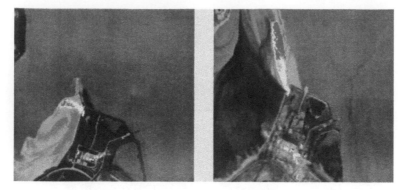

FIGURE 14.23
Airborne thermal imagery of hot water discharged into tidelands. Ebb tide right, flood tide left. Colors indicate temperature differences (0.5°C resolution) with red being hot and black being cool. (Acquired by Clyde Surveys.)

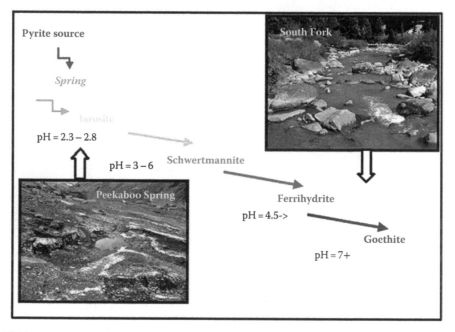

FIGURE 14.24
Iron minerals observed in the natural-ARD impacted streams of the Lake Creek watershed. (From Hauff, P.L. et al. 2006. Hyperspectral sensing of acid mine drainage—two Colorado case studies. In R.I. Barnhisel (ed.), *Proceedings of 7th International Conference on Acid Rock Drainage,* American Society of Mining and Reclamation: 738–762.)

(Peters et al., 2004; Hauff et al., 2006). A cooperative project between NASA and the Colorado Geological Survey set out to map the distribution of natural acid drainage in this watershed.

The acidic sections of these streams contain jarosite and copiapite mineral coatings on river gravels. In the Lake Creek watershed, the waters precipitate sulfates (jarosite, copia- pite, melanterite), oxides (lepidocrocite, goethite, maghemite, ferrihydrite), and hydroxide minerals. Neutral to alkaline sections of the watershed contain goethite and lepidocrocite mineral coatings. Since specific minerals are only stable in certain pH ranges, these min- erals are indicative of changing water pH (Figure 14.24). The streams become neutral to slightly alkaline downstream as they are mixed with and diluted by other tributaries.

The Advanced Spaceborne Thermal Emission and Reflection Radiometer (ASTER) satellite sensor has 15 m (visible) and 30 m (infrared) resolution. ASTER was used to identify classes of minerals, but not specific minerals. Some iron oxides could be identified in wider sections of the streambed. The airborne AVIRIS hyperspectral sensor was flown at an altitude that provided 4 m resolution. The 225 bands with 12–15 nm bandwidth allowed this imagery to identify jarosite and other high-acid minerals (Figures 14.25 and 14.26). The HST-1 hyperspectral sensor, built and operated by SpecTIR Corp., was also flown. With 0.5 m to 2.5 m resolution and 240 bands (10 nm bandwidth), the HST imagery was able to identify zoning of minerals, from jarosite-schwertmannite (pH 3–6) to ferrihydrite (pH 4.5) to goethite (pH 7+). Imagery, combined with mineral and water sampling, allowed characterization of water quality in the upper Arkansas River basin (Figure 14.27). The description and mapping of water acidity is being used for planning and prioritization of remediation.

Hyperspectral reflectance is widely used to assess water quality (Govender et al., 2007). This includes classifying the trophic status of lakes and estuaries, and characterizing algal blooms. Hyperspectral imagery is useful in determining the total suspended matter, chlorophyll content (Hakvoort et al., 2002; Vos et al., 2003), and phosphorus (Koponen et al., 2002) in surface waters. Estimation of chlorophyll (in blue-green algae) from imagery is a useful proxy for monitoring water quality. Hyperspectral imagery can detect chlorophyll-α and phycocyanin (algal pigments) if it includes narrow spectral bands between 450 and 650 nm.

Case History: Monitoring Algal Blooms and Toxins, Nebraska

Communities that depend on surface waters, especially those near feedlots and/or runoff from fertilized fields, have been motivated for years to inventory and measure the degree

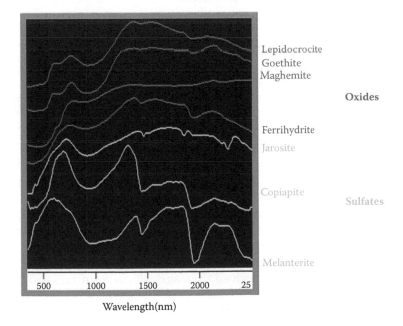

Wavelength(nm)

FIGURE 14.25
Spectral curves of iron oxides and sulfates found as precipitates in the Lake Creek watershed. (From Hauff, P.L. et al. 2006. Hyperspectral sensing of acid mine drainage—two Colorado case studies, In R.I. Barnhisel (ed.), *Proceedings of 7th International Conference on Acid Rock Drainage*, American Society of Mining and Reclamation: 738–762.)

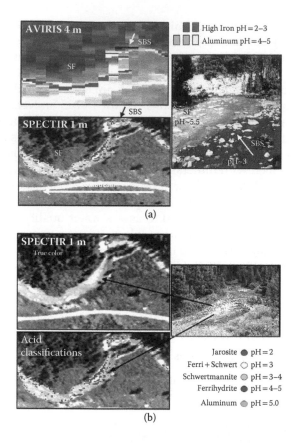

FIGURE 14.26

(a) Comparison of AVIRIS (4-m resolution) and SpecTIR HST-1 (1-m resolution) imagery at the confluence of South Fork (SF) and Sayres Bowl Stream (SBS). Changes in precipitate mineralogy on the streambed alluvium are indicative of stream pH. (b) The better spatial resolution of the HST-1 imagery allows mineralogical zoning to be observed on gravel bars, indicating changing stream pH conditions from high-flow to low-flow stream stages. (From Hauff, P.L. et al. 2006. Hyperspectral sensing of acid mine drainage—two Colorado case studies. In R.I. Barnhisel (ed.), *Proceedings of 7th International Conference on Acid Rock Drainage*, American Society of Mining and Reclamation: 738–762.)

of pollution in local lakes and streams (Brownlee et al., 1980; Wagner, 1996). A cooperative project between the Nebraska Department of Environmental Quality and the University of Nebraska set out to determine whether hyperspectral imagery could detect and monitor the distribution and concentration of chlorophyll-α and phycocyanin in Nebraska lakes (Nebraska Department of Environmental Quality, 2011). Phycocyanins are found in cyanobacteria (blue-green algae).

Algal blooms in lakes become a problem when there are excessive amounts of nutrients, particularly nitrogen and phosphorous derived from agricultural fertilizers and livestock excrement. In addition to surface scum and a foul odor, algal blooms add a bad taste to drinking water, they cause oxygen depletion which kills fish, and the algae become a health risk to humans and livestock. Cyanobacteria produce water soluble toxins including microcystins. These toxins are released when the blooms die or the bacteria are consumed. The concentration of C-phycocyanin, a pigment in cyanobacteria, is a measure of blue-green algal productivity.

The optical properties of inland waters are controlled by various combinations of algae, suspended sediment, and colored dissolved organic matter. Fremont Lakes in eastern

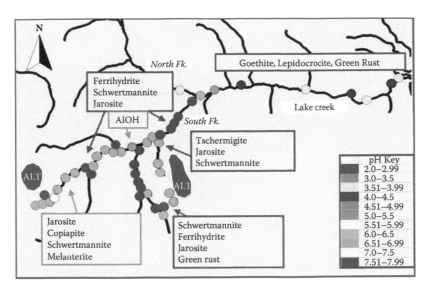

FIGURE 14.27

Schematic map of the Lake Creek watershed (Phase 1) correlating general stream pH with streambed precipitate mineral distribution. Hydrothermally altered areas, the sources of acid rock drainage, are labeled "ALT." (From Hauff, P.L. et al. 2006. Hyperspectral sensing of acid mine drainage—two Colorado case studies. In R.I. Barnhisel (ed.), *Proceedings of 7th International Conference on Acid Rock Drainage*, American Society of Mining and Reclamation: 738–762.)

Nebraska combine high turbidity with high nutrient content, which leads to algal blooms during the summer months (Figure 14.28). Fremont Lakes are popular for fishing and water sports. In the period from 2004 to 2006, the Nebraska Department of Environmental Quality had to close the lakes to the public for 25 weeks due to high levels of algal toxins. The purpose of this work was to determine the distribution of chlorophyll and total suspended solids concentration as well as presence of phycocyanin in the waters.

Chlorophyll-α has a reflectance peak near 560 nm, a reflectance trough at 665 to 675 nm, another peak at 685 nm related to solar-induced fluorescence, and a reflectance peak around 700–710 nm. *Phycocyanin* is an accessory pigment to chlorophyll that is light blue. It has an absorption near 620 nm, and fluoresces at about 650 nm. There is also a reflectance peak around 700 nm.

Two imaging systems were used to monitor algae and phycocyanin: the Medium Resolution Imaging Spectrometer (MERIS) on ENVISAT, and an AisaEAGLE airborne system built and operated by the Galileo Group of Melbourne, Florida. The AisaEAGLE was flown over Fremont Lakes at an altitude of 3 km above the ground in the summer and fall of 2008. The AisaEAGLE spectrometer has 488 bands in the range from 400–970 nm. The spectral resolution for this project was about 10 nm and spatial resolution was about 2 m.

Water samples were analyzed to determine chlorophyll and particulate concentrations and presence of phycocyanin. A correlation was established between the hyperspectral reflectance at sample locations and laboratory measurements (Gurlin, 2012). Once the correlation was established, it was applied to the entire lake area.

Concentrations of chlorophyll and presence of phycocyanin were successfully monitored (Figures 14.29 and 14.30). There was no single measure that could predict microcystin toxins with confidence. In the absence of directly measuring toxins, detecting chlorophyll-α and phycocyanin is considered indicative of potential microcystin problems.

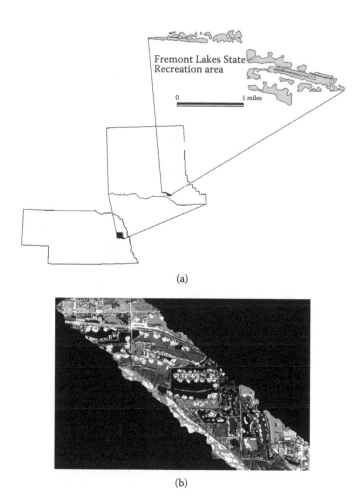

(a)

(b)

FIGURE 14.28

(a) Location map of Fremont Lakes State Recreation Area, Nebraska. (b) True color image of Fremont Lakes. From Nebraska Department of Environmental Quality, 2011. (Courtesy Nebraska Department of Environmental Quality, Water Quality Assessment Section.)

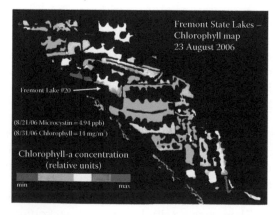

FIGURE 14.29

Chlorophyll-α concentration, Fremont Lakes. From Nebraska Department of Environmental Quality, 2011. (Courtesy Nebraska Department of Environmental Quality, Water Quality Assessment Section.)

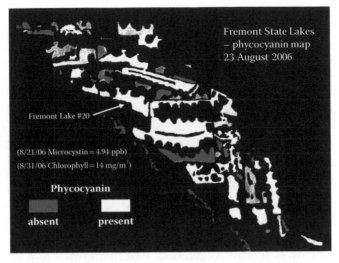

FIGURE 14.30

Phycocyanin concentration, Fremont Lakes. From Nebraska Department of Environmental Quality, 2011. (Courtesy Nebraska Department of Environmental Quality, Water Quality Assessment Section.)

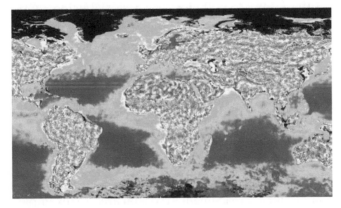

FIGURE 14.31

ENVISAT MERIS image showing global chlorophyll distribution. Reds and yellows show high values; violet to blue are low values; greens are mean values. (Courtesy of ESA.)

The MERIS instrument on the European Space Agency's (ESA) ENVISAT is currently mapping the distribution of chlorophyll, and potential pollution and fishing problems, on a global scale (Figure 14.31).

Groundwater Quality

Most remote sensing technology examines only the surface microlayer. It is not possible to determine the depth to the water table, or to locate water table contamination using conventional imaging remote sensing technology. Similarly, there are no conventional imaging techniques that will locate or track hydrocarbon accumulations on the groundwater surface. However, it is possible to map areas of surface moisture, springs, fracture zones, and surface cover related to groundwater.

In Chapter 12 under the section Locating Sources of Groundwater, we discussed using satellite imagery and airphotos to locate springs, seeps, and near-surface water. Thermal sensors can detect cool, moist soil in pre-dawn imagery and on daytime imagery. Radar flown from aircraft or satellites can penetrate dry sand and gravel up to 2 m under ideal (hyperarid) conditions. GPR, a non-imaging system, can detect the water table to a maximum depth of about 50 m. However, these systems do not tell us anything about the quality of the groundwater.

In arid climates, water is likely to be brackish to saline if white crusts or light-colored areas are seen on soils or in low spots (evaporites, including salt and gypsum, or alkali deposits). In temperate regions, plant density and diversity decreases, and plant growth is less vigorous where near surface water is not fresh. Few plants (with the exception of mangroves, some palms, coastal grasses and shrubs) can take large or even moderate amounts of salts in their water. They become stunted or die as salinity increases. In south Florida, marine salt water is encroaching on municipal and agricultural fresh water wells. The South Florida Water District is using satellite imagery to find the boundary between salt and fresh water by observing vegetation, like mangroves and saltgrass, that only grow in or near salt water (Figure 14.32) (Hough, 1995).

The potential for evaluating groundwater pollution is illustrated in a study done by the Baylor University Geology Department (Fritch et al., 2000). The Lower Cretaceous Paluxy aquifer in north-central Texas provides water for both domestic and agricultural purposes. The vulnerability of the aquifer to pollution was assessed using a GIS to integrate depth to the water table, net recharge, aquifer type, soil type, topography, impact of the vadose zone, and conductivity. These criteria were weighted based on their importance in determining regional groundwater quality. Land use and land cover derived from satellite imagery were combined with the other criteria to determine the final groundwater pollution potential of the area.

Airborne resistivity surveys can determine whether groundwater is fresh, brackish, or saline under certain conditions. A key to the success of groundwater resource studies is understanding that the resistivity is influenced by both sediment type and water quality. Sandy sediments are more resistive than clayey sediments, and fresh water is more resistive than saline water. Fresh groundwater in coarse clastic deposits has relatively high resistivity while saline water and clay-rich soils have relatively low electrical resistivity. This allows mapping of salt water encroachment in coastal area, for example, if the near-surface soil composition is known. Regional conductivity maps have been correlated with water

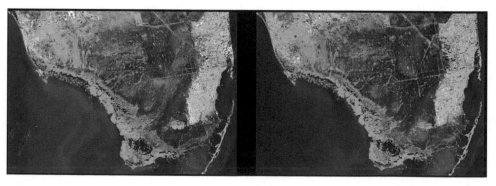

FIGURE 14.32
Mosaic of Landsat images over the south Florida Everglades showing agriculture and natural vegetation patterns. Left image is 1987–1988; right image is 2010–2011. (Courtesy of NASA.)

table recharge. Paleochannels have been detected where there is an electrical conductivity contrast with surrounding soil or rock (Fugro, online).

A joint project of the U.S. Geological Survey and several universities used frequency-domain helicopter-based EM surveys to collect resistivity data over the North Platte River valley in western Nebraska (Consortium of Universities for the Advancement of Hydrologic Science, Inc. 2010). The objectives were to map the aquifers and bedrock topography of the area to improve the understanding of groundwater–surface water relationships, leading to refined groundwater models and improved water management decisions. Numerical inversion was used to convert the measured data into a depth-dependent subsurface resistivity model. This inverted model, in conjunction with borehole data and surface geology maps, is used to characterize the groundwater system. This work improved the understanding of groundwater flow paths by providing the location of paleochannels and associated bedrock highs. The resulting groundwater models better represent the actual hydrology than previous data sets.

In a follow-up project, the U.S. Geological Survey undertook a groundwater resource investigation in the Sand Hills of western Nebraska using frequency domain and time domain airborne electromagnetic (AEM) surveys (Smith et al., 2012). The purpose, again, was to understand the hydrologic framework and provide input to groundwater management decisions. Dunes in the Sand Hills are constantly moving. Blue Creek had been blocked and partly buried by a natural sand dune dam. The dune dam is expressed as a resistive zone to a depth of 45 m and probably influences groundwater flow paths as well as surface water. Shallow lakes in the area have variable salinities. Some conductive zones suggest downward flow from saline lakes. The AEM surveys show the distribution of saline water that serves as a natural tracer for groundwater flow. Saline water above more resistive zones implies the presence of fresher water at depth. Interpreted flow paths are consistent with expected channels of the now buried ancestral Blue Creek. The groundwater flow paths determined in this project will help hydrologists understand how shallow groundwater is related to deep groundwater.

A three-frequency AEM system was developed by the Chinese Institute of Geophysical and Geochemical Exploration (Meng et al., 2006). The lower and middle frequencies (0.4 kHz and 1.5 kHz, respectively) are used to define conductors, and the upper frequency (8.3 kHz) is better suited for environmental and engineering applications with shallow targets. This instrument was flown in 2001 in Qian'an, Jilin Province, China, to map salinity, determine the freshwater/saltwater interface, and locate fresh water. The 8.3 kHz frequency was used for the salinity study. The location and intensity of saline groundwater was identified by comparing resistivities from the AEM data with available hydrological data. The lower frequencies were used to explore for fresh water. Two zones of fresh groundwater were found within the 100 m investigation depth of the system. Output from the three-frequency instrument provided a basis for detailed interpretations of the subsurface resistivity distribution.

Ackman et al. (2000) describes a collaborative effort by the U.S. Department of Energy National Energy Technology Laboratory and the U.S. Geological Survey to locate and determine the quality of groundwater in coal mining regions of Pennsylvania, West Virginia, and Maryland. They flew an airborne TIR system along with an electromagnetic conductivity (AEM) and very low-frequency (VLF) conductivity survey to identify groundwater flow paths and surface discharge locations. Surveys were flown over mines with known groundwater discharge. A GIS was used to georeference remote sensing data with existing data sets related to coal mining in the surveyed areas (mined-out areas, outcrops, mine workings, and so forth). A winter predawn TIR survey flown at 400 m had 1 m spatial resolution and 0.1°C spectral resolution. Results of this program indicate that TIR imagery is effective in identifying groundwater discharged sites. The TIR survey could not, however, determine groundwater quality or if the water was from a mine or a natural seep.

Airborne EM conductivity information was acquired using a five-frequency transmitter and receiver with an 8-m coil separation towed by helicopter over the surveyed areas. The instrument was designed to allow calculation of conductivity at five widely separated frequencies (400 Hz, 1600 Hz, 6400 Hz, 25 kHz, and 100 kHz). Because the frequencies are separated by a factor of four, the thickness of the strata being sensed decreases by a factor of two for each successively higher frequency, that is, a survey conducted at 400 Hz senses the conductivity of strata between the surface and a depth of ~100 m, whereas a survey conducted at 1600 Hz would only sense the conductivity of strata between the surface and 50 m depth. Multiple-frequency EM conductivity provided the most useful information on the location of mine pools and lateral groundwater flow paths.

The information obtained from VLF conductivity surveys is similar to that obtained from EM conductivity except that VLF is more sensitive to vertical conductors (usually water filled fractures, conductive ore bodies, or manmade features) oriented in the direction of the transmitter. Two U.S. Navy transmitters were used for these surveys: the one at Cutler, Maine (24 kHz) was ideally located for the detection of vertical, water-filled fracture zones oriented parallel to the major Northwest (NW) structural trend in the surveyed areas. A second transmitter at Seattle, Washington (24.8 kHz) was better located for the detection of water-filled fractures with Northeast (NE) trends. VLF conductivity was effective for the detection of vertical, water-filled fractures, which could be indicative of potential groundwater recharge zones and vertical flow paths.

The airborne remote sensing used in this study allowed the authors to model groundwater movements and address clean water issues. The integration of a TIR program, airborne EM, and a VLF survey with a GIS has enhanced their ability to manage this watershed.

Estuaries and Wetlands

Remote sensing of marine waters, estuaries, and wetlands is similar in most respects to sensing rivers, lakes, and ponds. It is often easier and less costly to use satellite imagery than airborne sensors because of the economies of scale involved in surveying large areas. Events such as spills, however, require detailed daily coverage by a variety of sensors to identify the spill, determine the volume of the spill, or track the spill. This coverage and detail are provided by both satellite and airborne instruments.

Estuaries, wetlands, and deltas are environmentally sensitive areas that may require periodic monitoring, especially if oil fields or chemical facilities are located nearby. For the purpose of monitoring or inventorying changes in these areas, it is appropriate to use color or color infrared satellite imagery or aerial photographs. These can be purchased at a variety of scales and resolutions, depending on the detail required.

The diversity of vegetation and range of water depths and turbidities make wetland areas particularly well suited to run supervised (if training areas are available) and unsupervised classifications on imagery.

Marine Environment

Algae

Some jurisdictions require that the amount of particulates discharged along with water from refinery or power plant cooling ponds must be below a specified level. Algae growing in the ponds are considered particulates, and the ponds must be periodically drained, cleaned, and painted with special algae-inhibiting paint. Marine algal blooms, also known as "red tides," are of concern to health officials and fishermen. They consume oxygen necessary for fish and other marine organisms, and release toxins potentially harmful to humans. It is possible to monitor blue-green algae by taking advantage of their chlorophyll reflectance peak between 0.7 and 0.75 μm (Figures 14.20 and 14.31). Algae may be distinguished from suspended sediment by subtracting the reflectance at 665 nm from the peak reflectance at 710 nm (Quibell, 1991).

The Fraunhofer Line Discriminator (FLD) has been described as an airborne imaging instrument that can detect oil refinery waste, sewage, feedlot runoff, sludge (wet or dry), algal blooms, and phosphate processing effluents by measuring the luminescence of these materials (Watson et al., 1978). Fraunhofer lines are absorption bands in the solar spectrum. Using the sun as a source, the FLD measures luminescence in these absorption bands under daylight conditions.

Discharges, Spills, and Seeps

Marine spills can occur as a result of ships flushing their bilges, tankers running aground or colliding, platform blowouts, ships colliding with platforms or sea ice, pipeline leaks, and so forth (Figure 14.22). Natural seeps exist in many areas, often slow and steady, in other cases intermittent and triggered by earthquakes or reservoir pressure changes. It is virtually impossible to tell the difference between natural and manmade slicks unless one can pinpoint the source and identify the type of hydrocarbon. Most manmade slicks originate at a point and spread outward from it; many natural seeps appear dispersed over large areas. The key difference is that manmade features tend to be of limited duration and the source location may move, whereas natural seeps tend to be stationary and are visible over long periods of time (Estes et al., 1985).

Operator's permits for power plants and refineries usually require monitoring of the temperatures of cooling water discharge. Thermal imagery is useful for monitoring these thermal plumes (Figure 14.23). Thermal imagery has been used since the mid-seventies to detect acid mine drainage into fresh water bodies. Aircraft systems can produce high-resolution imagery at frequent intervals, but costs are high relative to satellite imagery. It has been shown that Landsat TM thermal imagery has sufficient thermal and spatial resolution to map the cooling effluent from power plants such as the Diablo Canyon nuclear facility (Gibbons et al., 1989). When radiosonde measurements are used to adjust atmospheric models (such as LOWTRAN 6), the surface radiances can be converted to temperatures that are within 0.6°C of surface measurements.

In any spill or leak situation, there are three questions that must be answered as soon as possible:

1. Where is the spill located?
2. How much has been spilled?
3. Which direction and how fast is the spill moving?

Most companies have contingency plans in place. These plans have modeled oceanic current movements and wind directions at various times of year. They should also have emergency manpower and monitoring equipment (buoys, surveillance aircraft) on standby.

Daily repetitive flying with a suite of sensors that include radar, UV, and thermal scanners is the best way to answer the questions posed above. Radar can reveal variations of sea

FIGURE 14.33
RadarSat image of an oil spill at St. Ann's Head, Wales, taken February 22, 1996 (RADARSAT, 1996). Oil appears as dark streaks along the coast. (Processed by RadarSat International. RADARSAT Data © Canadian Space Agency/Agence Spatiale Canadienne 1996. All rights reserved.)

surface roughness and help pinpoint the location of objects on the surface, including drilling platforms, ships, and effluent. Radar imagery can locate a slick from high altitudes in any weather and at any time of day or night under normal conditions. Oil slicks appear on radar images as smooth areas due to the suppression of small surface waves (Figure 14.33) (Estes et al., 1985; RADARSAT, 1996). Radar cannot detect spills if there are storm waves, or if the sea surface is smooth due to a wind shadow, for instance. A UV scanner or an FLD can identify slicks at lower altitudes as an organic substance, and the thermal scanner can characterize the thickness of the spill. In general, slicks less than 300 µm thick appear cool because the emissivity of oil (0.94–0.97) is less than that of water (0.99). Slicks thicker than about 500 µm act as an absorber of solar radiation and tend to appear warmer than background (Goodman, 1989; Shih and Andrews, 2008; Distefano and Freni, 2012). However, if the oil has a high concentration of volatiles, these will evaporate quickly and make the oil appear cool. Ultimately, it is the contrast with a uniform background that allows a slick to be identified. All these sensors, as well as color or color-infrared images, photos, or video can provide the areal extent and movement direction of the spill. These systems can pinpoint the location of the spill when they are combined with inertial navigation systems or GPSs fixed to buoys that are dropped into the slick (Hielm, 1989; Meyers, 1989).

Repetitive coverage, combined with accurate positioning, provides movement vectors of a slick and allows the emplacement of booms and skimmers, the application of dispersants, and optimized oil collection for *in situ* burning. In high risk or heavy traffic areas, it is smart to know ahead of time the direction and speed of currents during various seasons so that in the event of a spill, local conditions can be quickly taken into account. Certain types of imagery, particularly true color or TIR, can assist in mapping currents if there are suspended sediments (near the coast) or if there is a temperature contrast between warm and cool currents (Figure 14.34). The location and extent of ocean currents can also be determined using radar altimetry.

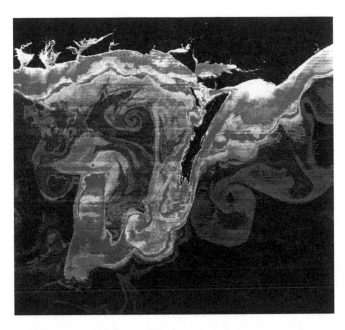

FIGURE 14.34
Landsat TM 3-2-1 image reveals current patterns near Kayak Island, Alaska. The image shows the circulation of sediments in coastal water including dipole eddies near the shore. (Image processed and analysed by Kristina Ahlnäs and Tom Royer, University of Alaska Fairbanks; Ahlnäs, K. et al. *J. Geophys. Res.* 92(C12), 13041–13047, 1987; Ahlnäs, K., T.C. Royer. *Adv. Space Res.* 9, 185–190, 1989a; Ahlnäs, K., T.C. Royer. *Rem. Sens. Environ.* 28, 85–93, 1989b.)

Hyperspectral imagery has been used to monitor oil spills in Chesapeake Bay (Salem and Kafatos, 2001). In the event of a spill such as the Deepwater Horizon blowout, satellite and airborne systems provide real-time data (McNutt et al., 2011). This imagery can provide information on the location, movement direction and speed, abundance, and composition of hydrocarbons (heavy versus light or type of refined product) based on their spectral signature. False positives caused by jellyfish or algal blooms can be eliminated. Knowing the location and type of pollutant will help direct cleanup efforts.

Coastal Change Mapping

For some purposes, it is important to know where the coastline is. Although this sounds obvious and trivial, there are areas where the mapped coastline can move laterally by tens of meters or even kilometers depending on the tides. The same sensors mentioned above can be used to map the extent of high and low tides for the building of jetties and marinas, breakwaters, for erosion control along beaches, for siting property developments, and for coastal heavy mineral and placer mining.

In the aftermath of a hurricane like Katrina or Sandy, there are changes not only to the coastline. Beaches are eroded, homes are destroyed, barrier islands are swept away or reshaped, lagoons and waterways are rearranged, low areas are flooded, and forests are damaged (Klemas, 2009). Detailed imagery used to record damage has come from low altitude "post-disaster visualization" footage and high-resolution satellites, among others (Ghosh et al., 2007); regional quick-look imagery has been provided by Landsat

ETM+ and RadarSat (Rykhus and Lu, 2008); vegetation change has been recorded by Landsat TM and the Moderate Resolution Imaging Spectroradiometer on the AQUA and TERRA satellites (Wang and Qu, 2009; Wang and Xu, 2010). Rising sea level coupled with large storms makes coastal change a topic of ongoing interest to the public.

Atmospheric and Air Pollution Monitoring

Atmospheric monitoring includes identification and tracking of industrial emissions such as smoke, particulates, and gases from coal-fired power plants, smelters, incinerators, and forest fires. Naturally occurring gases, such as carbon dioxide, which are implicated in global warming, and ozone that screens the Earth from harmful UV radiation are also monitored. Acid rain is a uniquely human problem. The burning of fossil fuels (mainly coal and oil) in power plants and some industries releases sulfur into the air that combines with oxygen to form sulfur dioxide (SO_2). Vehicle exhaust contributes nitrogen oxides. These gases combine with atmospheric moisture to form airborne sulfuric acid (H_2SO_4) and nitric acid (HNO_3). Acid-rain gases are carried hundreds of miles by winds. Although an environment can adapt to low amounts of acid rain, and slightly basic (limestone) rocks and soils naturally buffer the acid precipitation, forests and lakes can be damaged when acid rain exceeds some threshold. Low pH rain can harm forests (particularly conifers; Figure 14.35) and lakes, including some of the more sensitive creatures that live in those environments (fish, frogs). As early as 1983, research was conducted into identifying and tracking acid rain (Steffes, 1983). It was noted that SO_2, H_2S, COS, N_2O, O_3, H_2CO, CO, and gaseous HNO_3 exhibit significant microwave opacity. They also examined the microwave opacity of acid-bearing clouds. The magnitude of the absorption is largely dependent on cloud bulk density and physical temperature, but the frequency dependence is directly related to acidity in the cloud. It was concluded that radiometric remote sensing of the opacity of the cloud-related gases, and of the clouds themselves, could provide information on the production of acid rain. This work led directly to satellites that monitor atmospheric gases (Tables 14.1 and 14.2).

FIGURE 14.35
Acid rain stressed spruce (brown to red forest) near Long Lake, Adirondack Park, New York. (True color image courtesy of GeoEye and Google Earth, © 2012 Google.)

TABLE 14.1

Partial List of Satellites that Monitor Ozone, CFCs, Nitrogen Compounds, and Sulfur Dioxide

Platform	Sponsor	Launched	Gases Monitored							
			O_3 (ozone)	NO	NO_2	N_2O	HNO_3	NO_y (total reactive nitrogen)	SO_2	CFCs
ADEOS	Japan	1996	x							
Terra	United States	1999	x							
ODIN	ESA	2001	x		x					
Aqua	United States	2002	x						x	
EnviSat	ESA	2002	x	x	x		x			
Midori-2	Japan	2002	x		x	x				x
SORCE	United States	2003	x							
SCISAT-1	Canada	2003	x		x					
NOAA-N	United States	2005	x							
Suomi	United States	2011	x							
Discover-AQ (aircraft)	United States	2011	x	x	x		x	x		

TABLE 14.2

Partial List of Satellites that Monitor Greenhouse Gases and Aerosols

Platform	Sponsor	Launched	Greenhouse Gases Monitored							Aerosols/ Particulates (e.g., smoke)
			Water Vapor	CO_2	CO	CH_4	CH_2O	N_2O	O_3	
GOES series	United States	1974 to present	x						x	
ERS-2	ESA	1995	x							x
ADEOS	Japan	1996	x							
Terra	United States	1999			x	x			x	x
ODIN	ESA	2001							x	x
Aqua	United States	2002			x	x			x	x
EnviSat	ESA	2002	x	x	x	x			x	
Midori-2	Japan	2002	x			x		x	x	x
IceSat	United States	2003								x
SORCE	United States	2003							x	
SCISAT	Canada	2003							x	x
Parasol	France	2004								x
NOAA-N	United States	2005	x						x	x
Cloudsat	United States	2006	x							x
Calipso	United States	2006	x							x
GOSAT	Japan	2009	x	x		x				x
OceanSat	India	2009								x
Suomi	United States	2011	x						x	x
Discover-AQ (aircraft)	United States	2011	x	x	x	x	x		x	
OCO-2	United States	~2015		x						

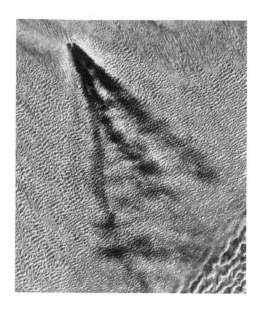

FIGURE 14.36
Smoke plume from an oil well flare in the Ar Rab' al Khali, Saudi Arabia. Although well flares are not that common, this Landsat MSS image dramatically demonstrates the ability of satellite imagery to reveal smoke plumes and some types of industrial stack emissions. (True color image processed by ERIM [now Michigan Tech Research Institute of Michigan Technological University].)

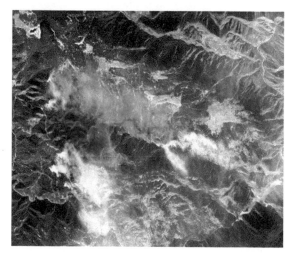

FIGURE 14.37
Color infrared image of the March 2012 fires above Boulder, Colorado. (Image courtesy of DigitalGlobe, Inc and GeoEye © 2012.)

B/W, color, or infrared images can detect smoke plumes, and color photography is particularly sensitive to atmospheric haze (blue wavelengths) such as vehicle-derived smog and smoke (Figures 14.36 and 14.37). Such images show, for example, areas most prone to acid rain as a result of coal-burning mills, copper smelters, or power plants. None of the sensors mentioned so far, however, can detect or identify clear gaseous emissions such as CO_2 or volatile organic compounds emitted from factories or plants. At least one company in the late 1970s used satellite VNIR images to map "fuzzy" or "hazy" tonal anomalies associated with naturally occurring photochemical smog caused by oil and gas fields leaking methane at the surface. It is not known whether this technique was ultimately successful, but there are several atmospheric conditions, such as ground fog or wind-blown dust, which could thwart such an approach.

In the United States, the Clean Air Act Amendments of 1990 and Clean Air Act of 1992 require ambient air monitoring by industry. As an example of such monitoring, a TIR video camera was used to record near source plumes from a turbine stack at an oil gathering center at Prudhoe Bay. Because the ambient temperatures were very cold, and the plume was hot (270–331°C), this method of monitoring plumes was shown to be more effective and more accurate than standard plume rise models such as the Industrial Source Complex model (Rickel et al., 1990).

Gas and Aerosol Detectors

There are instruments available and others in development that can detect and possibly identify fugitive emissions (Grant et al., 1992). "Nondispersive Infrared Spectrometry" is used to detect carbon monoxide and carbon dioxide in vehicle exhaust. An infrared source (heating element) is placed on one side of a road, and an IR detector is placed on the other side. When a vehicle passes through the IR beam, the instrument measures the strength of the absorption for carbon monoxide at 4.3 μm and for carbon dioxide at 4.6 μm (Bishop et al., 1989; Lawson et al., 1990; Bishop and Stedman, 2008). The ratio of carbon monoxide to carbon dioxide can be used to calculate emissions in terms of percent or grams per gallon. Similar absorption bands exist for hydrocarbon vapors such as benzene, toluene, xylene, and other aromatics. Low signal-to-noise is the major problem in development of these hydrocarbon detectors.

A related technology, "FTIR Spectroscopy" is being used to detect industrial hydrocarbon and nitrous oxide emissions. FTIR is essentially an air "grab sample" chemical analyzer that can be adjusted to measure gas species concentrations over some threshold. Gases that have been measured include H_2O, CO_2, CO, methane (CH_4), ethyne (C_2H_2), ethene (C_2H_4), propene (C_3H_6), formaldehyde (HCHO), formic acid (HCOOH), methanol (CH_3OH), acetic acid (CH_3COOH), phenol (C_6H_5OH), hydrogen cyanide (HCN), nitrous acid (HONO), ammonia (NH_3), peroxyacetyl nitrate (CH_3COONO_2), and ozone (O_3) (Burline et al., 2011). In volcanism studies, FTIR has been used to quantify emission rates of volatile gases including CO, HCl, and HF (McGee et al., 2005). Several companies have developed portable and airborne versions that are used to detect emissions as diverse as volcanic plumes, forest fire smoke, and airborne ethyl ether, a byproduct of illegal drug production.

Lidar has been used to measure movement and concentration of urban air pollution and to determine the composition of emissions near industrial plants. The technique consists of projecting a short laser pulse followed by reception of reflected or re-emitted radiation from atmospheric constituents such as gas molecules, aerosols, clouds, or dust. The incident radiation interacts with these constituents and causes a change in intensity and wavelength. There are several types of Lidar systems. The most commonly used systems for pollution detection include Atmospheric Backscatter Lidar (ABL), Differential Absorption Lidar (DIAL), and Fluorescence Lidar (FL) (see Chapter 3, section on Instrument Systems).

Atmospheric backscatter Lidar is used to track turbid effluent and gas plumes, including monitoring stack emissions from power plants, smelters, and refineries. DIAL has been used to detect molecules such as SO_2, NH_3, O_3, CO, CO_2, HCl, NO, N_2H_4, NO_2, and SF_6. FL has been used to detect atmospheric trace metals including Na, K, Li, and Ca, and the hydroxyl ion (Considine, 1988).

FIGURE 14.38

Remote imaging of gaseous emissions shows concentrations of SF_6 and NH_3 at distances up to 1.5 km. (Courtesy of Telops.)

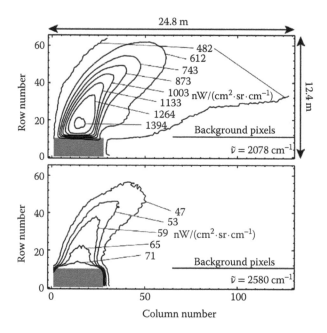

FIGURE 14.39

Top: a contour map of the time-averaged spectral radiance at 2078 cm^{-1} (CO_2 emission line). Bottom: a contour map of the spectral radiance at 2580 cm^{-1} corresponding to continuum emission from particulates in the plume. The gray rectangle is the top of the stack. The horizontal line at row 12 between columns 64 and 128 indicates the pixels used to estimate the background spectrum. (Courtesy of Telops.)

Infrared hyperspectral imaging has been used to locate toxic industrial compounds (TICs) in plant emissions. Remote imaging of gaseous emissions shows concentrations of SF_6 and NH_3 at distances up to 1.5 km using the Telops Hyper-Cam imaging spectrometer (Figure 14.38). This instrument has demonstrated capability to detect and identify multiple TICs at distances up to 5 km and with concentrations as low as a few ppm. A Hyper-Cam-derived contour map of time-averaged spectral radiance corresponding to CO_2 stack emissions can be seen in the upper panel in Figure 14.39. A contour map of the spectral radiance corresponding to particulates in the plume can be seen in the bottom panel.

Most countries require continuous monitoring of emissions from power plants, hazardous waste incinerators, cement plants, natural gas pipelines, and other industrial sources. A tripod-based infrared hyperspectral scanner developed by Telops remotely obtains a complete image of emissions using FTIR Spectroscopy. This passive remote sensing technique can be performed in a matter of hours from distances ranging from 100 m to 1 km. Spectral resolution is selectable over the 8.0–11.8 μm range. Good agreement has been found between measurements of smokestack plume temperature, average speed, and concentration of H_2O, CO_2, and SO_2 emissions and *in situ* sample analyses (Gross et al., 2010). An airborne helicopter survey flown at 450 m above a pipeline in a 10 km/h wind was able to detect methane leaking at a rate of 20 L/min (Telops, 2012). This is one of several sensors being developed to detect atmospheric gases. The primary advantage of hyperspectral imaging is that, because an entire spectrum is acquired, no prior knowledge of the sample is needed, and postprocessing allows all available information from the data set to be used.

Since the 1970s, industrial emissions have been monitored using correlation spectroscopy (COSPEC). COSPEC uses absorption bands to remotely detect SO_2 emissions (Moffat and Millan, 1971; Stoiber et al., 1983). The instrument contains a photodetector calibrated to SO_2. Diffuse skylight is used as the UV source. The instrument measures concentration path lengths of SO_2 along the atmospheric path viewed by the instrument. Concentration profiles are integrated to derive emission rates. Measurements can be made at a distance of several kilometers. While COSPEC can provide gas emission rates, only a few rates could be calculated in a day (Nadeau, 2011).

Differential Optical Absorption Spectrometers (DOAS) can measure emission rates of SO_2 on the order of minutes (Edmonds et al., 2003). These spectrometers provide better spectral, temporal, and spatial resolution than COSPEC measurements of SO_2 emissions. Imaging DOAS (I-DOAS) provides relatively low-spatial resolution two-dimensional plume imagery over a wide spectral range. It requires a scan through the plume and thus takes tens of minutes to acquire. Wide field-of-view (FOV) spectrometers can provide readings at a rate of 1 per second, but lack the imaging ability required to evaluate plume dynamics. Multiple wide angle DOAS instruments can be used to derive plume speed (Nadeau, 2011).

Similar instrumentation comes from the field of volcano observations. SO_2 flux rates appear to be intimately related to magma movement. UV remote sensing cameras have been successfully employed to detect and measure SO_2 flux from vents, fumaroles, and craters (Mori and Burton, 2006, 2009; Tamburello et al., 2011a,b,c). The UV camera system used by Tamburello et al. (2011a) consisted of two Apogee Instruments Alta U260 cameras with cooled charge-coupled device (CCD) array detectors. The cameras are tripod mounted and aligned to ensure identical pointing. Each camera had a Pentax B2528-UV lens with a 24° FOV. Each lens had a band-pass filter centered on 310 and 330 nm, respectively. The 310 nm filter is affected by plume SO_2 absorption, and the 330 nm filter falls outside the absorption band. This permits a qualitative measure of absorbance. Calibration is performed immediately prior to each acquisition, viewing SO_2-free sky at a 45° angle adjacent and away from the fumaroles and vents. Taking a suitable row of pixels, they calculated the integrated column amount across the plume and multiplied this by the plume speed to compute the SO_2 flux (see Figure 16.21). In this study, measurements were made ~300 m from the gas vents and the plume was highly transparent, so errors are considered to be minimal. As the plume transport speed is well defined, flux errors are estimated to be ±15%.

In the same way, volcanic CO_2, H_2O, and H_2S plume fluxes can be determined by using UV cameras in conjunction with a multigas sensor (a portable field instrument for sampling gases). Since H_2O, H_2S, and CO_2 are proportional to the SO_2 emission rates, the SO_2 rates are multiplied by gas molar ratios from the multigas sensor to arrive at emission rates of H_2O, H_2S, and CO_2 (Aiuppa et al., 2011a,b).

None of these instruments are traditional remote sensing systems in the sense that they do not generate map-like images. They do, however, generate images that show the distribution of gases and are effective standoff methods for monitoring atmospheric pollutants. These optical systems all share certain advantages and limitations. Advantages include the following:

1. No additional sampling is required for many of these gases.
2. Path-averaged measurements are more representative than geochemical or gas "grab" samples.
3. Exposure of the analyst to toxic gases is limited and minimized.

Some of the limitations are as follows:

1. The possibility of spectral interference from nonpollutants (water vapor, dust).
2. Spectral databases are required.
3. The minimum resolution may not be adequate for regulatory purposes, and they may not be readily quantifiable.

Satellite and Airborne Atmospheric Monitoring

Monitoring of the atmosphere began with the first the National Oceanographic and Atmospheric Administration (NOAA) weather satellites in 1974. At that time, there were two atmospheric constituents of interest: water vapor and ozone. Water vapor is an obvious target of weather satellites, since it helps forecast precipitation. Ozone, however, was another matter.

In 1973, chemists Frank Rowland and Mario Molina, of the University of California, Irvine, began studying the impact of chlorofluorocarbons (CFCs) on Earth's atmosphere. They determined that CFC molecules in the stratosphere are broken down by UV radiation and release chlorine atoms. Rowland and Molina proposed that these chlorine atoms could cause the breakdown of ozone (O_3) in the stratosphere. Their argument was based on work by Paul Crutzen and Harold Johnston, which showed that nitric oxide (NO) also caused the destruction of ozone. Since stratospheric ozone absorbs most of the UV-B radiation reaching the surface of the planet, the environmental consequence of depleting the ozone layer would be an in increase in UV-B radiation at the surface. Increased UV-B radiation would likely result in an increase in skin cancer, damage to crops and marine phytoplankton, and could have other as yet unknown effects. This work and other studies led directly to the Vienna Convention for the Protection of the Ozone Layer (1985) and the Montreal Protocol on Substances that Deplete the Ozone Layer (1989). The protocol acts as a framework for international efforts to protect the ozone layer and includes legally binding reduction goals for the use of CFCs. It also provides an incentive for monitoring global ozone (Table 14.1; Figure 14.40).

A second factor driving the monitoring of the atmosphere is concern about global warming. Satellite missions were established to help support and administer the Kyoto protocol (1997), which was dedicated to a reduction in greenhouse gases, particularly

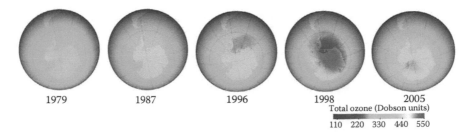

1979   1987   1996   1998   2005

Total ozone (Dobson units)

110   220   330   440   550

FIGURE 14.40
The size of the ozone hole over the Antarctic fluctuates from year to year, but it may be growing. Monthly maps from December 1979, 1987, 1996, 1998, and 2005, illustrate the fluctuations. (Courtesy of NASA.)

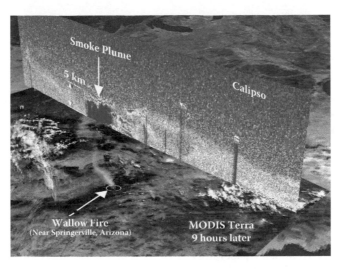

FIGURE 14.41
A CALIPSO vertical profile shows the smoke plume on June 3, 2011, from wildfires in Arizona. The profile is overlaid on an image taken by the Moderate Resolution Imaging Spectroradiometer (MODIS) instrument on the Terra satellite 9 hours later. Data show that the Wallow Fire plume reached 5 km high. (Courtesy of NASA/Kurt Severance, Jason Tackett and CALIPSO Team.)

CO_2. This international convention called for a reduction in greenhouse gas emissions by signatories to 5% below their 1990 levels by 2012. Several satellites and aircraft missions were designed to help enforce the protocol, as well as get a better understanding of the distribution and concentration of greenhouse gases (water vapor, carbon dioxide, carbon monoxide, methane, nitrous oxide, ozone, and particulates/aerosols; Figures 14.41 and 14.42). This is in addition to satellites that are monitoring ice cover, sea level changes, storm activity, and other factors related to the effects of global warming. Some satellites that monitor atmospheric constituents related to global warming are listed in Table 14.2.

Many future satellite systems are being promoted as climate change monitoring missions, among them NASA's ATTREX mission, the NASA/JAXA GPM mission, and the ESA/JAXA EarthCARE mission (see Chapter 3). Suffice it to say that our ability to remotely monitor the atmosphere and its constituents is improving.

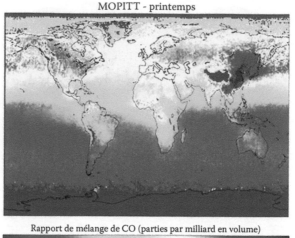

FIGURE 14.42

MOPITT image showing global average springtime carbon monoxide concentrations over four years (March 2000 to February 2004). (Courtesy of the Canadian Space Agency and NASA; Measuring atmospheric pollutants. http://www.asc-csa.gc.ca/eng/satellites/mopitt.asp. Canadian Space Agency, 2007. Reproduced with the permission of the Minister of Public Works and Government Services Canada, 2013.)

References

Ackman, T.E., G.A. Veloski, R.A. Dotson, Jr., R.W. Hammack, J.I. Sams. 2000. *An Evaluation of Remote Sensing Technologies for Watershed Assessment.* Pittsburgh, PA: U.S. Department of Energy, National Energy Technology Laboratory.

Aiuppa, A., M. Burton, P. Allard, T. Caltabiano, G. Giudice, S. Gurrieri, M. Liuzzo, G. Salerno. 2011a. First observational evidence for the CO_2-driven origin of Stromboli's major explosions. *Solid Earth.* 2: 135–142.

Aiuppa, A., M. Burton, P. Allard, T. Caltabiano, G. Giudice, S. Gurrieri, M. Liuzzo, G. Salerno. 2011b. *Linking plume CO_2 flux emissions and eruptive activity at Stromboli volcano (Italy)* (abs.). In 11th Gas Workshop, Kamchatka, Russia, 1–10 Sept. 2011, Commission on the Chemistry of Volcanic Gases (CCVG)—IAVCEI: 6.

Ahlnäs, K., T.C. Royer, T.H. George. 1987. Multiple dipole eddies in the Alaska coastal current detected with Landsat Thematic Mapper data. *J. Geophys. Res.* 92(C12): 13041–13047.

Ahlnäs, K., T.C. Royer. 1989a. Evaluation of the ability of various remote sensors to map distributions of suspended sediments in the Gulf of Alaska. *Adv. Space Res.* 9: 185–190.

Ahlnäs, K., T.C. Royer. 1989b. Application of satellite visible band data to high latitude oceans. *Rem. Sens. Environ.* 28: 85–93.

Andreoli, G., B. Bulgarelli, B. Hosgood, D. Tarchi. 2007. *Hyperspectral analysis of oil and oil-impacted soils for remote sensing purposes.* Institute for the Protection and Security of the Citizen, European Commission Joint Research Centre: 36 p.

Bandyopadhyay, T.K., T.K. Mukherjee, B.B.P. Shrivastava. 1994. Magnetic response and remote sensing for mapping coal seam fire-A study over Jharia coalfield, India. In *Proceedings of ERIM 10th International Conference on Geologic Remote Sensing. II.* Ann Arbor, MI: ERIM: II576–II584.

Benson, R.C., R.A. Glaccum, M.R. Noel. 1984. *Geophysical Techniques for Sensing Buried Waste and Waste Migration.* US Department of Commerce National Technical Information Service Report. EPA-600/7-84-064 Las Vegas, NV: Environmental Protection Agency: 18 p.

Bhattacharya, A., C.S. Reddy, T. Mukherjee. 1991. Multi-tier remote sensing data analysis for coal fire mapping in Jharia coalfield of Bihar, India. In *Proceedings of the 12th Asian Conference on Remote Sensing, Singapore*, October 30–November 5, 1991. Singapore: National University of Singapore: 22-1–22-6.

Bishop, G.A., D.H. Stedman. 2008. A decade of on-road emissions measurements. *Environ. Sci. Technol.* 42: 1651–1656.

Bishop, G.A., J.R. Starkey, A. Ihlenfeldt, W.J. Williams, D.H. Stedman. 1989. IR long-path photometry: a remote sensing tool for automotive emissions. *Analyt. Chem.* 61: 671–677.

Blum, E.D. 2008. *Love Canal Revisited: Race, Class, and Gender in Environmental Activism*. Lawrence, KS: University Press of Kansas: 194 p.

Boardman, J.W. 1993. *Automated spectral unmixing of AVIRIS data using convex geometry concepts*. Fourth JPL Airborne Geoscience Workshop 93–26 I. Pasadena: Jet Propulsion Lab: 11–14.

Brownlee, R.H., J.P. Scherz, J.M. Dennis. 1980. Remote sensing that motivated community action. In R.W. Kiefer (ed.), *Civil Engineering Applications of Remote Sensing*. In Proceedings of the Specialty Conference of the Aerospace Division of the American Society of Civil Engineers, August 13–14, 1980. Madison: University of Wisconsin: 60–77.

Bryson, B. 1998. *A Walk in the Woods*. New York: Broadway Books: 180–184.

Burline, I.R., R.J. Yokelson, S.K. Akagi, S.P. Urbanski, C.E. Wold, D.W.T. Griffith, T.J. Johnson, J. Reardon, D.R. Weise. 2011. Airborne and ground-based measurements of the trace gases and particles emitted by prescribed fires in the United States. *Atmos. Chem. Phys.* 11: 12197–12216.

Canadian Space Agency. 2012. Measuring atmospheric pollutants. http://www.asc-csa.gc.ca/eng/satellites/mopitt.asp (accessed 22 Dec. 2012).

Chin, L.Y., T.A. Tomberlin, G.G. Ramos, R.J. Chalaturnyk. 2012. Evaluation of caprock stability by coupled modeling of geomechanics and reservoir simulation under steam injection for producing oil sands reservoirs. In *Proceedings of the 46th U.S. Rock Mechanics/Geomechanics Symposium*, June 26–29, 2012, Chicago: American Rock Mechanics Association: 16 p.

Considine, D.M. 1988. *Van Nostrand's Scientific Encyclopedia*. New York: Van Nostrand Reinhold: 1678–1679.

Consortium of Universities for the Advancement of Hydrologic Science, Inc. 2010. http://www.thewaterchannel.tv/en/videos/categories/viewvideo/984/groundwater/using-airborne-geophysical-surveys-to-improve-groundwater-resource-management-models (accessed 26 Oct. 2012).

Cundill, S.L. 2008. Investigation of soil reflectance for detecting hydrocarbon pipeline leakages. MSc. Thesis. ITC, Enschede, The Netherlands: 66 p.

Davids, C., A.N. Tyler. 2003. Detecting contamination-induced tree stress within the Chernobyl exclusion zone. *Rem. Sens. Environ.* 85: 30–38.

Distefano, G., F. Freni. 2012. Dynamic thermograms processing for the optimization and the automation of hydrocarbons pollution on sea surface detection. In 11th International Conference on Quantitative InfraRed Thermography, June 11–14, 2012, Naples, Italy: 4 p.

Edmonds, M., R.A. Herd, B. Galle, C.M. Oppenheimer. 2003. Automated, high time-resolution measurements of SO_2 flux at Soufrière Hills Volcano, Montserrat. *Bull. Volcanol.* 65: 578–586.

EOSAT. 1991. *EOSAT* (calendar). Lanham, MD: Earth Observation Satellite Co. (now GeoEye/Digital Globe, Longmont, CO).

Erb, T.L., W.R. Philipson, W.L. Teng, T. Liang. 1980. Analysis of landfills with historic airphotos. In R.W. Kiefer (ed.), *Civil Engineering Applications of Remote Sensing*. In Proceedings of the Specialty Conference of the Aerospace Division of the American Society of Civil Engineers, August 13–14, 1980. Madison: University of Wisconsin: 15–23.

ERCB. 2010. ERCB Staff Review and Analysis: Total E&P Canada Ltd. Surface Steam Release of May 18, 2006. *Joslyn Creek SAGD Thermal Operation*. Alberta: Energy Resource Conservation Board: 177 p.

ERIM. 1991. *Earth* (calendar). Ann Arbor, MI: ERIM Image Processing Facility.

Estes, J.E., R.E. Crippen, J.L. Star. 1985. Natural oil seep detection in the Santa Barbara Channel, California, with Shuttle Imaging Radar. *Geology.* 13: 282–284.

Fischer, C., D. Hermsmeyer. 1999. *Digital photogrammetry and GIS in tailings and mine waste management*. In Proceedings of the 6th International Conference on Tailings and Mine Waste '99, Balkema, Rotterdam: 77–86.

Fritch, T.G., C.L. McKnight, J.C. Yelderman, J.G. Arnold. 2000. An aquifer vulnerability assessment of the Paluxy aquifer, central Texas using GIS and a modified DRASTIC approach. *Environ. Manage.* 25: 337–345.

Fugro. http://www.fugrowaterservices.com/downloads/mapping/AirborneGroundwater Surveys.pdf (accessed 26 Oct. 2012).

Gangopadhyay, P.K. 2000. Surface temperature mapping and detection of surface & subsurface coal-mine fires of the Raniganj Coal belt (West Bengal) using Remote Sensing and GIS. Dissertation, IIRS: Dehradun, India.

Ghosh, S., B.J. Adams, J.A. Womble, C. Friedland, R.T. Eguchi. 2007. *Deployment of remote sensing technology for multi-hazard post-Katrina damage assessment*. 2nd International Conference on Urban Disaster Reduction, Taipei, Nov. 27–29: 8 p.

Gibbons, D.E., G.E. Wukelic, I.P. Leighton, M.J. Doyle. 1989. Application of Landsat Thematic Mapper data for coastal thermal plume analysis at Diablo Canyon. *Photogram. Eng. Rem. Sens.* 55: 903–909.

Goodman, R.H. 1989. Application of the technology in the remote sensing of oil slicks. In A.E. Lodge (ed.), *The Remote Sensing of Oil Slicks*. Chichester, UK: John Wiley and Sons: 39–65.

Govender, M., K. Chetty, H. Bulcock. 2007. A review of hyperspectral remote sensing and its application in vegetation and water resource studies. *Water SA.* 33: 145–151.

Grant, W.B., R.H. Kagann, W.A. McClenny. 1992. Optical remote measurement of toxic gases. *J. Air Waste Manag. Assoc.* 42: 18–30.

Gross, K., K. Bradley, G.P. Perram. 2010. Remote identification and quantification of industrial smokestack effluent via imaging Fourier transform spectroscopy. *Environ. Sci. Technol.* 44: 9390–9397.

Gurlin, D. 2012. Near-infrared-red models for the remote estimation of chlorophyll-α concentration in optically complex turbid productive waters: from *in-situ* measurements to aerial imagery. Dissertation & Thesis in Natural Resources Paper 46, University of Nebraska: 133 p.

Hakvoort, H., J. DeHaan, R. Jordans, R. Vos, S. Peter, M. Rijkeboer. 2002. Towards airborne remote sensing of water quality in the Netherlands-validation and error analysis. *ISPRS J. Photogramm. Rem. Sens.* 57: 171–183.

Hauff, P.L., N. Lindsay, D.C. Peters, G.A. Borstad, W. Peppin, L. Costick, R. Glanzman. 1999. Hyperspectral evaluation of mine waste and abandoned mine lands: NASA- and EPA-sponsored projects in Idaho. In R.O. Green (ed.), *Summaries of the Eighth JPL Airborne Earth Science Workshop*, February 9–11, 1999: Pasadena, CA: JPL Publication 99-17: 229–238.

Hauff, P.L., D.C. Peters, D.W. Coulter, M.A. Sares, D.A. Bird, F.B. Henderson III, E.C. Prosh. 2006. Hyperspectral sensing of acid mine drainage—two Colorado case studies. In R.I. Barnhisel (ed.), *Proceedings of 7th International Conference on Acid Rock Drainage*, American Society of Mining and Reclamation: 738–762.

Hielm, J.H. 1989. NIFO comparative trials. In A.E. Lodge (ed.), *The Remote Sensing of Oil Slicks*. Chichester: John Wiley and Sons: 67–76.

Hook, S.J., T.J. Cudahy, A.B. Kahle, L.B. Whitbourn. 1998. Synergy of active and passive airborne thermal infrared systems for surface compositional mapping. *J. Geophys. Res.* 103: 18269–18276.

Hörig, B., F. Kühn, F. Oschütz, F. Lehmann. 2001. HyMap hyperspectral remote sensing to detect hydrocarbons. *Int. J. Rem. Sens.* 22: 1413–1422.

Hough, H. 1995. Coastal monitoring: more than a day at the beach. *Earth Obs. Mag.* 4:22–24.

Huber, T.P., K.E. Casler. 1990. Initial analysis of Landsat TM data for elk habitat mapping. *Int. J. Rem. Sens.* 11: 907–912.

Isaacson, D.L., D.A. Leckenby, C.J. Alexander. 1982. The use of large-scale aerial photography for interpreting Landsat digital data in an elk habitat-analysis project. *J. Appl. Photogr. Eng.* 8: 51–57.

Kahle, A.B., A.F.H. Goetz. 1983. Mineralogic information from a new airborne thermal infrared multispectral scanner. *Science.* 222: 24–27.

Klemas, V.V. 2009. The role of remote sensing in predicting and determining coastal storm impacts. *J. Coastal Res.* 25: 1264–1275.

Koltun, J. 1993. Satellite imagery classification for the development of wildlife habitat types. *Earth Observation Magazine.* 2. 32–40.

Koponen, S., J. Pulliainen, K. Kallio, M. Hallikainen. 2002. Lake water quality classification with airborne hyperspectral spectrometer and simulated MERIS data. *Rem. Sens. Environ.* 79: 51–59.

Kremer, R.G., J. Zack, A. Davis, S. McLeod, C. Byrns. 1998. Contribution of landscape attributes in determining revegetation potential: a geographical information system approach. In *Tailings and Mine Waste '98.* Colorado State University, CO: Rotterdam: Balkema: 789–793.

Kuehn, F., T.V.V. King, B. Hoerig, D.C. Peters (eds.). 2000. *Remote Sensing for Site Characterization.* Federal Institute for Geoscience and Natural Resources. Berlin: Springer: 252 p.

Kühn, F., K. Oppermann, B. Hörig. 2004. Hydrocarbon index—an algorithm for hyperspectral detection of hydrocarbons. *Int. J. Rem. Sens.* 25: 2467–2473.

Lawson, D.R., P.J. Groblicki, D.H. Stedman, G.A. Bishop, P.L. Guenther. 1990. Emissions from in-use motor vehicles in Los Angeles: a pilot study of remote sensing and the inspection and maintenance program. *J. Air Waste Manag. Assoc.* 40: 1096–1105.

Lyon, J.G. 1987. The use of maps, aerial photographs and other remote sensing data for practical evaluation of hazardous waste sites. *Photogram. Eng. Rem. Sens.* 53: 515–519.

Marsh, S.E., J.L. Walsh, C.T. Lee, L.A. Graham. 1991. Multitemporal analysis of hazardous waste sites through the use of a new bi-spectral video remote sensing system and standard color-IR photography. *Photogram. Eng. Rem. Sens.* 57: 1221–1226.

McGee, K.A., T. Elias, A.J. Sutton, M.P. Doukas, P.G. Zemek, T.M. Gerlach. 2005. Reconnaissance gas measurements on the East Rift Zone of Kilauea volcano, Hawai'i by Fourier transform infrared spectroscopy. U.S. Geological Survey Open File Rept. 2005-1062.

McNutt, M., R. Camilli, G. Guthrie, P. Hsieh, V. Labson, B. Lehr, D. Maclay, A. Ratzel, M. Sogge. 2011. *Assessment of Flow Rate Estimates for the Deepwater Horizon/Macondo Well Oil Spill.* U.S. Dept. Interior Flow Rate Technical Group report to the National Incident Command, Interagency Solutions Group: 21 p.

Meng, Q., H. Hu, Q. Yu. 2006. The application of an airborne electromagnetic system in groundwater resource and salinization studies in Jilin, China. *J. Environ. Eng. Geophys.* 11: 103–109.

Meyers, R.J., Associates. 1989. *Oil Spill Response Guide.* Park Ridge, IL: Noyes Data Corp.: 46–69.

Moffat, A.J., M.M. Millan. 1971. The applications of optical correlation techniques to the remote sensing of SO_2 plumes using sky light. *Atmos. Environ.* 5: 677–690.

Mori, T., M.R. Burton. 2006. The SO2 camera: A simple, fast and cheap method for ground-based imaging of SO_2 in volcanic plumes. *Geophys. Res. Lett.* 33(24): L24804.

Mori, T., M.R. Burton. 2009. Quantification of the gas mass emitted during single explosions on Stromboli with the SO2 imaging camera. *J. Volcanol. Geoth. Res.* 188: 395–400.

Mukherjee, T., T.K. Bandyopadhyay, S.K. Pande. 1991. Detection and delineation of depth of subsurface coal mine fires based on an airborne multispectral scanner survey in a part of the Jharia Coalfield, India. *Photogramm. Eng. Rem. Sens.* 57: 1203–1207.

Nadeau, P.A. 2011. Ultraviolet digital imaging of volcanic plumes: implementation and application to magmatic processes at basaltic volcanoes. PhD Dissertation, Michigan Technological University: 181 p.

Nebraska Department of Environmental Quality. 2011. Microcystin Toxin Migration, Bioaccumulation, and Treatment, Fremont Lake #20, Dodge County, Nebraska. Lincoln, NB: Nebraska Department of Environmental Quality: 44p.

Peters, D.C., D.A. Bird, P.L. Hauff, M.A. Sares, D.W. Coulter, E.C. Prosh, F.B. Henderson III. 2004. Assessment of water quality in a watershed impacted by natural ARD using mineralogy and remote sensing (extended abs.). In *Proceedings of the 2004 National Meeting of the American Society of Mining and Reclamation*: Lexington, KY: AMSR: 1480–1483.

Peters, D.C., P.L. Hauff. 2000. Multispectral remote sensing to characterize mine waste (Cripple Creek and Goldfield, USA). In F. Kuehn, T.V.V. King, B. Hoerig, D.C. Peters (eds.), *Remote Sensing for Site Characterization.* Berlin: Springer: 113–164.

Quibell, G. 1991. The effect of suspended sediment on reflectance from freshwater algae. *Int. J. Rem. Sens.* 12: 177–182.

RADARSAT. April 1996. RadarSat image of St. Ann's Head, Wales. *Earth Obs. Mag.* 5: 62.

Rickel, C., B. Lamb, A. Guenther, E. Allwine. 1990. An infrared method for plume rise visualization and measurement. *Atmos. Environ.* 24A: 2835–2838.

Rykhus, R., Z. Lu. 2008. Hurricane Katrina flooding and oil slicks mapped with satellite imagery. *U.S. Geol. Surv.* 4 p. http://pubs.usgs.gov/circ/1306/pdf/c1306_ch3_f.pdf

Sabins, F.F., Jr. 1987. *Remote Sensing Principles and Interpretation.* New York: W.H. Freeman and Co.: 198–199.

Salem, F., M. Kafatos. 2001. Hyperspectral image analysis for oil spill mitigation. In *Proceedings of the 22nd Asian Conference on Remote Sensing,* November 5–9, 2001, Singapore: Centre for Remote Imaging, Sensing, and Processing: 6 p.

Schoer, B. 1999. *Geophysics for Civil Engineers, an Introduction.* Briefing Sheet, London, UK: The Institution of Civil Engineers, 5 p.

Schwartz, G., E. Ben-Dor, G. Eshel. 2012. Quantitative analysis of total petroleum hydrocarbons in soils: comparison between reflectance spectroscopy and solvent extraction by 3 certified laboratories. *Appl. Environ. Soil Sci.* 2012: 11, Article ID 751956.

Shih, W.C., A.B. Andrews. 2008. Modeling of thickness dependent infrared radiance contrast of native and crude oil covered water surfaces. *Opt. Express.* 16: 10535–10542.

Slonecker, T., G.B. Fisher, D.P. Aiello, B. Haack. 2010. Visible and infrared remote imaging of hazardous waste: a review. *Rem. Sens.* 2: 2474–2508.

Smith, B.D., J. Cannia, J.D. Abraham, D.O. Rosenberry, A. Prikhodko, P.A. Bedrosian. 2012. Hydrologic implications from airborne resistivity mapping of the Sand Hills of western Nebraska (abs.): *GSA Ann. Mtg. and Exposit.* Paper No. 31922–210987.

Steffes, P.G. 1983. *Remote Sensing of Clouds Bearing "Acid Rain."* National Science Foundation Grant ECS-8205472: Arlington, VA: NSF: 28 p.

Stoiber, R.E., L.L. Malinconico, S.N. Williams. 1983. Use of the correlation spectrometer at volcanoes. In H. Tazieff, J.C. Sabroux (eds.), *Forecasting volcanic events.* New York: Elsevier: 424–444.

Tamburello, G., E.P. Kantzas, A.J.S. McGonigle, A. Aiuppa. 2011a. Vulcamera: a program for measuring volcanic SO2 using UV cameras. *Ann. Geophys.* 54: 219–221.

Tamburello, G., E.P. Kantzas, A.J.S. McGonigle, A. Aiuppa, G. Guidice. 2011b. UV camera measurements of fumarole field degassing (La Fossa crater, Vulcano Island). *J. Volcanol. Geotherm. Res.* 199: 47–52.

Tamburello, G., A.J.S. McGonigle, E.P. Kantzas, A. Aiuppa. 2011c. Recent advances in ground-based ultraviolet remote sensing of volcanic SO_2 fluxes. *Ann.Geophys.* 54: 199–208.

Telops. 2012. *Airborne detection of methane leaks using hyperspectral imaging.* www.telops.com (accessed 26 Jan. 2013).

Tuominen, J., T. Lipping. 2011. *Detection of environmental change using hyperspectral remote sensing al Olkiluoto repository site.* Working Rept. 2011–26, Posiva Oy: 51 p.

Voigt, S., A. Tetzlaff, J. Zhang, C. Künzer, B. Zhukov, G. Strunz, D. Oertel, A. Roth, P. van Dijk, H. Mehl. 2004. Integrating satellite remote sensing techniques for detection and analysis of uncontrolled coal seam fires in North China. *Int. J. Coal Geol.* 59: 121–136.

Vos, R.J., J.H.M. Hakvoort, R.W.J. Jordans, B.W. Ibelings. 2003. Multiplatform optical monitoring of eutrophication in temporally and spatially variable lakes. *Sci. Total Environ.* 312: 221–243.

Wagner, M.J. June 1996. The view from down under: a small end-user community in Canberra, Australia strives to provide remote sensing solutions for a series of water and forestry applications. *Earth Obs. Mag.* 5: 51–54.

Wang, W., J. Qu. 2009. Assessment of post-hurricane forest damage using optical remote sensing. *SPIE Newsr.* 3 p. http://spie.org/x35463.xml

Wang, F., J. Xu. 2010. Comparison of remote sensing change detection techniques for assessing hurricane damage to forests. *Environ. Monit. Assess.* 162: 311–326.

Watson, R.D., M.E. Henry, A.F. Theisen, T.J. Donovan, W.R. Hemphill. 1978. Marine monitoring of natural oil slicks and manmade wastes utilizing an airborne imaging Fraunhofer line discriminator. In *Proceedings of the American Chem. Soc. 4th Joint Conference on Sensing of Environment Pollutants*: Washington, D.C.: American Chemical Society: 667–671.

White, D.C., M. Williams, S.L. Barr. 2008. Detecting sub-surface soil disturbance using hyperspectral first derivative band ratios of associated vegetation stress. *Int. Arch. Photogram. Rem. Sens. Spatial Inform. Sciences.* XXXVI B7, Beijing: 243–248.

Williams, R.S., Jr. 1983. Geologic applications. In R.N. Colwell (ed.), *Manual of Remote Sensing*, 2nd ed. Falls Church, VA: American Society of Photogrammetry: 1667–1954.

Additional Reading

Gitelson, A.A., B.C. Gao, R.R. Li, S. Berdnikov, V. Saprygin. 2011. Estimation of chlorophyll-a concentration in productive turbid waters using a hyperspectral imager for the coastal ocean—the Azov Sea case study. *Environ. Res. Lett.* 6: 6 p.

Gitelson, A.A., Y.Z. Yacobi, J.F. Schalles, D.C. Rundquist, L. Han, R. Stark, D. Etzion. 2000. Remote estimation of phytoplankton density in productive waters. *Arch. Hydrobiol. Spec. Issues Advanc. Limnol.* 55: 121–136.

Green, G.W., R.M. Moxham, A.H. Harvey. 1969. Aerial infrared surveys and borehole temperature measurements of coal mine fires in Pennsylvania. In *Proceedings of the 6th International Symposium on Remote Sensing of Environment.* Ann Arbor, MI: ERIM: 517–525.

Kühn, F., T.V.V. King, B. Hoerig, D.C. Peters (eds.). 2000. *Remote Sensing for Site Characterization.* Berlin: Springer: 268 p.

Ritchie, J.C., P.V. Zimba, J.H. Everitt. 2003. Remote sensing techniques to assess water quality. *Photogram. Eng. Rem. Sens.* 69: 695–704.

Sengpiel, R.E. 2007. Using airborne hyperspectral imagery to estimate chlorophyll A and phycocyanin in three central Indiana mesotrophic to eutrophic reservoirs. MSc. Thesis. Indiana University: 149 p.

Slavecki, R.J. 1964. Detection and location of subsurface coal fires. In *Proceedings of the 3rd Symposium on Remote Sensing of Environment.* Ann Arbor, MI: ERIM.

15

Geobotanical Remote Sensing

Chapter Overview

When considering using plants for mapping geology or helping to find groundwater, mineral deposits, or petroleum seeps, it is useful to understand that plants are living organisms and that they respond to many environmental factors. Among many environmental influences, the most important are the rock, soil, and regolith on which they are growing. These determine the nutrients they can take up, the amount of soil moisture, and soil salinity. Equally important are roughness of the topography, the particular slope angle (steepness) and the slope azimuth (mostly facing the sun versus mostly in shadow). Also important are the season, how much precipitation, heat, or cold has affected them in the past years. It helps to know whether the plants in an area have been affected by burns, landslides, avalanches, or if they are a stable and mature plant cover. What is the impact of plant diseases, insect infestations, or overpopulation of native grazers and browsers? Then there are manmade influences such as overgrazing, pollution, timber harvesting, and groundwater withdrawal, to name a few. Plants respond to these stimuli in several, not always predictable ways. Among the responses are changes in density, vigor (health), height, species distribution, species replacement, growth stage, timing of leafout and senescence, and die out.

From the aforementioned list, one can see that there are a number of influences on plant communities other than geology. If all else were held constant, perhaps we could readily predict the significance of certain vegetation responses. This rarely being the case, under certain circumstances, it should still be useful to observe plant patterns to help map the distribution of rock types, soils, mineral concentrations, alteration, seeps, near-surface groundwater, and so forth. Areas with heavy vegetation cover, whether tropical or temperate rainforest or boreal forest, will benefit from being able to map geology by observing changes in the plant cover.

Stress is the primary reaction of plants to mineralization or hydrocarbons in their environment. Stress can be imaged in at least three ways: (1) thermal response of plants, (2) fluorescence response, and (3) reflected light response. Thermal response is a function of stomatal conductance, that is, leaf temperature is a result of evaporation or transpiration. Leaf temperature at any given time is also related to air temperature, humidity, wind speed, and net radiation (Jones and Schofield, 2008).

Fluorescence imaging detects changes in a plant's response to ultraviolet (UV)-A radiation by measuring emission from green leaves at 440, 520, 690, and 740 nm. The fluorescence peaks in the red and near-infrared are associated with chlorophyll-*a*, which is directly related to photosynthesis. Shorter wavelength emissions are related mainly to phenolic compounds in the leaves. Fluorescence intensity is related mainly to concentration of chlorophyll, phenols, and other compounds. Since the absolute fluorescence magnitude is an arbitrary measure, it is helpful to normalize the data and use ratios. Fluorescence is used to distinguish early stages of infection by viruses, fungi, and bacteria before symptoms are detectable in reflected light (Jones and Schofield, 2008). Multicolor fluorescence has been used to detect malnutrition on leaves. The low absolute intensity of plant fluorescence hinders its use as an airborne remote sensing tool.

Reflected light is used mainly to detect changes in chlorophyll, a good indicator of plant stress. Decreases in chlorophyll due to lack of water, disease, or other factors lead to yellowing or browning of leaves. Color infrared imagery is best for detecting changes in reflectance as a result of stress to groups of plants. Many diseases and mineral deficiencies are detectable by color patterns on leaves, for example, zinc deficiency causes browning between veins whereas sulfur deficiency leads to purpling of the veins (Jones and Schofield, 2008). Detecting these patterns is beyond the resolution of most airborne sensors at this time.

Other instruments that can detect plant stress include radar and Lidar, which measure canopies, and can detect changes in plant morphology (e.g., stunting), plant diversity, and plant density.

Geobotany and Geological Mapping

Soil and climate, altitude, latitude, moisture, and slope are the main natural factors in determining which plants grow in an area and how well they grow. Plant communities are intimately related to the substrate on which they are rooted. They take their nutrients and moisture from the soil, and the soil is usually derived locally from underlying rocks. Plants may be endemic (restricted to) and thrive in a given soil, just tolerate a difficult soil, or they may be part of a larger community that is influenced by the soil as far as vigor, morphology, density, and diversity. For example, serpentinite-derived soils are rich in magnesium and heavy metals such as iron, cobalt, and nickel, and low in calcium. These soils are clay rich and low in nutrients that plants need. Some plants, like Interior Silk Tassel (*Garrya congdonii*) and Leather Oak (*Quercus durata*) are endemics on serpentinite soils in northern California. Others, such as Knobcone Pine (*Pinus attenuata*) are typical of sepentinite soils. Some, like Chamise (*Adenostoma fasciculatum*), are not endemic but tolerate serpentinite soils (Schoenherr, 1992). The same goes for soils developed on basalts, limestones, and other rock types. Because endemics vary with locale, the analyst should become familiar with the plant community in the area of interest.

When we talk about remote sensing of vegetation, other factors enter into how well we can divine what is going on. The percent canopy cover will determine how much soil or rock is included in the averaged pixel, as will the amount of lichen cover. Early work at JPL showed that dead or dry vegetation does not greatly alter the spectral reflectance of underlying rock, whereas a mere 10% green vegetation cover can alter low reflectance rock or soil beyond recognition and render some Landsat ratios ineffective (Siegal and Goetz, 1977). Studies by the Geological Survey of Canada indicate that up to 50% lichen cover on carbonates will obscure (but not eliminate) the carbonate absorption signature, whereas 100% cover does (Harris et al., 2012). Instrument factors such as the spatial resolution will affect how many plant reflections are mixed (averaged) within a pixel, and spectral bandwidth and spectral wavelength determine which absorptions or reflection peaks are included in the analysis and how well resolved these features are. Any artifacts introduced by the instrument during acquisition must be detected and removed. Environmental noise removal, particularly atmospheric corrections, influences how well spectral unmixing and classification works.

Vegetation pattern mapping on remote sensing imagery or maps fall into two categories: (1) mapping rock units and (2) mapping structure. VLF-EM has been used to identify silicate versus carbonate substrate, particularly near-vertical contacts, in a project in the Bükk Mountains of northeast Hungary (Németh and Pethő, 2009). Airborne gamma ray surveys were used along with Landsat TM images to map surface geology in a heavily forested area near Manitouwadge, Ontario, Canada (Kettles et al., 2000). A supervised classification of TM data over Archaean greenstone belts created a usable surface geologic map, but it did not have much rock type information. Rather it outlined categories such as rock, thin till, thick till, sand/gravel, organic-rich soil, water, clear cuts, and unclassified. Likewise, Singhroy and Barnett (2003) made a surficial materials map of the Trout Lake area, Northern Ontario, Canada, for use in planning forest roads and other civil engineering programs. They merged information from a digital elevation model (DEM), Landsat TM bands 4, 5, and 7, and IRS-1c panchromatic imagery to make a terrain classification map based on the following four components: material, landform, relief, and moisture (Figures 15.1 and 15.2).

Workers at the Geological Survey of Brazil and Brazilian universities have demonstrated how to map geologic units in heavily vegetated terrain (Perrotta et al., 2008). They used optical hyperspectral imagery from the Advanced Spaceborne Thermal Emission and Reflection Radiometer (ASTER) along with airborne gamma ray data to map geology in the Guapore Valley region of the Amazon forest. Geology in the area consists of Cretaceous to Cenozoic sediments over metamafic-ultramafic units interspersed with banded iron formations and calc-silicate gneisses. A semi-deciduous forest is developed on clay-rich eutrophic soil and rocky outcrops. Trees heights reach to 30 m, and dominant plant families

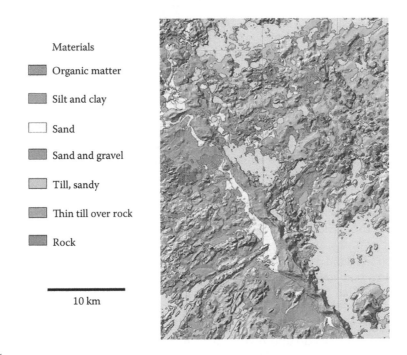

FIGURE 15.1
Surficial materials map of the Trout Lake area, northern Ontario, Canada, for use in planning forest roads and other civil engineering programs. (From Singhroy, V., P. Barnett, Satellite based standardized terrain maps: A case study. In D.T. Hansen et al. (eds.), *Spatial Methods for Solution of Environmental and Hydrologic Problems—Science, Policy, and Standardization* (STP 1420), West Conshohoken, PA, ASTM, 148–160, 2003; courtesy of Dr. Vern Singhroy.)

FIGURE 15.2
A DEM, Landsat TM bands 4, 5, and 7, and IRS-1c panchromatic imagery were merged to make a terrain classification map. Labels indicate various combinations of material, landform, relief, and moisture. For example, "sgME/HD" indicates "sand and gravel, end moraine, high and dry." (From Singhroy, V., P. Barnett, Satellite based standardized terrain maps: a case study. In D.T. Hansen et al. (eds.), *Spatial Methods for Solution of Environmental and Hydrologic Problems—Science, Policy, and Standardization* (STP 1420). West Conshohoken, PA, ASTM, 148–160, 2003; courtesy of Dr. Vern Singhroy.)

include Euphorbiaceae, Moraceae, Burseraceae, Fabaceae, and Momosaceae, along with palms that occur in humid valleys. Distribution of plants in this area is strongly influenced by soil type, topography, amount of sun, and climate gradients. Band ratios were generated by dividing ASTER band 3 (centered at 807 nm) by all the other bands (1 through 9; 556 nm through 2.40 μm, respectively). Principle components (PCs) for the ratios were generated and a minimum noise fraction (MNF) analysis was performed on the ratios to remove noise and maximize information content. High-frequency information in the PC and MNF components was smoothed using low-pass filtering. A transform was used to merge the first PC (contains most albedo and texture information) with the color composites consisting of PC and MNF components. The best combinations contained PC1 or PC3 with PC2 and PC8. It was found that different lithologies can have the same spectral signature, and that a signature can represent more than one lithology. However, when combined with tonal contrast, textural and topographic information, and field observations, the resulting images were useful for mapping lithologies (Figure 15.3). The airborne gamma ray data

FIGURE 15.3
Work by the Geological Survey of Brazil in the Guapore Valley region of the Amazon demonstrates how to map geologic units in heavily vegetated terrain. A transform was used to merge the different image components. (a) PC1, PC8, PC2 in RGB. (b) PC1, PC3, PC2 in RGB. (c) (-MNF1), (-MNF5), MNF4 in RGB. (d–f) Aerogeophysics grids sharpened by a Hue-Saturation Value transform with ASTER original bands PC1 in the value channel. (d) *e*Th channel, (e) K channel, (f) ternary (K-U-Th) radiometric map. (From Perrotta, M.M. et al., *Intl. Arch. Photogram. Rem. Sens. Spat. Inform. Sciences*, XXXVII part B8, 2008; courtesy of Dr. Mauricio G. M. Santos.)

flown at 500 m line spacing was processed to equivalent Thorium and Potassium grids and to a ternary K–U–Th (red–green–blue) radiometric map. These grids were likewise sharpened by fusion with the PC1 of the original ASTER bands. The final interpretive map shows the distribution of granites, monzonites, metagabbro, metasediment, and aluminous schist, among others (Figure 15.4).

An example of geobotany used to map structure is provided by work done in Scotland (Cracknell and Saraf, 1988). The areas studied, near Assynt and Loch Tay in northwest

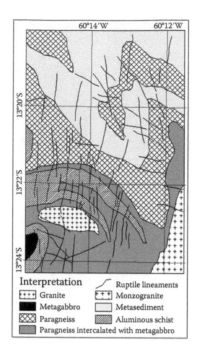

FIGURE 15.4
Interpretive map shows the distribution of granites, monzonites, metagabbro, metasediment, and aluminous schist, among others. (From Perrotta, M.M. et al. *Intl. Arch. Photogram. Rem. Sens. Spat. Inform. Sciences.* XXXVII part B8, 1275–1280, 2008; courtesy of Dr. Mauricio G. M. Santos.)

Scotland, have Cambro-Ordovician quartzites intruded by basic dikes. Other than some bare rock outcrop and scree, the area is completely covered by a mixture of ericaceous heaths, *Molinia* and *Juncus* sedges, fescue grasses, thyme bushes, and *Sphagnum* mosses. Grass dominates in well-drained areas over carbonate, alluvium, and fluvio-glacial deposits. Brackens are confined to coarse scree. Fresh green turf is associated with feldspathic claystones and carbonates. Rock is covered by lichens. The workers used color composites of Airborne Thematic Mapper Simulator (TMS) bands 5, 7, and 9 along with Landsat multispectral scanner (MSS) bands 4, 5, and 7 (RGB), the band 7/4 ratio, and PC 1 of the MSS bands. They recognized that the TMS imagery in particular was useful for mapping faults and dikes, but found it less sensitive to variations in lithology. Linear vegetation depressions were related to previously unrecognized faults, whereas positive lineaments (ridges) were related to ultrabasic dikes. Color composites of MSS band ratios 4/5, 5/7, and 7/4 were helpful in identifying linear vegetation features. Multitemporal imagery indicated that vegetation can be seasonally stressed and that September imagery (in the northern hemisphere) shows stress due to early dieback. They noted that gneisses had grass cover whereas overlying sedimentary rock supported dense heather. The MSS 3/7 ratio allowed identification of the contact between quartzites and quartz sandstones. They surmised that the plants growing on the quartzites are stressed. It was found that soil developed over mica schists supports European larch, whereas soil developed over epidiorite and hornblende schist has Scots pine.

The association of plants to substrate is somewhat unique in each area. Once the correlations are made, however, it should in many cases be possible to map major units on the basis of their plant cover. Vegetation alignments reveal the location of faults, dikes, contacts between near-vertical beds, and other linear geologic features.

Vegetation Suppression Techniques

Before leaving a discussion of geobotanical mapping, we should briefly review image processing techniques known as vegetation suppression. "Vegetation suppression" minimizes the effect of reflection from plants in order to enhance geologic information in densely vegetated terrain. The main requirement for this technique is that there is enough of the bedrock exposed that some light is reflected from the rocks and soil and is available to the remote sensing instrument (Yu et al., 2009).

These algorithms are versions of the "forced invariance" method, which splits out components of plant reflectance that are invariant with respect to some index such as the Normalized Difference Vegetation Index. Using band statistics, Crippen and Blom (2001) suggested a forced invariance method that minimizes vegetation. Such algorithms usually involve the following steps: (1) perform atmospheric correction, (2) mask barren areas using a vegetation index threshold, (3) calculate a statistical relationship between the vegetation index and digital number (DN) reflectance values for each band, (4) calculate a best-fit curve for each of these relationships, (5) select a target average DN value and scale all pixels at each vegetation index level by an amount that shifts the curve to the target DN, and (6) combine the processed areas with the masked areas. The results can be quite stunning (Figure 15.5).

Radar and Lidar can have an effect similar to vegetation suppression in that they record the top of the forest canopy, and/or the ground beneath the canopy when the vegetation is not dense. In many cases, the canopy mimics the ground and can reveal contacts and structures. In other cases, the radar or Lidar reflections can be processed such that the farthest reflections in a given area are accepted and the shortest reflections are deleted. This, in effect, eliminates reflections from the canopy and uses only ground reflections. A DEM can be generated, and geomorphology can be evaluated.

FIGURE 15.5
Standard true color hyperspectral image with vegetation in green (left) and MDA vegetation suppression image (right). (Courtesy of MDA Federal.)

Botanical Indicators of Mineralization and Alteration

Mineral deposits and the alteration that accompanies them are usually expressed in vegetation as some kind of stress. Some plants are poisoned by heavy minerals and acid soils associated with ore deposits, such as the birch-pine forest growing over sulfide copper deposits at Karasjok, Norway (Låg and Bølviken, 1974), or by ultramafic soils such as those associated with chromium, nickel, manganese, and mercury near New Almaden, California (Oze et al., 2003). Stress can take the form of stunted growth (dwarfism), premature leaf senescence, chlorosis (yellowing of leaves), lower plant density, wilting, and death (Carlisle and Cleveland, 1958; Cannon, 1960, 1971; Canney, 1969; Ayeni et al., 2010).

Different plant species concentrate certain metals in their leaves, stems, and roots to varying degrees, and toxicity levels differ among species. The key is whether the metal concentration affects photosynthesis, which is the vital energy-generation and food-production process in plants. Where some plants are stressed, others appear to thrive. Either situation can be an indicator of mineralization, that is, a vegetation anomaly.

Although specific indicator plants may not be identifiable on imagery, areas of stunted or dying vegetation should be recognizable as less vigorous on color infrared and hyperspectral imagery. We must not assume that stressed vegetation is always growing in mineralized soil: drought, infestation, lack of nutrients, slope, elevation, and other factors must be considered.

Lower plant densities can be detected using radar. In the Amazon, C band and L band radar has been effective in mapping microtopography (surface roughness) through a sparse canopy over iron laterites (de Morais et al., 2011). Although the Brazilian-German MAPSAR (Multi-Application Purpose Synthetic Aperture Radar) is a proposed L band SAR that is still in the planning stage, a MAPSAR simulation program used multipolarization L band airborne radar to collect data with 11 m resolution over a deposit in Carajás Province. Geology in this area includes basalts, dolerites, rhyolites, and ironstones consisting of various iron oxides. A distinctive low-density *campos rupestres* plant assemblage over the laterite is in sharp contrast to the high-density equatorial background forest. Radar signals penetrated the sparse cover and were returned from the duricrust surface. A Gray Level Co-occurrence Matrix was used to characterize texture, followed by a supervised classification using nine defined classes. The best classification was done using both HH and VV polarizations. The HH polarized signal had the least attenuation due to vegetation. Despite, or perhaps because of a drought prior to the survey (an extra stressor), they got good penetration of the canopy. The duricrust, containing limonite blocks and hematite ore blocks with subordinate specularite and cemented with hydrous ferric oxides, was classified as "smooth." Radar wavelength is considered the most important factor in sensing these laterites, and the authors suggest that multiwavelength SAR images are best for discriminating rock classes using texture in forested areas.

One spectral response of vegetation growing on mineralized soil may be changing reflectance intensity. Reflectance intensity most likely corresponds to stress or plant vigor. In northeastern Ontario (Kelvin Township), the Ontario (Canada) Geological Survey found that almost pure stands of Trembling aspen (*Populus tremuloides*) have brighter visible and near-infrared reflectance in the range 680–792 nm when growing in soils with elevated levels of nickel, calcium, chromium, and copper (Beswick et al., 1991). This echoes the pioneering work of Yost and Wenderoth (1971), which showed that both Red Spruce (*Picea rubens*) and Balsam Fir (*Abies balsamea*) are more reflective in the visible range when growing in copper–molybdenum mineralized soil as opposed to background soil. In Kelvin Township, nickel and calcium were preferentially enriched in the leaves and roots of the aspens. The trees are growing on mafic to felsic metavolcanics intruded by diabase dikes. The high soil metal content is associated with basaltic and serpentinized komatiites with high nickel and chrome values. The standard vegetation cover is dense stands of mixed Trembling aspen, White birch, Balsam fir, and spruce. Rocky ridges tend to have Jack pine, while swampy areas contain Speckled Alder and White Cedar.

Other workers (Singhroy et al., 2000) found it difficult to see any difference in reflection between mineralized and background sites in the boreal forest. They examined a copper–nickel mineralized site containing breccias of the Sudbury Nickel Igneous Complex, Sudbury basin, Ontario, and compared it with a nonmineralized granitic background site. Reflectance spectra in the range 400 to 944 nm were taken for both White birch (*Betula papyrifera*) and Trembling aspen. Leaves were collected and spectra were recorded within 2 hours of sampling. The 81 birch and aspen spectra were averaged for both the mineralized and background site. For plants growing on mineralized soil, the red edge of the reflectance curve, usually around 720 nm, shifted to longer wavelengths (a "red shift") by about 0.35 nm for the birch and 1.46 nm for the aspen. Yost and Wenderoth (1971) observed this effect in Balsam fir growing on copper-moly mineralized soil at Catheart Mountain, Maine. Yet, they observed a shift to shorter wavelengths (a "blue shift") for conifers growing in mineralized soil. This inconsistent effect may indicate that different plants react differently to metal stress, that soil metal content is not great enough to either increase or decrease reflectance, or that other environmental factors are more important to the plants. The authors conclude that the effect would not be detectable using airborne or satellite sensors.

On the other hand, airborne gamma ray surveys and Landsat TM imagery were combined in a supervised classification that mapped geology using vegetation patterns in mineralized and unmineralized training sets in Ontario (Kettles et al., 2000). The TM imagery provided an estimate of surficial cover while the gamma ray measurements showed anomalies that corresponded to the location of known mineral occurrences.

Almeida et al. (2009) described a project to assess geobotanical anomalies related to high-sulfidation gold mineralization in almost virgin tropical rain forests of the Tapajos gold province, northern Brazil. No rocks or soil are exposed at the surface. Landsat-5 TM imagery was processed using algorithms designed to enhance both specific and anomalous vegetation spectra. Their four-stage method consists of (1) preparing spectral indexes to enhance responses of vegetation pigments and leaf moisture, (2) a principal component analysis (PCA) applied to these indexes, (3) another PCA is applied to the original TM 4, 5, and 7, creating a PC1 that is a texture-rich albedo image, and (4) a low-pass proportional convolution filter applied to the principal component images derived in step 2. The final product (Figure 15.6) shows a near-perfect spatial correlation between color classes highlighted by this Landsat image product and hydrothermal alteration identified in outcrops. Although the product does not directly indicate rock type, minerals, or presence of hydrothermal alteration, it was suitable to outline geobotanical anomalies associated with mineralogical variations. The method should be an important aid for exploration in terranes covered by equatorial forests.

Perhaps because there has not been any resounding success using plants to locate mineral deposits, there have not been many articles reporting on these techniques in the past decade. Both Brazilian and Canadian workers, however, report some success mapping geology using gamma ray surveys along with multispectral or hyperspectral imagery. Interest in this topic continues because so much of the Earth is covered by plants.

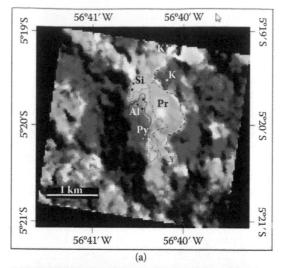

(a)

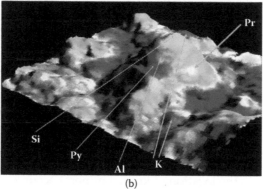

(b)

FIGURE 15.6
An area where essentially no rocks or soil are exposed at the surface, Tapajos gold province, Amazon rain forest. (a) Integrated image comprising a false color composite of PC4, PC7, and PC8 in RGB. Sampled outcrops are plotted on the imagery. Areas with similar tonal characteristics are outlined. Al = advanced argillic alteration with alunite; K = granites affected by potassic alteration; Py = advanced argillic alteration with pyrophyllite; Pr = propylitic alteration; Si = silica cap. (b) False color composite of PC4, PC7, and PC8 in RGB merged with an ASTER digital terrain model. The map shows a near-perfect correlation between interpreted classes and hydrothermal alteration identified in the field. (From Almeida, T.I.R. et al., Application of remote sensing to geobotany to detect hydrothermal alteration facies in epithermal high-sulfidation gold deposits in the Amazon region. In: Bedell, R. et al. (eds.), *SEG Reviews in Economic Geology*, Vol 16, Chapter 11: 135–142, 2009; courtesy of Dr. Teodoro Almeida and Dr. Carlos R. de Souza Filho.)

Botanical Responses to Hydrocarbons

Vegetation has been used as an indicator of seeping hydrocarbons, which can occur both over structural traps and stratigraphic traps. Some types of vegetation have been classified as "hydrocarbon indicator" plants, for example, *Anabasis salsa*, *Salsela* sp. (saltwort), and *Allium* sp. (onion) (Siegal, 1974). Hydrocarbon seepage may be detrimental to plant communities in areas of natural seepage over oil/gas fields, or over natural gas pipeline leaks. Sagebrush is stunted, and prairie grasses are less dense over the gas cap at Patrick Draw field, Wyoming (Richers et al., 1982; Jacobson and Khan, 2006). At Lost River field, West Virginia, high soil pH coincides with high soil gas and causes an inability of plants to absorb micronutrients, especially zinc, which has led to stunting (Rock, 1983). Changing the chemistry of soil, such as at Cement field, Oklahoma, can cause anomalous species to develop.

Many plants are harmed by the presence of hydrocarbons in the soil. Laboratory experiments on marsh plants show that *Spartina patens* (saltmeadow cordgrass) suffered increasing damage with increasing amounts of oil and that heavier crudes were more toxic than lighter oils. It has been noted by pipeline inspectors that natural gas leaks affect trees and shrubs near buried pipeline leaks. This has been explained (Davis, 1977) as a result of methane driving out soil air and contributing to the growth of methane-oxidizing bacteria that deplete soil oxygen. Flower et al. (1981) showed that trees growing above urban landfills were killed when leaking methane decreased the soil oxygen and increased the carbon dioxide in the soil root zone. Parrish and Rock (1983) and Rock (1984) studied vegetation patterns over the Lost River gas field. They found that mycorrhizal fungus on plant root hairs (that help with nutrient uptake) are inhibited on Chestnut Oaks (*Quercus prinus*) in areas of high soil manganese and/or methane concentrations.

Smith et al. (2004) demonstrated that soil oxygen displaced by natural gas, argon, nitrogen, and waterlogging caused an increase in visible reflectance. Bammel and Birnie (1994) found that the red edge at 700–740 nm was shifted to shorter wavelengths in sagebrush due to hydrocarbons in the soil. The "red edge" or "vegetation shoulder" is the abrupt increase in healthy plant reflectance that occurs between 650 and 750 nm. Crawford (1986) also observed this "blue shift" in the red edge for Douglas fir trees (*Pseudotsuga menziesii*) growing in an area of hydrocarbon seepage. In contrast, Yang et al. (1999) found a shift in the red edge toward longer wavelengths (red shift) in wheat (*Triticum aestivum*) growing over hydrocarbon microseepage. Work by Noomen et al. (2006) and Noomen (2007) also noted a shift of reflectance to longer wavelengths in the region 400–750 nm for wheat and maize growing in ethane-charged soil. The largest shift occurred in the yellow region near 580 nm. Noomen et al. (2005) found that corn (maize, *Zea mays*) subjected to varying amounts of methane and ethane became stunted.

Maracci et al. (1991) found that the photosynthetic efficiency of maize plants could decrease while chlorophyll content remained the same. To test whether this was the case in the hydrocarbon-treated plants, the *physiological reflectance index* (PRI) was calculated:

$$PRI = \frac{[R(570) - R(531)]}{[R(570) + R(531)]}$$

where R is reflectance at 570 and 531 nm.

PRI correlates with photosynthetic efficiency (Gamon et al., 1992). An increasing PRI indicates decreasing photosynthetic activity (Strachan et al., 2002) as a result of water stress, and indeed plants growing in ethane-charged soils had the highest PRI. The hydrocarbon gases reduce water uptake by plants, which causes a higher reflectance in water absorption regions (Noomen et al., 2006). Presumably, the mechanism is displacement of soil oxygen and water by the hydrocarbon gases.

Noomen et al. (2006) expected that hydrocarbon-treated plants would be stressed and show lower photosynthetic efficiency. For ethane-treated plants, this was indeed the case (Figure 15.7). Natural gas (methane plus heavier hydrocarbon gases) and pure methane caused decreased photosynthetic efficiency, but because of the low concentrations, there was no statistically significant effect. This work showed that natural gas leakage might be detected using vegetation reflectance, and suggested that this is the case when concentration of gas in the soil is in the range 4–14%.

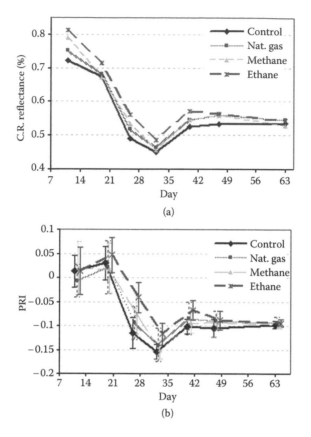

FIGURE 15.7

(a) Continuum removed (C.R.) reflectance at 580 nm over time. Except for the last date, the control plant reflectance is always lower than plants growing in natural gas and methane-charged soil, which in turn always have a lower reflectance than plants growing in ethane-charged soil. (b) The physiological reflectance index (PRI) over time. The maximum contrast between the control PRI and hydrocarbon treatments occurs at about 37 days after gas is introduced. Bars indicate the standard deviation. (From Noomen, M.F. et al., *Rem Sens Environ*, 105, 262–270, 2006; courtesy of Dr. Marleen Noomen.)

In later experiments, gas was delivered in large quantities, with the effect that the soil experienced oxygen shortage (Noomen, 2007; Noomen et al., 2008). The effect of oxygen deficiency (resulting from high concentrations of hydrocarbon gas) is easier to measure; any vegetation index should clearly indicate the difference between healthy and gassed vegetation (Noomen, email 1/8/2013). This work indicated that the effect should be detected in vegetation growing up to 0.5 m from the source.

The effect on agricultural vegetation of a pipeline leaking benzene in The Netherlands was investigated by van der Werff et al. (2008). Benzene condensate was leaking from pipeline joints at 9 m intervals. The buried pipeline crosses various agricultural fields. In June 2005, a HyMap airborne imaging spectrometer was flown at an altitude of 3000 m resulting in 4 m resolution. The sensor covers the spectral region from 436 nm to 2.485 μm with a bandwidth of 15–20 nm. A "red edge" image was generated showing the wavelength of the red edge (Figure 15.8). Although a shift toward shorter wavelengths can indicate vegetation stress, the variation in red edge position was greater between fields than within fields. Thus, the red edge position *per se* could not be related to hydrocarbon stress. A normalization process examined red edge values relative to neighboring pixel values to highlight pollution-related anomalies (Figure 15.9). Indications are that

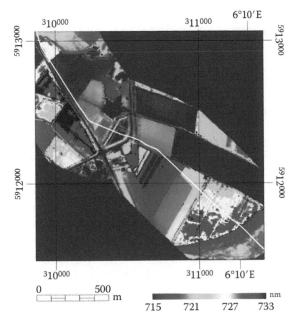

FIGURE 15.8

Background red edge values for each pixel. The values in each field are effectively homogeneous while boundaries between these fields are sharp. This shows that the average red edge value within a field can be calculated. (From van der Werff, H. et al., *Sensors*, 8, 3733–3743, 2008.)

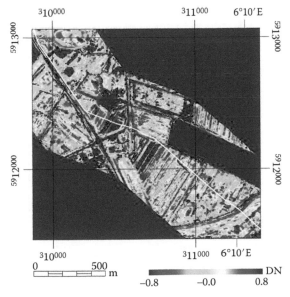

FIGURE 15.9

Normalized red edge values for each pixel. This image shows high-frequency changes within fields while contrasts between fields are minimized. The location of the pipeline is indicated by the white line. (From van der Werff, H. et al., *Sensors*, 8, 3733–3743, 2008.)

gaseous hydrocarbons leaking from a point source effect an area with a horizontal extent of roughly 4 m in sandy soils and 1 m in clayey soils. The influence of liquid hydrocarbons in a clayey soil was not expected to exceed 4 m from the pipeline. Image interpreters were asked to interpret the potential anomalies, resulting in a final classification. Comparison of the distribution of anomalies with geochemical data obtained by drilling showed that 8 of the 10 polluted sites were identified correctly (Figure 15.10).

Similar experiments using multispectral field radiometers and satellite sensors are being carried out on natural oil seeps in Slovakia and the Czech Republic (Smejkalová and Bujok, 2012). At the Pico anticline, near San Fernando, California, an oil, tar, and methane seep reportedly has harmful effects on orange and walnut groves, as well as on indigenous oak and chaparral.

It appears that some plants do well in the presence of hydrocarbons in soil. Riparian woodland vegetation (including mulefat, royal willow, *populus*) are healthy in and around the Pico anticline seep. At Lost River, field maples tolerate high soil manganese and methane, apparently because they can tolerate anaerobic root bacteria, and thus replace the natural oaks. The marsh plant *Scirpus olneyi* (a flowering sedge) not only grew well, but had enhanced survival in increasing concentrations of crude oil (Gebhart and Chabreck, 1975). A study by Warner (1997) determined that forest reflectance on early fall (September) Landsat TM imagery over the Volcano oil field in West Virginia actually increases in the near and mid-infrared range related to soil gas anomalies, suggesting that healthier plants are growing in the soil gas anomalies. Microseepage over an oil field 8 km north of Changyi in the Weibei area, southeast China, confirmed by soil gas measurements, causes the red edge to shift to longer wavelengths in crops of spring wheat (van der Meer et al.,

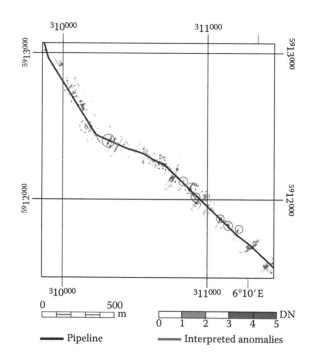

FIGURE 15.10
Weighted anomalies along the pipeline. White areas are where no anomaly has been found and are interpreted to represent the normal (background) state of vegetation. The pixels with yellow, orange, or red shade indicate areas that are interpreted as having vegetation stress. (From van der Werff, H. et al., *Sensors*, 8, 3733–3743, 2008.)

2000). This indicates stronger chlorophyll production and is interpreted as microseepage acting as a fertilizer for the wheat crop.

Whether a plant does well, just tolerates, or is harmed by hydrocarbons in the soil depends on the plant and on local soil and climatic conditions. When looking for vegetation anomalies, it is necessary to know what plants are natural in a given area, what environmental factors (moisture, nutrients, soil gasses, soil pH) they can tolerate, and what factors they cannot.

Vegetation as a Groundwater/Geothermal Resource Indicator

This topic is treated in greater depth in Chapter 12, Hydrology. Suffice it to say here that the presence or absence of groundwater in the near surface, and the degree of soil moisture is a significant factor in determining the type and vigor of vegetation cover developed on the surface. Some plants have tap roots well adapted to deep water tables, whereas others have shallow root systems and rely on shallow or perched water tables. Drought tolerant plant varieties will give way to phreatophytes when soil moisture is high enough. Healthy plants in moist soil appear bright green on true color imagery, or bright red on color infrared imagery (Figure 15.11). Moist soils host dense and diverse plant communities compared with dry soils. Near-surface groundwater in semi-arid to arid environments may be a result of fracturing of otherwise impermeable bedrock that allows water to percolate into the zone of broken rock and support a vibrant plant community. Under these conditions, the vigorous vegetation is an indicator of geologic structure. In other cases, near-surface water may be the result of a perched water table, suggesting that there are impermeable clay zones that trap water in the near surface where there would otherwise be only permeable sands or gravels.

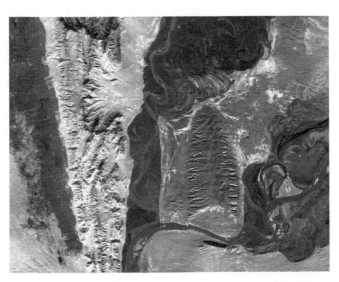

FIGURE 15.11
Color infrared RapidEye image of the Afghanistan–Tajikistan border area acquired October 16, 2011. Note the intense red color of the cultivated areas and phreatophyte vegetation along the rivers relative to the arid surrounding areas. Note as well the breached, north-plunging anticline at the junction of the Vakhsh and Amu Darya rivers (bottom center). (Data from RapidEye, Copyright © RapidEye.)

Likewise, there are vegetation clues to geothermal resources that range from clay-tolerant species to plant stress and die-offs due to high heat flow or CO_2 flux in the soil. All these plant indicators are detected primarily in the visible and near-infrared by multispectral or hyperspectral instruments, although some textural/dielectric information may also be available from microwave and radar sensors.

References

Almeida, T.I.R., C.R. deS. Filho, C. Juliani, F.C. Branco. 2009. Application of remote sensing to geobotany to detect hydrothermal alteration facies in epithermal high-sulfidation gold deposits in the Amazon region. In: Bedell, R., Crósta, A.P., Grunsky, E. (eds.), *SEG Reviews in Economic Geology*, Vol 16, Chapter 11: Tulsa, OK: Society of Exploration Geophysicists: 135–142.

Ayeni, O.O., P.A. Ndakidemi, R.G. Snyman, J.P. Odendaal. 2010. Chemical, biological and physiological indicators of metal pollution in wetlands. *Sci. Res. Essays.* 5: 1938–1949.

Bammel, B.H., R.W. Birnie. 1994. Spectral reflectance response of big sagebrush to hydrocarbon-induced stress in the Bighorn basin, Wyoming. *Photogramm. Eng. Rem. Sens.* 60: 87–96.

Beswick, A.E., P.J. Beckett, G.M. Courtin, G.O. Tapper. 1991. Evaluation of geobotanical remote sensing as an aid to mineral exploration in northeastern Ontario. *Ontario Geological Survey Open File Rept.* 5757: 22 p.

Canney, F.C. 1969. Remote detection of geochemical soil anomalies. In *Second Annual Earth Resources Aircraft Program Status Review 1*. Houston, TX: NASA: 7-1–7-8.

Cannon, H.L. 1960. Botanical prospecting for ore deposits. *Science.* 132: 591–598.

Cannon, H.L. 1971. The use of plant indicators in ground water surveys, geologic mapping, and mineral exploration. *Taxon.* 20: 227–256.

Carlisle, D., G.B. Cleveland. 1958. Plants as a guide to mineralization. *Calif. Div. Mines Spec Rept.* 50: 31 p.

Cracknell, A.P., A.K. Saraf. 1988. Improvement of geological mapping in north-west Scotland. Proc. 16th Intl. Congress of Photogrammetry and Remote Sensing. *ISPRS.* 27: 132–141.

Crawford, M.F. 1986. Preliminary evaluation of remote sensing data for detection of vegetation stress related to hydrocarbon microseepage: Mist gas field, Oregon. In *Proceedings of the 5th Thematic Conference, Remote Sensing for exploration geology*. Reno: ERIM: 161–177.

Crippen, R.E., Blom, R.G. 2001. Unveiling the lithology of vegetated terrains in remotely sensed imagery. *Photogramm. Eng. Rem. Sens.* 67: 935–943.

Davis, S.H. Jr. 1977. The effect of natural gas on trees and other vegetation. *J. Arbor.* 3: 153–154.

de Morais, M.C., P.P. Martins, Jr., W.R. Paradella. 2011. Mapping iron-mineralized laterite environments based on textural attributes from MAPSAR image simulation—SAR-R99B (SIVAM/SIPAM) in the Amazon region. *Rev. Braz. Geofís.* 29: 99–111.

Flower, F.B., E.F. Gilman, I.A. Leone. 1981. Landfill gas, what it does to trees and how its injurious effects may be prevented? *J. Arbor.* 7: 43–52.

Gamon, J.A., J. Penuelas, C.B. Field. 1992. A narrow-waveband spectral index that tracks diurnal changes in photosynthetic efficiency. *Rem. Sens. Environ.* 41: 35–44.

Gebhart, J.L., R.H. Chabreck. 1975. Effects of various levels of crude oil on Olney bulrush (*Scirpus olneyi*) and marsh hay cordgrass (*Spartina patens*). In *Proceedings of the 29th Annual Conference on SE Association of Game and Fish Comm*, St. Louis: 567–577.

Harris, J.R., J. Peter, L. Wickert, P.H. White, M. Maloley, R. Gauthier, P. Budkewitsch. 2012. *Hyperspectral remote sensing for geological mapping and mineral exploration—a review of activities at NRCAN*. 12th International Circumpolar remote sensing symposium, May 14–18, 2012, Levi, Finland. http://alaska.usgs.gov/science/geography/CRSS2012/abstracts2012.pdf (accessed October 26, 2012).

Jacobson, S., S. Khan. 2006. Identifying Surface Alterations Caused by Hydrocarbon Microseepages in the Patrick Draw Area of Southwest Wyoming, Using Image Spectroscopy and Hyperspectral Remote Sensing (abs). *AAPG Search Discov.*, #90052: 1 p.

Jones, H.G., P. Schofield. 2008. Thermal and other remote sensing of plant stress. *Gen. Appl. Plant Physiol. Spec. Issue.* 34: 19–32.

Kettles, I.M., A.N. Rencz, S.D. Bauke. 2000. Integrating Landsat geologic, and airborne gamma ray data as an aid to surficial geology mapping and mineral exploration in the Manitouwadge area, Ontario: Photogram. *Eng. Rem. Sens.* 66: 437–445.

Låg, J., B. Bølviken. 1974. Some naturally heavy-metal poisoned areas of interest in prospecting: soil chemistry, and geomedicine. *Nor. Geol. Unders. Bull.* 304: 73–96.

Maracci, G., G. Schmuck, B. Hosgood, G. Andreoli. 1991. Interpretation of reflectance spectra by plant physiological parameters. In *Proceedings of the International Geoscience and Remote Sensing Symposium (IGARSS '91), Remote Sensing: Global Monitoring for Earth Management.* New York: IEEE: 2303–2306.

Németh, N., G. Pethő. 2009. Geological mapping by geobotanical and geophysical means: a case study from the Bükk Mountains (NE Hungary). *Cent. Eur. J. Geosci.* 1: 84–94.

Noomen, M.F., A.K. Skidmore, F.D. van der Meer, H.H.T. Prins. 2006. Continuum removed band depth analysis for detecting the effects of natural gas, methane, and ethane on maize reflectance. *Rem. Sens. Environ.* 105: 262–270.

Noomen, M.F., F.D. van der Meer, A.K. Skidmore. 2005. Hyperspectral remote sensing for detecting the effects of three hydrocarbon gases on maize reflectance. In Proc. Univ. Utrecht, United Academics Publishers. Utrecht, The Netherlands: 4 p. http://www.cpnt.ru/userfiles/Hyperspectral_remote_sensing_for_detecting_three_HC_gases.pdf (accessed October 26, 2012).

Noomen, M.F. 2007. Hyperspectral reflectance of vegetation affected by underground hydrocarbon gas seepage. ITC Dissertation 145, International Institute for Geo-information Science & Earth Observation, Enschede: Netherlands: 136 p.

Noomen, M.F., K.L. Smith, J.J. Colls, M.D. Steven, A.K. Skidmore, F.D. Van der Meer. 2008. Predicting soil oxygen concentrations using indices based on hyperspectral reflectance of maize and wheat canopies. *Int. J. Remote Sens.* 29: 5987–6008.

Oze, C.J., M.J. LaForce, C.M. Wentworth, R.T. Hanson, D.K. Bird, R.G. Coleman. 2003. Chromium geochemistry of serpentinous sediment in the Willow core, Santa Clara County, CA. *USGS. Open-File Rept.* 03–251: 9 p.

Parrish, J.B., B.N. Rock. 1983. The effect of soil methane on in vitro growth and vigor of *Quercus prinus* seedlings. *Botan. Soc. Am.* 70: 52.

Perrotta, M.M., T.I.R. Almeida, J.B.F. Andrade, C.R. Souza Filho, G.J. Rizzotto, M.G.M. Santos. 2008. Remote sensing geobotany and airborne gamma-ray data applied to geological mapping within Terra Firme Brazilian Amazon forest: a comparative study in the Guapore Valley (Mato Grosso State, Brazil). *Intl. Arch. Photogram. Rem. Sens. Spat. Inform. Sciences.* XXXVII part B8, Beijing: 1275–1280.

Richers, D.M., R.J. Reed, K.C. Horstman, G.D. Michels, R.N. Baker, L. Lundell, R.W. Marrs. 1982. Landsat and Soil-Gas Geochemical Study of Patrick Draw Oil Field, Sweetwater County, Wyoming. *Bull. Am. Assn. Pet. Geol.* 66: 903–922.

Rock, B.N. 1983. Vegetation at Lost River, West Virginia. in Fourth GEOSAT Workshop. Flagstaff.

Rock, B.N. 1984. Remote detection of geobotanical anomalies associated with hydrocarbon microseepage. In *Proceedings of the International Symposium on Remote Sensing of Environment.* Ann Arbor, MI: ERIM, 16–19 April 1984, Colorado Springs: 183–195.

Schoenherr, A.A. 1992. *A Natural History of California.* Berkeley, CA: University of California Press: 100–101.

Siegal, F.R. 1974. *Applied Geochemistry.* New York: John Wiley and Sons: 353 p.

Siegal, B.S., A.F.H. Goetz. 1977. Effect of vegetation on rock and soil type discrimination. *Photogram. Eng. Rem. Sens.* 43: 191–196.

Singhroy, V., P. Barnett. 2003. Satellite based standardized terrain maps: a case study. In D.T. Hansen, V.H. Singhroy, R.R. Pierce, A.E. Johnson (eds.), *Spatial Methods for Solution of Environmental and Hydrologic Problems—Science, Policy, and Standardization* (STP 1420). West Conshohoken, PA: ASTM: 148–160.

Singhroy, V., R. Saint-Jean, J. Levesque, P. Barnett. 2000. Reflectance spectra of the boreal forest over mineralized sites. In *Geoscience and Remote Sensing Symposium, 2000. IGARSS 2000. IEEE 2000 International* v. 4. New York, NY: IEEE: 1379–1381.

Smejkalová, E., P. Bujok. 2012. Remote sensing methods in the identification of oil contaminations. *GeoSci. Eng.* LVIII: 24–33.

Smith, K.L., M.D. Steven, J.J. Colls. 2004. Spectral responses of pot-grown plants to displacement of soil oxygen. *Int. J. Rem. Sens.* 25: 4395–4410.

Strachan, I.B., E. Pattey, J.P. Boisvert. 2002. Impact of nitrogen and environmental conditions on corn as detected by hyperspectral reflectance. *Rem. Sens. Environ.* 80: 213–224.

van der Meer, F., P. van Dijk, S. Kroonenberg, Y. Hong, H. Lang. 2000. *Hyperspectral hydrocarbon microseepage detection and monitoring: potentials and limitations.* 2nd EARsel Workshop on Imaging Spectroscopy, Enschede: 9 p.

van der Werff, H., M. van der Meijde, F. Jansma, F. van der Meer, G.J. Groothuis. 2008. A spatial-spectral approach for visualization of vegetation stress resulting from pipeline leakage. *Sensors.* 8: 3733–3743.

Warner, T. A. 1997. Geobotanical and lineament analysis of Landsat satellite imagery for hydrocarbon microseeps. Natural Gas Conference, March 24–27, 1997, Houston, Texas; http://www .fetc.doe.gov/publications/proceedings/97ng/ng_pro.html (accessed October 26, 2012)

Yang, H., J. Zhang, F. van der Meer, S.B. Kroonenberg. 1999. Spectral characteristics of wheat associated with hydrocarbon microseepages. *Int. J. Remote. Sens.* 20: 807–813.

Yost, E., S. Wenderoth. 1971. The reflectance spectra of mineralized trees. *Proceedings of the Seventh International Symposium on Remote Sensing of Environment, v. 1.* Ann Arbor, MI: ERIM: 269–284.

Yu, L., A. Porwal, E.J. Holden, M. Dentith. 2009. Suppression of vegetation in LANDSAT ETM+ remote sensing images. *Geophys. Res. Abstr.* 12, EGU2010-510, 2010: 1 p.

Additional Reading

Brooks, R.R. 1983. *Biological Methods of Prospecting for Minerals.* New York: John Wiley & Sons: 322 p.

Collins, W., S.H. Chang, G. Raines, F. Canney, R. Ashley. 1983. Airborne biogeophysical mapping of hidden mineral deposits. *Econ. Geol.* 78: 737–749.

Sabins, F.F. 1999. Remote sensing for mineral exploration. *Ore. Geol. Rev.* 14: 157–183.

16

Environmental Hazards and Global Warming

Chapter Overview

On the morning of August 29, 2005, Hurricane Katrina made landfall at Grand Isle, LA. Wind speeds were measured at 200 km/h (125 mph, a strong category 3), with gusts up to 224 km/h (140 mph). Rainfall exceeded 19–24 cm (8–10 in.) along much of the hurricane's path, causing extensive flooding. Katrina also spawned tornadoes, causing further damage. At least 1833 people died in the hurricane and subsequent flooding. Total property damage is estimated at $81 billion (2005 USD).

The loss of life and property damage was worsened by breaks in the levees protecting New Orleans. At least 80% of New Orleans was under up to 6 m of water on August 31, 2005. The levee failures were considered the worst civil engineering disaster in U.S. history, with responsibility for the failures and flooding laid squarely on the Army Corps of Engineers by the U.S. District Court.

Before it was over, more than 1.7 million people had lost power in the Gulf Coast states as a result of the storm. Drinking water was unavailable in New Orleans due to a broken water main. Disruption of the oil industry in the Gulf of Mexico provided a shock to the nation's economy. The Mineral Management Service estimated that oil production in the Gulf of Mexico was reduced 95% (by 1.4 million barrels per day). Gasoline reached record highs on August 30, with concerns over refinery capacity driving the increase (NOAA, online; Blake and Gibney, 2011).

Much of what we know about Hurricane Katrina comes from remote sensing data. Satellites imaged the clouds, measured wind speed, rainfall intensity, sea surface temperature, and wave height, and monitored subsequent flooding (Figures 16.1 through 16.3). Whether a result of geology (subsiding delta, low-lying coastal plain), engineered structures (levees), or climate change (ocean temperature, sea level rise, loss of wetlands), the dangers could be foreseen using integrated remote sensing analysis. Integration of several systems, including weather and land monitoring satellites, was able to detect the storm, project its path, issue early warnings, and monitor its progress. In the event, preparations failed to prevent this disaster, but remote sensing was able to monitor the impacts and assist in the response (Ghosh et al., 2007; Klemas, 2009).

Predicting and accommodating environmental hazards is a key role of integrated remote sensing analysis. Building a nuclear power plant on an active fault or building an oil pipeline across an active or a potential landslide are examples of the types of activities that are all too common and yet could be averted with the careful use of remote sensing imagery as a first step in a hazard assessment. The construction of roads and facilities within river floodplains costs industry billions of dollars each year. Drilling wells or building roads on landslides, and building facilities or pipelines across faults or in areas of active subsidence

FIGURE 16.1
NOAA image of Hurricane Katrina, August 28, 2005. http://www.ncdc.noaa.gov/special-reports/katrina.html.

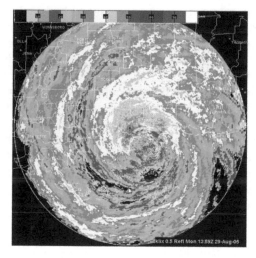

FIGURE 16.2
Radar image showing heaviest precipitation from Hurricane Katrina at landfall. (Courtesy of NOAA. http://www.ncdc.noaa.gov/special-reports/katrina.html.)

are also costly and often avoidable. Likewise, flood control can be assisted by monitoring snowpack and the historical extent of flooding. Both floods and normal runoff can cause erosion, as can wave action and storm tides along coastlines. The progress of erosion, which can cost millions of dollars in property damage annually, can be observed using multitemporal remote sensing. Global warming is revealed through changing coastlines, the retreat of ice caps, decreasing sea ice, desertification, forest fires, and the progress of the pine bark beetle.

Detecting environmental or geologic hazards is straightforward: telltale indicators can be seen on many types of aircraft or satellite imagery, including black-and-white (B/W), color, and color-infrared (CIR) film and imagery, multispectral and hyperspectral data, and radar and thermal images.

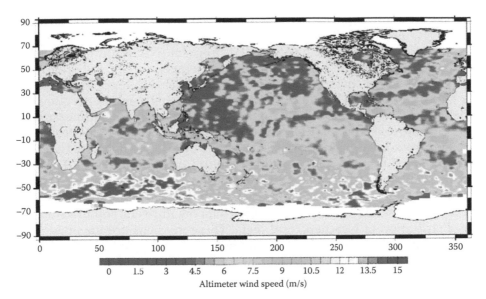

FIGURE 16.3
Global ocean wind speeds as determined by the Jason-2 altimeter for July–Aug, 2008. (Image courtesy of NASA.)

Monitoring the Effects of Global Warming

Remote sensing is particularly well suited for monitoring the effects of global warming (sometimes referred to ambiguously as "climate change"). Earth scientists recognize that the Earth has been warming since the end of the last ice age approximately 10,000–12,000 years ago, and warming may have accelerated with the introduction of industrial CO_2 in the past 100 years. In any case, remote sensing has a role in monitoring changes related to global warming.

Global change mapping must be "grounded" by geographic locations and scales, and have a time frame. Satellite systems provide consistent, continuous data for atmospheric, ocean, and land studies at a variety of spatial and temporal scales. From a geoscience perspective, the important aspects of global warming include the following:

1. Global temperature and its effect on land cover
2. Shifts in ecosystems
3. Changing land use
4. Global moisture distribution and its effect on precipitation, water supplies, flooding, and erosion
5. Changes in ice sheets, sea level, and ocean circulation
6. Changes in ocean temperatures and the effect on local weather, storm systems, and ecosystems
7. Changing global concentration and distribution of atmospheric gases and particulates

The European Space Agency has a Climate Change Initiative dedicated to "provide an adequate, comprehensive, and timely response to the requirement for ... long-term satellite-based products for climate. The initiative ... is based on the delivery of climate variables derived from satellite data and includes data acquisition, calibration and validation..." (ESA, online). NASA has a similar program (NASA, online) that "provides information on solar activity, sea level rise, the temperature of the atmosphere and the oceans, the state of the ozone layer, air pollution, and changes in sea ice and land ice." As of 2007, NASA/NOAA had 17 space missions collecting Earth climate-related data. Instruments on the Terra and Aqua satellites provide global measurements of atmospheric aerosols from natural sources such as volcanoes, dust storms, and forest fires, and from man-made sources such as the burning of fossil fuels. Instruments on the Aura satellite study processes that control ozone abundance. Data from the GRACE and ICESat missions and from spaceborne radar show changes in the Earth's ice sheets, while the Jason-1 and OSTM/Jason-2 missions record indications of sea level rise. Other satellites are available to monitor the flow in rivers, soil moisture changes, and the distribution of bleached coral reefs. These missions are intended to monitor and measure the following parameters and how they change over time. They are considered essential to understanding global warming and are needed as inputs to climate models.

- Cloud properties (height, water and ice content, temperatures)
- Carbon dioxide, methane, water vapor, and other greenhouse gases (Figure 16.4)
- Ozone, nitrous oxides, and chlorofluorocarbons (Figure 16.5)

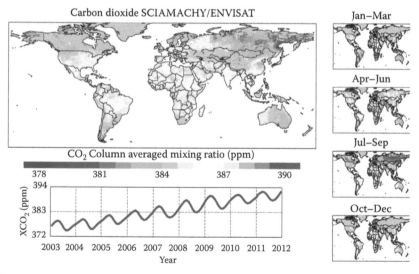

FIGURE 16.4

Global CO_2 concentrations averaged over 2006 to 2011 as measured by the SCIAMACHY instrument on ENVISAT. Maps on the right show the seasonal variations. CO_2 is low (blue) during the summer months at mid and high latitudes over the northern hemisphere as this is the growing season when plants take up atmospheric CO_2, thereby reducing the amount in the atmosphere. CO_2 varies during the year with a maximum around April and a minimum around August. This is more clearly shown in the red curve at the bottom, which also shows that atmospheric CO_2 is increasing by about 1.8 ppm per year. (Courtesy of ESA, DLR, and Dr. M. Buchwitz, Univ. Bremen. http://www.iup.uni-bremen.de/sciamachy/NIR_NADIR_WFM_DOAS/wfmd_image_gallery_co2.html.)

- Aerosols and forest fire smoke (Figure 3.34a)
- Sea surface temperature, surface roughness, salinity, ocean color, phytoplankton, and currents (Figure 16.6)
- Land surface temperatures
- Ocean wind speed and direction (Figures 3.63 and 16.3)

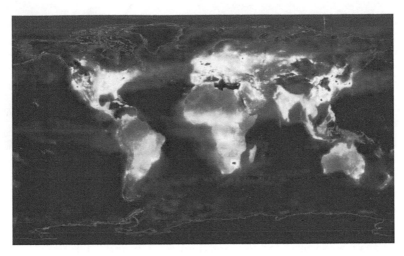

FIGURE 16.5
Global mean NO_2 concentrations for 2008 as measured by SCIAMACHY. High concentrations are shown by warm colors (reds, yellows) that appear to cluster in industrial areas. (Courtesy of ESA.)

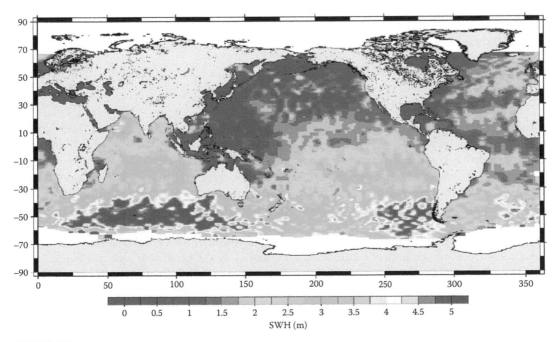

FIGURE 16.6
Jason-2 altimeter cycle 3 (2008-07-31 to 2008-08-10) maps of significant wave heights (0–5 m). (Image courtesy of NASA, CNES/CLS, and EUMETSAT. http://www.aviso.oceanobs.com/en/news-storage/news-detail .html?tx_ttnews%5Btt_news%5D=320&cHash=8752d8a4d1343572edffe94bf88b9dca.)

FIGURE 16.7
Global land cover between December 2004 and June 2006 as determined by the MERIS instrument on ENVISAT. (Courtesy of ESA.)

- Sea level changes
- Extent of sea ice, glaciers, and ice caps (Figures 3.28 and 3.43)
- Land cover and deforestation (Figure 16.7)

In addition to general Earth observation satellites with visible-near IR imaging systems, Table 14.2 lists some satellites that monitor greenhouse gases. Table 16.1 provides a partial list of satellite programs launched by international space agencies to detect and measure key global climate parameters.

Landslides and Avalanches

Landslides are among the most costly and damaging of natural hazards, both in the destruction of life and property. They damage infrastructure (powerlines, phone lines and cables, highways, railroads) as well as housing.

Hazard assessment can be performed using geomorphic techniques as well as statistical methods that analyze multiple variables such as proximity to active faulting, slope azimuth and steepness, vegetation cover, soil and rock type, dip of bedding, rainfall amounts, snowmelt, forest and grassfire history, and historical landslide patterns (Mantovani et al., 1996). For example, timing of grass senescence is found to be correlated with soil thickness variations on slide-prone hillsides: senescence is delayed over deeper, wetter soils that are likely debris flow source areas (McKean et al., 1991). Likewise, on forested slopes, slow slide movement disrupts timber canopy, exposes understory vegetation and soils, and alters a site's textural and spectral characteristics.

The primary sensors involved in monitoring slopes are airphotos, visible-near IR satellite images (detect scarps, hummocky topography), interferometric radar (detect downslope movement), and Lidar-derived digital elevation models (DEMs; provide slope inclination and azimuth). These instruments measure surface morphology, vegetation patterns, and changes in surface elevation.

TABLE 16.1

Global Warming-Related Missions

Mission	Agency	Launched	Instrument	Monitors
AcrimSat	NASA	1999	Active Cavity Irradiance Monitor	Sun's energy reaching Earth's atmosphere
ADEOS	JAXA/ NASA	1996	Scatterometer (NSCAT)	Ocean wind speed and direction
			Ocean Color Temperature Scanner	Ocean temperature
			POLDER	Ocean color, vegetation, water vapor
			ILAS	Measure trace gases
			RIS and IMG	Trace gases and greenhouse gases, respectively
ADM-AEOLUS	ESA	2013	Doppler Lidar	Wind dynamics
Aqua	NASA	2002	Atmospheric Infrared Sounder	Measure greenhouse gases, water vapor, air and surface temperatures
Aura	NASA	2004	Microwave Limb Sounder	Monitor thermal radiation from Earth's atmosphere; measure atmospheric gases, pressures. Understand ozone changes, pollution.
			Tropospheric Emission Spectrometer	Monitor ozone
CALIPSO	NASA	2006	CALIOP Lidar	Aerosols and clouds
			Imaging Infrared Radiometer	Cloud emissivity
CloudSat	NASA	2006	Radar	Internal structure of clouds; monitor tropical cyclones
CRYOSAT	ESA	2010	Interferometric Radar	Ice thickness
EarthCare	ESA/ JAXA	2016	Backscatter Lidar	Distribution of clouds, aerosols
			Cloud-profiling Radar	Distribution of clouds, aerosols
			Broadband Radiometer	Thermal radiation balance
ENVISAT	ESA	2002	AATSR	Sea surface temperature
			Microwave Radiometer (MWR)	Atmospheric water, soil moisture, surface temperature
			GOMOS	O_3, NO_2, O_2, H_2O, aerosols
			MIPAS	NO_2, NO, CH_4, O_3, H_2O, nitric acid
			SCIAMACHY	O_3, O_2, NO, CO, CO_2, CH_4, H_2O
			Radar altimeter	Changing sea level, sea ice monitoring
ERS-1 and ERS-2	ESA	1992, 1992	Active Microwave Instrument, Wind Scatterometer	Wind speed and direction; ocean surface state
			Radar Altimeter	Sea surface elevation; wave heights, sea ice
			Microwave Sounder	Sea temperature, cloud top temperature, atmospheric water vapor

(*Continued*)

TABLE 16.1 (*Continued*)

Global Warming-Related Missions

Mission	Agency	Launched	Instrument	Monitors
			Infrared Radiometer	Sea surface and cloud top temperature
			Global Ozone Monitor Experiment (GOME)	Measure O_3, NO_2, aerosols
GOCE	ESA	2009	Gravity Gradiometer	Ocean circulation
GOSAT	JAXA	2009	TANSO-FTS	Cloud and aerosol measurements
GRACE	NASA	2002	K band Microwave	Monitor water and ice sheets
Himawari 1 to 5	JAXA	1977–1995	Infrared Radiometer	Cloud motion, sea surface temperature
ICESat	NASA	2003	Geoscience Laser Altimeter System (GLAS)	Monitor clouds and aerosols
Jason-1	NASA/ CNES	2001	Altimeter, MWR	Ocean circulation, monitor El Niño, ocean-atmosphere ties
Jason-3	NASA/ CNES	2014	Altimeter, MWR	Sea level, ocean circulation
Midori-2	JAXA	2002	AMSR	Surface temperature
			ILAS-II	O_2, NO_2, N_2O, CH_4, H_2O, CFCs, aerosols, atmospheric temperature, pressure
			POLDER	Ocean color, vegetation, water vapor
			SeaWinds	Ocean wind speed and direction
OceanSat-2	ISRO	2009	Scanning Scatterometer	Monitor ocean winds
OCO-2	NASA	2014	Spectrometer	Measure CO_2
ODIN	ESA	2001	OSIRIS	Monitor O_3, NO, aerosols
			Sub-millimeter Radiometer	Measure atmospheric gases
OSTM/Jason-2	NASA/ CNES	2008	Altimeter, Advanced MWR	Ocean surface topography, circulation, monitor El Niño, ocean-atmosphere ties
PARASOL	CNES	2004	POLDER	Monitor clouds, aerosols
QuikScat	NASA	1999	Active Radar Scatterometer (SeaWinds)	Ocean winds, cloud cover, weather forecasting
Satélite de Aplicaciones Científicas	Argentine Space Agency	2011	Aquarius—SAC-D	Sea surface salinity
Soil Moisture Ocean Salinity (SMOS)	ESA	2009	Microwave Imaging Radiometer	Soil moisture, ocean salinity, currents
Soil Moisture Active Passive (SMAP)	NASA	2014	Advanced Radar and Radiometer	Soil moisture, freeze-thaw state, flood prediction, drought monitoring
Surface Water and Ocean Topography	NASA/ CNES	2019	KA-band Radar Interferometer, Nadir Altimeter	Monitor changes in surface waters
Terra	NASA	1999	Advanced Spaceborne Thermal Emission and Reflectance Radiometer (ASTER)	Earth surface temperature

TABLE 16.1 (*Continued*)

Global Warming-Related Missions

Mission	Agency	Launched	Instrument	Monitors
			Multiangle Imaging Spectroradiometer (MISR)	Atmospheric particulates, clouds, surface cover. Monitors hurricanes, floods, air pollution
TerraSAR-X	DLR	2007	X-band synthetic aperture radar (SAR)	Monitor ocean currents; create global digital elevation model
Topex/ Poseidon	NASA/ CNES	1992	Radar Altimeter, Microwave Altimeter	Ocean circulation

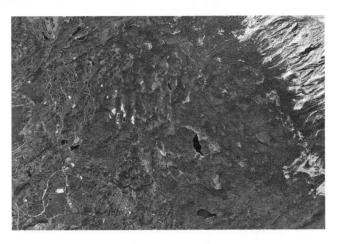

FIGURE 16.8
The Dugout Creek landslide south of Glenwood Springs, Colorado. This landslide is characterized by hummocky topography, numerous small lakes and poorly developed drainage, and narrowing of the East Muddy Creek valley along the toe of the slide. (Courtesy of U.S.D.A. Farm Service Agency, Digital Globe, and Google Earth, © 2012 Google.)

FIGURE 16.9
Landslide above Guinsaugon, the Philippines, 2006. This slide is characterized by abrupt scarps, light colored rubble, and an alluvial fan at the base. (Courtesy of NASA; M. Kennedy photo.)

Landslides are characterized by concave escarpments and generally lobe-shaped (in map view) hummocky topography (Figures 16.8 and 16.9). Trees are often knocked down or tilted at unnatural angles. Many forest-covered slides have groves of aspen or other colonizer trees surrounded by the old-growth forest. Valleys can be dammed behind old debris flows, forming lakes, and rivers will narrow where they are downcutting through old slides that have partially blocked their flow. The hummocky ground on landslides can have many small ponds with no well-developed drainage system. On gentle, grass-covered slopes that are water saturated and prone to slump, one often sees patterns of downslope soil creep revealed by ridges and depressions parallel to slope contours. As the surface particles creep downslope slightly faster than deeper soils, the tops of trees with deep root systems are seen to tilt slightly downslope.

Many of the nongeomorphic aspects of detecting and predicting landslides can be evaluated with geographic information systems that include a digital topographic database and surface materials database. Detailed stereo airphotos at scales from 1:15,000 are most useful for mapping landslide features, although satellite imagery has also been used. Global positioning systems can be used in conjunction with a geographic information system to monitor movements, as can interferometric radar imagery. A landslide hazard map should contain information such as slope stability, bedrock versus soils and unconsolidated units, past slides and slumps, the most likely areas of failure, and the probable areas downslope that would be affected.

Aerial imaging for pipeline management has been used for decades (Jadkowski, 1994). An example of mapping geohazards related to pipeline right-of-ways was provided by Henschel et al. (2012). They used interferometric synthetic aperture radar (InSAR) along with automated amplitude change detection and polarimetric change detection to determine where the landscape had been altered. The InSAR measures local ground movement. Polarimetric radar determines where the ground cover has changed. A scattering model is calculated for each pixel that reveals changes in the type of object providing the radar return. This is usually vegetation change, but can also be man-made changes such as cut-and-fill, roads, and so forth. In automated change detection, two or more images are compared for changing amplitude of the backscatter on the order of tens of meters. These methods are meant to cover large areas quickly and inexpensively and to predict where hazards exist before they become problematic.

Avalanche chutes are easily seen on moderate- to fine-resolution imagery. They are funnel shaped and characterized by steep slopes with rock swept clean of vegetation at higher levels and by bent, broken, or young trees and colonizer vegetation (willows, berry bushes,

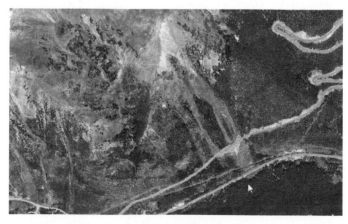

FIGURE 16.10
Avalanche chute (arrow) southwest of Berthoud Pass, Colorado. Note the bright green colonizer vegetation in the lower parts of the chute. (True color image courtesy of Google Earth, © 2012 Google.)

aspen trees) at lower levels (Figure 16.10). Predicting whether an avalanche is imminent requires monitoring snowpack and weather patterns. The combination of heavy snowpack and alternating warm and cold weather (soft and hard snow) makes ideal conditions for avalanches.

Earthquakes and Tsunamis

Many regions of the world have large populations living in tectonically active settings. Certainly, this is true for the Pacific Rim; it is also true in areas considered tectonically quiet such as the coastal plain of Texas, which is prone to growth faulting. Sites subject to liquefaction during earthquakes are generally in flat, low-lying areas along a coastline or lake that contain either natural or man-made landfills and are water saturated (e.g., the margins of San Francisco Bay). Construction in places such as Kobe, Japan, the Los Angeles area, California, or Anchorage, Alaska, whether for roads and bridges or high-rise offices and refineries, must take into consideration earthquakes and faulting (Figure 16.11). Some areas have regulations concerning the siting of power plants that require, for example, that faults in the vicinity have not been active during the past 10,000 years or more (see Chapter 13, Kakrapar power plant site selection case history). This requires trenching and age dating the soils to determine the youngest strata cut by a fault. In the initial stages of an evaluation, however, it is convenient to use remote sensing imagery to map the indications of faulting in an area. There are a number of criteria used to recognize faults on imagery, including linear escarpments, offset drainage patterns, and abrupt changes in rock or soil types (see Chapter 5).

Radar interferometry (see Chapter 3) provides an almost real-time method for monitoring creep along active faults. Rosen et al. (1998) used ERS-1 radar images taken 14

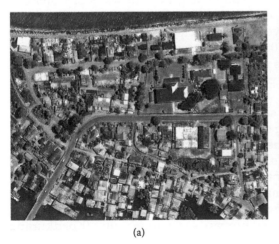 

(a)   (b)

FIGURE 16.11
Aerial views of coastal Banda Aceh, Indonesia, before and after the magnitude 9.2 Indian Ocean earthquake on December 26, 2004. (a) Before the tsunami. (b) After the earthquake and resulting tsunami devastated low-lying areas. Images such as these are not only dramatic, but can assist in determining the magnitude of the earthquake and help prioritize aid efforts. (Images courtesy of DigitalGlobe, Inc and GeoEye © 2012.)

months apart to map creep along the San Andreas fault near Parkfield, California. The radar observed the fault from roughly 800 km. Offset along a 170 km section of the fault was estimated from the imagery to be approximately 32 mm/year, decreasing to zero in the locked section southeast of Parkfield. Any change in motion from parallel to the fault would indicate an increase in strain across the fault. Although strain patterns related to a change from locked to creeping are not clear, the decrease in creep from northwest to southeast along the fault suggests a potentially unstable situation.

A similar study examined nearly instantaneous ground surface displacements near Kobe, Japan (Ehrismann et al., 1996). A large and destructive earthquake struck Kobe on January 17, 1995, resulting in over 5000 deaths and 40,000 injured. The Japanese National Institute for Earth Science and Disaster Prevention, which is responsible for investigating such disasters, analyzed the deformation to better prepare for future occurrences. Two JERS-1 synthetic aperture radar images collected one and a half years apart were used, along with a DEM, to measure surface deformation. The DEM was used to subtract topographic elevation information from the radar interferogram generated from the two images. The resulting differential interferogram revealed a maximum horizontal deformation of about 1.5 m toward the southwest as a result of movement along a northeast-southwest fault in the area. The maximum vertical displacement occurred at the northern tip of Awaji Island, where the surface moved up nine color cycles, or about 1 m (Figures 16.12 and 16.13).

The humanitarian response to the Haitian earthquake of January 12, 2010, made extensive use of remote sensing and geographic information systems (Hall-Beyer, 2010). In 1999, the European Space Agency initiated the International Charter on Space and Major Disasters. Most nations with space programs participated. Governments or relief agencies responding to a disaster can request imagery and interpreters trained to extract required information. The Charter was activated during the Haitian crisis, providing radar and high-resolution visible-near IR imagery. Among other things they were used to locate collapsed buildings and blocked access roads and to plan relief efforts.

Coastal communities should be concerned with the effects of tsunamis. Although less common than major earthquakes, these earthquake-generated walls of water can destroy property and kill large numbers of people. The December 26, 2004, Indian Ocean tsunami claimed about 300,000 lives. The March 11, 2011, Japan earthquake displaced 400,000 people, killed about 20,000, and destroyed the Fukushima Dai'ichi Nuclear Power Station. Remote sensing has a role in outlining areas in danger, predicting the magnitude of the tsunami, and in post-flood relief efforts and damage assessments (Figure 16.14).

People living in low-lying coastal areas should be aware of the risk of a tidal wave or tsunami. Depending on the angle of the coast with respect to the epicenter, and the distance, any areas near sea level could be in danger. Historical imagery should be reviewed for evidence of recent tsunamis (e.g., debris deposits at the high water mark), the extent of areas affected, and the amount and type of damage inflicted. When combined with a DEM and bathymetric information, the maximum area in danger can be defined and areas at risk can build warning towers, strengthen building codes, shield power systems, construct shelter buildings, and prepare evacuation routes (Koedkurang et al., 2005; Chrysoulakis et al., 2008; Dewi et al., 2010).

Near-real-time measurements from a global differential GPS network, combined with information on sea surface height from satellite altimetry, and the location and magnitude of an earthquake allows government agencies to estimate the amount of damage and where and when it will occur (Song, 2007; Wang and Li, 2008; Blewitt et al., 2008).

FIGURE 16.12
JERS-1 OPS color infrared image of Osaka and northeast Awaji Island, Japan, taken October 31, 1992. (Image processed by ERSDAC; courtesy of METI/Japan Space Systems.)

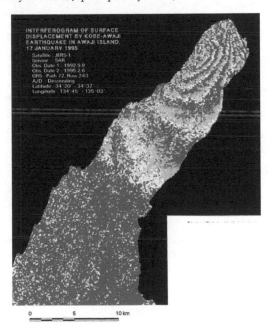

FIGURE 16.13
Radar interferogram derived from JERS-1 synthetic aperture radar (SAR) over northeast Awaji Island, Japan. Displacement is a result of the Hyogo-Nanbu earthquake of January 17, 1995. Vertical displacement of about nine color cycles represents approximately 1 m. (Image processed by ERSDAC; courtesy of METI/Japan Space Systems.)

In the aftermath of a tsunami, remote sensing helps with site and structural engineering impact assessment (Figure 16.14). High-resolution imagery was used to help coordinate the response by directing and prioritizing aid and reconstruction efforts in the 2010 Haiti earthquake and 2011 Japan earthquake (Messinger et al., 2010, 2011). Change detection algorithms highlight the tsunami-induced damage (Yamazaki et al., 2006; Roemer et al., 2010; Yoshikawa et al., 2012).

FIGURE 16.14

Digital Globe image of the Fukushima Daiichi atomic power plant, Japan, acquired March 14, 2011, showing earthquake and tsunami damage to the facility. (Image courtesy of DigitalGlobe, Inc and GeoEye © 2012.)

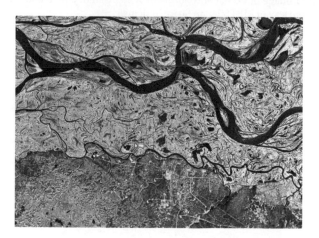

FIGURE 16.15

Image showing a section of the Ob River, Russia. The Priobskoye oil field extends into the floodplain (arrow) and wells must be drilled from gravel islands. The floodplain is revealed by low-lying areas with sandbars and lacking trees. (Image courtesy of TerraMetrics and Google Earth, © 2012 Google.)

Flooding

River floodplains are identified as those low-lying flat zones on either side of a river that extend outward until one encounters bluffs and higher ground (Figure 16.15). Floodplains contain swampy depressions and abandoned meanders known as oxbow lakes (Figure 16.16). Construction in these areas must account for both periodic flooding and unstable soil due to near-surface water tables. Roads and railroad grades may need to be elevated. Facilities may have to be raised on pilings, and oil wells may have to be placed on artificial gravel islands. Shifting of the river channel and erosion along the banks must be taken into consideration. In cool temperate and arctic regions, one must consider the effects of ice jam-induced flooding as well as the effects of river ice on pilings and abutments during spring breakup. Dams subject to failure must be inventoried, especially if failure and

FIGURE 16.16
Oxbow lake on the Yazoo River floodplain southwest of Yazoo, Mississippi. (True color image courtesy of USDA Farm Service Agency and Google Earth, © 2012 Google.)

subsequent flooding would lead to loss of life and property (Blystra and Dumas, 1980). Low-lying coastal areas may be exposed to storm surges and associated flooding.

The best way to determine the extent of areas affected by flooding is to look for indications of recent sandbars and/or a lack of mature shrubs and trees. Identification of zones affected by major floods requires examination of multitemporal imagery over a period of several years to decades. These images show those locations where channels are most prone to shifting, and provide an estimate of the direction and rate of channel migration.

Surface waters (reservoirs, lakes, rivers, streams) can be monitored for flood prediction and public safety, irrigation planning, fisheries, and for recreational activities. The main remote sensing contributions include monitoring snowpack, impending storms, and runoff from snowmelt and storms. Weather services use dedicated geostationary satellites like the NOAA, GOES, and Meteosat series. Sensors such as Hyperion on the EO-1 satellite provide near real time flood detection (Felipe et al., 2006). Automated spacecraft technology reduces the time to detect and react to flood events to a few hours. NASA and the U.S. Geological Survey are using satellite observations of rainfall, rivers and surface topography to create early warning systems (Brakenridge et al., 2006). Satellite microwave sensors can determine river flow by measuring changes in river widths and provide satellite-based estimates of rainfall as input to these warning systems. For example, RadarSat was used for near real-time monitoring of the 2009 flooding along the Red River in Minnesota, North Dakota, and Manitoba (Brennan, 2010).

Satellite imagery is particularly well suited for mapping the extent of snowpack over time (Figure 16.17). Airborne radio echo sounding (see Chapter 12) can determine snowpack thickness. If ground control points exist, it should be possible to use imagery to extrapolate snow thicknesses over the area of known snow cover. Additional information on slope steepness and azimuth (north-facing slopes retain more snow in the northern hemisphere) derived from a digital terrain model should further refine the calculations. After several years of imagery, a historical database can be compiled showing the volume of water associated with a given extent and thickness of snow. Flood control reservoir managers need this information to plan when to lower water levels in anticipation of higher than normal runoff. Farmers, particularly those cultivating rich bottom lands, want to know the chance of flooding prior to planting. Farmers dependent on irrigation want to know if water will be available in a given year. Downstream fisheries can use information on water salinity and turbidity related to higher or lower than usual river volume.

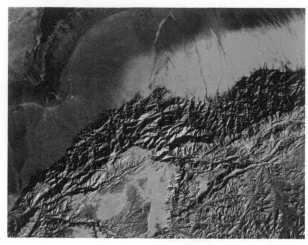

FIGURE 16.17
Snowpack is shown as blue on this Landsat TM 5-4-3 (RGB) image of the Altun Mountains, China. Snowpack can be used to predict runoff and flooding. (Image processed by EOSAT [now GeoEye/DigitalGlobe, Longmont, CO].)

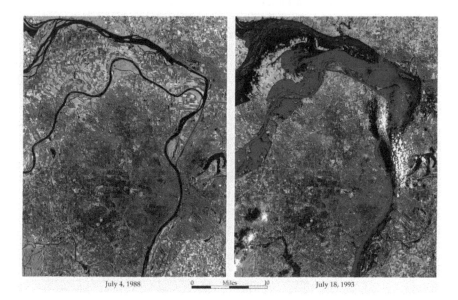

July 4, 1988   0   Miles   10   July 18, 1993

FIGURE 16.18
Landsat images of the St. Louis area showing the confluence of the Missouri and Mississippi rivers. The first, taken on July 4, 1988, shows normal flow; the second, from July 18, 1993, shows the rivers in flood stage. (Images processed by EOSAT [now GeoEye/DigitalGlobe].)

Late snows combined with long or heavy rains and rapidly warming temperatures can cause catastrophic runoff and flooding. Other floods are purely the result of long periods of intense rainfall. Monitoring floods may not do much for those affected by the flood-waters, but it can show which areas are prone to flooding (Figures 16.18 through 16.20). This allows those areas to be declared off limits for development, and permits prudent planning for development of the floodplain. Insurance companies are interested in the

FIGURE 16.19
Flooding at New Madrid, Missouri, 5th May, 2011. (False color image courtesy of DigitalGlobe, Inc and GeoEye © 2012.)

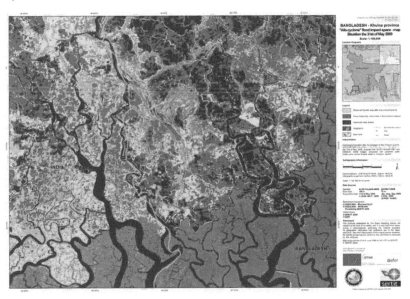

FIGURE 16.20
Flood impact maps such as this one for southwestern Bangladesh can aid in planning infrastructure and agriculture. Dark blue indicates permanent water bodies; progressively lighter colors indicate less chance of flooding. Generated from the PALSAR (Wide) instrument on the ALOS satellite. (Produced by JAXA/PALSAR; courtesy of USGS, SERTIT; copyright JAXA and METI. http://www.disasterscharter.org/web/charter/activation _details?p_r_p_1415474252_assetId=ACT-256.)

limits of flooding, so they can set flood insurance rates, and cities and county governments need this information to prepare storm sewers and drainage projects that will minimize damage and control runoff. In the United States, the National Flood Insurance Act of 1968 established the Federal Emergency Management Agency (FEMA) to manage a national flood insurance program. As part of the program, the act requires identification of all floodplains and flood-prone areas. FEMA has prepared a series of Flood Insurance Rate Maps that are used to establish premiums and to help with flood relief operations. Input to these maps comes from topographic maps and historical flood records. These are derived from DEMs and multitemporal remote sensing images (Grzeda, 1996). Agencies with similar agendas exist in most nations.

Erosion

Erosion is a function of several factors including precipitation, currents, slope, dip of strata, vegetation cover, surface materials, and frequency of storms and storm surges. Imagery, whether airphoto or satellite derived, can contribute to mapping erosion hazards. Digital terrain models provide a quantitative measure of slope, whereas image interpretation gives only a qualitative measure (i.e., a slope is more or less prone to fail). Mapping scours and cutbanks along rivers, and headlands and barrier bars along a coastline can also furnish information on where, historically, erosion has occurred. Effort should be taken not to build roads, dams, and other structures where they are likely to be undercut by running water or waves, where the slope is greater than the angle of repose for saturated soils, or where the dip of bedding is parallel to the slope.

An example of an erosion mapping project was described by Šúri et al. (2002) for the Slovak Republic. Both potential and actual soil erosion risks were evaluated on a regional scale (1:500,000) using the following five criteria: (1) erosional potential of rainfall, (2) potential soil erosion risk, (3) actual soil erosion risk, (4) slope, and (5) land cover. Remote sensing was able to contribute to mapping of soil erodabililty (soil texture and type), slope, and land cover (vegetation density, structures, soil conservation practices). Potential erosion risk indicates the tendency of an area toward erosion. Actual erosion risk takes into account land cover and conservation practices. They found that the loamy and silty soils developed on strongly dissected slopes of the Carpathians put 75% of the country in the high potential soil erosion risk category. The actual erosion risk in the forested mountains is lower due to the plant cover.

Another approach was used by Rao and Adinarayana (1994) to assess erosional potential in the Kukadi River watershed of Maharashtra, India. They determined that soil erosion is a function of three factors: (1) energy provided by falling rain, runoff, and topographic slope, (2) soil resistance, and (3) vegetation cover. Satellite imagery was used to map land use and modify drainage maps derived from topographic mapping. This information was then combined with soil maps derived from field observations and laboratory measurements to develop a database that describes the relative erosion status for different watersheds in a quantitative way.

Shoreline erosion has always been a concern of coastal communities. The kind of information required to understand and predict coastal erosion includes a knowledge of surface materials, elevation, vegetation cover, man-made cover, frequency and direction of storm attack, water circulation patterns, wave activity, and storm intensity (Figure 8.11). Some or all these features can be determined, at least in part, by remote monitoring (Welby, 1980). A study of coastal erosion was undertaken at Chesapeake Bay by the Maryland Geological Survey (Hennessee et al., 1997; Maryland Department of Natural Resources, 2000). They mapped the extent of shoreline change by examining historical maps and airphotos. Then, they calculated the rate of erosion along transects perpendicular to the coast.

They combined this information with sediment type and topographic relief to determine the volume of sediment eroded. This knowledge helped determine set-backs for construction and flood insurance. It also allowed them to calculate the volume of sediment available for silting up other areas of the bay, assess turbidity and nutrients in bay fisheries, and determine the factors that contribute the most to coastal erosion.

Volcanoes

There are parts of the world where volcanism has been and will continue to be a concern. Parts of Hawaii and Iceland, for example, have active lava flows that destroy property. The slopes of Mt. Rainier, in Washington State, are prone to mudflows when eruptions melt glaciers on the volcano. These mudflows threaten the suburbs of Seattle. The outskirts of Mexico City, like ancient Pompeii, are vulnerable to ash flows from nearby Mt. Popocatepetl. The eruption of Eyjafjallajökull in Iceland in 2010 shut down air traffic over Europe for a week.

Villages, camps, facilities, and mines should not be, but often are located on the slopes of active volcanoes or in areas where explosive eruptions are possible. They ought not be located in low-lying areas that are prone to volcanic eruptions or debris and mud flows. Such areas, and those affected by previous eruptions, can be readily identified on imagery. Steam eruptions and heightened seismic activity may be the best indications of an impending eruption, but it is also possible to monitor fissures and melting icepack using thermal imagery to detect increasing heat flow in the near surface (Figure 3.23). Radar interferometry has been used to map deformation (subsidence, doming) on and near volcanoes in an effort to understand the stress regime before and after eruptions. In a study of Mt. Etna, Briole et al. (1997) were able to map areas with a maximum subsidence of 47 mm/year and attributed this to a combination of compaction of recent lava flows and relaxation of the substrate up to 3.5 years after an eruption. In a similar study of the area around Ito City, Japan, swelling and uplift up to 4 cm is indicated by JERS-1 radar interferometry and is attributed to magma movement at depths between 5 and 10 km (Fujiwara et al., 1998). Wicks et al. (1998) used ERS-1 and ERS-2 SAR images to create six interferograms of the Yellowstone Caldera between 1992 and 1997. They were able to demonstrate as little as 20 mm inflation and 60 mm of subsidence, apparently moving between resurgent domes. They concluded that this indicates magmatic migration at depths of approximately 8 km beneath the caldera.

SO_2 flux rates may be a reliable predictor of imminent eruptions. Ultraviolet (UV) remote sensing cameras are being used to measure SO_2 flux from fumaroles on Volcano Island, Italy (Figure 16.21; Tamburello et al., 2011a,b,c). UV camera systems are also being used to monitor CO_2 plume fluxes at Stromboli volcano, Italy (Figure 3.62; Aiuppa et al., 2011a,b). Increasing CO_2 emissions have been shown to be precursors to explosive volcanism.

When El Chichón Volcano in Mexico erupted in 1982 imaging satellites detected the ash flow (Figure 8.15) and the Total Ozone Mapping Spectrometer (TOMS) on the Nimbus 7 satellite detected the SO_2 cloud emitted by the eruption. Other satellite instruments including the Ozone Monitoring Instrument (OMI), Atmospheric Infrared Sounder (AIRS), Advanced Spaceborne Thermal Emission and Reflection Radiometer (ASTER), Moderate Resolution Imaging Spectroradiometer (MODIS), and Global Ozone Monitoring Experiment (GOME)

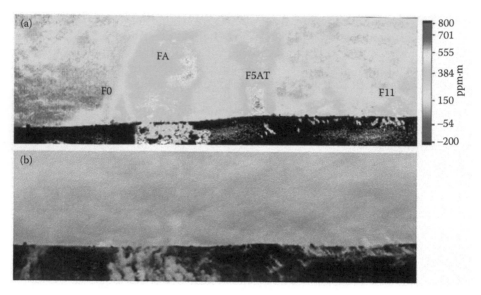

FIGURE 16.21
(a) UV SO_2 flux (slant column density, ppm m) from La Fossa crater, Volcano Island, measured remotely using a ground-based UV instrument. (b) Visible light photo of the same scene. (From Tamburello, G. et al., *J. Volcanol. Geotherm. Res.* 199, 47–52, 2011; courtesy of Dr. G. Tamburello, University of Palermo.)

FIGURE 16.22
SO_2 plume from Mt. Etna observed by the Atmospheric Infrared Sounder (AIRS) instrument on the AQUA satellite on October 28, 2002. The image clearly shows a sulfur dioxide plume (dark blue) emitted by Mt. Etna. The image was created by comparing data from two channels and creating an image that highlights the differences between them. (Courtesy of NASA.)

have been able to measure SO_2 and other gases emitted from volcanoes by using absorption bands in UV and IR wavelengths (Figure 16.22). The coarse resolution and long repeat times of satellite systems precludes detection of small SO_2 plumes or short-lived events. The Spinning Enhanced Visible and Infrared Imager (SEVERI) on Meteosat does provide SO_2 data every 15 minutes, but is limited to coverage of Europe and Africa (Nadeau, 2011).

Case History: Monitoring Volcanic Unrest in Kenya

The following work is being carried out (2012) by Elspeth Robertson at the Department of Earth Sciences, University of Bristol, Wills Memorial Building, Bristol, United Kingdom. The project described is a part of Robertson's PhD research entitled "Volcano Deformation in Kenya" under the supervision of Dr. Juliet Biggs (Bristol), Dr. Marie Edmonds (Cambridge), and Dr. Kathy Cashman (Bristol). The remote sensing and fieldwork component of this project has benefited from the collaboration and support of Charlotte Vye, Kay Smith, and Sue Loughlin from the British Geological Survey. Fieldwork has been funded by a Royal Geographical Society Geographical Fieldwork Grant and an NERC studentship.

In the Kenyan Rift, where volcanoes are numerous yet there are no documented historical eruptions, satellite observations have identified unrest at volcanic centers. Radar images reveal that shallow magma systems may be active under at least four of the volcanoes in Kenya, but whether the signals are driven by an influx of new magma remains to be determined.

Evidence for extensive magmatism is rife throughout the Rift Valley (Figure 16.23). Effusive lava flows cover the rift floor, and ash fall deposits from explosive eruptions cover great distances. Yet, in Kenya no eruptions have been recorded in human history. Consequently, no monitoring infrastructure is in place and the 15 Quaternary volcanoes along the rift axis remain understudied.

InSAR observations have recently revealed periods of ground deformation at four Kenyan volcanoes. Between 1997 and 2008, uplift of 9 and 21 cm occurred at Longonot and Paka, while subsidence was seen at Menengai and Suswa of 3 and 4.6 cm, respectively (Biggs et al., 2009). This discovery implies that the magmatic systems under these Kenyan volcanoes may be more active than previously thought. This work is focused on understanding the dynamics of these volcanoes, and more specifically, teasing apart the mechanisms of the observed ground deformation (Figure 16.24).

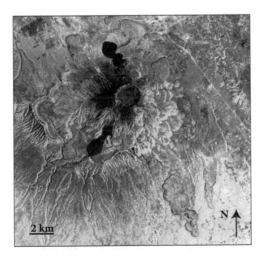

FIGURE 16.23
SPOT-5 color infrared image of Longonot volcano, Kenyan Rift, being used to produce a geological and volcanic hazards map. Band combination RGB 321; red color represents vegetation, predominately growing on the Longonot Ash deposit, and green colors indicate exposed trachyte deposits. The most recent lava flows, mentioned in the text, appear almost black, reaching northwest and southwest from the caldera. The morphological features of lava flows, particularly on the eastern flank, are clearly defined in the imagery. (Courtesy of Elspeth Robertson and Dr. Juliet Biggs, Department of Earth Sciences, University of Bristol.)

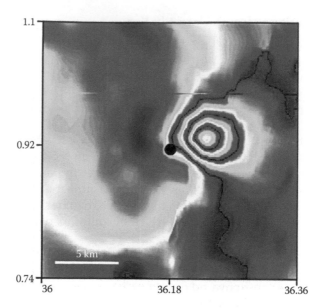

FIGURE 16.24
Radar interferogram of ground deformation observed at Paka volcano between July and August 2006. Paka is located in the central region of the Kenyan Rift. The caldera is represented by the central black dot and the fringes, each representing 2.83 cm of uplift, are centered on the northeastern flank of the volcano. The interferogram is processed using ENVISAT imagery with ROI_PAC software. (Courtesy of Elspeth Robertson and Dr. Juliet Biggs, Department of Earth Sciences, University of Bristol.)

Of the four volcanoes that have shown volcanic ground displacements, Longonot poses a large risk to Kenyan society and its economic exports, especially the flower farms that are located 10 km from the summit. Longonot is a small stratovolcano 70 km northwest from Nairobi and just south of Lake Naivasha. The volcanic flanks are piled high with viscous lava flows and the surrounding area is covered in an ash fall deposit, the Longonot Ash, dating back to 3200 years ago. The explosive eruption that produced this ash fall was likely concurrent with the formation of the circular pit caldera seen today (Figure 16.25). Distinctive black lava flows, clearly overlying the ash deposits, mark the Longonot's last eruptive activity. Although the lava flows have not been dated, they are estimated to be less than 1000 years old.

The Longonot Ash likely resulted from an explosive Plinian event (Pyle, 1999) producing a plume that distributed ash beyond the neighboring volcano, Olkaria. Studies on Longonot concentrated on the geochemistry of the lavas, and very little is known about the emplacement processes and the physical detail of recent eruptives, both ash fall and lavas. A detailed analysis of the deposits can be used to reconstruct past eruptions. For example, by mapping the Longonot Ash, we can determine physical parameters such as eruptive volume, eruptive mass, plume height and a mass discharge rate.

Understanding the physical volcanology of the ash fall and effusive deposits provides a constraint on the magmatic conditions that can lead to an eruption. To investigate this, we use a combination of remote sensing mapping techniques and fieldwork. The optical imagery over Kenya is spectacular, largely due to the dry climate and lack of cloud cover. We are creating a detailed geological map of the deposits using SPOT 5 and ASTER imagery, and are using spectral techniques to discriminate between the chemical compositions of the deposits.

FIGURE 16.25
Panoramic view of Longonot's caldera looking southeast. Measuring 2 km in diameter at 2776 m above sea level, the caldera has steep sided walls that lead into the flat-bottomed caldera floor that is covered in vegetation. (Courtesy of James Hammond, Imperial College, London.)

As always, a geological map requires field work. Two of the volcanoes, Longonot and Suswa, were visited during a reconnaissance trip. Both have shown ground deformation in the last decade. A longer field season is planned. During the reconnaissance program, rocks were sampled, fumaroles were identified and their temperatures were measured. Comprehensive sampling is anticipated during future field seasons, as well as mapping the Longonot Ash deposit and collecting rock samples to validate the geological map created from satellite images. The preliminary findings will guide a soil CO_2 survey at Longonot. A diffuse survey will be conducted to map the overall pattern of degassing at the volcano, followed by a more focused survey to accurately quantify fluxes around areas such as faults and fractures. The pattern of soil degassing will provide clues to understanding the underlying magmatic system.

Subsidence and Sinkholes

Subsidence can be natural or culturally induced. In low-lying carbonate platforms, such as southern Florida, it is important to know where karst solution collapse has occurred and will occur (Figure 16.26). Karst collapse features tend to form along fracture zones due to groundwater movement and dissolution of the carbonates along those zones (Casper et al., 1980). For this reason, they often form narrow, linear caverns and subsequent collapse areas. These depressions often have better-developed soils and near-surface groundwater, making them ideal areas for cultivation. They can sometimes be detected prior to collapse by using thermal surveys to map areas where cool caverns within a few meters of the surface appear as thermal anomalies (Marino et al., 1984). Not only are caverns usually cooler than the surrounding rock, but their temperature tends to be constant whereas the surrounding rocks and soils will warm up during the day and cool off at night.

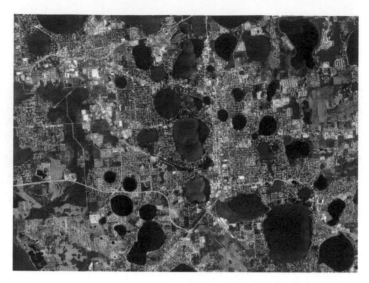

FIGURE 16.26
Sinkholes, mainly water-filled, in and around Winter Haven, Florida. (True color imagery courtesy of Google Earth, © 2012 Google.)

Subsidence occurs when oil (e.g., San Pedro, California) or groundwater (e.g., south of Houston, Texas) is withdrawn from underground reservoirs. In such cases, structural foundations will tilt or crack, and surface water can encroach on part of the area.

Abandoned underground coal mines occur in areas such as the Silesian basin of Poland or along the foothills south and west of Denver, Colorado (Turney, 1985). In these areas, it is important to know where coal mine collapse has occurred, and where it may occur in the future. It is possible to map areas that have subsided in the past, but is nearly impossible to predict those areas that will subside. Subsidence is indicated by topographically low areas distinct from natural drainage patterns, often containing ponds or indications of mining activity such as tailing piles or rectilinear depressions that contradict the natural slope.

Case History: Subsidence over Abandoned Coal Mines, Colorado Springs, Colorado

The U.S. Bureau of Mines has been interested in mapping subsidence over abandoned mines since at least the early 1980s (Elifrits and Barr, 1982; Dames and Moore, 1985; Amundson et al., 2009). Subsidence over mines can occur during mining, or may not happen for decades after a mine has been abandoned. An area of abandoned coal mines in northern and northeastern Colorado Springs, Colorado, was investigated by Doug Peters et al. (1996) using remote sensing techniques, primarily lineament analysis of satellite images and airphotos. The subsidence hazard was mapped by interpreting the surface expression of geologic structures, chiefly faults and joint systems. It is proposed that subsidence should occur where the mine roof is weakened by fractures. It is also possible to map actual subsidence using large-scale stereo airphotos and stereoplotters, airborne laser profiling, or radar interferometry. The U.S. Bureau of Mines has investigated locating underground mines using ground-based resistivity, where the depth of penetration is a function of electrode spacing and resistivity contrast is a function of the size of the void and whether it is filled with air or water (Burdick et al., 1986). This application may be adaptable to airborne resistivity surveys.

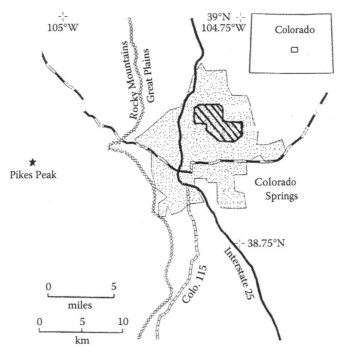

FIGURE 16.27

Location map showing area affected by subsidence (crosshatched) over abandoned coal mines, Colorado Springs, Colorado. (From Peters, D.C. et al., *Environ. Geosci.* 3, 11–20, 1996.)

Coal was mined in the Colorado Springs area starting in the 1870s and continued until 1957 (Figure 16.27). The mines occur at depths up to 150 m, and most of the mine locations are known from old maps. Coals of the Upper Cretaceous Laramie Formation are up to 6 m thick and dip 2° to 5° northeast. The mining method was room and pillar. Overburden consists of sandstones, siltstones, and claystones of the Cretaceous-Tertiary Dawson Formation, and scattered eolian deposits exist on the surface. No faults had been previously mapped in the mined areas. Prior to 1950, most of the surface was used for agriculture and no one was concerned with subsidence. Since that time urban development has modified the surface through grading and the construction of roads, homes, and commercial buildings. Occasionally, an old mine collapses and surface subsidence destroys property, as happened after heavier than usual spring rains during May 1995.

Landsat TM (28 m resolution), SPOT XS (20 m resolution), and SPOT P (10 m resolution) satellite images were acquired. In addition, high-altitude color and B/W airphotos at scales of 1:58,000 and 1:80,000 were used, as were large-scale B/W airphotos taken in 1947 (1:23,600) and 1969 (1:28,000). The SPOT images and high-altitude airphotos were not considered helpful due to the clutter of cultural features. The coarser resolution of the Landsat TM made structural features more continuous and subdued the noise of cultural features. It was felt that the larger fractures identified using Landsat would be more likely to extend to depth and control subsidence. Large and subtle features are also more difficult to recognize on detailed airphotos, because the faults are composed of multiple smaller segments. The contrast of the photos and images was enhanced to maximize their spatial and spectral detail, and airphotos were then examined in stereo.

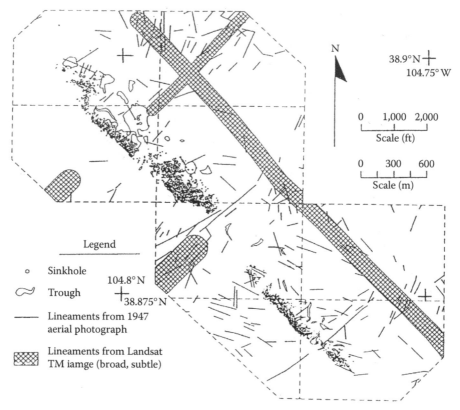

FIGURE 16.28
Fracture systems and subsidence features mapped in an attempt to locate collapsed, abandoned coal mines in the Colorado Springs area. (From Peters, D.C. et al., *Environ. Geosci.*, 3, 11–20, 1996.)

Linear patterns of topography, vegetation, soil tones, and outcrops were interpreted as controlled by faults and/or joint zones. Interpreted structures were then compared with the location of known mines, known subsidence features, and published isopach and structure contour maps.

Among the factors controlling subsidence are type of mining, groundwater movement, and thickness and lithology of the overburden. For example, gaps in subsidence could be correlated to areas that are buried deeper or where thick sandstone channels lie between the mines and the surface. Although it could not be confirmed that the northwest-trending lineaments are indeed fractures, evidence exists that probable structures identified through this analysis have an effect on the stability of the mined areas. Previously unrecognized northeast and northwest fracture systems in this area appear to influence the location of subsidence over abandoned coal mines (Figure 16.28).

Marine Hazards

Natural marine hazards include submarine landslides or slumps that can disrupt pipelines and platforms, shoals that can disrupt or impede shipping, and sea ice that can damage ships, platforms, and shallow subsea pipelines or facilities. Man-made hazards include collisions between ships or between ships and platforms, and anchor damage to pipelines, for example. In shallow water, it is possible to use true color images to examine the ocean surface or seafloor. This is because blue and green light penetrates water more than other wavelengths. Pulsed blue-green Lidar is being used for bathymetric mapping. Radar and satellite imagery is useful for mapping sea ice movements, particularly in arctic areas. In deeper water, it is necessary to use side-scan sonar to image ocean-bottom hazards.

Shoals

Shoals are best mapped using the visible blue and green light channels of Landsat, SPOT, and other satellites, or any true color aerial photography (Figure 16.29). Visible light penetrates water up to 20–30 m, depending on water clarity. Laser bathymetry using blue and green pulsed lasers is useful for mapping water depths to 40 or 50 m (see Bathymetric Mapping, Chapter 12). Shoals have been observed on radar imagery, because currents moving through a channel cause the water to accumulate over the shoals, and this slight rise in the water surface is detectable on radar imagery. Sandidge and Holyer (1998) used AVIRIS hyperspectral imagery and a neural network to map shallow bathymetry in Tampa Bay off the west coast of Florida and in the Florida Keys. The spectral radiance of coastal waters was determined to be a function of optical properties of the water column (scattering, absorption, and fluorescence of suspended sediment, organic matter, and chlorophyll), illumination conditions, and bottom type as well as depth. Hydrographic maps existed for control. Only 36 AVIRIS bands in the range 400 to 742 nm were used, as these have the water penetration required for mapping. An additional five bands between 867 nm and 1.203 μm were used for atmospheric corrections. Maps were generated to depths of 6 m with an RMS error of 0.39 to 0.48 m. Maximum correlation of radiance to depth occurred at the band centered at 608 nm.

FIGURE 16.29
Landsat TM 7-4-2 RGB image of shoals near Abu Dhabi City. (Image generated by Environmental Remote Sensing Unit, British Aerospace Space System Ltd., Bristol, for EOSAT [now GeoEye/Digital Globe, Longmont, CO].)

Shoreline Change

Shorelines move due to longshore drift, the action of tides, storm wave erosion, delta subsidence, and tectonic uplift or subsidence. Coastal change can also be a result of human development such as building groins to slow longshore drift of sand, or building marinas and draining swamps. Changing shorelines can be mapped using historical imagery, and change detection algorithms can be applied to multitemporal digital imagery. Appropriately timed imagery can be used to show the difference between coastlines at high and low tides.

Submarine Slumps

Submarine slumps have cut trans-ocean phone cables and can cause damage to other man-made structures such as offshore drilling platforms. Their surface manifestation appears similar to terrestrial landslides: arcuate escarpments, hummocky topography, and lobate shape are characteristic. In areas of unstable slopes, such as on the Mississippi or Mackenzie River deltas, it is fairly easy to identify recent slumps (Figure 16.30). These areas should be avoided if possible when planning pipelines, platforms, or subsea well completions.

Sea Ice and Sea Ice Scour

Sea ice can be mapped using any visible or near-IR sensor system, and using radar, which also provides data at night or during Arctic/Antarctic winter. Imagery provides information on the average and maximum extent of sea ice, ice floe distribution and movements, and timing of ice formation and breakup (Figure 16.31). RadarSat was launched specifically for monitoring ice in the Canadian arctic. ICESat and CryoSat are designed to map the extent of sea ice. This information is necessary for supply ships, ore barges, and tanker traffic to avoid collisions and for safe operation of drill ships and platforms (Kostilainen,

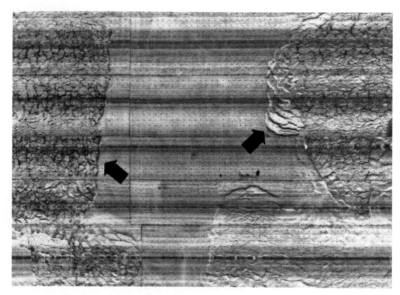

FIGURE 16.30
Sonar image of submarine slumps (arrows). (Courtesy of John Chance and Assoc., Lafayette, LA [now Fugro GeoServices, Inc, LA .].)

FIGURE 16.31
The Molikpaq platform was used in the Beaufort Sea and off Sakhalin Island. It was possible to build it to withstand the movement of sea ice based on a knowledge of the kinds of forces it would be subjected to.

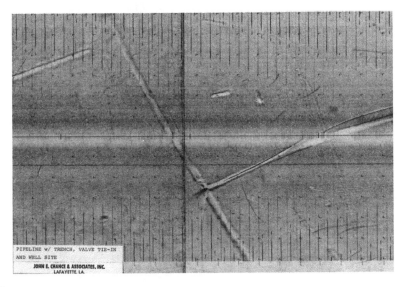

FIGURE 16.32
Side-scan sonar image of a pipeline gas leak (where the two pipes join) in the Gulf of Mexico. The pipeline on the sea bottom was snagged by a ship's anchor and pulled out of its trench. (Courtesy of John Chance and Assoc., [now Fugro GeoServices, Inc.].)

1990). Side-scan sonar may be necessary to identify areas with sea ice bottom scour so as to avoid them when planning subsea well completions and the laying of pipe. Sonar can also detect areas where pipelines have been uncovered by currents or damaged by shipping (Figure 16.32).

Subsea structures such as pipelines in areas with sea ice are subject to bottom scours. When scours exist, it is necessary to bury structures well below the depth of scouring. Sea ice as such generally will not scour the seafloor. When floes collide, they form pressure

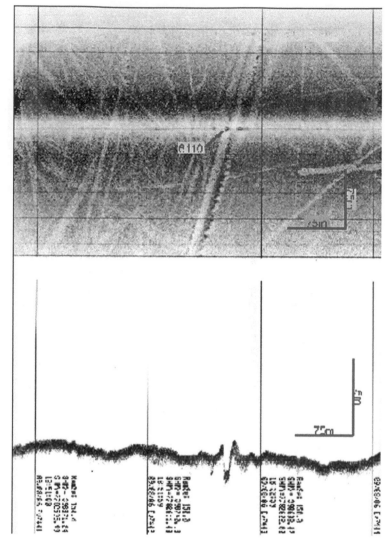

FIGURE 16.33
Top: side-scan sonar image of sea ice scour, Beaufort Sea. Bottom: cross section of sea ice scour in the soft sea-bed. Scour is approximately 1 m deep. (Courtesy of Dr. S. Blasco, Geological Survey of Canada.)

ridges that rise above the ice surface and extend below the ice. Scours as a result of ice keels have been seen up to a meter deep and in water depths to 150 m (Figure 16.33) (Blasco et al., 2010). Multibeam sonar is the preferred remote sensing technology for mapping sea ice scours.

Hydrates, Gas Vents, and Craters

Gas vents can indicate unstable areas where natural gases escape to the surface. Sometimes these vents occur as domes consisting of bottom mud, tar, and chemosynthetic communities of organisms. At other times, there are craters where gas has escaped suddenly and violently (Figure 13.12). Gas bubbles and oily slicks may be visible on the sea surface. Not only is the gas potentially explosive, but rapid escape of large volumes of gas can pose hazards to ships by decreasing their buoyancy. Drill rigs prefer to avoid encountering shallow sub-bottom gas for the same reasons. These vents may be aligned along recent fault scarps. Multibeam side-scan sonar or sparker (shallow seismic) surveys are the best ways to map shallow and venting gases. Photography, multispectral/hyperspectral imagery, and radar can map surface slicks (Jauer et al., 2011).

References

Aiuppa, A., M. Burton, P. Allard, T. Caltabiano, G. Giudice, S. Gurrieri, M. Liuzzo, G. Salerno. 2011a. First observational evidence for the CO_2-driven origin of Stromboli's major explosions. *Solid Earth*. 2: 135–142.

Aiuppa, A., M. Burton, P. Allard, T. Caltabiano, G. Giudice, S. Gurrieri, M. Liuzzo, G. Salerno. 2011b. *Linking plume CO_2 flux emissions and eruptive activity at Stromboli volcano (Italy)* (abs.). In 11th Gas Workshop, Kamchatka, Russia, 1–10 Sept. 2011, Commission on the Chemistry of Volcanic Gases (CCVG)—IAVCEI: 6.

Amundson, A., C. Greenman, B.K. Stover. 2009. *Subsidence above Inactive Coal Mines: Information for the Homeowner*. Denver: Colorado Dept. Natural Resources Spec. Pub. 26: 36 p.

Biggs, J., E.Y. Anthony, C. Ebinger. 2009. Multiple inflation and deflation events at Kenyan volcanoes. East African Rift. *Geology*. 37: 979–982.

Blake, E.S., E.J. Gibney. 2011. The Deadliest, Costliest, and Most Intense United States Tropical Cyclones from 1851 to 2010 (And Other Frequently Requested Hurricane Facts) NOAA Technical Memorandum NWS NHC-6. Miami: National Weather Service, National Hurricane Center: 49 p.

Blasco, S., K. MacKillop, V. Kostylev. 2010. Seabed geohazard constraints, offshore Beaufort Sea hydrocarbon development research results 2009, plans for 2010: Geol. Survey Canada PowerPoint, 29 slides.

Blewitt, G., W.C. Hammond, C. Kreemer, H.P. Plag, S. Stein, E. Okal. 2008. GPS for real-time earthquake source determination and tsunami warning systems. *J. Geod*. 83: 335–343.

Blystra, A.R., R.R. Dumas. 1980. Application of Landsat to the inventory of dams. In R.W. Kiefer (ed.), *Civil Engineering Applications of Remote Sensing*. Proceedings of the Specialty Conference of the Aerospace Division of the American Society of Civil Engineers, August 13–14, 1980. Madison: University of Wisconsin: 54–59.

Brakenridge, R., E. Anderson, S.V. Nghiem. 2006. Satellite microwave detection and measurement of river floods. NASA Spring Annual General Conference www.nasa.gov/vision/earth/lookingatearth/springagu_2006.html (accessed October 26, 2012).

Brennan, M. 2010. Satellite monitoring of Red River flooding in near real-time. http://nrcan.gc.ca/science/story/3585 (accessed October 29, 2012).

Briole, P., D. Massonnet, C. Delacourt. 1997. Post-eruptive deformation associated with the 1986-87 and 1989 lava flows of Etna detected by radar interferometry. *Geophys. Res. Lett*. 24: 37–40.

Burdick, R.G., L.E. Snyder, W.F. Kimbrough. 1986. *A Method for Locating Abandoned Mines*. U.S. Department Interior, Bureau of Mines Report. of Investigations 9050: Pittsburgh, PA: 27 p.

Casper, J.W., B.E. Ruth, J.D. Degner. 1980. Remote sensing study of sinkhole occurrence. In R.W. Kiefer (ed.), *Civil Engineering Applications of Remote Sensing*. Proceedings of the Specialty Conference of the Aerospace Division of the American Society of Civil Engineers, August 13–14, 1980. Madison: University of Wisconsin: 1–14.

Chrysoulakis, N., E. Flouri, E. Diamandakis, V. Dougalis, C. Synolakis, S. Foteinis. 2008. *Remote sensing in support of tsunami mitigation planning in the Mediterranean*. Proc. of the First International Conference in Disaster Management and Emergency Response in the Mediterranean Region, Zadar, Croatia, September 22–24, 2008: 10 p.

Dames and Moore. 1985. *Colorado Springs Subsidence Investigation, El Paso County.* Completed for the Colorado Inactive Mine Reclamation Program, text and maps Los Angeles, CA: Dames and Moore.

Dewi, R.S., N. Salam, S. Suwadi. 2010. Remote sensing for disaster mitigation: case study for tsunami evacuation route modeling in Cilacap—Central Java, Indonesia. Intl. Archives of the Photogrammetry, Remote Sensing and Spatial Information Science, v. XXXVIII pt. 8, Kyoto, Japan: 281–286.

Ehrismann, J., B. Armour, M. van der Kooij, H. Schwichow. April 1996. Mapping a broken land: using repeat pass, space-based SAR interferometry, researchers measure effects of the Kobe, Japan earthquake. *Earth Observ. Mag.* 5: 26–29.

Elifrits, C.D., D.J. Barr. 1982. Applications of remote sensing to subsidence detection at a room-and-pillar coal mine. In C.J. Johannsen and J.L. Sanders (eds.), *Remote Sensing for Resource Management.* Ankeny, IA: Soil Conservation Society: 347–361.

EOSAT. 1991. *EOSAT 1991* (calendar). Lanham, MD: Earth Observation Satellite Co. (now GeoEye/ Digital Globe, Longmont, CO).

ESA Climate change initiative. 2012. http://www.esa-sst-cci.org/?q=node/1 (accessed 26 October 2012).

Felipe, I.P., J.M. Dohm, V.R. Baker, T. Doggett, A.G. Davies, R. Castano, S. Chien et al. 2006. Flood detection and monitoring with the autonomous science craft experiment onboard EO-1. *Rem. Sens. Environ.* 101: 463–481.

Fujiwara, S., P.A. Rosen, M. Tobita, M. Murakami. 1998. Crustal deformation measurements using repeat-pass JERS 1 synthetic aperture radar interferometry near the Izu Peninsula, Japan. *J. Geophys. Res.* 103: B2–2411–2426.

Ghosh, S., B.J. Adams, J.A. Womble, C. Friedland, R.T. Eguchi. 2007. Deployment of remote sensing technology for multi-hazard post-Katrina damage assessment. Second International Conference on Urban Disaster Reduction, Taipei, November 27–29: National Science and Technology Center for Disaster Reduction (NCDR), Taiwan; Institute of Social Safety Sciences (ISSS), Japan; and Earthquake Engineering Research Institute (EERI), United States: 8 p.

Grzeda, S. 1996. FEMA digital flood mapping program. *Earth Observ. Mag.* 5: 22–25.

Hall-Beyer, M. 2010. Remote sensing for geologists IV: disasters, change detection, and becoming RS literate. *CSPG Reservoir.* 3: 20–25.

Hennessee, L., R. Kerhin, J. Isoldi, R. Gast. 1997. *Coastal Erosion, Chesapeake Bay, Maryland: A Digital Approach.* Carlisle, PA: Maryland Geological Survey, and Robertson, M., Dickinson College.

Henschel, M.D., G. Robert, B. Deschamps, R.W. Gailing. 2012. Geohazard mapping and identification along pipeline right-of-ways using space-borne synthetic aperture radar. Proceedings of the 2012 9th International Pipeline Conference IPC2012. September 24–28, 2012, Calgary: 5 p.

Jadkowski, M.A. March 1994. EOCAP funding develops aerial imaging system for pipeline infrastructure management. *Earth Observation Magazine* 3: 29–32.

Jauer, C.D., P. Budkewitsch, G.N. Oakey, D. Lavoie, K. Dewing. 2011. A survey of Canadian Arctic offshore basins with satellite radar for reconnaissance mapping of natural seep occurrences (abs.). AAPG Search and Discovery Article #90130, 3P Arctic: 1 p.

Klemas, V.V. 2009. The role of remote sensing in predicting and determining coastal storm impacts. *J. Coastal Res.* 25: 1264–1275.

Koedkurang, K., P. Phetprayoon, P. Aphaphant. 2005. Application of remote sensing and GIS for tsunami warning system (abs.). Proceedings of 26th Asian Conference on Remote Sensing: 1 p.

Kostilainen, V. 1990. *Environmental Problems in Ice-Covered Waters.* Houston, TX: Offshore Technology Conference: 217–224.

Mantovani, F., R. Soeters, C.J. Van Western. 1996. Remote sensing techniques for landslide studies and hazard zonation in Europe. *Geomorphology.* 15: 213–225.

Marino, C.M., G. Del Pero, A.M. Tonelli. 1984. Examples of airborne thermal infrared analysis applied to engineering geology and civil works site investigation. *Proceedings of the 18th International Symposium on Remote Sensing of Environment.* Paris: ERIM III: 1905–1914.

Maryland Department of Natural Resources. 2000. Shore erosion task force final report. Annapolis, MD: Maryland Dept. of Natural Resources, Shore Erosion Control: 65 p.

McKean, J., S. Buechel, L. Gaydos. 1991. Remote sensing and landslide hazard assessment. *Photogram. Engr. Rem. Sens.* 57: 1185–1193.

Messinger, D., D. McKeown, N. Raqueño. 2011. Remote sensing aid to the Japanese earthquake and tsunami relief effort. *SPIE Newsroom,* doi: 10.1117/2.1201107.003785, http://spie.org/x51497.xml

Messinger, D.W., J. van Aardt, D. McKeown, M. Casterline, J. Faulring, N. Raqueño, B. Basener, M. Velez-Reyes. 2010. High resolution and LIDAR imaging support to the Haiti earthquake relief effort. *Proc. SPIE* 7812, doi:10.1117/12.862090: 78120L.

Nadeau, P.A. 2011. Ultraviolet digital imaging of volcanic plumes: implementation and application to magmatic processes at basaltic volcanoes. PhD dissertation. Michigan Technological Univ.: 181 p.

NASA, *Global climate change,* http://climate.nasa.gov/NasaRole/ (accessed 26 Oct. 2012).

NOAA. *Hurricane Katrina.* http://www.ncdc.noaa.gov/special-reports/katrina.html (accessed 21 Dec. 2012).

Peters, D.C., K.E. Livo, P.L. Hauff. 1996. Remote sensing for analysis of mine subsidence and mine wastes. *Environ. Geosci.* 3: 11–20.

Pyle, D. 1999. Widely dispersed Quaternary tephra in Africa. *Global Planet. Change.* 21: 95–112.

Rao, K.G., J. Adinarayana. 1994. Qualitative assessment of soil erosion for watershed prioritization—A database approach. *Asian-Pac. Remote Sens. J. Bangkok: ESCAP.* 7: 147–154.

Roemer, H., G. Kaiser, H. Sterr, R. Ludwig. 2010. Using remote sensing to assess tsunami-induced impacts on coastal forest ecosystems at the Andaman Sea coast of Thailand. *Nat. Hazards Earth Syst. Sci.* 10: 729–745.

Rosen, P., C. Werner, E. Fielding, S. Hensley, S. Buckley, P. Vincent. 1998. Aseismic creep along the San Andreas Fault northwest of Parkfield, CA measured by radar interferometry. *Geophys. Res. Lett.* 25: 825–828.

Sandidge, J.C., R.J. Holyer. 1998. Coastal bathymetry from hyperspectral observations of water radiance. *Rem. Sens. Environ.* 65: 341–352.

Song, Y.T. 2007. Detecting tsunami genesis and scales directly from coastal GPS stations. *Geophys. Res. Lett.* 34, L19602. doi: 10.1029/2007GL031681.

Šúri, M., T. Cebecauer, J. Hofierka, E. Fulajtár jun. 2002. Soil erosion assessment of Slovakia at a regional scale using GIS. *Ecology (Bratislava).* 21: 404–422.

Tamburello, G., E.P. Kantzas, A.J.S. McGonigle, A. Aiuppa. 2011a. Vulcamera: a program for measuring volcanic SO2 using UV cameras. *Ann. Geophys.* 54: 219–221.

Tamburello, G., A.J.S. McGonigle, E.P. Kantzas and A. Aiuppa. 2011b. Recent advances in ground-based ultraviolet remote sensing of volcanic SO_2 fluxes. *Ann. Geophys.* 199–208.

Tamburello, G., E.P. Kantzas, A.J.S. McGonigle, A. Aiuppa, G. Guidice. 2011c. UV camera measurements of fumarole field degassing (La Fossa crater, Vulcano Island). *J. Volcanol. Geotherm. Res.* 199: 47–52.

Turney, J.E. 1985. *Subsidence above Inactive Coal Mines: Information for the Homeowner.* Denver: Colorado Department of Natural Resources, Colorado Division of Reclamation, Mining, and Safety, and Colorado Geological Survey. 36 p.

Wang, J.F., L.F. Li. 2008. Improving tsunami warning systems with remote sensing and geographical information system input. *An Int. J.* 28: 1653–1668.

Welby, C.W. 1980. Satellite imagery and shoreline erosion prediction. In R.W. Kiefer (ed.), *Civil Engineering Applications of Remote Sensing.* Proceedings of the Specialty Conference of the Aerospace Division of the American Society of Civil Engineers. Madison: University of Wisconsin, Aug. 13–14, 1980: 78–87.

Wicks, C. Jr., W. Thatcher, D. Dzurisin. 1998. Migration of fluids beneath Yellowstone caldera inferred from satellite radar interferometry. *Science:* 282:458-462.

Yamazaki, F., K. Kouchi, M. Matsuoka. 2006. Tsunami damage detection using moderate-resolution satellite imagery. Proceedings of the 8th U.S. National Conference on Earthquake Engineering, April 18–22, 2006, San Francisco, California, Paper no. 465: 10 p.

Yoshikawa, K., Y. Okajima, S. Takagishi. 2012. Disaster monitoring using remote sensing for the great east Japan earthquake. Special Session on Catastrophic Disaster of East Japan Earthquake and Tsunami, Rome, Italy, 6–10 May 2012: 8 p.

Additional Reading

Bierwirth, P.N., T.J. Lee, R.V. Burne. 1993. Shallow sea-floor reflectance and water depth derived by unmixing multispectral imagery. *Photogramm. Eng. Remote Sens.* 59: 331–338.

Govender, M., K. Chetty, H. Bulcock. 2007. A review of hyperspectral remote sensing and its application in vegetation and water resource studies. *Water SA.* 33: 145–151. http://www.wrc.org.za (accessed 26 October 2012).

Khan, M.A., Y.H. Fadlallah, K.G. Al-Hinai. 1992. Thematic mapping of subtidal coastal habitats in the western Arabian Gulf using Landsat TM data—Abu Ali Bay, Saudi Arabia. *Int. J. Remote Sens.* 13: 605–614.

Kuehn, F., G. Trembich, B. Hoerig. 1997. Multisensor remote sensing to evaluate hazards caused by mining. *Proceedings of the 12th International Conference on Applied Geologic Remote Sensing.* Ann Arbor: ERIM: 425–432.

Rosen, P., C. Werner, E. Fielding, S. Hensley, S. Buckley, P. Vincent. 1998. Surface deformation and coherence measurements of Kilauea volcano, Hawaii, from SIR-C radar interferometry. *Geophys. Res. Lett..* 25: 825–828.

Section V

Astrogeology

What is a section on astrogeology doing in a book on remote sensing in the geosciences? Besides being a topic of interest to many geologists, astrogeology is the ultimate in remote sensing: the analysis of geologic systems from afar. This is all about understanding planetary processes remotely: using gravity to determine mass and internal composition, using magnetics to determine whether there is a rotating iron core, and using the electromagnetic spectrum to determine whether there is an atmosphere, whether there is water ice, and to determine the composition of the surface. Earth-based analogs have been used to suggest explanations for various planetary landforms. On the other hand, planetary landforms provide almost pure end-members for dunes, channels, mass wasting, impact structures, fold structures, and other erosional, depositional, and structural features found on Earth. Features on Earth are often obscured by the atmosphere, vegetation, soil, snow, ice, and water. On the planets, for the most part, they are not.

Whether or not humans will ever land on these bodies, planetary surface geology has been a source of animated discussion. At one time there was heated debate about whether lunar craters are the result of impacts or volcanism (Vita-Finzi, 2005). Scientists wondered whether the lunar surface was solid rock or consisted of rock powder that would swallow up a lunar lander (Hess, 1965). Is there water on the moon? These questions at least have been answered. Did Mars ever have an Earth-like atmosphere and an ocean? Did life emerge on Mars? Is Earth-based life derived from Mars? Does Venus have plate tectonics? Did Venus ever *not* have a CO_2 atmosphere and if so, what happened? Is the climate of Venus a result of runaway global warming? These questions are yet to be answered.

Thirty-three years ago the results of the first extraterrestrial remote sensing program, the Apollo lunar orbital missions, were published by the U.S. Geological Survey (Moore et al., 1980). They state that "remote sensing and measurements of the Moon from Apollo orbiting spacecraft and Earth form a basis for extrapolation of Apollo surface data to regions of the Moon where manned and unmanned spacecraft have not been and may be used to discover target regions for future lunar exploration which will produce the highest scientific yields." Up to that point, counting craters was the best (and only) way to estimate the age of planetary surfaces. Now, there are gravity measurements from orbiting spacecraft that show mass concentrations in lunar mare basins; remnant magnetism was detected that suggests the Moon once had a molten core; orbital geochemistry has revealed Al:Si and Mg:Si ratios and how they vary across the surface; radar returns show surface roughness and dielectric constants; reflected sunlight provides microscale physical properties, chemical composition, and maturity of the regolith; and thermal infrared measurements have found "hot spots" associated with young craters. Seismic software is being used to process 2D shallow radar (SHARAD) recordings of Martian icecaps into a 3D data set that reveals sub-ice geology and is helping interpret lithospheric thickness and mantle viscosity (Lukosavich, 2013).

Chapters 17 through 19 will review what we know about the terrestrial planets and moons derived from Earth-based and satellite remote sensing.

References

Lukosavich, N. 2013. Celestial seismic: 3D interpretation unlocks Martian mysteries. *World Oil.* 234: 27.

17

Mapping Planetary Structure

Chapter Overview

Geologic structure is interpreted largely by its geomorphic expression and context, taking into account that each planet or moon has a different atmosphere and that none have liquid water at the surface (although some have had in the past). All of the planets have extremes of temperature. An area of horsts and grabens implies extension and normal faulting. Lateral offset of linear ridges indicates strike-slip faulting. Arcuate escarpments suggest thrust faulting, especially if folding is seen and shortening is implied by the local context.

Sensors used to map surface structures range from digital elevation data to radar, from visible light to thermal infrared emittance. Gravity provides a connection between surface features and crustal or mantle processes. Most of the images and much of the text in this section come from the NASA Images web site and are gratefully acknowledged.

This chapter is meant to serve only as a brief introduction to the topic of planetary structure. For more detailed discussions and reviews see Vita-Finzi (2005), Schultz et al. (2010), and Melosh (2011).

Faulting

Faults are identified on planets much as they are on Earth, that is, by their geomorphic expression. Normal faults tend to be linear, near-vertical, and have scarps associated with grabens, horsts, and relay ramps (Figures 17.1 through 17.3). Strike-slip faults are linear or branching and anastomosing, with secondary features such as en echelon faults and folds that bend into the fault system (Figure 17.4). Offsets on strike slip faults suggest lateral displacement (Figure 17.5). Thrust faults are suggested by curving or lobate scarps (Figures 17.6 and 17.7) that offset crater rims and other features with a distinctive reverse fault appearance. Visible light and radar images combined with topographic data allow vertical and lateral offsets to be determined (Watters et al., 2000).

With the return of the first Lunar Reconnaissance Orbiter Camera (LROC) images in 2010, analysts began to identify signs of contraction on the lunar surface in the form of lobe-shaped scarps. The scarps appeared to be thrust faults, and provided evidence that the moon may be shrinking globally, as these scarps are widely distributed across the moon. But identification of extensional faults shows that this shrinking, probably due to thermal cooling and contraction, is not homogeneous everywhere. Researchers analyzing high-resolution LROC images saw small, narrow grabens typically much longer than they

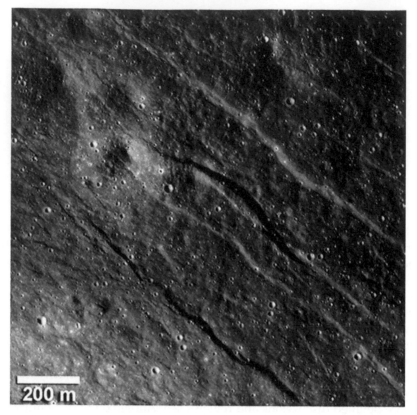

FIGURE 17.1
The largest lunar graben observed with the Narrow Angle Camera (NAC) of the LRO spacecraft is about 500 m wide and nearly 20 m deep. (Courtesy of NASA/Goddard/Arizona State University/Smithsonian Institution. http://www .structuralgeology.org/2012/02/grabens-shrinking-expanding-moon.html.)

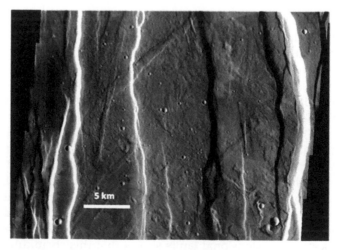

FIGURE 17.2
THEMIS image of normal faults, grabens, and relay ramps in the Alba Patera region, Mars. (Taken by the Mars Odyssey spacecraft; courtesy of NASA/JPL-Caltech/Arizona State University.)

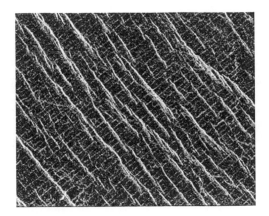

FIGURE 17.3

Gridded fractured plains of Sedna Planitia in the Lakshmi region of Venus. Fainter lines are spaced at about 1 km. North is at the top of the image. (Magellan image courtesy of NASA.)

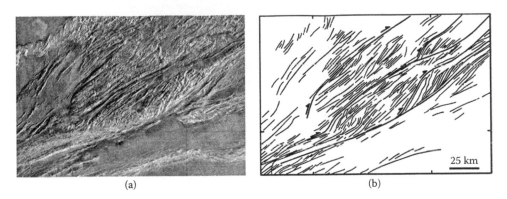

(a)   (b)

FIGURE 17.4

(a) Magellan SAR image of eastern segment, Thetis Boundary Shear Zone , Venus. (b) Structural interpretation. NASA image. (Data from Kumar, P.S. *J. Geophys. Res.* 110, EO7001: 10, 2005.)

FIGURE 17.5

Ridges in icy crust of Europa are offset by strike-slip faults. Image is approximately 12 × 15 km with 26 m resolution. North is at the top. (Image acquired by the Galileo spacecraft; courtesy of NASA.)

FIGURE 17.6
Discovery Rupes probable thrust scarp (arrows), Mercury, taken by Mariner10. (Courtesy of NASA/JPL/ Northwestern University. http://www.nasaimages.org/luna/servlet/detail/nasaNAS~4~4~16508~120145: Discovery-Rupes-Scarp#.)

FIGURE 17.7
Viking Orbiter image of Amenthes Rupes, a probable thrust scarp, Mars. This thrust cuts the walls and floor of impact craters. (Courtesy of USGS/NASA.)

are wide. A handful of these graben systems have been found across the lunar surface. NASA estimates that these faults are no older than 150 million years (NASA, online).

Using measured fault topography, mechanical modeling of fault displacements, and three-dimensional displacement–scaling relations, workers suggest that grabens such as those in Alba Patera on Mars (Figure 17.2) resulted from extensional deformation of a crustal section 2–3 km thick that may correspond to the thickness of sedimentary cover in this area (Polit et al., 2009). Note that these normal faults display many of the same features as normal faults on Earth: they are paired in horsts and grabens, have straight to slightly curving scarps, relay ramps, and displacement transfer zones at their terminations.

The faults of Sedna Planitia, Venus (Figure 17.3), may be normal faults caused by uplift-generated tensile stress. This image shows an area of the Lakshmi region about 37 km wide and 80 km long. Based on data from the Pioneer Venus Orbiter and the ground-based Arecibo Radar Observatory, it is known that this region is located on a low rise that separates Sedna Planitia and Guinevere Planitia. Two sets of parallel lineations are seen intersecting almost at right angles. It is not clear whether the two sets of lineations are faults or jointing. While we do see regional and orthogonal fracture sets on Earth, this type of "plaid" terrain has not been seen elsewhere on Venus, on Earth, or on other planets.

A striking example of a dextral wrench-fault system can be seen along the Thetis Boundary Shear zone of Venus (Figure 17.4). Strike-slip ductile duplexes are present at the restraining bends. Sigmoidal folds bending into the master faults suggest dextral displacement. Note that Tuckwell and Ghail (2003) interpreted this as a sinistral strike-slip zone. The debate continues.

A high-resolution image of the icy crust of Europa, one of the Jupiter's moons, reveals a surface criss-crossed by multiple sets of ridges and fractures (Figure 17.5). The large ridge in the lower right corner of the image is approximately 2.5 km across and is one of the youngest features on the surface, as it cuts across many of the other features. Note that one ridge (a piercing point) is offset in a right-lateral sense.

One of the most prominent lobate scarps on Mercury, the Discovery Scarp, is about 350 km long and transects two craters 35 and 55 km in diameter (Figure 17.6). The maximum height of the scarp is about 3 km just south of the larger crater. Arcuate scarps are similar to thrust faults that are convex toward the foreland. Thus, these lobate scarps are thought to represent reverse or thrust faults (foreland in the lower left of the figure) and suggest either local or planetary shortening.

The lobate scarp seen in Figure 17.7 is in the Amenthes region of Mars. The part shown here includes heavily cratered highlands. This lobate scarp is considered a thrust fault (the foreland is in the lower left of the figure) and suggests local shortening.

There is active debate whether plate tectonics functions on other planets, and if it takes forms other than that on Earth. There are some tantalizing clues. Spreading centers appear to be evident on the icy moons Ganymede and Europa, sometimes associated with oblique extension. Figure 17.8 shows a same-scale comparison between Arbela Sulcus on Jupiter's moon Ganymede (a) and an unnamed band on Europa (b). Arbela Sulcus is one of the smoothest zones of bright terrain identified on Ganymede and shows subtle striations along its length. Lightly cratered Arbela is in stark contrast to the surrounding heavily cratered dark terrain.

On Europa, dark bands appear to have formed by tectonic crustal spreading. Bands have completely separated preexisting features in the surrounding bright-ridged plains. The scarcity of craters on Europa illustrates the relative youth of its surface compared to

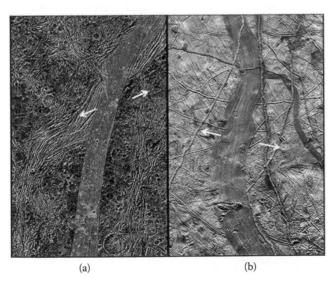

(a)   (b)

FIGURE 17.8
Possible icy spreading centers. (a) Arbela Sulcus on Ganymede shows about 65 km strike-slip offset. (b) A band on Europa (right) apparently formed by crustal spreading. Both images were acquired by the Galileo spacecraft and are at the same scale. (Courtesy of NASA/JPL.)

Ganymede's. It is possible that Arbela Sulcus has, like bands on Europa, formed by complete separation of the icy crust. Prominent fractures on either side of Arbela appear to have been offset about 65 km, suggesting that oblique extension was an important component of this spreading center. The Ganymede image covers an area of approximately 258 × 116 km with a resolution of 133 m. The Europa image covers an area of approximately 275 × 424 km. Resolution is 220 m.

Mantle convection and plume activity on Mars may have created Tharsis Montes on the margin of a postulated Boreal Austral plate (Sleep, 1994). The apparent hot spot volcanism may be analogous to the Hawaiian volcanic chain (Figure 17.9). However, the lack of a magnetic field on Mars suggests that Noachian (3.5–4.5 billion years ago) plate tectonics ended when dynamic heat flux and convection ended.

Plate tectonics is possible on Venus, which has active heat flow. Sandwell et al. (1997) suggested that Venusian tectonics is driven by isostacy and that areas of positive geoid are generally in a state of extension (rifting, grabens) while areas of negative geoid are in compression (folding, wrinkle ridges). One recent interpretation of Venusian landforms is that limited plate tectonics may be taking place along segments of "chasmata" (linear troughs) on the margins of Artemis, Latona, and other "coronae," elevated circular uplifts perhaps located above diapiric mantle plumes (Figure 17.10) (Vita-Finzi, 2005). Ghail (2002) suggested that, with a surface temperature near 750°K, the Venusian lithosphere is likely buoyant and therefore tectonically more similar to Earth's continents than to Earth's oceanic crust. Thus, the Venusian version of plate tectonics would encompass mainly buoyant vertical movements rather than lateral plate migration. Buoyancy, which inhibits subduction, may encourage obduction in areas such as southeastern Thetis Regio, for example.

Another form of plate tectonics, combining vertical and lateral motion, can be seen on the surface of Europa. Ice rafts appear to have tilted, bobbed up and down, and moved sideways in response to impacts or currents in the underlying liquid ocean. The blocks are from a few kilometers to a few tens of kilometers across. After moving and allowing upwelling of subsurface liquid, the surface again freezes solid. Figure 17.11 shows the ice-rich crust of Europa. Crustal plates ranging up to 13 km across, which have been broken apart and rafted into new positions, resemble multi-year pack-ice on polar seas on Earth. The size and geometry of these features suggest that motion was enabled by an "asthenosphere" of semi-frozen water or soft ice in the subsurface.

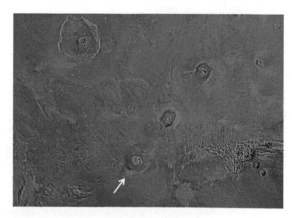

FIGURE 17.9
Possible hot spots in the Tharsis Montes region, Mars (arrow). (Courtesy of NASA.)

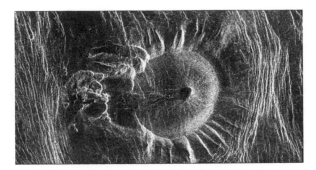

FIGURE 17.10
The corona at Thetis Regio, Venus, a volcanic uplift over a probable diapiric mantle plume. The uplift is bounded by thrust faults. (Magellan image courtesy of NASA.)

FIGURE 17.11
Ice rafts on Europa suggest mixed vertical and lateral plate tectonics. The area shown is about 34 × 42 km, and resolution is 54 m. Taken by Galileo spacecraft. (Courtesy of NASA/JPL/ASU.)

Folding

Folding on the planets is considered to be a result of shortening and compression. Some folds are high amplitude and short wavelength, whereas others are broad, gentle upwarps. "Wrinkle ridges" are long, sinuous, and generally narrow ridges seen on the Moon and other planets. They are thought to result from cooling and contraction in lavas or may develop over blind thrusts. The following examples illustrate these fold types.

The Magellan image in Figure 17.12 shows part of Ovda Regio, one of the highlands ringing the Venusian equator. Several tectonic events formed this complex terrain. An underlying fabric of ridges and valleys strikes northeast-southwest. These ridges are spaced 10 to 20 km apart and appear to have been caused by shortening of the crust at right angles to this trend. These structures are cut by thoroughgoing extension fractures trending northwest-southeast, suggesting a syntectonic or later episode of northeast southwest extension. Finally, the largest valleys (probable synclines), particularly the 20-km-wide valley extending diagonally across the image, were filled with dark material that looks like lava. The complex structure of Ovda Regio suggests a long deformation history.

The northern boundary of Ovda Regio can be seen in Figure 17.13. The scene contains low-relief, rounded linear ridges. These ridges, 8–15 km wide and 30–60 km long, lie along a 100–200 km wide slope where the elevation drops 3000 m from Ovda Regio to the surrounding plains. Some of the ridges have been cut at right angles by normal faults.

Dark material, either lava or windblown sediment, fills the synclines. The curvilinear, layered nature of these ridges and abrupt terminations of probable bedding at apparent thrust fronts suggests roughly north-south crustal shortening.

The High-Resolution Imaging Science Experiment on NASA's Mars Reconnaissance Orbiter shows stream valleys cut into the flanks of a gently sloping mountain range in the center-right of Figure 17.14. A prominent ridge that extends from the top to the bottom of the image is thought to have formed above a thrust fault. On planetary surfaces, these ridges are termed "wrinkle ridges." The wrinkle ridge imaged here is Ogygis Rupes. The area has a number of small, young impact craters, which are distinguished by sharp crater rims and bright or dark halos of ejected material. A large, old crater can also be seen in the center of the dip slope. It has a smooth rim and is partly filled with sand or soil. Wrinkle ridges are thought to be a result of local compression or perhaps developed over buried thrusts (Figure 17.15) (Goudy et al., 2005).

The smooth dark areas on the Moon's surface are called *maria* (Latin for seas). These volcanic plains are made up of basalt, highly fluid lava that covered a large percent of the Moon's lowlands. The oblique photograph in Figure 17.16 looks north across the southern part of Mare Imbrium. The low sun angle and long shadows accentuate details

FIGURE 17.12
Ovda Regio on the margin of Venus' equatorial highlands. Northeast fold axes are 10 to 20 km apart and are cut by northwest-oriented extension fractures (grabens). The fold interpretation is by the author. This image measures 225 × 150 km. (Taken by the Magellan spacecraft; courtesy of NASA.)

FIGURE 17.13
Ovda Regio, Venus. This image measures 300 × 225 km. Ridges (folds) are 8 to 15 km wide and 30–60 km long. Compare to Figure 5.75 showing thrusted folds on the Tibetan Plateau. (Taken by the Magellan spacecraft; courtesy of NASA.)

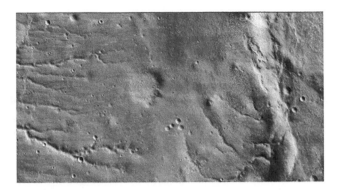

FIGURE 17.14
Ogygis Rupes, a wrinkle ridge in Bosporus Planum, Mars, is thought to be developed over a buried thrust fault. Note the asymmetry of the ridge, with a steep right flank and gentle dip slope on the left. High-Resolution Imaging Science Experiment image, Mars Reconnaissance Orbiter. This image is 50 km wide, 24 km long, and has 2.5 m resolution. (Courtesy of NASA/JPL/University of Arizona.)

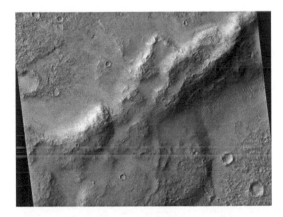

FIGURE 17.15
Wrinkle ridge at Solis Planum, Mars. High-Resolution Imaging Science Experiment image, Mars Reconnaissance Orbiter. Resolution is 50 cm. (Courtesy of NASA.)

FIGURE 17.16
Mare Imbrium wrinkle ridge, Moon. (Apollo 15 photo courtesy of NASA.)

of the surface structure. The surface in this area is *mare* basalt. The prominent ridges running from upper left to lower right are *en echelon* wrinkle ridges, folds that probably formed when the mare surface sagged under the weight of several kilometers of basalt. Similar wrinkle ridges are seen in other *mare* regions, including Mare Serenitatis and Mare Humorum.

Astypalaea Linea on the Jovian ice moon Europa is the broad smooth region running through the Galileo image in Figure 17.17. The criss-crossing ridges are believed to result from upwelling of new material through cracks in the surface ice. Evidence that stress is folding Europa's surface is shown by the presence of smaller cracks and wrinkles. These span the width of the broad swells suggestive of anticlines and synclines on Earth. Although ice covered, the crust of Europa is thought to be geologically active, riding over an ocean of icy water.

Impact Structures

The connection between impacts on Earth and mineral deposits has been recognized for some time. It has been speculated that the large layered intrusions at Sudbury, Bushveld, and Skaergaard have their origins as impact structures (Guilbert and Park, 1986; Golightly, 1994; Krogh et al., 1996). More recently, the importance of impact in the evolution of life has become evident, especially with the discovery of the Chicxulub crater in Yucatan and its presumed association with K-T extinctions. There are even a few oil fields, such as Steen River, Alberta and Red Wing, North Dakota, that are thought

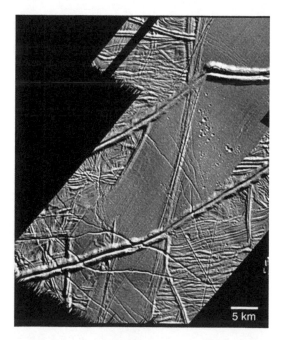

FIGURE 17.17
Galileo image of wrinkle ridges on Europa. (Courtesy of NASA/JPL, L. M. Prockter [JHU APL], and R. T. Pappalardo [Brown University].)

to produce from impact craters. In the case of Chicxulub, Steen River, and Red Wing, there are virtually no surface indications of the structure: they are evident primarily on gravity, magnetics, and seismic data.

Impact structures typically are rounded depressions, although they can be elliptical depending on the angle of impact. They commonly have a central peak due to rebound, and sometimes have concentric rings of ridges and fractures. Depending on their size, impacts can cause local melting and volcanism. They are responsible for blankets of ejecta that are splashed out of the crater and contribute to sedimentation on many planets and moons. We will first look at a few examples of terrestrial impact structures, then review typical planetary structures.

Meteor Crater, near Winslow, Arizona, is one of the best known impact craters on Earth (Figure 17.18). The crater is 1.6 km wide, and up to 200 m deep. This was the first crater recognized to be caused by an impact rather than a volcanic eruption. Modern research indicates that the responsible meteorite weighed 272 million kilos and was composed of nickel and iron. Its age is estimated at 50,000 years. It is estimated to have been about 40 m across and traveling over 42,000 km per hour. For comparison, the asteroid or comet that created the Chicxulub crater in Mexico 65 million years ago, and is thought to have caused the extinction of the dinosaurs, was between 9.5 and 19 km in diameter.

Chicxulub crater, shown in Figure 17.19, is revealed by the near-circular gravity low in the center of the figure. The negative gravity anomaly corresponds to low-density breccias, melt sheet, and Tertiary sediments filling the crater. The double-humped central gravity high corresponds to central uplift. This 65 million years old, 180-km-wide ring structure is still detectable under 0.7 km of younger rock in the Yucatan Peninsula region of Mexico. The outlines of the structure, called the Chicxulub crater (named for a local village), are visible in the gravity and magnetic field data from the region.

A shaded relief image of Mexico's Yucatan Peninsula (Figure 17.20) shows a subtle indication of the Chicxulub impact crater. Most of the peninsula is visible here, along with the island of Cozumel off the east coast. The Yucatan plateau is composed mostly of limestone and is a very low relief area with elevations varying less than a few hundred meters. In this image, the topography has been greatly exaggerated to highlight a semicircular trough (arrows). This trough is only about 3 to 5 m deep, is about 5 km wide, and is the surface expression of the crater's outer boundary. Color coding is directly related to topographic

FIGURE 17.18
Barringer Crater (Meteor Crater), Arizona. (Shane Torgerson photo courtesy of of http://en.wikipedia.org/wiki/File:Meteorcrater.jpg.)

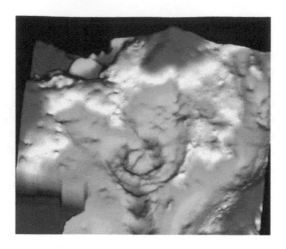

FIGURE 17.19
Bouguer gravity map over the Chicxulub crater. North is up. (Reproduced with the permission of Natural Resources Canada, 2013; courtesy of Alan Hildebrand and Michael Higgins, Geological Survey of Canada. http://miac.uqac.ca/MIAC/chicxulub.htm.)

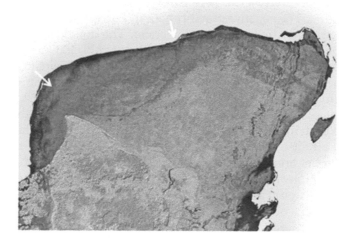

FIGURE 17.20
Surface manifestations of the Chicxulub structure are primarily arcuate cenotes (sinkholes) and a subtle arcuate topographic ring seen for the first time on Shuttle radar images. (Shuttle Radar Topography Mission elevation model courtesy of NASA/JPL/NIMA.)

height, with green at the lower elevations, rising through yellow and tan, to white at the highest elevations.

The Richat structure lies in the barren Gres de Chinguetti Plateau in central Mauritania, northwest Africa (Figure 17.21). It represents sedimentary rocks that have been domally uplifted by a large meteor impact, and were subsequently eroded by water and wind. The 40-km-wide structure is not apparent to a geologist on the ground, but is obvious from orbit.

As mentioned, some terrestrial impact structures produce oil and gas. Among these are the Red Wing Creek field of North Dakota (Figure 17.22), the Ames structure in Oklahoma, and Steen River in Alberta. Red Wing Creek was discovered by True Oil in 1972. They encountered an 820 m oil column in highly fractured and brecciated carbonates of the

FIGURE 17.21
Oblique view toward the east over the Richat impact structure, Mauritania. Banding is due to Paleozoic sedimentary layering. (International Space Station image courtesy of NASA.)

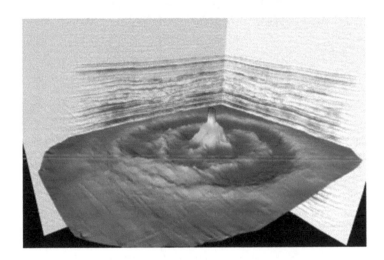

FIGURE 17.22
3D seismic image of the Red Wing Creek field structure, North Dakota. Central depression is about 2 km in diameter; uplifted rim is about 8 km in diameter. Structure on Mississippian Mission Canyon carbonates at about 2100 m depth. (From Burton, R. et al. *Oilfield Rev.* 21: 14–29 ,2010. Reprinted with permission from Ben Herber, Roger Barton, and True Oil Company.)

Mississippian Mission Canyon Fm. Although the average Mission Canyon net pay was 6–12 m in surrounding areas, they found 490 m net pay. The structure consisted of a circular central uplift about 1.5 km in diameter, with the Mission Canyon about 1000 m above its regional stratigraphic level. Outward from the central zone was a ring-shaped depression more than 1.5 km wide, then an uplifted ring more than 8 km across. The structure is buried under 2100 m of cover. Shatter cone fragments and shocked quartz in cuttings confirmed the nature of the structure, which is thought to be between 200 and 220 million years old. Till 2010, the Red Wing Creek field has produced 16.6 mmbo and 25 bcf gas, with an estimated 70 mmbo and 100 bcf recoverable remaining in the structure (Burton et al., 2010).

Impact craters on the Moon and other planets take a variety of forms. The ejecta deposits around Mars' Yuty crater (18 km in diameter) consist of many overlapping lobes

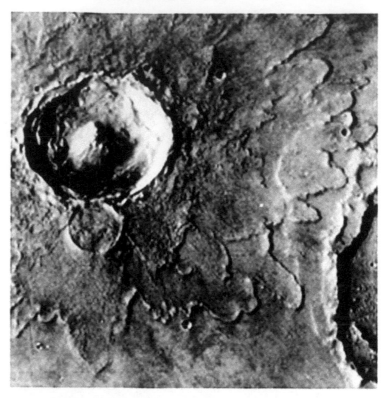

FIGURE 17.23
Yuty impact structure showing a central high and ejecta blanket. Note the similarity to the Red Wing structure, Figure 17.22. (Mars Viking 1 image, courtesy of NASA.)

(Figure 17.23). Craters with this type of ejecta deposit are known as rampart craters. This ejecta morphology is characteristic of craters at equatorial and middle latitudes on Mars, and it is believed to form when a meteorite impact rapidly melts ice in the subsurface. The presence of liquid water in the ejected material allows it to flow along the surface, giving the ejecta blanket its fluidized appearance.

The Mars Global Surveyor (MGS) Mars Orbiter Camera image in Figure 17.24 shows a pedestal crater in the Promethei Terra region. Rocks in the ejecta help shield the underlying terrain from being eroded by wind. Thus, over time, some craters appear to be raised on pedestals defined by their ejecta blankets, because wind has stripped away the surrounding material.

Clementine images acquired through three spectral filters (415, 750, and 1000 nm) were processed and combined into a multispectral mosaic of the Aristarchus region of the Moon (Figure 17.25). A color-ratio composite was created with the 750/415 ratio red, 415/750 blue, and 750/1000 green. Color ratios emphasize differences related to composition or mineralogy. The Aristarchus plateau, an elevated crustal block about 200 km across, is surrounded by the mare lava plains of Oceanus Procellarum. The plateau has experienced intense volcanic activity, both effusive and explosive. It includes vents and channels for low-viscosity lavas that formed deposits that cover the plateau. These deposits probably consist of iron-rich pyroclastics or cinders and have a deep red color on this image. Rather than forming cinder cones as on Earth, the lower gravity and vacuum of the Moon allows

FIGURE 17.24
The Promethei pedestal crater and ejecta, Mars Global Surveyor 2. (Courtesy of NASA/JPL/Malin Space Science Systems. For more information, contact Malin Space Science Systems at wwrv.msss.com. Copyright MSSS. http://photojournal.jpl.nasa.gov/catalog/pia08029.)

FIGURE 17.25
Aristarchus crater and ejecta, Moon. (Clementine 2 visible-NIR mineralogy image courtesy of USGS/NASA.)

the pyroclastics to travel much greater heights and distances, thus depositing an extensive ejecta blanket. The sharp rims suggests that the Aristarchus impact occurred relatively recently in geologic time. The 42 km diameter crater and its ejecta reveal two different stratigraphic sequences: that of the plateau, and that of the Oceanus Procellarum. This is apparent in the colors of the ejecta as seen in this image, where reddish colors indicate excavated mare lava, and bluish colors are caused by excavation of highland materials.

The extent of the continuous ejecta blanket appears asymmetric: it extends about twice as far northwest than in other directions. These ejecta lobes could have been caused by an oblique impact from the southeast.

Some of the most important results of the MGS include what we have learned about the presence and nature of the sedimentary rock record on Mars. The meteor crater in northwestern Schiaparelli Basin (Figure 17.26) shows amazing details of the layered, sedimentary rock affected by the impact. The 2.3-km-wide crater may have once been completely filled with sediment; the crater was later eroded to its present form. Many layers of similar thickness and physical properties are exposed in the bottom of the crater. The uniform physical properties and bedding of these layers might indicate that they were originally deposited in a lake (it is possible that the crater was at the bottom of a large lake that filled the Schiaparelli Basin); alternatively, the layers may have been deposited by settling out of the atmosphere in a dry environment.

The picture in Figure 17.27 shows Tyre, a large impact structure on Jupiter's icy moon, Europa. Tyre is unusual in that, while the crater is about 40 km across, the entire structure is much larger. The concentric rings display relatively little relief. One hypothesis for such characteristics is that the impactor that formed Tyre penetrated through an icy crust into a less-brittle layer.

Two large meteorite impact craters, with diameters that range from 37 to 50 km, are seen in the Magellan image of the Lavinia region of Venus (Figure 17.28). The craters and their ejecta are located in a region of fractured plains and show many features typical of impact craters, including rough (bright) ejecta around the rim, terraced inner walls, and central peaks.

So far we have not been able to visit any objects beyond our Moon. However, the range of structures on the terrestrial planets and larger moons mimics those on Earth. We use our knowledge of the expression of structure on Earth to help us understand what we are seeing elsewhere. In turn, we use what we see on other planets and moons to understand what the earliest Earth may have been like during the late heavy bombardment (3.8–4.1 billion years ago).

FIGURE 17.26
Schiaparelli crater, Mars, showing layered sediments. Note the similarity to Figure 17.21, the Richat impact structure, Mauritania. (Mars Global Surveyor image courtesy of NASA/JPL/Malin Space Science Systems. For more information, contact Malin Space Science Systems at wwrv.msss.com. Copyright MSSS.)

FIGURE 17.27
Tyre multi-ring crater, Europa. Multiple rings may be due to low compressive strength of ice over water. Vertical black line in the Tyre mosaic indicates a gap in the data received for this image. (Galileo 2 image courtesy of NASA/JPL.)

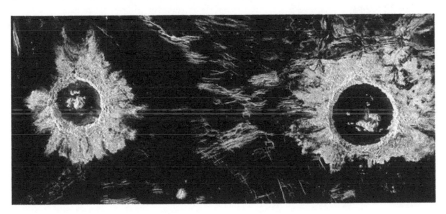

FIGURE 17.28
Impact craters in the Lavinia region, Venus, showing a central high and ejecta blankets. (Magellan radar image courtesy of NASA/JPL.)

References

Burton, R., K. Bird, J.G. Hernández, J.M. Grajales-Nishimura, G. Murillo-Muñetón, B. Herber, P. Weimer et al. 2010. High impact reservoirs. *Oilfield Rev.* 21: 14–29.

Ghail, R.C. 2002. Structure and evolution of southeast Thetis Regio. *J. Geophys. Res.* 107(E8), 5060: 1–7.

Golightly, J.P. 1994. The Sudbury igneous complex as an impact melt: evolution and ore genesis. In P.C. Lightfoot, A.J. Naldrett (eds.), Proceedings of the Sudbury-Norl'sk Symposium, Ontario Geological Survey Spec., v. 5: 105–117.

Goudy, C.L., R.A. Schultz, T.K.P. Gregg. 2005. Coulomb stress changes in Hesperia Planum, Mars, reveal regional thrust fault reactivation. *J. Geophys. Res.* 110, E10005: 12 p.

Guilbert, J.M., C.F. Park, Jr. 1986. *The Geology of Ore Deposits.* New York: W.H. Freeman: 985 p.

Hess, W.D. 1965. *Nature of the Lunar Surface.* Proceedings of the 1965 IAU-NASA Symposium. Conference on the Nature of the Surface of the Moon, 1965, Greenbelt, MD: 328 p.

Hildebrand, A., M. Michael Higgins. Gravity measurements over the Chicxulub crater. http://miac.uqac.ca/MIAC/chicxulub.htm (accessed October 28, 2012).

Krogh, T.E., S.E. Kamo, B.F. Bohor. 1996. Shock metamorphosed zircons with correlated U-Pb discordance and melt rocks with concordant protolith ages indicate an impact origin for the Sudbury structure. In A. Basu and S. Hart, (eds.) Geophysical Monograph 95, *Earth Processes: Reading the Isotope Code*. Washington, DC: American Geophysical Union: 343–353.

Kumar, P.S. 2005. An alternative kinematic interpretation of Thetis Boundary Shear Zone, Venus: evidence for strike-slip ductile duplexes. *J. Geophys. Res.* 110, EO7001: 10.

Lukosavich, N. 2013. Celestial seismic... 3D interpretation unlocks Martian mysteries. *World Oil*. 234: 27.

Melosh, H.J. 2011. *Planetary Surface Processes*. Cambridge: Cambridge University Press: 500 p.

Moore, H.J., J.M. Boyce, G.G. Schaber, D.H. Scott. 1980. *Lunar Remote Sensing and Measurements*. U.S. Geological Survey Professional Paper 1046-B: Washington, D.C.: U.S. Government Printing Office: B1–B78.

NASA. *Images Home Page*. http://www.nasaimages.org/ (accessed October 27, 2012).

NASA. *Lunar Reconnaissance Orbiter*. Headquarters release No. 12-055 http://www.nasa.gov/mission_pages/LRO/news/lunar-graben.html (accessed October 27, 2012).

Polit, A.T., R.A. Schultz, R. Soliva. 2009. Geometry, displacement-length scaling, and extensional strain of normal faults on Mars with inferences on mechanical stratigraphy of the Martian crust. *J. Struct. Geol.* 31: 662–673.

Sandwell, D.T., C.L. Johnson, F. Bilotti, J. Suppe. 1997. Driving forces for limited tectonics on Venus. *Icarus*. 129: 232–244.

Schultz, R.A., E. Hauber, S.A. Kattenhorn, C.H. Okubo, T.R. Watters. 2010. Interpretation and analysis of planetary structures. *J. Struct. Geol.* 32: 855–875.

Sleep, N.H. 1994. Martian plate tectonics. *J. Geophys. Res.* 99: 5639–5642.

Tuckwell, G.W., R.C. Ghail. 2003. A 400-km strike-slip zone near the boundary of Thetis Regio, Venus. *Earth. Planet. Sci. Lett.* 211(1–2): 45–55.

Vita-Finzi, C. 2005. *Planetary Geology: An Introduction*. Harpenden, UK: Terra Publishing: 168 p.

Watters, T.R., R.A. Schultz, M.S. Robinson. 2000. Displacement-length relations of thrust faults associated with lobate scarps on Mercury and Mars: comparison with terrestrial faults. *Geophys. Res. Lett.* 27: 3659–3662.

Additional Reading

Balme, M.R., S. Gupta, C. Gallagher, A. Bargery. 2011. *Martian Geomorphology*. London: Geological Society Special Publication 356: 237 p.

Christensen, P.R., B.M. Jakosky, H.H. Kieffer, M.C. Malin, H.Y. McSween, Jr., K. Nealson, G.L. Mehall et al. 2004. The Thermal Emission Imaging System (THEMIS) for the Mars 2001 Odyssey Mission. *Space Sci. Rev.* 110: 85–130.

Ghail, R.C., C. Wilson, M. Galand, D. Hall, C. Cochrane, P. Mason, J. Helbert et al. 2011. EnVision: taking the pulse of our twin planet. *Exp. Astron.* (published online), Springer Science+Business Media: 29: 27 p.

Goudy, C.L., R.A. Schultz, T.K.P. Gregg. 2005. Coulomb stress changes in Hesperia Planum, Mars, reveal regional thrust fault reactivation. *J. Geophys. Res.* 110: 12.

Hansen, V.L. 2006. Geologic constraints on crustal plateau surface histories, Venus: the lava pond and bolide impact hypothesis. *J. Geophys. Res.* 111: E11010.

Klimczak, C., R.A. Schultz, A.L. Nahm. 2010. Evaluation of the origin hypothesis of Pantheon Fossae, central Caloris basin, Mercury. *Icarus*. 209: 262–270.

Nahm, A.L., R.A. Schultz. 2010. Evaluation of the orogenic belt hypothesis for the formation of the Thaumasia Highlands, Mars. *J. Geophys. Res.* 115: 14 p.

NASA, Jet Propulsion Lab. *Malin Space Science Systems*. http://photojournal.jpl.nasa.gov/catalog/pia08029 (accessed January 25, 2013).

Okubo, C.H., R.A. Schultz. 2004. Mechanical stratigraphy in the western equatorial region of Mars based on thrust fault-related fold topography and implications for near-surface volatile reservoirs. *Geol. Soc. Am. Bull.* 116: 594–605.

Okubo, C.H., R.A. Schultz. 2006. Variability in Early Amazonian Tharsis stress state based on wrinkle ridges and strike-slip faulting. *J. Struct. Geol.* 28: 2169–2181.

Phillips, R.J., M.C. Malin. 1984. Tectonics of Venus. *Annu. Rev. Earth. Pl. Sci.* 12: 411–443.

Schultz, R.A., J.M. Moore, E.B. Grosfils, K.L. Tanaka, D. Mège. 2007. The Canyonlands model for planetary grabens: revised physical basis and implications. In M. Chapman (ed.), *The Geology of Mars: Evidence from Earth-based Analogs*. Cambridge: Cambridge University Press: 371–399.

18

Mapping Planetary Stratigraphy

Chapter Overview

Sedimentation on the planets and moons consists primarily of lava flows, ash falls, and impact ejecta. In rare cases such as Mars, there appears to have been a surface liquid ocean and running water with attendant sedimentation. On Venus and the Moon, there are channels resulting from flowing lava. On Mars, there are winds that carry dust and perhaps some coarser material. The dominant agent of erosion on planets appears to be gravity, although in some cases running or seeping water, moving ice, wind, impacts, and lava can erode into the substrate.

Multispectral and hyperspectral visible to infrared imagery and radar are the primary sensors used to map sedimentary layers. Compositional analyses (mineralogy) are accomplished using spectrometers.

Erosion and Deposition

There is evidence of layering on other planets, which means that some sort of depositional process must be active. Sediments on Venus or the Moon may be purely volcanic flows, ash, or impact ejecta. Sediments on Mars appear to include, in addition to the above, windblown dust and sand, stream channel deposits, deltas, shoreline deposits, and glacial deposits. Wind and water erosion, mass wasting, and glacial processes are active on Mars.

The best evidence for active depositional processes is layering. In 1972, images from Mariner 9 revealed that some of the mesas in Valles Marineris (Mars) have layers. Speculation as to the origin of these layered materials ranged from volcanic ash deposits to sediments laid down in lakes. Mars Orbiter Camera (MOC) images have confirmed the presence of layered outcrops within Valles Marineris and many other locations as well. The layered rock is visible because of impacts, faulting, and most importantly, erosion.

The picture in Figure 18.1 shows a 1.5 × 2.9 km area in Candor Chasma. Over 100 beds have been counted in this area, and each has about the same thickness, estimated to be about 10 m. Each layer has a relatively smooth upper surface, and together they are good cliff formers. This layering could indicate that the sediments were deposited in a lake or shallow sea. Other MOC images suggest that these layers were deposited in a crater or other basin that existed before Candor Chasma was eroded into the surrounding terrain.

Eolian Erosion and Deposition

Since wind requires an atmosphere, of our immediate neighbors, this pretty much rules out Mercury and our Moon. The atmosphere on Venus is so thick, and the surface is so

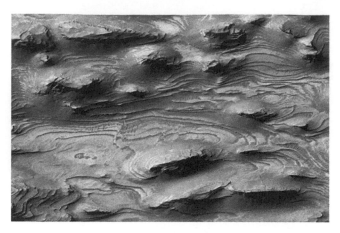

FIGURE 18.1
Layered Rock in Candor Chasma, Mars. These patterns could very well indicate that the materials were deposited in a lake or shallow sea. (Mars Global Surveyor image courtesy of NASA/JPL-Caltech.)

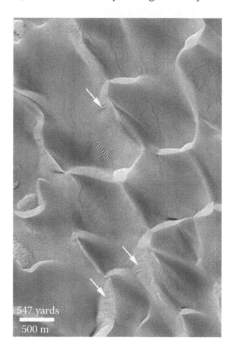

FIGURE 18.2
Dune field in Rabe crater, Mars. Arrows indicate wind direction. (Mars Global Surveyor image courtesy of NASA, Jet Propulsion Lab, and Malin Space Science Systems. For more information, contact Malin Space Science Systems at wwrv.msss.com. Copyright MSSS. http://photojournal.jpl.nasa.gov/catalog/pia02354.)

hot, that the wind would have a hard time moving loose material, assuming that there is any. That leaves Mars as the best candidate for wind erosion. Despite the thin atmosphere (relative to Earth), there is a lot of wind on Mars, and a lot of eolian deposits and landforms (Figure 18.2). In fact, sand dunes are a common feature on the Martian surface.

Sandy and dust-covered surfaces frequently have linear to curving light and dark streaks. Many of these streaks are straight and narrow, while others exhibit twists and loops. They cross hills, run straight across dunes, and go through fields of boulders. The example shown in Figure 18.3 shows dark streaks on the rippled flats of Argyre Planitia and covers an area of 3 × 5 km. These streaks indicate relatively recent (i.e., in the past few years or less) disruption of the surface. Streaks seen criss-crossing the dunes are thought to be the result of passing dust devils. Each dust devil leaves a dark streak by removing bright dust from the terrain in its path, revealing a darker surface underneath. The streaks are indicative of an active atmosphere and relatively mobile surface materials.

The image of "yardangs", features carved by windblown sand seen in Figure 18.4 near Olympus Mons, was obtained by the High-Resolution Stereo Camera (HRSC) on the European Space Agency (ESA) Mars Express spacecraft. This image has a resolution of 20 m. The scene shows a structure south of Olympus Mons that was probably formed by wind erosion. Loose sand blasted the bedrock, slowly eroding the surface till it resembles the fins of

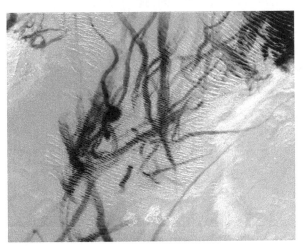

FIGURE 18.3
Dark streaks are probable dust devil tracks. (Mars Global Surveyor, Mars Orbiter Camera image of Argyre Planitia courtesy of NASA/JPL/MSSS. For more information, contact Malin Space Science Systems at wwrv. msss.com. Copyright MSSS.)

FIGURE 18.4
Oblique view of yardangs near Olympus Mons acquired by Mars Express satellite. These are long sharp ridges cut by wind erosion that may have accentuated regional fractures. The flat regions (the first in the foreground on the left, and the others top right), each measure about 9 × 17 km. (Courtesy of ESA.)

wind-eroded rock seen in Jurassic sandstone near Lake Powell, Utah (Figure 5.46). The three flat regions (in the left foreground and top right) may have been more resistant to erosion.

Oceans and Lakes

There is evidence that suggests that there was once a liquid water ocean on Mars. This is based on the paucity of craters in the essentially flat lowlands as well as canyons and channels flowing into the lowlands, deltas, and (subtle indications of) beach or shoreline deposits. The lack of cratering suggests that any craters in the area were buried by more recent sedimentation.

Figure 18.5 is an MOC high-resolution image that shows the contact between the Lycus Sulci uplands and Amazonis Planitia lowlands, Mars. In this picture, Amazonis and Lycus Sulci are separated by a subtle rise that runs diagonally across the scene from the lower left toward the upper right. None of the features seen in this image look like typical coastal landforms found on Earth: there are no beaches, windblown dunes, or wave-cut terraces. Instead, this could be a faulted contact, or lava flows could have filled a structural or topographic depression. Alternatively, this could be a dry seabed where evidence of the coastline has been buried by windblown deposits.

On the elevation map of the Valles Marineris region (Figure 18.6), one can see the canyon system, adjacent Chryse outflow channels and deltas, and Chryse basin. The basin has fewer craters, and some of those in the basin are partially buried by windblown or water-deposited sediment.

Rivers and Deltas

One of the most important results from the Mars Orbiter Camera on NASA's Mars Global Surveyor (MGS) has been the discovery of an ancient delta (Figure 18.7). Located

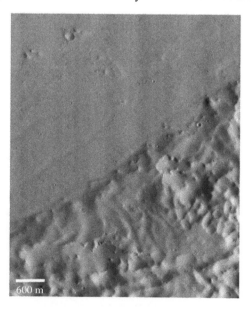

FIGURE 18.5

The contact between the Lycus Sulci highlands and Amazonis Planitia lowlands is a possible ancient shoreline. Mars Global Surveyor MOC. (Courtesy of NASA/JPL/MSSS. For more information, contact Malin Space Science Systems at wwrv.msss.com. Copyright MSSS.)

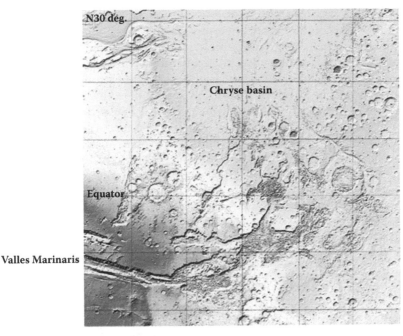

FIGURE 18.6
Valles Marineris canyon and others empty into the Chryse basin, which might once have held a Martian sea. Mars Global Surveyor (MGS) laser altimeter image. (Courtesy of NASA.)

FIGURE 18.7
The Eberswalde delta complex was discovered by the Mars Global Surveyor. Distributary fans suggest a long-lived river system. The delta is approximately 13 km long. This fossil delta appears to be an eroded remnant of a larger deposit, and some channels have inverted topography. The image covers an area about 3 × 3 km with a ground resolution of 50 cm. (Courtesy of NASA/JPL/Malin Space Science Systems. For more information, contact Malin Space Science Systems at wwrv.msss.com. Copyright MSSS.)

northeast of Holden Crater, the Eberswalde delta system includes at least two avulsing deltas. The north end of a looping, inverted, meandering channel can be seen feeding distributary channels in the delta to the right. The curving channel at some point became more indurated than the surrounding rock and, after wind erosion, the topography was inverted such that the channel fill is now higher than the surrounding ground.

The scene in Figure 18.8 is about 4 × 8 km. A series of anastomosing channels and streamlined islands reveals fluid erosion of layered bedrock that produced terraces and abandoned spillways, cutbanks, and point bars. Rib-like bed forms and hydrodynamic stream channel erosion (islands are pointed to rounded upstream and taper gently downstream) indicate that the direction of flow is from top to bottom of this image. The lack of impact craters on the flood-scoured surfaces indicates a relatively young age for the last flow.

Channels scoured by ancient flood waters are seen in this orbital view (Figure 18.9) from Odyssey's Thermal Emission Imaging System (THEMIS). The channels may be billions of years old and have likely been affected by multiple processes over time, including water and wind erosion. Two stream channels, Tiu Vallis on the left and Ares Vallis on the right, flow northward (bottom to top) from the highlands of the Martian southern hemisphere. The streamlined islands in the channels indicate flow direction.

The false-color image (Figure 18.10) of gullies in a crater in the southern highlands of Mars was taken by the High-Resolution Imaging Science Experiment camera on the Mars Reconnaissance Orbiter. The gullies emerging from the rocky cliffs near the crater's rim (upper left) show meandering and braided patterns typical of water-carved channels. Resolution is 26 cm per pixel.

Figure 18.11 shows a Magellan radar mosaic with a 200 km long section of a dark, sinuous channel on Venus. The channel is approximately 2 km wide. Such channels are common on the plains of Venus. They appear to have been formed by flowing lava that may have melted or eroded a path across the surface of the plain. They resemble terrestrial rivers in

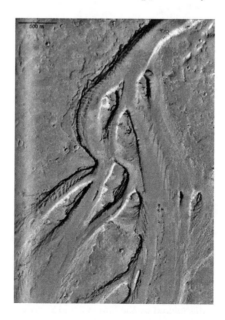

FIGURE 18.8

Channels in the Athabasca part of the Marte Vallis system. (Mars Orbiter Camera south of Cerberus Rupes courtesy of NASA/Malin Space Science Systems. For more information, contact Malin Space Science Systems at wwrv.msss.com. Copyright MSSS.)

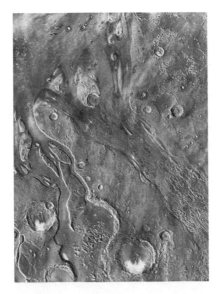

FIGURE 18.9
Flood channels Tiu Vallis (left) and Ares Vallis (right). The image is from Odyssey's Thermal Emission Imaging system. (Courtesy of NASA/JPL-Caltech/ASU.)

FIGURE 18.10
Gullies highly suggestive of water erosion are seen along the margin of a crater in the southern highlands of Mars. Acquired by the High-Resolution Imaging Science Experiment camera on the Mars Reconnaissance Orbiter. (Courtesy of NASA/JPL/University of Arizona.)

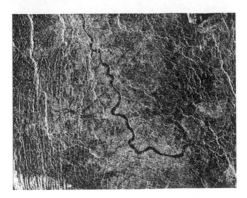

FIGURE 18.11
Meandering lava channel on Venus. (Magellan radar image courtesy of NASA/JPL.)

that they have meander belts, cutoff oxbows, and abandoned channel segments. Channel meanders on Venus have longer wavelengths and lower amplitudes than terrestrial rivers. In many cases, they appear to be partly buried by younger lavas, making it difficult to identify their sources. A few empty into vast plains, suggesting large flow volumes and fluid lavas. These channels are crossed by fractures and wrinkle ridges, and appear to run both upslope and downslope, suggesting that after eruption and channel formation, the lavas were warped by regional tectonism. Resolution of the Magellan data is about 120 m.

The radar image in Figure 18.12, of Sedna Planitia, Venus, shows not only orthogonal fracturing that seems to exert some control on erosion, but also a meander cutoff of the lava channel that may reflect underlying structural control.

Methane rain, ethane lakes, intermittent streams, and water–ice volcanoes may all exist on Saturn's moon Titan, according to preliminary analyses of images taken during the Cassini mission. A riverbed is identified by the dark channel near the top of the image in Figure 18.13, while a smooth lakebed can be seen at the bottom of this figure. Both the river and lake are thought to be dry when the image was taken, but contained a flowing

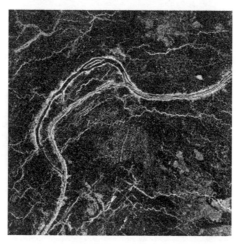

FIGURE 18.12
Meander cutoff at Sedna Planitia, Venus. (Magellan image courtesy of NASA.)

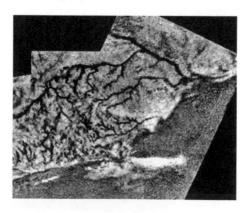

FIGURE 18.13
Mosaic of three frames from the Huygens Descent Imager/Spectral Radiometer on the Cassini-Huygens mission. (Courtesy of NASA/JPL/ESA/University of Arizona.)

(liquid—likely methane—in the recent past. Titan's surface appears strangely similar to Earth even though it is 94°K (–179°C).

A network of river channels is located in the Xanadu region of Titan. The radar image in Figure 18.14 was acquired by the Cassini Radar Mapper. These meandering rivers appear to emerge from rolling highlands at the top of the image and flow toward the bottom. Streams are thought to consist of methane and/or ethane.

Mass Wasting

Studies of slope failure on Mars suggest that three mechanisms may be at work: (1) artesian fluid pressures, (2) ground accelerations due to Marsquakes or impacts, and (3) gravitational collapse due to oversteepened or undercut slopes (Neuffer and Schultz, 2006). Failure of rock walls probably required quakes or impacts: softer layered deposits may collapse under their own weight.

Slopes collapse under the weight of gravity whenever they are unstable. This happens along valley rims, on the flanks of volcanic domes (see Figure 18.21 later in the chapter, Olympus Mons), or at the margins of impact craters, especially if they intersect a water table and there is active seepage. Debris flows lubricated by water or ice can also happen along the margins of glaciers, craters, or volcanoes and in steep canyons.

FIGURE 18.14
Meandering rivers in the Xanadu region of Titan. Rivers are thought to be methane or ethane. Image is roughly 230 × 340 km. Resolution is 500 m. (Cassini Radar Mapper image courtesy of NASA and JPL.)

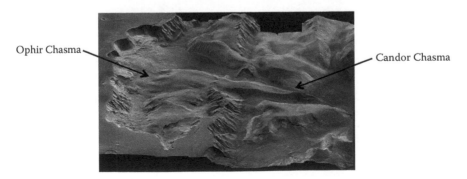

FIGURE 18.15
Oblique view of mass wasting in Ophir Chasma (left) and Candor Chasma (right), Valles Marineris, Mars. (Viking mission image produced by U.S. Geological Survey.)

Figure 18.15 is an oblique, color image of central Valles Marineris, Mars, showing Ophir and Candor Chasmata. The figure is a composite of Viking high-resolution images in black and white and low-resolution images in color. Ophir Chasma on the left is approximately 300 km across and as deep as 10 km. The connected chasma (valleys) of Valles Marineris may have formed from a combination of erosional collapse and structural activity. Tongues of interior layered deposits on the floor of the chasmata can be observed as well as young landslide material along the base of Ophir Chasma's north wall.

Glaciation

Glacial scouring and erosion are suggested by images of landforms in the north and south polar regions of Mars. The layered, ice-rich material appears to have flowed away from the Martian poles.

The part of the Martian south polar icecap that lasts through each summer is known as the residual cap. The landscape of the south polar residual cap is dominated by layered, frozen carbon dioxide ("dry ice") that has been eroded into a variety of pits, troughs, buttes, and mesas. The darkened slopes of the pits and troughs suggest that dust is mixed in with the ice. The ice is layered, indicating many cycles of deposition. Continuing MGS images indicate that the scarps are retreating an average of 3 m per Martian year. As more carbon dioxide is released into the atmosphere each southern spring and summer, the atmospheric pressure of Mars may increase. The mesa shown in Figure 18.16 exposes reddish-brown strata in an area covered by snow and ice.

Figure 18.17 is an MGS image that shows polygons formed in ice-rich material in the north polar region of Mars. The bright surfaces in this image are covered by a thin water ice frost. The patterned ground is analogous to frost polygons found the arctic on Earth (see Figure 13.2).

The Martian north pole (Figure 18.18) shows glacial tongues (with very few craters) extending outward into the lowlands. The large valley in the center has what appears to be glacial debris, including moraines and outwash. In some cases, ice appears to have flowed over or around craters. The scarcity of craters in the ice indicates flow and constant resurfacing.

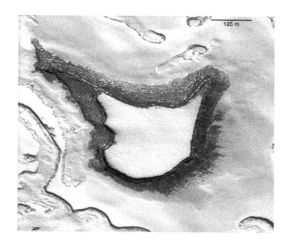

FIGURE 18.16
Mars Global Surveyor image of a layered mesa near the Martian south pole. (Courtesy of NASA/JPL/MSSS. For more information, contact Malin Space Science Systems at wwrv.msss.com. Copyright MSSS.)

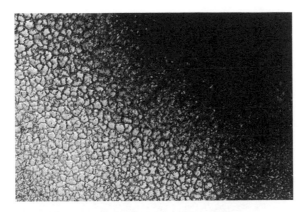

FIGURE 18.17
Permafrost polygons, or patterned ground on Mars. (Courtesy of NASA.)

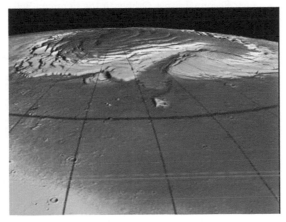

FIGURE 18.18
Mars Orbiter Laser Altimeter false color image of the north pole. Red color indicates highlands; blue indicates lowlands. (Courtesy of NASA/GSFC.)

Volcanism

Volcanic surface materials can occur as viscous or fluid flows, layered flows, cinder cones, strato-volcanic domes, and ash-fall deposits.

The smooth and rounded volcanic cone seen on the Moon in see Figure 18.19 may be either a cinder cone, because of the rubble on top, or consist of a silica-rich lava, since steep slopes suggest a viscous magma. Gamma-ray mapping of this area shows it to be rich in thorium (NASA LRO, online).

At the center of the area on the Moon seen in Figure 18.20 is an irregular depression that might be a caldera. At its edges are domes with features that suggest that they were formed by the intrusion of high-viscosity silicic lava, relatively rare on the Moon. Most lunar volcanism is basaltic, so this suggests that lunar volcanism may have evolved (differentiated) over time. The area is known as the Compton-Belkovich region. These volcanoes are estimated to be around 800 years old (The Daily Galaxy, online).

Figure 18.21 is a Viking image mosaic of Olympus Mons draped over topography. The volcanic dome is 600 km in diameter and rises 18 km above the surrounding plains. The mountain consists of overlapping basalt flows. Mars Orbiter Laser Altimeter's regional

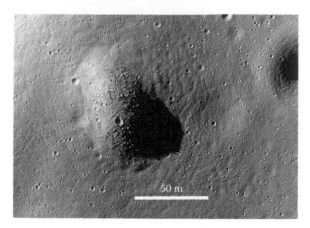

FIGURE 18.19
Volcanic cone imaged by the Lunar Reconnaissance Orbiter in the Compton-Belkovich region of the far-side highlands. (Courtesy of NASA/GSFC/Arizona State University.)

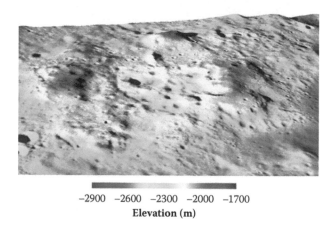

FIGURE 18.20
Siliceous volcanic deposits on the Moon. Lunar Reconnaissance Orbiter imagery draped over a DEM. (Courtesy of F. Scholten, DLR. NASA/GSFC/ASU/WUSTL.)

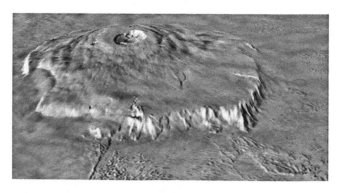

FIGURE 18.21
Mars Global Surveyor oblique image of Olympus Mons, Mars. Vertical exaggeration is 10:1. (Courtesy of NASA/Mars Orbiter Laser Altimeter Science Team.)

topography clearly shows the volcano's fault scarp in the foreground and the large land-slide deposits (left) caused by flank collapse.

The Magellan mosaic in Figure 18.22 shows an area of 160 × 250 km in the Eistla region of Venus. The prominent circular features are volcanic domes, the largest being about 65 km in diameter, having broad, flat tops up to 1 km above the surrounding plains. Known as "pancake" domes, they represent viscous extrusives with extremely steep margins. The fractures and pits are thought to result from cooling and withdrawal of lava.

The color image in Figure 18.23 shows two volcanic plumes on Jupiter's moon Io captured by the Galileo spacecraft. One plume on the edge of the moon was erupting over a caldera named Pillan Patera. This plume is 140 km high. The second plume, seen near the boundary between day and night, is called Prometheus. The shadow of the Prometheus plume can be seen extending down from the vent. (The vent is near the center of the bright and dark rings.) Plumes on Io have a blue color, and the plume shadow is reddish brown. The Prometheus plume can be seen in every Galileo image with the appropriate geometry, as well as every such Voyager image acquired in 1979, suggesting that it has been continuously active for over 18 years (Galileo mission home page, online).

The margin of the lava field associated with the Prometheus volcanic plume on Io can be seen in Figure 18.24. The darker lava has margins similar to those formed by fluid lava flows on Earth. This entire area is under the active plume of Prometheus, which is

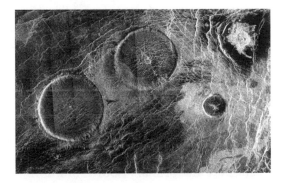

FIGURE 18.22
"Pancake" volcanic domes, Eistla region, Venus. (Magellan radar image courtesy of NASA.)

FIGURE 18.23
Caught in the act: active volcanism on Io. The resolution is about 2 km per pixel. This composite uses images taken with green, violet, and near-infrared bands. North is to the right. (Galileo spacecraft image courtesy of NASA/JPL.)

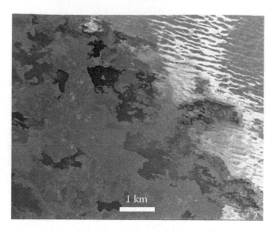

FIGURE 18.24
Lava flows and ridged plains in Prometheus region, Io. Image resolution is 12 m. (Galileo image courtesy of NASA/JPL-Caltech.)

constantly raining bright material onto the surface. Galileo mission scientists interpret the darkest flows as being the most recent since they are not yet covered by bright plume fallout. The flows may be too warm for sulfur dioxide gas to condense. The older plains are covered by bright ridges that may have formed by folding of a surface layer or as wind deposits. Bright streaks across the ridged plains radiate from the lava flow margins. This suggests that the bright material is being ejected at a low angle because it only coats the lava-facing sides of the ridges.

Composition and Mineral Mapping

The composition of planets and their moons is determined using a variety of instruments. Visible, near infrared, and thermal infrared spectrometers, configured as multispectral or hyperspectral instruments, are the primary source of information. Gamma ray spectrometry has been used. Gravity is used to determine composition of the core, thickness of sedimentary cover, thickness of the crust, and whether or not there is a sub-crust ductile or fluid asthenosphere. Magnetic measurements tell us whether there is a rotating liquid core that is generating a magnetic field. In most cases, however, it is the spectrometer that provides the most information about a planet's surface.

The lunar crust, based mainly on Apollo sampling, is dominantly anorthosite, believed to be a result of early magmatic differentiation. Plagioclase-rich magmas rose to the surface and olivine-ilmenite-rich magmas sank into the mantle (Vita-Finzi, 2005).

One of the best (and least known) sources of information on the distribution of minerals on the Moon is the Moon Mineralogy Mapper (M3) on the Indian Space Agency's Chandrayan-1 spacecraft. M3 is a visible-near infrared imaging spectrometer (420 nm to 3.0 μm) that mapped lunar mineralogy at relatively high spatial and spectral resolution. It mapped 95% of the lunar surface from an altitude of 100 km at 140 m resolution and 40 nm spectral sampling (86 spectral channels), with selected targets mapped at 80 m resolution and 10 nm bandwidths (260 channels). The mission, launched in October 2008, lost communication in August 2009. Targets of lunar mineral mapping included outcrops exposed at the walls and central peaks of large craters, complex volcanic terrain, boundaries where different kinds of rocks converge, unusual or rare compositions, and polar regions (Pieters et al., 2009).

The M3 acquired the image seen in Figure 18.25, a three-color composite of reflected and near-infrared radiation. Small amounts of water were detected on the surface of the moon at various locations. This image illustrates the distribution of small amounts of water at high latitudes near the poles. Blue shows the indications of water, green shows the brightness of the surface as measured by reflected infrared radiation, and red shows surface material with the iron-bearing mineral pyroxene.

Detailed maps of the lunar surface, including the location of water, if any, will be of use to future astronauts who will live and work on the Moon for extended periods. These maps are being derived from images such as Figure 18.26, acquired by the Galileo mission.

The Martian crust is thought to be mainly weathered basalts, although some sites appear to be richer in SiO_2. The image of Candor Chasma (Figure 18.27a) combines exposures taken by Odyssey's THEMIS at three different wavelengths of infrared light: 6.3, 7.4, and 8.7 μm. Note the landslide (hummocky terrain) on the left side of the image. The false-color image of Melas Chasma shown in Figure 18.27c was created using a decorrelation stretch, which emphasizes spectral differences to highlight compositional variations. It is difficult

FIGURE 18.25
Moon Mineralogy Mapper image, north half of the Moon. (Courtesy of ISRO/NASA/JPL-Caltech/Brown University/USGS.)

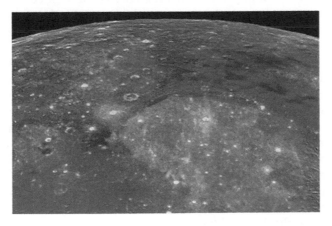

FIGURE 18.26
This false-color mosaic of part of the Moon was taken by the Galileo spacecraft. Images were processed to map titanium-rich regolith (blue) and soils lower in titanium (orange). Most of the lunar highlands appear red, indicating their low titanium and iron composition. (Courtesy of JPL/NASA.)

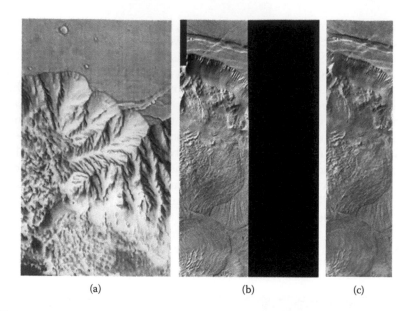

(a)   (b)   (c)

FIGURE 18.27

Compositional differences on Mars. (a) Color differences in a daytime infrared image taken by the Thermal Emission Imaging System (THEMIS) instrument on the Mars Odyssey spacecraft represent differences in the mineral composition on the surface. The image covers a part of Candor Chasma within Valles Marineris that is approximately 30 × 175 km. (b) Topographic image of Melas Chasma, Mars. (c) THEMIS image draped over topography, Melas Chasma. Colors indicate composition, for example, dark purple represents olivine-rich material. (Courtesy of NASA/JPL/Arizona State.)

to interpret some compositions due to the high variability of sunlit and shaded surfaces in this area, which cause a wide range of temperatures within each pixel (100 m resolution).

The composition of the crust of Venus appears similar to Earth's oceanic crust, that is, basaltic. Russian Venera landers discovered surface rocks that are alkali basalts (Vita-Finzi, 2005). Figure 18.28 shows a volcano near Phoebe Regio, Venus. The Magellan image shows emissivity in the radio-thermal range. Red represents high emissivity and blue represents low emissivity. The image is 590 km on a side. The unnamed volcano is about 2 km high and shows low emissivity at the summit, which could indicate the presence of pyrrhotite or pyrite, minerals that may not be stable at lower altitudes (Magellan press release P-40698).

The Mariner 10 flyby measured the albedo and spectrum of Mercury's crust. This data and ground-based spectroscopy suggest a surface rich in alkali feldspar, as in terrestrial alkali granites (Vita-Finzi, 2005). Updated calibration and mosaicing have vastly improved the Mariner 10 color image data. The color composite shown in Figure 18.29 highlights differences in opaque minerals (such as ilmenite), iron content, and soil maturity. Kuiper crater (lower left) shows color consistent with fresh material excavated from a subsurface unit that may have an unusual composition. Relatively dark and blue units are enriched in titanium. Bright red units may represent primitive crustal material.

The false color image of Europa (Figure 18.30) was obtained by Galileo's near-infrared mapping spectrometer. It reveals the presence of materials with differing compositions on Europa's surface: in this image, blue areas represent the cleanest, brightest icy surfaces, while the reddest areas have the highest concentrations of darker, non-ice materials. The mixture of colors is most likely the result of variation in the age and composition of surface materials. This area is unusual on Europa because of its concentration of fresh-appearing

FIGURE 18.28
Volcanic units southeast of Phoebe Regio, Venus. (Magellan emissivity image courtesy of NASA.)

FIGURE 18.29
Mariner 10 image mosaic of Kuiper crater area (lower left), Mercury. (Courtesy of NASA/JPL/Northwestern University.)

bright ridges and fractures. On other parts of Europa, the darker areas appear to be the most recently formed, but here the ridges and fractures appear to "overprint" the underlying darker mottled terrain. Scientists disagree about the chemistry of the dark areas; both sulfuric acid and salts, perhaps from a subsurface ocean, have been suggested. The area imaged is about 400 × 400 km.

Surface ice on Callisto and Ganymede contains both clays and organic molecules. Titan may have methane lakes in a water–ice landscape. The crust of Io appears to consist of an upper, primarily sulfur-based unit and a lower silicate basement. Pluto is thought to have four kinds of ice on its surface, including H_2O, CO, N_2, and CH_4 (Vita-Finzi, 2005).

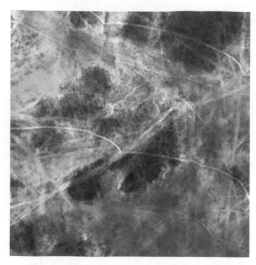

FIGURE 18.30
Blue represents clean ice; red areas have non-ice materials. This composite image of the Jupiter-facing hemi-sphere of Europa was obtained by two instruments on NASA's Galileo spacecraft. (Courtesy of NASA/JPL.)

FIGURE 18.31
The icy surface of Ganymede in visible (center) and infrared light (left). The image on the right combines the visible and infrared renditions. (This New Horizon spacecraft montage is courtesy of NASA/Johns Hopkins University Applied Physics Laboratory/Southwest Research Institute.)

Figure 18.31 shows images of Ganymede from the New Horizons spacecraft's Long-Range Reconnaissance Imager (LORRI) and its infrared spectrometer, the Linear Etalon Imaging Spectral Array (LEISA). LEISA captures more than 200 wavelength bands in the infrared, allowing detailed analysis of surface composition. The color LEISA image shown here combines three of these wavelengths (1.3 μm, 1.8 μm and 2.0 μm) to highlight relatively clean water ice (blue), while brown colors show regions contaminated by a dark material. The image on the right combines the high-resolution grayscale LORRI image with color-coded compositional information from the LEISA image to produce a composite that combines both data sets. With a diameter of 5268 km, Ganymede is the largest satellite in the solar system.

Hyperion, eighth largest of Saturn's moons, is 300 km in its longest dimension and is covered in craters. Distributed over the surface is a thin layer of organic dust, which appears to have been concentrated in the bottoms of some of the craters. In Figure 18.32, blue indicates frozen water, red denotes carbon dioxide ice, magenta indicates regions of

FIGURE 18.32

A Hyperion composition map shows evidence of organic molecules. Visual and Infrared Mapping Spectrometer image from the Cassini spacecraft. (Courtesy of NASA/JPL/University of Arizona.)

water plus carbon dioxide, and yellow is a mix of carbon dioxide and an unidentified material. A surface composition map is overlain on a Cassini image of Hyperion taken with the Imaging Science Subsystem.

In summary, then, we continue to learn about the composition of the terrestrial planets and asteroids, to the benefit of future astronauts as well as the astrogeologists and astrophysicists studying the solar system.

References

Galileo Mission home page. 2010. http://galileo.jpl.nasa.gov (accessed October 28, 2012).

Magellan press release P-40698.1992. *Volcano southeast of Phoebe Regio, Venus with emissivity data.* http://nssdc.gsfc.nasa.gov/imgcat/html/object_page/mgn_p40698.html (accessed October 28, 2012).

NASA. Jet propulsion lab, Malin space science systems. http://photojournal.jpl.nasa.gov/catalog/pia02354 (accessed January 25, 2013).

NASA LRO. 2011. *Farside highlands volcanism!* http://lroc.sese.asu.edu/news/?archives/426-Farside-Highlands-Volcanism!.html (accessed October 28, 2012).

Neuffer, D.P., R.A. Schultz. 2006. Mechanisms of slope failure in Valles Marineris, Mars. *Quart J. Eng. Geol. Hydrol.* 39: 227–240.

Pieters, C.M., J. Boardman, B. Buratti, A. Chatterjee, R. Clark, T. Glavich, R. Green et al. 2009. The moon mineralogy mapper (M3) on Chandrayaan-1. *Curr. Sci.* 96: 500–505.

The Daily Galaxy. 2011. *NASA finds volcanoes on dark side of the moon.* http://www.dailygalaxy.com/my_weblog/2011/07/-weekend-feature-nasa-finds-volcanoes-on-dark-side-of-the-moon.html (accessed October 28, 2012).

Vita-Finzi, C. 2005. *Planetary Geology: An Introduction.* Harpenden, UK: Terra Publishing: 168 p.

Additional Reading

Bell, J.F. III., B.A. Campbell, M.S. Robinson. 1999. Planetary geology. In A.N. Rencz (ed.), *Remote Sensing for the Earth Sciences, Manual of Remote Sensing*, 3rd ed., Vol 3. New York: John Wiley and Sons: 509–563.

Hartman, W.K. 2003. *A Traveler's Guide to Mars.* New York: Workman Publishing: 468 p.

19

Planetary Resources

Chapter Overview

In this admittedly brief chapter, we review the search for resources on planets and moons. Because of the extreme inhospitable environment of most bodies, the focus must be on our Moon and Mars.

Water

The search for Dilithium crystals, used to contain and regulate the annihilation reaction of matter and antimatter in a starship's warp core, was a plot driver in *Start Trek IV: The Voyage Home*. Unobtainium, found only on Pandora, has a unique magnetic field and properties of superconductivity that cause it to levitate and make it worth $40 million/kg refined. In the movie *Avatar*, humans are willing to wipe out innocent native populations to get this Unobtainium.

Although this may be what most people think about when they consider extraterrestrial exploration or mining, we should probably get a bit more "down to Earth." The resource most in demand when humans go into space will be water. Water is needed for everyday human functions and to grow food. It will likely be a source of both oxygen and hydrogen fuel. At this time, it is the only resource that NASA and other space agencies are seriously interested in finding on other planets and moons because it will be essential in making human habitation even partly self-sufficient.

We now are fairly certain that there is water ice on the Moon in permanently shadowed regions (PSRs). In 1996, data from NASA's Clementine spacecraft suggested that there is water ice in the bottom of craters near the Moon's south pole. In 2009, NASA's Lunar CRater Observation and Sensing Satellite (LCROSS) sent a projectile into Cabeus Crater and measured the spectra of the plume of material ejected. LCROSS found evidence that as much as 20% of the material in this permanently shaded area consisted of volatiles including water, methane, ammonia, hydrogen gas, carbon dioxide, and carbon monoxide. Instruments also detected relatively large amounts of sodium, mercury, and perhaps silver (Ambrose et al., 2012).

Spectrometers on several spacecraft, including the LRO, Moon Mineralogy Mapper (M3) on Chandrayan-1, the VIMS instrument on Cassini, and the Epoxi mission (to comet Hartley 2) have since confirmed discovery of water ice and the hydroxyl ion (OH^-) on the Moon. As much as 1000 ppm of water molecules in lunar soil are thought to be disseminated in the upper

40 cm of regolith in permanently shadowed areas of near-polar craters. The Lyman Alpha Mapping Project (LAMP) on the Lunar Reconnaissance Orbiter examined PSRs and found that up to 2% of the surface material consists of ice crystals, likely imported from comets and asteroids early in Moon's history (Kluger, 2012). Most ice has since been lost to space due to the Moon's low gravity and lack of atmosphere. Estimates of the amount of remaining water range from 10 to 300 million metric tons (about 3–90 billion gallons).

The images in Figure 19.1 show part of the interior of a young lunar crater on the far side of the moon captured by the M3 on the Indian Space Research Organization's Chandrayaan-1 spacecraft. On the left is an image showing brightness at short infrared wavelengths. On the right, the distribution of water-rich minerals (light blue) is shown. Both water- and hydroxyl-rich materials were found in the crater.

Evidence for water on Mars comes from hyperspectral imagery, thermal imagery, and geomorphology. Water today is probably all in ice form unless heated by near-surface geothermal activity. The source of this ice is likely from comets or volcanic outgassing.

High-resolution images of the Martian surface strongly suggest that there was once running water on the surface and that lowlands may have contained lakes and oceans (see Figures 18.6 through 18.10). Evidence for liquid water in the shallow subsurface of contemporary Mars is based on geomorphic evidence in the form of seeps and gullies. Crater morphology that implies impact into soft ground (Figure 17.23) also suggests that there are regions of permafrost near the Martian poles.

In 2002, the Mars Odyssey spacecraft used thermal inertia data to determine that there are large amounts of water ice in soil near the Martian poles (Bandfield, 2007). It is estimated that up to 80% by volume of the upper 0.6 m of soil near the Martian poles is water ice.

Color coding in Figure 19.2 indicates the change in nighttime ground-surface temperature between summer and fall. This site has water ice mixed with soil near the surface. The ice is probably in a frozen layer beneath a few centimeters of looser, dry soil. The amount of temperature change at the surface should indicate how close the icy material is to the surface. Since the ice layer retains heat better than the loose soil above it, the surface temperature changes more slowly when the ice layer is closer to the surface and faster when the ice layer is deeper. Areas that cooled more slowly between summer and autumn (ice closer to the surface) are shown as blue and green, whereas

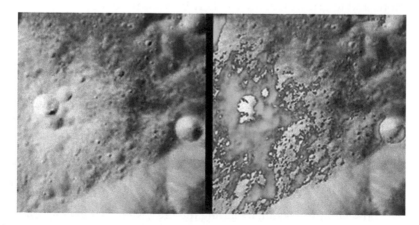

FIGURE 19.1

The Moon Mineralogy Mapper on Chandrayan-1 shows infrared reflectance (left) and distribution of water-rich minerals in blue (right). (Courtesy of Phillips, T. *NASA Science News*, 2009, http://science.nasa.gov/science-news/science-at-nasa/2009/24sep_moonwater.)

areas that cooled more quickly are red and yellow. The depth to the top of the icy layer is estimated to be as little as 5 cm.

The false color image in Figure 19.3 shows water ice and frost in a crater on northern Mars. Water ice has also been found in abundance in the north and south polar ice caps (see Chapter 18).

Indirect evidence for water and water ice can be seen on Europa: large craters show signs of relaxation that point to a warm, weak lithosphere (Figure 17.27). Multiring impact structures appear to have penetrated a low-viscosity layer at 6–15 km that could be warm ice (Vita-Finzi, 2005). Gravity data gathered by the Galileo spacecraft suggest that Europa has a 100-km-thick anhydrous silica mantle over a metallic core. This would leave room

Ice depth   132 E 67.5 N

Shallow   Deep

FIGURE 19.2
Map showing the relative depth to water ice in soil near the Martian north pole. (Courtesy of NASA/JPL/ASU.)

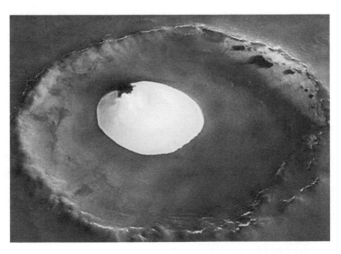

FIGURE 19.3
Water ice in an unnamed impact crater on Vastitas Borealis, northern Mars. The image, acquired by the THEMIS VIS camera, has a resolution of 19 m. (Courtesy of ESA/DLR/FU Berlin [G. Neukum]. http://www.nasa.gov/vision/earth/everydaylife/jamestown-water-fs.html.)

FIGURE 19.4
Ice erupting from the surface of Enceladus. (Cassini image courtesy of NASA/JPL/ESA.)

for a 60-km-thick subsurface ocean. Galileo magnetometer data show that Europa has an induced magnetic field consistent with a shallow layer of salty water. Extensional structures on Europa look like spreading centers on Earth and suggest an asthenosphere consisting of water or ductile ice.

Callisto and Ganymede also have induced magnetic fields that suggest that subsurface water bodies are underlain by ice rather than rock (Vita-Finzi, 2005).

Ice geysers erupt on Enceladus, an inner moon of Saturn. Figure 19.4 is a back-lit false-color image of the moon's limb. The (water) ice plumes were discovered by instruments on the Cassini Spacecraft in November 2005. The source of these geysers has been identified as fractures in the Moon's south polar region. The geysers are thought to arise from near-surface pockets of liquid water. The cryovolcanism probably contributes to Saturn's faint E ring.

From the foregoing discussion, it is apparent that water or ice derived from water (known as "water-ice") is fairly common throughout the solar system, and we have the means to identify and map it.

Minerals and Metals

As on Earth, mineral exploration on the moon will use hyperspectral remote sensing to find hydrothermal alteration minerals around volcanic centers. Hyperspectral imagery may also help locate mineral deposits associated with impacts such as the Sudbury district on Earth. LRO has mapped the distribution of metals in lunar basalts and has shown indications of thorium and uranium associated with silicic volcanic domes. High concentrations of rare earth elements, potassium, and phosphorus have been found in regolith associated with magmas in Oceanus Procellarum.

The Moon may have valuable reserves of helium-3, currently being considered as a power source in fusion reactors. The LRO has detected concentrations of helium-3 (Ambrose et al., 2012).

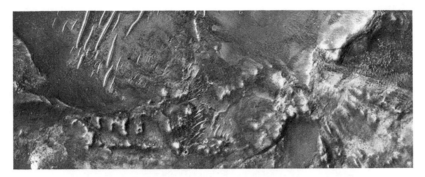

FIGURE 19.5

The floor of Toro Crater, Mars. Warm colors in this HiRISE image may indicate hydrothermal alteration. (Courtesy of NASA/JPL/University of Arizona.)

The best places on Mars to find valuable ores are volcanoes, lava flows, and impact craters (O'Hanlon, 2010). Hyperspectral imagery will likely direct Martian mineral exploration to areas with key indicator minerals. Harrison Schmidt (2004), the only geologist to walk on the Moon, gave a lecture on Martian resources in which he proposed that water-deposited spherules in Martian sediments may be copper sulfate or copper carbonate (malachite) concretions, and that iron-bearing minerals in the soil may include hematite and jarosite. In combination, these suggest the possibility of hydrothermal, volcanogenic, fumarole, or hot spring sulfide deposits. He further states that copper, lead, zinc, and precious metals might be found in pyroclastic and hot spring deposits, and that bauxite, iron oxides, manganese oxides, cobalt, and nickel deposits may all result from ancient chemical weathering. Iron and nickel may be found in meteorite debris near impact craters. Magmatic differentiation (in large intrusions or due to impact melting) could have formed layered intrusives containing titanium, chromium, iron, nickel, copper, and platinum group minerals. Martian sand also is an excellent source of silica.

Toro Crater on Mars may have experienced hydrothermal alteration (McEwen, 2012). Potential alteration minerals appear as diverse colors on HiRISE images, especially when enhanced as in Figure 19.5. In this image, the blue and green colors indicate unaltered minerals like pyroxene and olivine, whereas the warmer colors (yellow to red) indicate alteration to clays and other minerals. The linear north-south trending features are windblown dunes that are much younger than the bedrock.

The Nili Fossae region of Mars (Figure 19.6) has an assortment of minerals that include mafics and phyllosilicates (Ehlmann, 2008). The mineral assemblage suggests widespread liquid water activity and a variety of alteration caused by surface weathering and hydrothermal processes (Mangold et al., 2007).

Hydrocarbons

The widely accepted organic origin of hydrocarbons leads one to expect that oil will not exist on other planets. On the other hand, methane is a well-known product of outgassing of planetary interiors, including on Earth. Indications are that ethane may also exist on some extraterrestrial bodies.

Detection of plumes of seasonal methane venting into the Martian atmosphere suggests that natural gas hydrates may exist in the regolith. This methane could provide fuel and

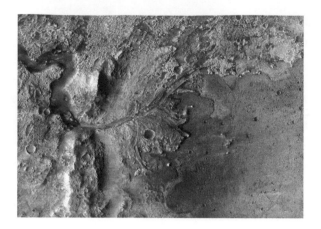

FIGURE 19.6
Nili Fossae region, Mars. This Compact Reconnaissance Imaging Spectrometer for Mars (CRISM) infrared spectrometer image is a composite of wavelengths 2.38 μm (red), 1.80 μm (green), and 1.15 μm (blue). Resolution is approximately 5 m. Geomorphology and minerals in this area suggest surface water was present and weathering and alteration were active. (Courtesy of NASA and Brown University.)

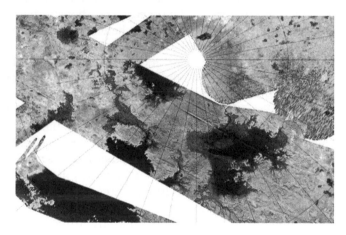

FIGURE 19.7
Cassini-Huygens radar image of Saturn's moon Titan. Blue-black lakes consist of liquid methane. (Courtesy of NASA/JPL/USGS.)

chemical industry feedstocks (Johnson et al., 2012). The NASA Curiosity spacecraft, which landed a rover in Gale Crater on Mars during August 2012, is using hyperspectral instruments to search for organic carbon associated with evaporites in a section estimated at close to 5000 m thick (Friedman, 2012).

Methane is abundant in the atmosphere of Saturn's moon Titan (Figure 18.14). Cassini-Huygens DISR images suggest that methane, and perhaps ethane, springs and lakes exist on the surface of Titan (NASA, online). A Cassini false-color mosaic (Figure 19.7) shows all synthetic-aperture radar images to date of Titan's north polar region. About 14% of the mapped region is covered by what is interpreted as liquid hydrocarbon lakes. Features thought to be liquid are shown in blue and black, and the areas likely to be solid surface are tinted brown. These seas are most likely filled with liquid ethane, methane, and dissolved nitrogen (NASA/JPL, online).

References

Ambrose, W.A., D. Beike, B.L. Cutright. 2012. Lunar energy and mineral resources: New insights from the Lunar Reconnaissance Orbiter (abstract). *AAPG Search Discov Article* #90142: 1 p.

Bandfield, J.L. 2007. High-resolution subsurface water-ice distributions on Mars. *Nature* 447: 64–67.

Ehlmann, B. 2008. Deltaic sediments on Mars. IAG Planetary Geomorphology Working Group, http://www.psi.edu/pgwg/images/oct08image.html (accessed November 3, 2012).

Friedman, B. April 2012. Mars explorer to look for organic carbon. *AAPG Explorer.* 66–67.

Johnson, A.H., M.D. Max, S.M. Clifford. 2012. Gas hydrate as part of the petroleum system of Mars (abstract). *AAPG Search Discov. Article* #90142: 1 p.

Kluger, J. 2012. Lunar liquid: more water than ever found on the surface of the moon. *Time Space Science,* http://www.time.com/time/health/article/0,8599,2104483,00.html#xzz1qS5PJ9wm (accessed October 28, 2012).

Mangold, N., F. Poulet, J.F. Mustard, J.-P. Bibring, B. Gondet, Y. Langevin, V. Ansan et al. 2007. Mineralogy of the Nili Fossae region with OMEGA/Mars Express data: 2. Aqueous alteration of the crust. *J. Geophys. Res.* 112: E08S04, doi: 10.1029/2006JE002835.

McEwen, A. 2012. The floor of Toro Crater. *SpaceREF,* http://hirise.lpl.arizona.edu/ESP_025067_1970 (accessed November 3, 2012).

NASA. *Cassini Solstice Mission.* http://saturn.jpl.nasa.gov/science/index.cfm?SciencePageID=73 (accessed October 28, 2012).

NASA/JPL. 2007. *Titan north polar region.* http://photojournal.jpl.nasa.gov/catalog/?IDNumber= PIA10008 (accessed November 3, 2012).

O'Hanlon, L. 2010. Mining Mars? Where's the Ore? *Discovery News,* http://news.discovery.com/space/mars-prospecting-ores-gold.html (accessed October 28, 2012).

Phillips, T. 2009. Water molecules found on the moon. *NASA Science News.* http://science.nasa.gov/science-news/science-at-nasa/2009/24sep_moonwater/ (accessed October 28, 2012).

Schmidt, H.H. 2004. Potential resources of Mars. NASA/JPL Lecture 19 NEEP 533, online PowerPoint: 53 p. fti.neep.wisc.edu/neep533/SPRING2004/lecture19.pdf (accessed October 28, 2012).

Vita-Finzi, C. 2005. *Planetary Geology: An Introduction.* Harpenden, UK: Terra Publishing: 168 p.

Additional Reading

Baldridge, A.M., J.D. Farmer. 2004. Mars remote sensing analog studies in the Badwater Basin, Death Valley, California. *JGR* 109, E12006, 18 p.

Section VI

Remote Sensing, Geoscience, and the Public

Remote sensing has a role to play in educating the public regarding exploration and development in extractive industries. It has a regulatory function in locating, monitoring, and measuring the impact of development. It can monitor the availability of surface water, the degradation or improvement of surface water quality, and can help predict erosion, flooding, and other natural hazards. It can illuminate resource-related, hazard and health-related, and climate-related matters of public interest. Monitoring of ice caps and sea level is helping to measure the impact of global warming. Imagery can help mobilize community action to eliminate or mitigate adverse environmental impacts.

The development of geographic information systems, global positioning systems, and web-based maps has revolutionized the way the public interacts with their environment. The geoscientist can now look over an area before starting a field mapping and sampling program. You do not have to be a skilled analyst to use Google Earth to check out an area while planning a vacation or an off-road weekend. We can generate topographic maps when interested in a property, and find out what the local seismicity is. A real estate agent can determine whether there are indications of unstable slopes when buying or selling a home. Imagery can help governments plan disaster relief, and it can help news organizations show the public where flooding will occur or has occurred. In these and many other ways, remote sensing has become just another part of every-day modern life.

20

Public Relations, the Media, and the Law

Chapter Overview

It is not going too far afield to say that remote sensing can influence the public, the media, and legal proceedings related to geosciences. Whether in a university setting, government, or industry, everyone has to deal with the public at some point. Remote sensing has a role to play in convincing the public that development can be monitored, controlled, and made safe. Imagery can show areas that historical development has or has not adversely affected, and can demonstrate that processes are in place to locate, measure, and mitigate adverse impacts.

The media can affect public opinion by attacking a development or by pointing out the extent of environmental impact in a negative or positive way. Print and television news programs tend to emphasize the sensational, but can also educate. Social media can overturn governments if enough people are aroused to action. Pictures are indeed worth thousands of words. Imagery can and does provide a "good visual" to get a point across.

Regulatory hearings and legal claims can be supported or refuted using historical and present-day imagery. Due diligence can be exercised at least in part by reviewing past imagery. Liability can be assigned based on evidence from archival images.

In these and many more ways, remote sensing can influence outcomes. In this chapter, we examine some ways in which this happens.

Public Relations

Government agencies (a geological survey, a space agency, an environmental protection agency) have to show the results of their efforts, how they help the public, and why they are worth funding. Educational institutions need to support their research programs. Oil companies, mining companies, and geotechnical firms all have to deal with the public, whether at stockholder meetings or environmental impact hearings. Images put things in a spatial and temporal context that people can relate to. Base maps derived from images or digital elevation models (DEMs) generated from radar or laser altimetry are the foundation for running surveys or starting developments (Figure 14.2). People can see what has been or is being affected by drilling a well, building an access road, or constructing a tailings pond (Figures 20.1, 20.2, and 21.1). They can also begin to see what might be affected over time and how it relates to them.

Everyone loves to see an image of their hometown and of places they have been (Figures 3.61, 20.3 and 20.4). The value of airphotos and satellite images as a means to

FIGURE 20.1
Eagle Ford play area, McMullen County, Texas. Natural color digital orthophotography with 30 cm resolution shows distribution and status of wells and mud pits. (Courtesy of Spatial Energy and GeoVantage Inc., Peabody, MA. All rights reserved.)

FIGURE 20.2
GeoVantage image, Ponderosa Park, Elbert County, Colorado. Natural color digital orthophotography with 15 cm resolution. (Courtesy of Spatial Energy and GeoVantage Inc., Peabody, MA. All rights reserved.)

familiarize the public with an area and program should not be underestimated. An image of a city, county, or province, with a prominent logo, in the office of a governor, environmental agency head, minister of natural resources, or in a local school can be a source of good will and convey important information about a project and its environment. Pictures go a long way toward generating favorable publicity. Producing high-quality imagery is a wonderful and effective way to promote the high-tech capabilities of a company through an annual report or media advertising.

Environmental groups are not the natural enemies of resource extraction industries and other types of development. Providing imagery over migratory bird nesting areas or caribou migration routes can educate, and perhaps convince groups that development is benign and environmental impact is minimal. Imagery can also convince companies why they should not develop in an area, or to modify their development to accommodate the environment. At the very least, imagery serves as a nonbiased starting point for discussion

FIGURE 20.3
GeoEye image of the Jefferson Memorial, Tidal Basin, and part of the Mall, Washington, D.C., 23 March 2012. Natural color image has 50 cm resolution. (Satellite images courtesy of GeoEye and Spatial Energy.)

FIGURE 20.4
Pléiades image of Paris showing the Eiffel Tower, River Seine, and Arc de Triomphe. 50 cm panchromatic commissioning phase data merged with color bands. (© CNES 2012 – Distribution Astrium Services. Courtesy of Astrium Services. Copyright Astrium Services 2013.)

and may help generate a cooperative atmosphere when projects of mutual concern are being planned.

Companies are always trying to recruit the best students from universities. The goodwill and publicity generated by providing imagery to a geology, engineering, or environmental sciences department can help make students aware of a company and impress them with its technical ability and environmental ethic. At the very least, it shows that a company is willing to use cutting-edge technology for solving exploration and environmental problems.

Media

There are always news stories that involve the geosciences. They involve global warming, earthquakes, volcanoes, flooding, erosion, and the search for water on Mars (Figures 16.18, 16.19, 20.5 and 20.6). Stories can be as direct as explaining why sinkholes are forming under a subdivision (Figure 16.26) or showing how tourists got caught in a flash flood; they can be as subtle as trying to identify the location of caves sheltering terrorists.

With public access to satellites that have resolution on the order of half a meter to a couple of meters, what was once a capability available only to spy satellites is now accessible to the media. There is no question that the media use the products of remote sensing. Neither

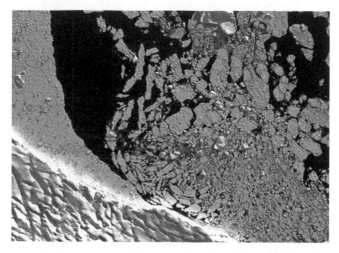

FIGURE 20.5
Ice breakup around Adelaide Island, Antarctic Peninsula. The media use images such as this to help explain global warming. GeoEye 50 cm resolution true color image acquired April 18, 2012. (Satellite images courtesy of GeoEye and Spatial Energy.)

FIGURE 20.6
NASA Aqua satellite image of Eyjafjallajökull volcano, Iceland. The eruption during May 2010 released an ash plume, seen clearly in this image, which disrupted air traffic over Europe for several weeks. (Courtesy of NASA.)

your evening news anchor nor Hollywood always explains imagery properly. It has been implied, for example, that thermal images can see through buildings to find victims of earthquakes, or that surface materials can in all cases be unambiguously identified. It is up to the remote sensing community to educate the media whenever possible as to what can and cannot be seen on, or interpreted from imagery.

Tutorials on remote sensing are available from many sources including NASA, the Canada Centre for Remote sensing, Penn State, and Dr. Nicholas Short. Links include http://landsat.gsfc.nasa.gov/education/tutorials.html, http://www.nrcan.gc.ca/earth-sciences/geography-boundary/remote-sensing/fundamentals/1430, https://wikispaces.psu.edu/display/RemSens597K/A and http://www.fas.org/irp/imint/docs/rst/Front/overview.html. (accessed June 2013).

Remote Sensing and the Law

Politics, scare tactics, and pseudoscience frequently determine how environmental issues are perceived by the public. Remote sensing offers those who use the technology an unbiased way to meet the requirement for cost-effective environmental monitoring. Imagery is not only efficient and effective, but it can help avoid alarmist responses that exacerbate real or perceived problems.

Regulations

The regulatory process in most countries is highly dependent on monitoring both the environment and development. For example,

> Sites contaminated with radioactive waste, mine tailings, or other material can have severe impacts on both ecosystem integrity and human health. In this regard remediation is aimed at diminishing the interaction between the concerned site and the impacted environment…. Overall site management (characterization, evaluation, risk assessment, remediation, monitoring) aims to protect human health and ecosystem condition…. Environmental monitoring of remediation management activities researched through the hyperspectral program at Canada Centre for Remote Sensing includes the evaluation of imaging spectrometer data (hyperspectral imagery) for characterizing environmentally contaminated sites and providing techniques to monitor their remediation.
>
> **White, online**

The permitting process usually requires an environmental impact statement that describes plants, animal habitats, breeding areas, and sensitive ecosystems, and how these will be affected by the proposed development. There will be further scrutiny if endangered plants or animals exist in the area. In the United States, the Federal Resource Conservation and Recovery Act (RCRA) of 1976 outlines a system of permitting, record keeping, and waste management for companies that generate, transport, treat, store, or dispose of hazardous wastes. These regulations define hazardous wastes to include certain tailings, drilling fluids, and produced water. This has an impact on mining and drilling operations for oil, gas, and geothermal exploration and production. Records are required showing the disposition of dumps, tailings, mud pits and drilling fluids. Imagery can provide such records.

In the United States, the Comprehensive Environmental Response, Compensation and Liability Act (CERCLA) of 1980 is designed to help clean up hazardous waste sites where no responsible party was available to pay the cleanup costs. CERCLA imposed retroactive liability on parties responsible for hazardous sites. Those responsible parties are defined as past and present owners or operators of a site, those responsible for disposal at a site, or those who selected a site. Those who purchase properties where contamination is later discovered routinely sue sellers under CERCLA. Courts have imposed liability on banks and other lenders who participated in management of the polluted site. "Bankers are becoming increasingly concerned about lending money for a property that may contain hazardous waste. Banks can end up as defendants in lawsuits arising from deals involving real estate with hazardous waste" (Bivens, 1990). An industrial park in Rochester, New York, was built on a sanitary landfill. New York's Environmental Conservation Department found radioactive thorium sludge on part of the property and listed it as a hazardous site. It is not considered a health threat, yet banks won't make loans there, so the business can't sell, refinance, or expand (Samuelson, 1990). In many countries, oil and mining companies have been told to clean up environmental hazards that existed prior to their acquisition of a lease (Figure 13.3). These factors have dramatically increased the demand for environmental audits. As mentioned in Chapter 14, imagery is an essential part of any environmental review, providing the historical evidence that is often lacking when the original property owners are uncooperative or their records are unavailable.

Remote sensing images serve as point-in-time documentation of environmental conditions in an area (Figure 14.35). This type of documentation is valuable both before and after an environmental problem occurs. Since most imagery is in the public domain, multitemporal records provide indisputable and unbiased evidence of the surface condition of an area. Imagery shows the extent of environmental damage that existed on a lease before a company acquired property, thereby limiting the liability of the present landowner (O'Donovan, 1989). Imagery shows the extent of damage, the timing of spills, leaks, trespass, or dumping, and may provide clues as to the party involved.

Law Suits

In recent years, there have been numerous cases involving contaminated groundwater and the oil and mining industries. Private landowners claim that past activity on or near their land has contaminated their surface or well water. The cause of action for lawsuits can be filed as a nuisance (seeping water, offensive smell, leaking fumes or gasses), negligence (unreasonable use of the surface owners' land), gross negligence (reckless disregard of the consequences of one's conduct, say, in reinjecting drilling fluids), and possibly trespass (unauthorized entry, including cracks from a frac job or subsidence due to mining). To be useful in a court of law imagery must meet certain criteria: 1) it must be clear, 2) it must be credible, and 3) it must be pertinent. The imagery must not be ambiguous: it should be clear to an unbiased observer what is being shown. The imagery must be credible, that is, from a credible source and not retouched or Photoshopped. Finally it must address the issue at hand.

It may be possible, using historical imagery, to show exactly where mud pits and tailings piles were, whether liners were or were not used, whether there were crude or fuel spills around a well, the disposition of drilling fluids, produced water, and cuttings, and the extent of actual subsidence related to abandoned mines or hydrocarbon or water withdrawal.

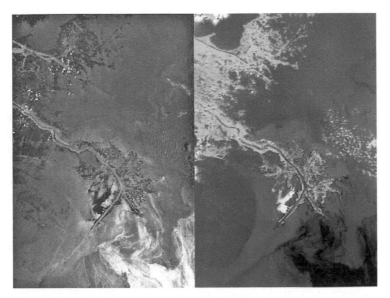

FIGURE 20.7

MISR images acquired in May 2010 show the encroachment of oil from the former Deepwater Horizon. Oil appears bright on the near-true color image (left); oil appears dark blue to black on the enhanced image (right). The images cover an area 110 × 119 km. (Courtesy of NASA/GSFC/LaRC/JPL MISR team.)

In the case of tanker or pipelines spills, the liability lawsuits will be based on negligence that can be assigned in three categories:

1. The grounding, collision, or source of the break
2. The containment
3. The cleanup

It is not likely that there will be image documentation of the accident itself. Immediately afterward, however, it will be possible to use imagery to map the distribution and direction of movement of a spill so that it is possible to contain and clean the pollution using appropriate techniques and adequate resources. The images shown in Figure 20.7 were acquired on May 24, 2010, by the Multiangle Imaging SpectroRadiometer (MISR) on NASA's Terra spacecraft. They show the encroachment of oil from the former Deepwater Horizon rig into Louisiana's wildlife habitats. The source of the spill is off the southeast (bottom right) edge of the images. Dark filaments of oil can be seen approaching the east edge of the Mississippi River delta. These areas are home to many varieties of fish, birds, and other wildlife. This image shows stringers of oil crossing the island barrier—which had been heavily eroded by Hurricane Katrina in 2005—and entering the Breton and Chandeleur Sounds. On the near-true color image (left), oil makes the surface appear brighter. To separate oil-related factors from other factors such as wind-related glint, the MISR imagery was enhanced using a combination of vertical and oblique views and several MISR channels. In the false-color view (right), oil appears in shades of dark blue to black; silt-laden water from the Mississippi River is orange, red and violet; and land and clouds appear as shades of cyan (NASA, online).

Images such as these are used to monitor spill-related environmental changes and as evidence in damage hearings. Most compensatory damage calculations depend in some way on the extent of damage to natural resources. It thus becomes essential to quantify the impact of a spill, broken tailings dam, or eroded mine dump. This means an environmental audit or damage assessment program. Such a program measures the extent of the injury to resources, assesses the impact such that full compensation will be provided, and determines how rehabilitation or restoration can be assured. Imagery can be used to help determine the extent of injury and guide rehabilitation efforts by monitoring the progress of the recovery program and comparing conditions to those existing before the accident.

Perhaps the simplest application of remote sensing is to show compliance with regulations. For example, thermal images of oil slicks related to shipping, or of a refinery's hot water outflow will show whether or not discharges are within environmental agency guidelines (Figures 14.22 and 14.23).

Images can be used in international disputes to demonstrate to a court, for example, that a nationalized lease is still producing, or to document illegal operations. One country may want to show that its neighbor is taking more water from a river than it has a legal claim to. In such cases, satellite imagery can clearly show how much water is being taken for irrigation, flares from wells, and piles of sulfur near the refinery where the oil is processed (Figure 13.5). This could help to direct a country into compliance or determine the amount of compensation due to a former lease owner.

We have seen how remote sensing can influence the public, the media, and regulatory and legal proceedings. Imagery provides a "good visual" for the media. Regulatory requirements have been met by monitoring and remediation. Damages have been assessed, legal claims have been supported (or refuted), and compensation has been determined, at least in part, using archival and present-day imagery.

References

Bivens, R. 1990. *Banks give boost to environmental exam business.* The Houston Chronicle (November 18, 1990).

NASA. NASA images show oil's invasion along Louisiana coast. http://www.nasa.gov/topics/earth/features/oil20100602.html (accessed November 4, 2012).

O'Donovan, P. 1989. The law and remote sensing evidence. In A.E. Lodge (ed.), *The Remote Sensing of Oil Slicks*. Chichester, IL: John Wiley and Sons: 105–113.

Samuelson, R.J. 1990. *Who cleans up the waste?* Newsweek, May 20: 49.

White, H.P. Contaminated site remediation monitoring. http://www.nrcan.gc.ca/earth-sciences/geography-boundary/remote-sensing/hyperspectral/1631 (accessed October 29, 2012).

21

Geographic Information Systems and Online Remote Sensing Resources

Chapter Overview

In the age of Google and smart phones, there is a lot that an individual can do using readily available public remote sensing resources. Many satellite images are available free from NASA or the U.S. Geological Survey (USGS) EROS Data Center. Individuals can download Google Earth at little or no cost and view any place on Earth, including the oceans. Many areas covered by Google Earth also have digital topography such that one can now do a virtual "fly through" of an area of interest. Google Earth and Google Maps® have layers of data, a virtual geographic information database, which can be turned on and off as needed. These auxiliary datasets include roads, political boundaries, buildings, services, even volcanoes, and snapshots contributed by the public. Yahoo!® Maps and Google Maps® offer not only maps but also satellite images of areas of interest, and map-satellite merged images at a variety of magnifications. These applications will take a starting and ending location and plot the shortest route between the two based on distance and/or travel time and provide directions. Yahoo!® Maps even provide traffic speeds at any given time. Google Maps® has a "Street View" feature that allows the user to see what is at a location at street level, and pan the camera 360°, over many areas of the world.

Applications (apps) have been developed that allow users to take advantage of remote sensing technology, geographic information systems (GIS), and global positioning systems (GPS). Your mobile phone and even your jogging watch may have GPS tracking that provides you with a printout of where you have been, your travel time, and your speed. The Earth Observer app ($0.99) developed by Colombia University (http://itunes. apple .com/us/app/earthobserver/id405514799?mt=8) provides iPhone, iPod Touch, and iPad users with a range of GIS layers that include imagery from Landsat 7, airphotos, Advanced Spaceborne Thermal Emission and Reflection Radiometer (ASTER) and Shuttle Radar Topography Mission (SRTM) digital elevation models, and NASA Visible Earth (http://visibleearth.nasa.gov/, updated monthly). Data types include onshore and offshore geologic maps, seismology, gravity, magnetic, snow cover, land and sea temperature, and weather maps. Since 2011 NASA's Visualization Explorer (NASA Viz) app has been providing two stories a week with an emphasis on Earth science (http://www.nasa.gov/ topics/technology/features/nasa-vis-update.html). Apps for the iPad include seismology (e.g., Quakewatch by LateNightProjects, $0.99), weather information (e.g., Weather+ by International Travel Weather Calculator Association, $0.99), star charts (e.g., Star Chart by Feel Great Publishing Ltd, $2.99), and "GeoEye" (free; http://www.geoeye.com/CorpSite/) that offers high-resolution GeoEye imagery from around the world. NASA provides free apps related to remote sensing for Android, iPad, and iPhone (http://www.nasa.gov/centers/ames/iphone/index.html). The DLR also offers free remote sensing apps for iPad, iPhone, and iPod Touch in both German and English (http://itunes.apple.com/us/app/dlr/id418202930?mt=8).

Research resources are offered online by many universities (e.g., University of California Berkeley: http://cluster3.lib.berkeley.edu/EART/aerial.html; University of Calgary: http://www.ucalgary.ca/f3gisci/) and their facilities may be available to the public. User groups such as the Geologic Remote Sensing Group (chairman@grsg.org.uk), AmericaView (http://www.americaview.org/), the (U.K.) Society of Remote Sensing and Photogrammetry (http://rspsoc.org/information-zones/public-outreach/), or Arizona State University's Open Topography (http://www.opentopography.org/index.php) can be accessed online. Remote sensing resources available to the public online are varied and increasing daily.

Geographic Information Systems and Mapping

GIS have become the way to manage data for all large projects. The GIS is simply a relational database that allows the user to call up and integrate several types of data based on geographic location. There are three attributes that make a GIS useful: (1) the ability to automatically locate an object based on a set of coordinates, (2) the ability to relate an object to other nearby objects, and (3) the ability to add descriptive text or tabular data linked to the object of interest. Generally speaking, remote sensing imagery or topographic maps are used as the base for overlaying graphical information including physical or administrative features. Physical information includes geology, roads or tanker lanes, river drainage networks, oil wells, mine pits and tailings, pipelines, and power lines. Administrative features include land use, wildlife habitat, rights-of-way, and land ownership. By specifying an area on the base map, one can be hot linked to all types of information related to that area, including historical photos, diagrams, video clips, and text such as lists of wells, soil samples, engineering reports, or legal land descriptions. The GIS allow the user to change scales for greater or lesser detail, and permits the interpreter or database manager to update all levels of the database instantly. It also provides the user instant access to various levels of data detail, which can be critical in the rapid and accurate analysis of diverse information. The user is able to point to a map feature and ask "what are you?" One can query the database to "show all the properties crossed by a pipeline right-of-way from point 'a' to point 'b'", or "show all the areas where bedrock is shale, dip is greater than 30 degrees, rainfall is greater than 40 cm/year, and vegetation cover is sparse."

It should be evident that a GIS can be very useful for engineering, exploration, and environmental purposes such as powerline construction (Wyvill, 1992), reclamation (Peters and Kirschner, 1996), project management (Ellis, 1997), pipeline inspection (Jadkowski, 1994), and open-pit slope stability analysis (Nelson et al., 2007). Two case histories are provided to illustrate applications of a GIS to engineering remote sensing.

Case History: Mine Management Using a Geographic Information System

Highland Valley Copper (HVC) had the largest base metal mine in northeast British Columbia, Canada, in 1996, shipping about 320 million pounds of copper (Speed, 1996). They used GIS technology to make and update maps on a daily basis. These maps support exploration for new ore bodies, the mining process, and restoration of the land. The company holds about 5900 hectares of disturbed land including two large open pits, facilities, dumps, and a tailings pond (Figure 21.1). Local landowners include the provincial government, forest companies, and private farmers and ranchers.

Historically, maps were made individually in a CAD system and were difficult to maintain, update, or scale up or down. Information existed on more than one map or computer file, and updates were not made consistently across all versions. With the introduction of a GIS, the engineers at HVC were able to take all existing maps and generate them in any coordinate

(a)

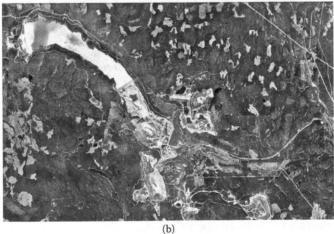

(b)

FIGURE 21.1
Revegetation is underway at the Highmont and Lornex pits, dumps, and tailings piles, Highland Valley, British Columbia. A GIS is coordinating the process (Speed, 1996). (a) Oblique view northwest over the pits toward the tailings in Highland Valley. (b) True color image of the Highland Valley mine pit and tailings. (Courtesy of Digital Globe, CNES/SPOT Image, and Google Earth, © 2012 Google.)

system and to any scale. They also added baseline data from aerial photography and topographic maps at 1:5000. This information is available to all the staff, from engineers to geologists and environmentalists, immediately after it is brought into the GIS environment.

During the mining process, it is important to monitor slope stability. Pit crews expect to be able to record field observations tied to GPS; to show rates of ground movement; the orientation, size, offset, and amount of movement on joints and faults; to map changes in hydrology (water table, springs); and to show the relationship of these features to buildings and other structures. This is accomplished using the GIS.

In the early 1990s, reclamation cost HVC about $2 million annually. About 1300 hectares had already been reclaimed. Reclamation involves resloping and contouring the land, capping waste dumps with soil, preparing the soil to seed, seeding and planting, then fertilizing and maintaining the area for up to 4 years. The GIS allow them to prepare maps of the reclamation status quickly and accurately. In addition to reclamation, the GIS show the land ownership situation in Highland Valley. There is an ongoing need to document and update leases, rights-of-way, and water rights, among other items. The GIS allows HVC to do this in a timely manner on an ongoing basis.

Case History: Pipeline Routing Using Remote Sensing and a Geographic Information Systems

Engineers at Bechtel planned a 700 km pipeline to bring oil from the Tengiz field in Kazakhstan on the Caspian Sea to Novorossiysk in Russia on the Black Sea (Feldman et al., 1995). The completed pipeline would be capable of delivering 1.5 million barrels of oil per day. A least-cost analysis was performed incorporating pipeline length, topography, geology, land use, and stream, wetland, road, and railroad crossings. Landsat Thematic Mapper (TM) and SPOT imagery was used as a base on which to display results and for land cover mapping.

Most of the pipeline route passes through low-relief terrain along the northern end of the Caucasus Mountains. Tectonic activity began in the Caucasus in Miocene time, with the most intense uplift and folding occurring in the Pliocene and Quaternary. Many faults are still seismically active. Although the shortest and generally least cost pipeline is a straight line, factors to be considered included avoiding population centers and environmentally sensitive areas, and technical issues such as slope steepness and stability, geology, and river crossings. Rugged topography requires more pump stations, whereas hard rocks require blasting, and river crossings require bridges.

Maps and high resolution imagery are needed to evaluate these factors and make routing decisions. Airphotos were not available, so 1:10,000 and 1:200,000 topographic maps and TM and SPOT imagery were used. A supervised classification was performed on the satellite imagery to generate a land use map (Figure 21.2). Topographic, drainage, geologic, and infrastructure (roads, cities) data were also input as separate layers to the GIS. A slope steepness map was derived from the topographic elevation information. The least-cost analysis was performed by assigning cost factors associated with crossing of steep versus gentle slopes, streams, wetlands, roads, railroads, outcrops, agricultural land, and urban and industrial areas. From this, it was possible to generate a "weighted cost surface," that is, a surface analogous to a topographic surface with peaks and troughs that correspond to high and low costs, respectively (Figure 21.3). The "least cost" path could then be traced. Although the straight line path was shorter, the GIS analysis indicated that the least cost path was 14% less expensive to build.

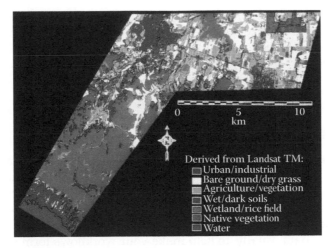

FIGURE 21.2
Land use map of the Caspian pipeline corridor derived from TM bands 3, 4, and 5. Image processed by Bechtel. (From Feldman et al., *Rem. Sens. Environ.* 53, 123–131, 1995.)

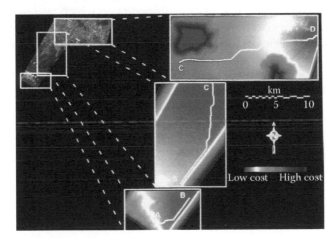

FIGURE 21.3
Cumulative cost surface and least cost path generated from a GIS database for the Caspian pipeline. Image processed by Bechtel. (From Feldman et al., *Rem. Sens. Environ.* 53, 123–131, 1995.)

Online Resources for Remote Sensing

Matt Hall, Agile Geoscience

Online resources for remote sensing and geology are being added on a daily basis. For example, the Utah Geological Survey recently created an online application that allows the user to overlay geologic maps on DEMs and virtual globes to create a "virtual field trip" (http://geology.utah.gov/geo_guides/st_george/). This is a brief review of some of the resources available online.

The following section is contributed by Matt Hall, Agile Geoscience. Matt Hall is a geoscientist in Nova Scotia, Canada. Find him online at agilegeoscience.com and on Twitter at twitter.com/kwinkunks.

Geospatial data analysis has been transformed by the Internet. The Internet hosts dozens of giant data and image repositories, many of which are free to download and use. There are software tools that could not have been developed or distributed without the Internet. On the social web, you will find hundreds of blogs and forums where people share their insights and workflows in remote sensing and GIS. Let us look at each of these resources in turn.

Data Repositories

Many remote sensing images are public data: Landsat images, for example, are copyright free and the entire archive is online. Amazingly, new Landsat 5 and 7 images of the United States are online within 24 hours of acquisition. This is just one of the dozens of freely available global datasets. To introduce some of the resources on the web, we look here at some places to find data, innovative ways people are finding new data sources, and some tools for publishing data. Finally, to help make your workflows format independent, you should be aware of some of the standard formats for data interchange.

There are so many data sources and tools for accessing them, that it can be a little overwhelming. Many of the repositories carry the same data, however, so you can narrow it down to a few tools. Which one you use just depends on which you prefer.

Repositories that are available to the public and allow browsing and ordering of data include the U.S. National Aeronautics and Space Administration (NASA), the USGS, and the European Space Agency (ESA).

The first three of the following are part of the USGS Land Processes Distributed Active Archive Center (LP DAAC, lpdaac.usgs.gov/lpdaac/get_data).

NASA Reverb (reverb.echo.nasa.gov)

Reverb houses about 2700 spatial and temporal Earth science datasets for NASA's Earth Observing System (EOS). It is particularly well suited to climate observation data. The online catalog allows you to select an area from a map, then apply filters based on various remote sensing platforms and instruments, including Landsat, Meteosat, and GOES. It is worth setting up a free user account to get access to more features.

Rapid Response (lance.nasa.gov/imagery/rapid-response/)

Rapid Response allows the user to browse and download NASA's Land Atmosphere Near Real-time Capability for EOS (LANCE) images by region or orbit. You can view and download Earth's observation imagery using the interactive Web Mapping Service (WMS). Multiple data products can be overlaid with population information and administrative boundaries from the Socioeconomic Data and Applications Center. A large number of geo-rectified images from around the world are available in GIS compatible formats.

USGS EarthExplorer (earthexplorer.usgs.gov)

EarthExplorer is intended to be a one-stop repository for Earth imagery. Like Reverb, it provides access to global satellite, aerial, and cartographic images via a Google Maps interface, then the user selects which datasets to search. It is perhaps better suited to land cover and other surface studies than Reverb. The list of data is very long indeed: there are raster data from NASA's Earth Observing orbiter (EO-1), ASTER, MODIS, and Landsat, as well as time series from global projects such as the Forest Carbon Tracking Task of the international Group on Earth Observations.

USGS GloVis (glovis.usgs.gov)

The *Global Visualization Viewer*, or *GloVis*, is a search tool for satellite data. You can browse images from Landsat, EO-1, ASTER, MODIS, and the Landsat Tri-Decadal project (1975–1990–2000). The map interface is not as smooth as EarthExplorer, but you can specify a maximum allowable cloud cover and date ranges, and you see the images immediately, making it easy to select the ones you need.

ESA EOLi (eoli.esa.int)

The *Earth Observation Link* tool, or *EOLi*, must be downloaded and installed to use, but is available for all major platforms. By means of a virtual globe interface (Figure 21.4), it provides download or ordering access to dozens of free and proprietary imagery catalogs, including Landsat, ASTER, ENVISAT, ERS, IKONOS, and SPOT. It also offers gravity data from GOCE, and greenhouse gas data from the Japanese GOSAT project.

Community Remote Sensing

Social media permeates the Internet (see sections on Networks and Forums). Beyond merely keeping in touch and up to date, however, researchers are getting creative with social media data. The basic idea is that social networks are themselves remote detectors.

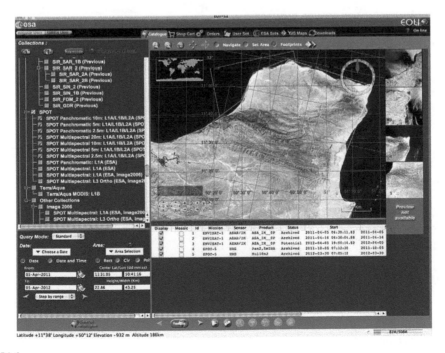

FIGURE 21.4
ESA's EOLi catalog tool provides access to dozens of online image repositories. Shown are some SPOT images from the Gulf of Aden rift zone, Somalia, Horn of Africa.

Bill Gail, a remote sensing scientist at Microsoft, coined the term "community remote sensing" for the collection of data from highly distributed, crowd-powered networks. The advantage of such data is that they are real-time or near-real-time, and close to the phenomenon or event in question, helping to close the "scale gap" of satellite-based observations (ref: goo.gl/O12bK). The disadvantage is that the data are bobbing along in a tide of unrelated, and mostly uninteresting, data.

Ongoing Research

The U.S. National Center for Atmospheric Research wants to build *IntelliDrive*, a vehicle communication network to collect real-time data from cars about air temperature, and the use of windshield wipers (ref: goo.gl/sAZOP).

Whatsinvasive.com educates users about invasive species in the United States and gathers geotagged images of possible sightings.

The USGS is experimenting with scanning Twitter for updates mentioning earthquakes, after a suggestion by a blogger. With appropriate filtering, this data mining could provide valuable information about apparent earthquake magnitude (Figure 21.5; goo.gl/CkEcX).

In a similar way, the Air Twitter project aims to collect air quality data (ref: goo.gl/PHInM).

The concept extends to natural phenomena in general: think about the likelihood of someone with a smart phone or digital camera witnessing once-in-a-century events today, and the

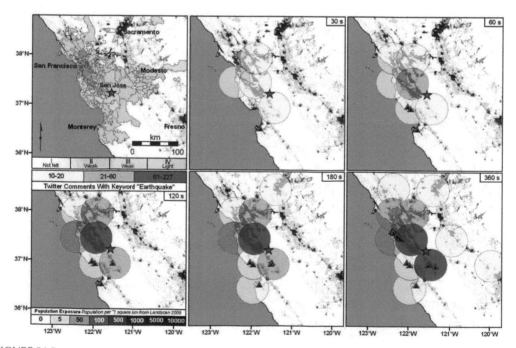

FIGURE 21.5

Geocoded tweet counts at discrete times in the 3 minutes after the March 30, 2009, magnitude 4.3 Morgan Hill earthquake in California. The color indicates the number of tweets. Compare with the map based on the "Did You Feel It" program (upper left). (From Earle et al., *Seismol. Res.* Lett. 81, 246–251, 2010.; maps courtesy of U.S. Geological Survey.)

"skin" of photographs covering the Earth at Flickr.com. The Tōhoku tsunami in Japan in March 2011 was the first event of its kind to be caught on video at multiple locations. University of Durham professor Dave Petley maintains landslideblog.org, with footage of recent landslides from all over the world. These observations quickly become data for geoscience researchers.

Publishing Data

As well as being easy to find and consume data and images on the Internet, it is also easy to publish your data and images, if you are able and inclined to do so. For example, it is simple to share a spreadsheet of data, or even a "Fusion Table" database, from the free Google Docs service. You can publish interpreted images on *Figshare*. You can publish papers in *Remote Sensing*, an open access journal published by Molecular Diversity Preservation International (MDPI) of Basel, Switzerland. You can write about your data and methods on a blog.

It is also possible to expose large image databases to a public or private audience through a map server. This requires some know-how and resources, but a basic setup is within anyone's reach. There is no shortage of technology options, from open-source options such as GeoServer, MapServer, and World Wind Server, to proprietary solutions such as ArcGIS Server, ArcIMS, and the web-ready WeoGeo.

Standards

The Open Geospatial Consortium, an international voluntary organization, has maintained standards for geospatial content and services since 1994. Some of the emerging standards in online geospatial information are as follows.

Geographic Markup Language, or GML, is a system of XML-based tags for two-dimensional (2D) and 3D geometries representing geographic information. It can encode almost any kind of data, including coverages and raster, and is accordingly large and complex.

Keyhole Markup Language, or KML, is in some ways a subset of GML, tailored for Google Earth. It is increasingly widespread on the web, and is a convenient way of interchanging data between, say, Google Earth and ArcGIS.

WMS is a server protocol for retrieving georeferenced images over the Internet. It is used inside most of the open-source and proprietary server-side and client-side map applications on the web.

As well as the standards maintained by the OGC, there are other protocols in use on the web. An important one is COLLADA, which defines an open XML schema for 3D data interchange. Files usually have a *.dae* (digital asset exchange) extension. Like much 3D computer graphics technology, its roots are in the entertainment industry, although tools such as ArcGIS, Google Earth, and NASA World Wind support it.

Software Tools

There are two classes of software tools that depend on the Internet: web applications, which run inside a browser or on a mobile device; and desktop applications with critical online components, usually data. An obvious shortcoming of such tools is that they are usually completely useless without a high-speed Internet connection, not to mention the usual hardware requirements and software dependencies. However, when Internet connectivity is guaranteed, there are increasingly powerful tools at our disposal.

Another class of tools that are enabled by the web is open-source software, or software whose code is available for inspection, reuse, and modification. Usually free of charge, open-source software is otherwise just like any other software and not necessarily dependent on the Internet to function. Nonetheless, the tools could not exist without the Internet—it would be hard to coordinate the community to develop it, and distribution would be difficult.

We can also classify online-based tools according to their function. There are three important functional types that are interesting to geoscientists:

1. Geographic information systems
2. Virtual globes
3. Image analysis tools and libraries

Let us look at some of these in more detail.

Geographic Information Systems

Open-source GIS are similar to proprietary tools like ArcGIS and MapInfo, so there is no need for an exhaustive treatment here. Only enabled by the Internet insofar as they are open source, these tools do not have any unusual degree of connectivity or awareness of online data repositories. The most commonly encountered open-source GIS are GRASS GIS, Quantum GIS, and SAGA GIS. You will find a complete list on Wikipedia (ref: goo.gl/AkKoE).

An emerging application type, related to virtual globes (next section), is the web-based GIS. There are no highly functional examples of this genre yet, but ArcGIS Online, ArcGIS Explorer, Google's extended software platform, and the direction of web applications in general all point toward more web-based GIS in the near future.

Virtual Globes

The concept of a virtual 3D representation of the Earth was first explored by Autometric, a small company spun out of Paramount Pictures in the mid-1990s. Computer graphics company SGI took the idea further, mainly as a promotional tool for their InfiniteReality graphics engine.

Two of the core technology concepts underlying virtual globes, which must process very large amounts of data rendered at a wide range of scales, are texture mapping and level of detail. Texture mapping adds surface detail, from satellite or other images, to a 3D model of the Earth. Level of detail means reducing the resolution of distant objects, for example, with so-called mipmaps (multiresolution images, akin to a set of spatially filtered images), which can be processed to adapt to the scene in real time.

There are at least a dozen virtual globe applications (ref: goo.gl/7dp7X), but three deserve special mention: Google Earth, NASA World Wind, and OpenWebGlobe.

Google Earth

This tool began life in 2001 as Earthviewer, a product of Keyhole, Inc., a private company funded by Sony, graphics hardware company NVIDIA, and the U.S. Central Intelligence Agency. After being acquired by Google in 2004, *Google Earth* went from occupying an exclusive niche to being free and indispensible in a matter of months. Geoscientists quickly realized that for them, this was one of the most important innovations of the decade.

Today, the tool is almost totally platform independent, highly functional, and approaches a 3D GIS in scope. It can display points, polygons, georeferenced images, and 3D objects such as buildings. It can import comma-delimited (CSV) data and GPS waypoints. It can read local and online data, including private image databases. The tool is not even restricted to the Earth: you can also explore Mars and the moon, or indeed any globe you have imagery for, as well as visualizing the night sky in its true geographic and temporal configuration.

As well as operating as a stand-alone application, it can be integrated into any web page via a browser plug-in and a JavaScript interface. The plug-in supports terrain, cultural layers, 3D buildings, and photographic overlays.

Unless you create a custom globe with proprietary data, Google Earth images are copyrighted and not licensed for reuse; this would require permission from Google, which you can apply for online (ref: goo.gl/Z75sv). If you think your use of a screenshot falls within "fair use," then one of the license terms is that you must legibly reproduce Google's copyright and attribution notice immediately adjacent to the image.

NASA World Wind

There are two substantial advantages to *World Wind*, compared to Google Earth: first, the software is open source, which means you are free to modify it; second, the default imagery is public domain, which means you can use it for any purpose without permission.

The World Wind application only runs on Windows and requires the .NET and DirectX libraries, but NASA maintains a free and open-source Java software development kit (SDK) for World Wind at goworldwind.org. This means that, although there is no platform-independent application, any number of platform-independent tools can be built from the

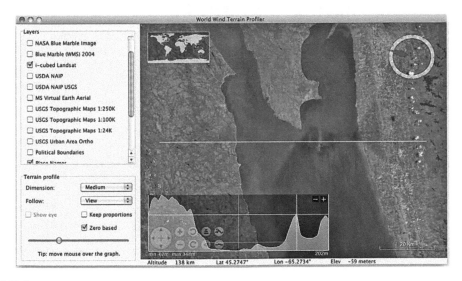

FIGURE 21.6
The World Wind Terrain Profiler displaying the elevation in a section across the Bay of Fundy between New Brunswick and Nova Scotia, Canada. (This is one of the demos of the World Wind Java SDK available at goworldwind.org/demos.)

SDK, as demonstrated by the various demos provided by NASA (Figure 21.6). Indeed, NASA is especially keen to see development on Google's Android operating system.

The tool displays low-resolution imagery from NASA's 2005 Blue Marble Next Generation image (a cloud-free MODIS image at 500 m per pixel), and high-resolution image from the Landsat GeoCover product (a pseudocolor, orthorectified image at 15 m per pixel) and aerial photography from the USGS. Several extraterrestrial datasets are also available, for example, from the several Martian imagers, and the Magellan mission to Venus.

A range of plug-ins extends the functionality of the Windows application, but since the software is open source, you can make World Wind do almost anything you like. The SDK supports all the functionality of Google Earth (image overlays, 3D buildings, and so on), but the implementation is up to you and the community at large. Find out more about it at the worldwindcentral.com wiki.

OpenWebGlobe

Like World Wind Java, *OpenWebGlobe* is not a downloadable application but an open-source SDK. Unlike World Wind, however, OpenWebGlobe is written in WebGL, a JavaScript library for web-based 3D graphics without plug-ins. It is maintained by the Institute of Geomatics Engineering at the Northwestern Switzerland University of Applied Sciences.

In addition to raster data and elevation models, the SDK supports points, vector data, and 3D objects such as buildings. Compared to Google Earth and World Wind, however, a little more is left to the user. Intended for programmers comfortable with Java programming, it is nonetheless easy to implement if you have a little programming experience. For example, this small JavaScript function, from the "Hello Globe" tutorial at openwebglobe.org, shows how simply an image and elevation model can be added to a virtual globe viewer:

```
function main()
{
var ctx = ogCreateContextFromCanvas("canvas", true); //create a
canvas and size fullscreen
var globe = ogCreateGlobe(ctx); //create the virtual globe
var imgBlueMarble500 = //select an image
{
url :["http://www.openwebglobe.org/data/img"],
layer :"World500",
service:"i3d"
};
var elvSRTM_CH = //select an elevation model
{
url :["http://www.openwebglobe.org/data/elv"],
layer :"SRTM",
service:"i3d"
};
og AddImageLayer(globe, imgBlueMarble500); //render the image on the
  globe
og AddElevationLayer(globe, elvSRTM_CH); //apply the elevation model
}
```

The result of the OpenWebGlobe "Hello Globe" JavaScript example can be seen in Figure 21.7. Another example is shown in Figure 21.8. Only demos are online today, but the entire SDK is available, enabling anyone to build their own WebGL virtual globe app.

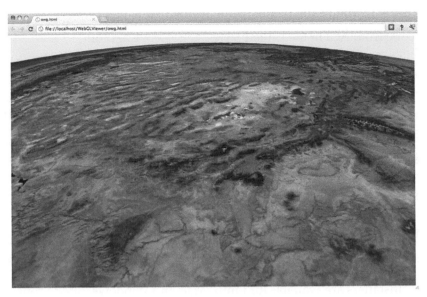

FIGURE 21.7

OpenWebGlobe "Hello Globe" JavaScript example, rotated to show a view over the Colorado Plateau from the Grand Canyon northwest to the Basin and Range Province of the western United States.

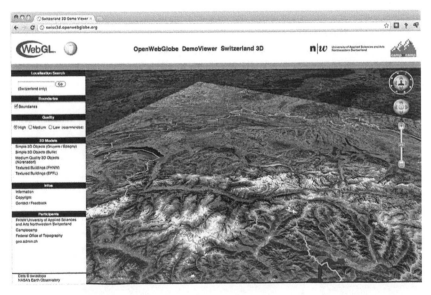

FIGURE 21.8

OpenWebGlobe view of 50 cm/pixel imagery over the Swiss Alps.

OpenWebGlobe is one of the several open-source WebGL-based virtual globes. This is relatively new technology, so at the time of writing (mid 2012), it is unclear which product will develop and build the strongest following. Here are some others:

- WebGL Earth from Klokan in Switzerland, at webglearth.org
- ReadyMap from Pelican Mapping in the United States, at readymap.com

- WebGL Globe from Google, at chromeexperiments.com/globe
- Geoscope from AGI in the United States, at geoscope.agi.com

Image Analysis

Alongside the wealth of tools for image viewing and organization, there is an equally rich ecosystem of image analysis software. Some are general image processing toolboxes with desktop applications (e.g., VISAT), others are more specialized software libraries, accessed via the command line or scripts (e.g., GDAL). What follows is a selection of software projects on the web. The best way to keep up with what is new and useful is via social media channels. There are many others besides those listed here: Opticks, Orfeo, PyShp, and ILWIS, to name a few.

ESA BEAM (www.brockmann-consult.de/cms/web/beam)

The *Basic ERS-ENVISAT(A) ATSR MERIS* toolbox, or *BEAM*, is an open-source project maintained by Brockmann Consult, a German environmental consultancy. Besides the source code, installers are available for Linux, Mac, and Windows. BEAM consists of a few components:

- VISAT, a desktop application for visualization and analysis of very large images (Figure 21.9). Its layer management system makes it easy to integrate imagery and shape files. You can perform arbitrary operations on rasters, change image projections, and apply bitmasks.

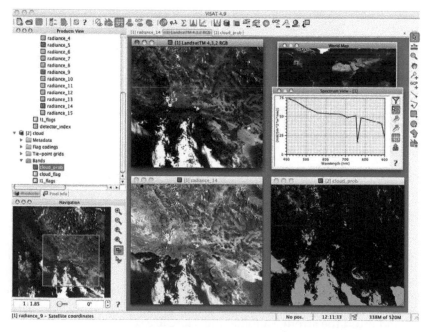

FIGURE 21.9

This VISAT example shows the result of cloud probability analysis in Landsat imagery over southern California. The dark shape just below right of center is the Salton Sea.

- A set of data processing tools for binning, mosaicing, collocation, clustering, and so on. The toolbox accepts images in a range of formats, including Envisat N1, HDF4, GeoTIFF, and NetCDF.
- A Java API supporting a range of user-developed plug-ins.
- Tutorials, sample data, and various support channels including a wiki.

GDAL (gdal.org)

The *Geospatial Data Abstraction Library*, or *GDAL*, is an open-source translator library with bindings for several widely used languages including Python, Java, and R. GDAL itself is for raster data processing, but distributions include the OGR Simple Feature Library for vector data (ESRI shape files, KML, and so on). GDAL is the main data access engine for open-source GIS applications GRASS and QGIS, and is used by Google Earth and ArcGIS. The libraries support GeoTIFF, ERDAS Imagine, MrSID, and many other image formats.

GeoTools (geotools.org)

An open-source Java library for geospatial programming, *GeoTools* is mature and full featured. It is closely tied to the geospatial community, so Open Geospatial Consortium specifications are implemented immediately. GeoTools is used by a number of other open-source projects including web services and desktop applications.

OSSIM (www.ossim.org)

The *Open Source Software Image Map*, or *OSSIM* (pronounced "awesome" by some people), is a suite of applications and C++ image processing libraries for remote sensing images and vector data. The U.S. government has invested substantially in the project and the tools are widely used in government. The main desktop application is ImageLinker, a mosaicing tool for histogram matching and image fusion.

Networks

It can be encouraging to find and meet others involved in geospatial image analysis and interpretation; happily there is an active and easy-to-find geospatial community on the web. You will find a lot of forum and blog activity focused on geospatial analysis, geoscience, planetary geology, and remote sensing. Getting involved can be entertaining and educational, and will certainly provide an informal support network.

Forums

The preeminent technical discussion community is at StackExchange, a question-and-answer site with a dedicated GIS channel (ref: gis.stackexchange.com). If you are careful about describing a problem, you will get helpful replies. Chit-chat is not allowed, so the signal-to-noise ratio is very high.

The GIS Café is a large and active community, with lots of discussion groups about topics concerning GIS in business and education. There is a group devoted to discussing remote sensing issues (http://www10.giscafe.com/wwwthreads-5.3/wwwthreads.php).

Another community is developing at LinkedIn. Conversations happen at a slower pace than Twitter, but are more organized and are archived. Try the following groups, all of which are quite large and have active discussions:

- GIS and Geospatial Technology, Innovation, Future Trends
- The Earth Observation Network
- GEOinformatics
- Hyperspectral Imaging Group
- Remote Sensing of Vegetation

Twitter

One of the best places to find people and resources is Twitter. You do not need an account to explore Twitter, but you do need to log in to easily return to the people you wish to read updates from. To get started, follow a few accounts like these (just go to twitter.com/NASA_EO for the Earth Observatory account, for example):

Organizations: NASA_EO, ESA, RSPSoc

Corporations: ESRI, DigitalGlobe, ERDAS, terraremote

Individuals: TheMapAddict, joefrancica, MattArtz, deepseadawn

Blogs

Compared with other channels, blogs are the place for the deepest conversations. Although a few are professional writers and journalists, most bloggers in technical geoscience topics are professionals, students, and researchers. To the newcomer, it can seem difficult to discover the gems in the rubble, but keeping your eyes on forums and Twitter is a surprisingly efficient filter. Here are some places to start looking:

- ESRI runs a number of blogs around technical subjects, including one focused on imagery (ref: http://blogs.esri.com/esri/arcgis/category/subject-imagery).
- James Fee of WeoGeo.com and PlanetGS.com writes a blog at SpatiallyAdjusted.com, covering a very broad range of topics, within and beyond geospatial science.
- Bill Dollins, a consultant, writes a blog at blog.geomusings.com with a focus on geospatial computing issues.

Should you wish to share your work online, starting a blog is easily done. The Blogger and Wordpress platforms, both of which are free and easy to use, are very popular. Wordpress is an open source and offers more flexibility, while Google's Blogger focuses on ease-of-use. Consider sharing your algorithms and interpretation methods for others to reuse, adapt, extend, and even improve. To make your desire to share unambiguous, it is advisable to use a clear license for your work, such as one of the Creative Commons licenses at creativecommons.org.

For a bit of fun, you can also play geospatial games, scouring satellite images for geological and cultural locations. For example, whereongoogleearth.com and whereon-googleearth.net are both for a general audience, while Where on (Google) Earth is strictly geoscientists only. This contest moves around, but Felix Bossert runs a blog that keeps track of it at woge-felix.blogspot.ca.

The following section provides a tutorial on how to find and apply a publicly available topographic database. It is contributed by David Haddad and colleagues at Arizona State University and U.C. San Diego.

OpenTopography: Tutorial for a Web-Based Public Digital Topographic Database and Tools

[1]*Active Tectonics, Quantitative Structural Geology, and Geomorphology Laboratory, School of Earth and Space Exploration, Arizona State University, Tempe, AZ 85287-6004.* E-mail: david.e.haddad@asu.edu
[2]*UNAVCO, 6350 Nautilus Drive, Boulder, CO 80301-5394.*
[3]*OpenTopography Facility, San Diego Supercomputer Center, University of California, San Diego, San Diego, CA 92093-0505*

Introduction

The drive for understanding geologic phenomena at submeter-resolution scales has led to a rapid increase in the demand for high-resolution digital topographic datasets. Many of these are collected by light detection and ranging (Lidar) systems. The richness of Lidar datasets makes them highly valuable beyond the original application that drove their acquisition. However, Lidar datasets and analyses tend to be project-specific, computationally intensive, and large, thus creating technical challenges in their processing and distribution to the geoscience community. Furthermore, the useful analysis of Lidar datasets requires special computing resources that are unavailable to many geoscientists. To help alleviate these challenges, the Arizona State University School of Earth and Space Exploration (http://sese.asu.edu) and the University of California (San Diego) San Diego Supercomputer Center (http://www.sdsc.edu), with support from the U.S. National Science Foundation (grants EAR 0930731 and EAR 0930643), have collaborated to create the OpenTopography Facility (http://opentopography .org). OpenTopography facilitates community Web access to high-resolution geoscience-oriented topographic data, related tools, and resources as a means to democratize Web-based access to topographic, and specifically Lidar datasets. Many thanks to Christopher Crosby for his contributions to the OpenTopography Lidar facility (www .OpenTopography.org) and for countless discussions about Lidar technology, data processing, and integration.

The purpose of this tutorial is to demonstrate how to access publicly available Lidar datasets using the OpenTopography Web facility. This tutorial is not meant to be comprehensive documentation of advanced OpenTopography functionalities. Instead, the reader is encouraged to explore OpenTopography on the Web via its extensive online documentation, video tutorials, and educational resources.

Computing Considerations

There are certain computing requirements to consider when visualizing and analyzing Lidar datasets. These depend on their intended use. For this tutorial, the basic computing needs include access to high-speed Internet, a Web browser, and the free edition of Google Earth. More advanced Lidar data analysis requires access to a GIS (e.g., Arc Info or Global Mapper).

Step 1: Set Up a Free OpenTopography Account

Go to http://opentopography.org (Figure 21.10). Click the "Overview Video" to watch a brief video about the OpenTopography facility.

Click the "myOpenTopo" button on the top-right corner of the homepage and "Create new account." Fill out your user information (Figure 21.11).

Once you activate your account, go back to the myOpenTopo log-in page and log into your account (Figure 21.12).

This page is divided into three main sections: (1) Data—where you can access past user-defined Lidar job requests (Figure 21.13), (2) Contribute—where you can contribute Lidar datasets or processing tools, and (3) User Account—where you can perform basic account management operations.

Step 2: Determine If Lidar Data Exist for Your Area of Interest

Click the "Data" tab near the top of the page. This will take you to a Google Maps interface that shows all publicly available Lidar datasets (Figure 21.14). The datasets are color coded by their data source. For example, red dots are datasets that are available directly via OpenTopography. Zoom to your area of interest to check the availability of Lidar data. In this example, you will access Lidar data that are located within the OpenTopography facility.

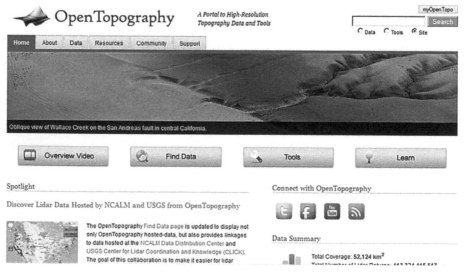

FIGURE 21.10
OpenTopography home page.

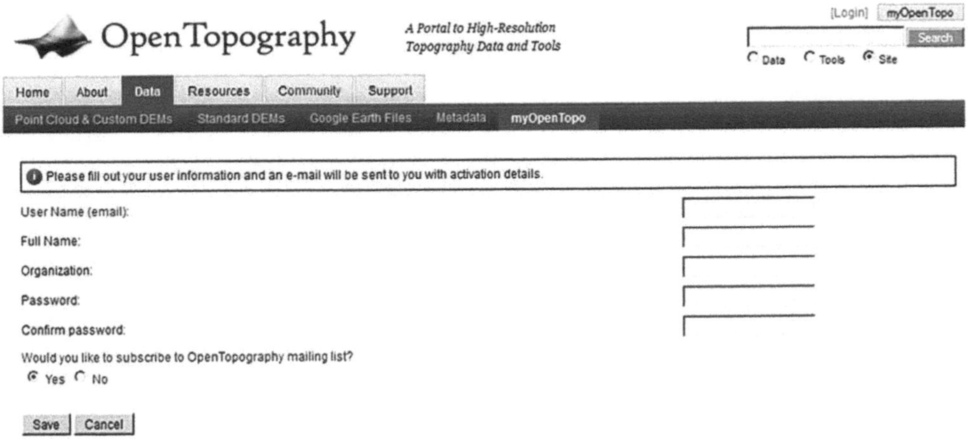

FIGURE 21.11
OpenTopograpy new user information form.

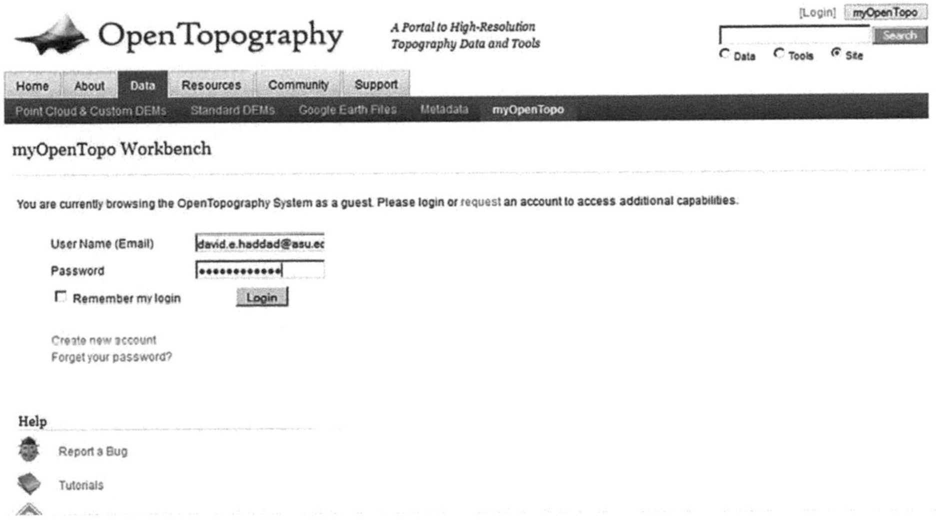

FIGURE 21.12
OpenTopography login page.

Zoom in to a portion of the Garlock fault in California (Figure 21.15). Left-clicking once on the data extent will display a pop-up balloon with some basic information about that dataset. Take note of the name of the dataset (this example shows the "EarthScope Southern & Eastern California Lidar Project"). We will use it to search for the actual data in OpenTopography.

Step 3: Access Lidar Data for Your Area of Interest

Click the "Point Cloud & Custom DEMs" link under the "Data" tab to go to the complete list of Lidar datasets that are available in OpenTopography (Figure 21.16). Scroll down to the dataset of interest (in this example it is "EarthScope Southern & Eastern California Lidar Project") and click the link.

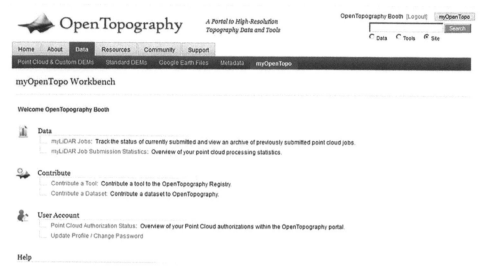

FIGURE 21.13
OpenTopography data page.

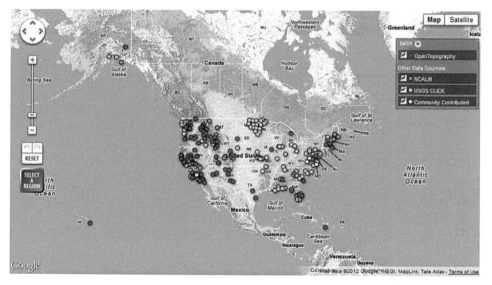

FIGURE 21.14
OpenTopography data selection page. Zoom to your area of interest to check data availability.

You will be taken to the dataset's page, which is divided into two sections (Figure 21.17). The top section contains general information about the dataset, the full metadata report that was generated by the data collector, information about the project funders, and DEM options. The contents of the lower section depend on which DEM option is selected.

Step 4: Choose a Method to Create/Download Lidar-Derived DEMs

There are three ways to generate DEMs in OpenTopography (Figure 21.18).

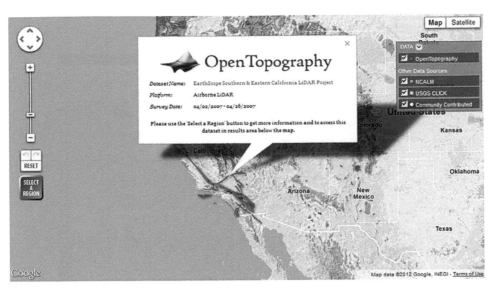

FIGURE 21.15
OpenTopography projects in southern California.

FIGURE 21.16
Lidar databases page in OpenTopography.

The first option ("Point Cloud Download & Processing") is the most advanced and allows you to control the parameters that go into creating DEMs. The second option ("Standard DEM Download") allows you to download DEMs that were already created by OpenTopography. Both these options require a GIS to display and process the DEMs (e.g., Arc Info or Global Mapper). See the supporting documents in the "Support" tab for details on how to do this. The third option ("Google Earth Imagery File") allows the user to download DEM imagery that was created

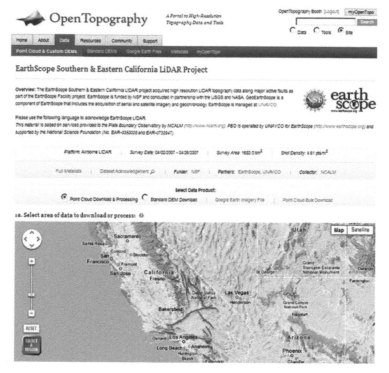

FIGURE 21.17
Lidar data page in OpenTopography.

FIGURE 21.18
"Select Data Product" menu, OpenTopography.

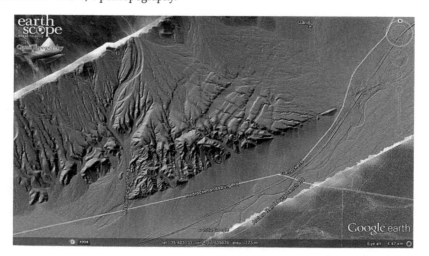

FIGURE 21.19
OpenTopography vertical DEM image of part of the Garlock fault, California, merged with Google Earth imagery.

FIGURE 21.20
OpenTopography oblique DEM image of part of the Garlock fault, California, merged with Google Earth imagery.

specifically for the Google Earth platform. The fourth option ("Point Cloud Bulk Download") provides the user with the ability to download the raw Lidar point data. For this tutorial, we use the third option of downloading DEM imagery that was created by OpenTopography to be displayed in Google Earth.

Step 5: Display DEM Imagery in Google Earth

Click the "Google Earth Imagery File" link to download the Google Earth file. Once the download is complete, double-click the Google Earth file and explore your area of interest (Figures 21.19 and 21.20).

This tutorial illustrates that a number of user interfaces and easy to use remote sensing products are available. Remote sensing has truly become a public enterprise.

References

Earle, P., M. Guy, R. Buckmaster, C. Ostrum, S. Horvath, A. Vaughan. 2010. OMG earthquake! Can Twitter improve earthquake response? *Seismol. Res. Lett.* 81: 246–251. Online at goo.gl/CkEcX. doi: 10.1785/gssrl.81.2.246.

Ellis, J.M. 1997. Business reasons for geologic remote sensing embracing GIS. 12th International Conference and Workshops on Applied Geologic Remote Sensing, Denver, 17–19 November 1997: I-75–I-81.

Feldman, S.C., R.E. Pelletier, E. Walser, J.C. Smoot, D. Ahl. 1995. A prototype for pipeline routing using remotely sensed data and geographic information systems analysis. *Rem. Sens. Environ.* 53: 123–131.

Jadkowski, M.A. 1994. EOCAP Funding Develops Aerial Imaging System for Pipeline Infrastructure Management. *Earth Obs. Mag.* 3: 29–32.

Nelson, E.P., K.A. Connors, C. Suárez S. 2007. GIS-based slope stability analysis, Chuquicamata open pit copper mine, Chile. *Nat. Resour. Res.* 16: 171–190.

Peters, D.C., F.E. Kirschner. February 1996. Midnight madness: reconciling site data from multiple and tangled coordinate systems. *GIS World*. 64–67.

Speed, V. October 1996. Highland valley copper mine. *Earth Obs. Mag.* 5: 22–25.

Wyvill, M. April 1992. Power routes laid in by GIS, remote analysis. NEPA requirements for impact studies being addressed quickly. *Earth Obs. Mag.* 1: 46–48.

22

Closing Thoughts

Remote sensing is a relatively new "science," involving technologies such as satellite multispectral and hyperspectral sensors, radar, thermal, potential fields, and acoustics (sonar). Geoscience remote sensing involves fields as diverse as geology, geomorphology, hydrology, geochemistry, geobotany, civil engineering, wildlife ecology, astrogeology, and image processing. However, the interpretation of images is not new. New applications are constantly emerging as industry struggles to succeed in an ever more competitive world, as universities advance research, and as government works to provide services to the public. The role of remote sensing analysis in mining, petroleum, ground water, engineering, and environment is to reduce exploration risk, minimize engineering uncertainty, improve public safety, and mitigate environmental impact. The driving force behind remote sensing has been and will continue to be the comprehensive and cost-effective nature of the data. It is the most efficient way to quickly survey and evaluate large areas of the Earth's surface.

Advances of the past decade are in the areas of commercialization, technology, integration, and access. Imagery is now offered by commercial enterprises as well as governments. Technology has provided improved spectral and spatial resolution. Satellite data are now available to the public via the internet, as are geographic information systems (GIS), and merged remote sensing/global positioning system (GPS) technology. GIS allows easy updating of existing maps using imagery and better evaluation of areas by integrating multiple diverse datasets. GPS provides the opportunity to make images into accurate base maps that can be registered and merged with other geographic databases. These technologies have revolutionized remote sensing by making it more accurate and easier to use when analyzing multiple datasets. In addition, hyperspectral remote sensing provides the ability to map an ever-greater suite of minerals, soils, and surface materials. Radar interferometry allows us to measure small changes in the Earth's surface related to earthquake deformation and surface changes over oil fields and abandoned mines. Lidar is being used to generate digital elevation models and make topographic maps. Earth now has nearly full radar coverage. The public has access to high-resolution imagery that used to be classified and available only to government agencies. Earth and environmental scientists now routinely visualize field areas by merging digital topography with imagery and displaying the three-dimensional (3D) representations of the surface on desktop workstations.

What can we expect from remote sensing in the next decade? In the short term, there will continue to be improvements in airborne and satellite multispectral scanners. Hyperspectral instruments will be used routinely. Private corporations will continue to launch satellites that provide sub-meter scale resolution, stereo coverage, and near real-time turnaround. Despite all the advances in remote sensing technology to date, stereo airphotos are still the most commonly used dataset. The availability of low cost distributed high-resolution stereo satellite and unmanned aerial vehicle (UAV) imagery will likely usurp the role of aerial photography. UAVs (drones) will become ubiquitous and provide high resolution, repeat coverage, and real-time data. Improved spatial resolution will enable geologists and engineers to focus on increasingly finer details of the surface, whereas coarser resolution will be used by environmental scientists to chart worldwide changes in plant cover, ocean currents, atmospheric ozone, sea level, and sea ice. Improvements will continue in spatial, spectral, and temporal resolution. Satellites will monitor atmospheric aerosols and gases, such as CO, CO_2, NO_2, and CH_6, as well as sea surface temperatures, wave height, salinity, and currents. Sea ice thickness data and the movement of glaciers will be charted on a regular basis. Farmers and urban planners will be able to measure soil moisture and plant health in real time and perhaps even determine the depth to groundwater. Interferometric radar and Lidar will allow identification and observation of ground movements and improved earthquake forecasting. 3D data visualization will become routine. 4D data (multitemporal fusion) will become routine for change detection. More web-based GIS applications will be developed and made available to the public.

In the not too distant, future geoscientists will apply spectral libraries and unmixing algorithms to generate ever more accurate and detailed maps of surface materials. Multisensor fusion will allow better feature recognition and identification of surface materials. *Data mining* is a process that attempts to discover patterns in large datasets. This will involve the analysis of data from different types of imagery which can then be summarized into useful information. Patterns, associations, or relationships among remote sensing data will provide information on the geology, vegetation, land use, and long-term changes over an area. The internet and mobile communication devices will provide the vehicle for increasingly interactive use of remote sensing by the public, from apps that locate real estate in areas without faults, flooding, or landslides, to maps that direct them to a restaurant for dinner.

There will always be a place for the image analyst who can explain the significance of what is being displayed. This is true whether one is searching for copper in Peru or for water on the moon. It is hoped that this text has served and will continue to serve as a useful reference for geoscience image analysts of the future.

Index